Electrochemical Biosensors

Electrochemical Biosensors

Edited by

Ali A. Ensafi, PhD

Professor
Department of Chemistry
Isfahan University of Technology
Isfahan, Iran

Elsevier
Radarweg 29, PO Box 211, 1000 AE Amsterdam, Netherlands
The Boulevard, Langford Lane, Kidlington, Oxford OX5 1GB, United Kingdom
50 Hampshire Street, 5th Floor, Cambridge, MA 02139, United States

ISBN: 978-0-12-816491-4

Publisher: Susan Dennis
Acquisition Editor: Kathryn Eryilmaz
Editorial Project Manager: Ruby Smith
Production Project Manager: Kiruthika Govindaraju
Cover Designer: Alan Studholme

Contents

Contributors.......... xi
Preface xiii

CHAPTER 1 An introduction to sensors and biosensors.......... 1
Ali A. Ensafi, PhD
1.1 Sensors.......... 1
1.2 Classification of sensors 1
1.3 Biosensors.......... 2
1.4 Electrochemical biosensors 3
1.5 Characteristics of an electrochemical biosensor 4
1.5.1 Linearity.......... 4
1.5.2 Linear dynamic range.......... 5
1.5.3 Sensitivity 5
1.5.4 Detection limit.......... 5
1.5.5 Selectivity 5
1.5.6 Response and recovery time 5
1.5.7 Ruggedness 6
1.5.8 Reproducibility and repeatability.......... 6
1.5.9 Accuracy.......... 6
1.5.10 Storage and operational stability.......... 6
1.6 Biosensor applications.......... 7
1.7 Electrochemical techniques.......... 7
1.7.1 Challenges facing biosensor research.......... 8
Acknowledgment.......... 9
References.......... 9
Further reading.......... 10

CHAPTER 2 Electrochemical detection techniques in biosensor applications.......... 11
Behzad Rezaei, PhD and Neda Irannejad
2.1 Electrochemical detection techniques.......... 11
2.1.1 Amperometric method.......... 13
2.1.2 Potentiometric devices.......... 14
2.1.3 Cyclic voltammetry.......... 15
2.1.4 Chronoamperometry and chronocoulometry 17
2.1.5 Chronocoulometry 18
2.1.6 Field-effect transistor 19
2.1.7 Electrochemical impedance spectroscopy.......... 25
2.1.8 Waveguide-based techniques and electrochemistry 26

2.1.9 Electrochemical-surface plasmon resonance 27
2.1.10 Ellipsometry and electrochemistry 30
2.1.11 Electrochemical atomic force microscopy 32
2.1.12 Electrochemical quartz crystal microbalance using dissipation monitoring 34
References 35

CHAPTER 3 Surface modification methods for electrochemical biosensors 45

N. Sandhyarani, PhD

3.1 Introduction 45
3.2 The need for surface modification 46
3.3 Specific versus nonspecific binding 46
3.4 Surface modification strategies 47
3.4.1 Self-assembled monolayers 47
3.4.2 Electrodeposition 49
3.4.3 Conducting polymers 50
3.4.4 Nanomaterials 50
3.4.5 Metal-organic frameworks 51
3.5 Immobilization techniques for biomolecules 53
3.5.1 Physisorption 53
3.5.2 Entrapment 54
3.5.3 Chemisorption 55
3.5.4 Active ester chemistry 56
3.5.5 Affinity immobilization 58
3.6 Types of immobilization strategies used in various categories of sensors 58
3.6.1 Enzymatic sensors 58
3.6.2 Affinity-based biosensors 62
3.6.3 Whole-cell biosensors 69
3.7 Conclusion 70
Acknowledgment 71
References 71

CHAPTER 4 Typically used carbon-based nanomaterials in the fabrication of biosensors 77

Esmaeil Heydari-Bafrooei, PhD and Ali A. Ensafi, PhD

4.1 Introduction 77
4.2 Properties of Carbon Nanomaterials 79
4.2.1 Properties of Carbon Nanotubes 79
4.2.2 Properties of graphene and graphene oxide 80
4.2.3 Properties of fullerene 81

4.3 Different synthesis methods of carbon nanomaterials for biosensing application....81
4.4 Modification of carbon nanomaterials....83
4.5 Application of carbon-based nanomaterials as biosensing....84
4.6 Biosensor using carbon nanotubes....85
4.7 Carbon nanomaterial–based aptamer and DNA biosensors....85
4.8 Carbon nanomaterial–based immunosensors....90
4.9 Conclusions....94
References....94

CHAPTER 5 Typically used nanomaterials-based noncarbon materials in the fabrication of biosensors....99
Fatemeh Farbod and Mohammad Mazloum-Ardakani, PhD
5.1 Introduction....99
5.1.1 Modifying electrotransducer surface and substrate for immobilization of biomolecules....100
5.1.2 Electrochemical signal amplification....100
5.1.3 Nanomaterials as signal-producing probes (i.e., as labels)....101
5.1.4 Mimicking the enzymatic behavior....102
5.2 Classification of noncarbon nanomaterials and their application in electrochemical biosensors....102
5.2.1 Metal and metal oxide....103
5.2.2 Core-shell....110
5.2.3 Quantum dots....113
5.2.4 Composites....116
5.2.5 Nanowires, nanofibers, and nanosheets....125
5.3 Conclusion....130
References....130

CHAPTER 6 Types of monitoring biosensor signals....135
Hamid R. Zare and Zahra Shekari
6.1 Introduction....135
6.2 Signal monitoring....135
6.3 Direct detection biosensors....136
6.3.1 Direct optical biosensors....137
6.3.2 Direct electrochemical biosensors....138
6.3.3 Direct piezoelectric (mechanical) biosensors....139
6.4 Indirect detection biosensors....140
6.4.1 Indirect optical biosensors....140
6.4.2 Indirect electrochemical biosensors....142
6.5 Signal monitoring in electrochemical DNA biosensors....144

6.5.1 Electrochemical DNA biosensors based on direct detection methods....144
6.5.2 Electrochemical DNA biosensors based on an indirect detection method....148
6.6 Conclusion....157
References....157

CHAPTER 7 Enzyme-based electrochemical biosensors........ 167
Aso Navaee and Abdollah Salimi
7.1 Introduction....167
7.2 What is biosensor?....168
7.3 Classification of enzymes-based bioreceptors....169
7.4 Oxidoreductase subclasses....176
7.5 Other flavoprotein-dependent enzymes....180
7.6 Methods for immobilization of enzymes....192
7.6.1 Adsorption....193
7.6.2 Covalent bonding....194
7.6.3 Entrapment/encapsulation....195
7.6.4 Cross-linking....196
7.7 Kinetics of immobilized enzyme....198
Conclusion....200
References....200

CHAPTER 8 Aptamer-based electrochemical biosensors........ 213
Seyedeh Malahat Shadman, MSc, Marzieh Daneshi, MSc, Fatemeh Shafiei, BSc, Maryam Azimimehr, MSc, Mehrdad Rayati Khorasgani, BSc, Mehdi Sadeghian, MSc, Hasan Motaghi, PhD and Masoud Ayatollahi Mehrgardi, PhD
8.1 Introduction....213
8.2 Electrochemical aptasensors against small molecules....214
8.2.1 Heavy metals ion....214
8.2.2 Antibiotics....214
8.2.3 Dopamine....218
8.2.4 Adenosine-5'-triphosphate....219
8.2.5 Toxins....221
8.2.6 Pesticides....221
8.2.7 Drugs....222
8.3 Electrochemical aptasensors against proteins....224
8.4 Electrochemical aptasensors against cancer cells and microorganisms....232
8.5 Conclusion....237
Abbreviations....237
Acknowledgments....239
References....239

CHAPTER 9 Nucleic acid–based electrochemical biosensors 253
Ayemeh Bagheri Hashkavayi and Jahan Bakhsh Raoof
9.1 Introduction 253
9.2 Hybridization and effective factors on this process 255
9.3 Probes and their immobilization on the electrode surface 255
9.3.1 Immobilization of the DNA probe through adsorption 255
9.4 Types of DNA interactions with molecules and ions 257
9.4.1 Intercalation of compounds between DNA base pairs 257
9.4.2 Attaching to the DNA molecule grooves 258
9.4.3 External junction via electrostatic bond 258
9.5 Mutations and damages in DNA 258
9.5.1 Types of mutations 258
9.6 Some applications of electrochemical DNA biosensors 259
9.6.1 Preparation of an electrochemical biosensor to detect DNA damage 260
9.6.2 Use of electrochemical nucleic acid–based biosensors for diagnosis of genetic defects and clinical applications 262
9.6.3 Preparation of electrochemical DNA biosensors for recognition of G-quadruplex structure and the study of its interaction with some stabilizing ligands 265
9.6.4 Preparation of electrochemical biosensors based on nucleic acid, to measure very small amounts of some heavy metal ions 268
9.7 Conclusion 271
References 271

CHAPTER 10 Peptide-based electrochemical biosensors 277
Mihaela Puiu and Camelia Bala
10.1 Introduction 277
10.2 Creating peptidic interfaces: coating strategies for electrodes 278
10.3 Peptide-modified surfaces and interrogation modes in electrochemical bioassays 279
10.4 Electron transfer across peptide bridges 280
10.5 Electrochemical techniques in peptide-based biosensing assays 283
10.5.1 Cyclic and linear sweep voltammetry 284
10.5.2 Square wave and differential pulse voltammetry 289

10.5.3 Alternating current voltammetry and electrochemical impedance spectroscopy.....292
10.5.4 Potentiometry.....294
10.6 Earnings and drawbacks of electrochemical peptide–based assays.....296
10.7 Conclusions and future trends.....300
Acknowledgments.....300
References.....300

CHAPTER 11 Receptor-based electrochemical biosensors for the detection of contaminants in food products.....307
Valérie Gaudin, PhD
11.1 Introduction.....307
11.1.1 Context.....307
11.1.2 Regulation.....310
11.1.3 Requirements for screening methods.....310
11.1.4 Conventional screening methods.....311
11.1.5 Biosensors.....312
11.2 Electrochemical biosensors for the detection of contaminants in food products.....312
11.2.1 Amperometry.....313
11.2.2 Chronoamperometry.....314
11.2.3 Potentiometry/field-effect transistors.....317
11.2.4 Voltammetry.....318
11.2.5 Electrochemical impedance spectroscopy.....321
11.2.6 Photoelectrochemical sensor.....327
11.2.7 Electrochemiluminescence sensor.....331
11.2.8 Electrochemical surface plasmon resonance.....335
11.2.9 Electrochemical quartz crystal microbalance.....336
11.3 Perspectives/future developments.....336
11.3.1 Developing new biosensing elements.....337
11.3.2 Miniaturization: micro/nanofluidic platforms and micro/nanoelectrodes.....337
11.3.3 Amplification strategies.....339
11.3.4 Smartphone-based biosensors for on field use ("point-of-care testing").....344
List of abbreviations.....346
References.....348

Index.....367

Contributors

Maryam Azimimehr, MSc
Department of Chemistry, University of Isfahan, Isfahan, Iran

Camelia Bala
R&D Center LaborQ, University of Bucharest, Bucharest, Romania; Department of Analytical Chemistry, University of Bucharest, Bucharest, Romania

Marzieh Daneshi, MSc
Department of Chemistry, University of Isfahan, Isfahan, Iran

Ali A. Ensafi, PhD
Professor, Department of Chemistry, Isfahan University of Technology, Isfahan, Iran

Fatemeh Farbod
Department of Chemistry, Faculty of Sciences, Yazd University, Yazd, Iran

Valérie Gaudin, PhD
Anses, Laboratory of Fougeres, European Union Reference Laboratory (EU-RL) for Antimicrobial and Dye Residue Control in Food-Producing Animals, Fougeres, France

Ayemeh Bagheri Hashkavayi
Electroanalytical Chemistry Research Laboratory, Department of Analytical Chemistry, Faculty of Chemistry, University of Mazandaran, Babolsar, Iran; Department of Chemistry, Faculty of Sciences, Persian Gulf University, Bushehr, Iran

Esmaeil Heydari-Bafrooei, PhD
Associate Professor, Department of Chemistry, Faculty of Science, Vali-e-Asr University of Rafsanjan, Rafsanjan, Iran

Neda Irannejad
Professor, Department of Chemistry, Isfahan University of Technology, Isfahan, Iran

Mehrdad Rayati Khorasgani, BSc
Department of Chemistry, University of Isfahan, Isfahan, Iran

Mohammad Mazloum-Ardakani, PhD
Department of Chemistry, Faculty of Sciences, Yazd University, Yazd, Iran

Masoud Ayatollahi Mehrgardi, PhD
Department of Chemistry, University of Isfahan, Isfahan, Iran

Hasan Motaghi, PhD
Department of Chemistry, University of Isfahan, Isfahan, Iran

Aso Navaee
Postdoctoral researcher, Department of Chemistry, Nanotechnology Research Center, University of Kurdistan, Sanandaj, Iran

Mihaela Puiu
R&D Center LaborQ, University of Bucharest, Bucharest, Romania

Jahan Bakhsh Raoof
Electroanalytical Chemistry Research Laboratory, Department of Analytical Chemistry, Faculty of Chemistry, University of Mazandaran, Babolsar, Iran

Behzad Rezaei, PhD
Professor, Department of Chemistry, Isfahan University of Technology, Isfahan, Iran

Mehdi Sadeghian, MSc
Department of Chemistry, University of Isfahan, Isfahan, Iran

Abdollah Salimi
Professor, Department of Chemistry, Nanotechnology Research Center, University of Kurdistan, Sanandaj, Iran

N. Sandhyarani, PhD
Professor, School of Materials Science and Engineering, National Institute of Technology Calicut, Kozhikode, Kerala, India

Seyedeh Malahat Shadman, MSc
Department of Chemistry, University of Isfahan, Isfahan, Iran

Fatemeh Shafiei, BSc
Department of Chemistry, University of Isfahan, Isfahan, Iran

Zahra Shekari
Dr, Department of Chemistry, Faculty of Science, Yazd University, Yazd, Iran

Hamid R. Zare
Professor, Department of Chemistry, Faculty of Science, Yazd University, Yazd, Iran

Preface

Ever since their first development in 1956 by Clark and later in 1962 by Clark and Lyons, biosensors distanced themselves from their original use an oxygen biosensor based on the electrochemical method for glucose detection in blood to witness a number of modifications, whereas new detection systems have been proposed for their further development. By definition, any analytical device composed of a bioreceptor and a transducer that combines a biological component and a physicochemical detector to detect an analyte is called a biosensor. They owe their high selectivity toward the target analyte to the specific interaction(s) of the bioreceptors present in their structure with the analyte (biorecognition). It is this specificity of the interaction that prevents the interference of signals because of other substances with the biosensor's signals. Biosensors have been fabricated using a variety of biological, or biorecognition, compounds such as antibodies, aptamers, enzymes, nucleic acids, and cells. These systems have reduced in size to allow more biosensors installed on a small surface thanks to the recent developments in electronics and microchips.

A transduction method widely used in biosensors is electrochemistry, which includes such varied techniques as voltammetry, conductometry, amperometry, impedance spectroscopy, potentiometry, and electrochemiluminescence. More recently, the biosensor field has transcended its conventional grounds by taking advantage of the immense developments in nanoscience and nanotechnology to entertain unprecedented research into novel materials and biomaterials of superior electrical, mechanical, biocompatible, and physical properties for the manufacture of ever more efficient electrodes. The novel electrochemical biosensors thus developed are being increasingly used in new areas. Among the increasingly adjustable biomaterials are especially included those that are nanostructured, which makes them suitable for fabricating electrodes of higher surface areas in the order of micrometer dimensions. Nanoobjects such as quantum dots, carbon nanotubes, and nanohybrids nowadays used in biosensors have given rise to hitherto nonexistent properties. Obviously, such superbly small electrodes can be used with trace amounts of analytes for in vivo applications and have additionally turned the field of biosensors into an economically flourishing and technically attractive one.

It is in the light of these developments that the present book strives to introduce the state-of-the-art field of electrochemical biosensors, their types, and surface modification methods. The topics covered in this volume include those of prominent importance for a wide readership. These include nanomaterial-based carbon and noncarbon materials as well as types of biosensor signal monitoring systems typically used in the fabrication of biosensors. Applications of electrochemistry for the detection of metabolites and exploration of physiological processes are illustrated by introducing a wide variety of electrochemical biosensors based on enzymes, aptamers, nucleic acids, proteins, and peptides together with receptor-based ones used for the detection of contaminants in food products. The book will appeal to a wide scientific audience as well as graduate students for it contains contributions by universally recognized scientists known for their expertise in different fields of electrochemical transduction for biosensors.

Ali A. Ensafi
Editor

CHAPTER

An introduction to sensors and biosensors

1

Ali A. Ensafi, PhD
Professor, Department of Chemistry, Isfahan University of Technology, Isfahan, Iran

1.1 Sensors

A sensor may be defined as a device to convert an input of physical quantity into a functionally related output usually in the form of an electrical or optical signal that can be read or detected either by human users or by electronic instruments. Sensors and their associated interface are used to detect and measure different physical and chemical properties of compounds including temperature, pH, force, odor, and pressure, the presence of special chemicals, flow, position, and light intensity (Ensafi and Kazemzadeh, 1999; Zhu et al., 2018).

A sensor is mainly characterized as one that (1) is solely sensitive to the chemical or physical quantity to be measured, whereas it is insensitive to all other parameters likely to be encountered in its application and (2) while in operation, does not influence the properties of the input chemical and/or physical quantities. The sensitivity of a sensor indicates the degree of variation of the output relative to the change in the measured chemical or physical property. Selection of a sensor should be based on such essential features as its selectivity (Fraden, 2004), sensitivity, accuracy, calibration range, resolution, cost-effectiveness, and repeatability as well as the prevailing environmental conditions (Vetelino and Reghu, 2010; Ensafi et al., 2011; Grandke and Ko, 2008; Gründler, 2007).

1.2 Classification of sensors

Depending on the properties of the substance or analyte to be measured, sensors may be broadly classified into physical and chemical types, with the physical one referring to the device detecting and/or measuring such physical responses as temperature, pressure, magnetic field, force, absorbance, refractive index, conductivity, and mass change (Gründler, 2007). Moreover, the devices do not have any chemical interface.

A chemical sensor has a chemically selective layer that responds selectively to a special analyte (Janata, 2009). It deals specifically with the chemical information obtained from the chemical reaction of the analyte or a physical property of the system being probed. Such information may include the concentration of a specific

Electrochemical Biosensors. https://doi.org/10.1016/B978-0-12-816491-4.00001-2

component or the analysis of a total composition, which is then transformed into such signals of analytical use as conductance change, light, voltage, current, or sound.

Chemical sensors are gaining a leading position among the presently commercially available ones with a wide array of clinical, industrial, environmental, and agricultural applications.

1.3 Biosensors

A biological component, a bioreceptor, and a physicochemical detector, and a transducer, may be combined to form a biosensor (Buerk, 1993). A biomolecule, such as an antibody, aptamer, enzyme, nucleic acid, or cell, capable of detecting or identifying the target analyte is used as the bioreceptor. These sensors offer such advantages as high selectivity to the target analyte mainly due to the specific interaction of the bioreceptor present in their structure with the target analyte (biorecognition) (Buerk, 1993). More important, this specific interaction prevents the interference of signals from other substances with the desired biosensor signal. Finally, the event recognized by the bioreceptor is transformed by a transducer into a measurable signal (Fig. 1.1).

The prerequisite to a stable biosensor is the immobilization of the bioreceptor at the surface of the transducer using a reversible or irreversible immobilization method. To achieve this, different strategies may be used, which are classified into surface adsorption, covalence binding, cross-liking, entrapment (beads or fibers), bioaffinity, and chelation or metal binding based on such criteria as a type of sample, desired selectivity, difficulty, and ranging (Buerk, 1995).

Being an element for converting one form of energy produced by a physical change accompanying a reaction into another, the transducer in a biosensor transforms the biorecognition event into a measurable signal in a process called "signalization." Transducers come in a variety of optical, electrochemical, quartz crystal piezoelectric, calorimetric (heat output or absorbed by the reaction), and thermal

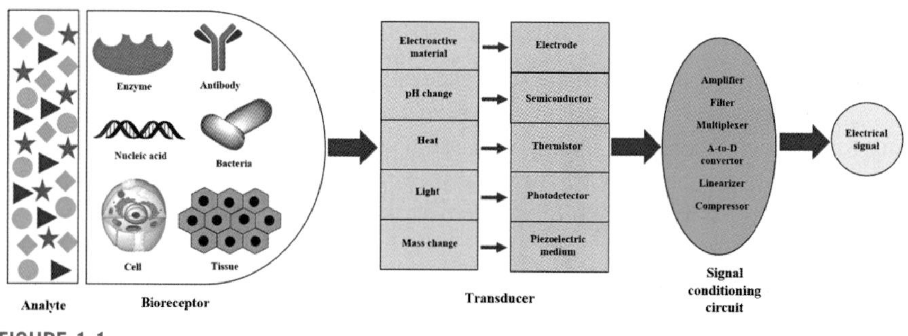

FIGURE 1.1

Schematic diagram of a biosensor.

types (Karunakaran et al., 2015). Most transducers, however, produce either optical or electrical signals in proportion to the analyte-bioreceptor interactions. A schematic diagram of the main components of a biosensor is shown in Fig. 1.1.

Originally, Clark and Lyons (1962) introduced the first biosensor in 1962. Using the enzyme glucose oxidase (GOx) as a recognition element, it was indeed an amperometric oxygen electrochemical sensor for detecting glucose. The term "biosensor" was coined as the shortened form of the so-called "bioselective sensor" proposed by Rechnitz et al. (1977) for arginine selective electrode that used living organisms as its recognition elements.

Biosensors have gone viral as analytical and diagnostic tools of widespread use, as they outperformance any other presently in use. Thanks to their operational simplicity, low cost, and no skills requirements, they have become the ordinary man's tools of everyday use. These advantages have won them increasingly wide applications in such varied areas as diabetic and cardiac self-monitoring, forensic investigations such as drug discovery, agricultural and environmental detection systems, the food industry, and biodefense (Scott, 1998). No doubt, further commercialization of biosensors rely much on such improved features as enhanced selectivity, sensitivity, stability, reproducibility, and portability, all at lower costs.

A variety of transducer-based output signals have been used in biosensors (Fig. 1.1). These include optical (such as absorbance, luminescence, chemiluminescence, and surface plasmon) (Ligler and Taitt, 2008), mass (piezoelectric and magnetoelectric) (Steinem and Janshoff, 2007), thermometric (Zhou et al., 2013), and electrochemical signals (Cosnier, 2013). From among these, the electrochemical ones are more prominently important, as they are not only economical and user-friendly but also allow robust, portable, and miniaturized devices to be fabricated for particular applications. Moreover, their excellent capacity for detecting and monitoring any changes in the electrical parameters of electrode potential, current, and charge transfer impedance or capacitance as a function of analyte concentration has made them suitable for many commercial applications (Ensafi et al., 2014). Based on the signal monitored, electrochemical biosensors may be classified into amperometric, potentiometric, voltammetric, impedimetric/conductometric, and capacitive sensors (Fig. 1.2).

1.4 Electrochemical biosensors

IUPAC defines an electrochemical biosensor as a self-contained and integrated device that uses a biological recognition element (biochemical receptor) in direct spatial contact with an electrochemical transducer to produce certain quantitative or semiquantitative analytical information on an analyte of interest (Theâvenot et al., 1999).

As already mentioned, a wide variety of electrochemical detection techniques are available that include amperometry, potentiometry, voltammetry, chronoamperometry, chronocoulometry, field-effect transistors, electrochemical impedance

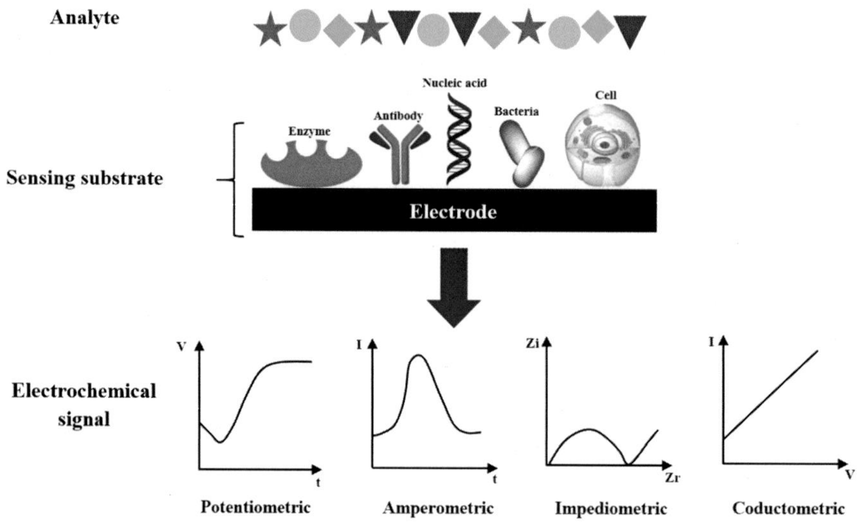

FIGURE 1.2

A schematic diagram of an electrochemical biosensor.

spectroscopy, electrochemical surface plasmon resonance, ellipsometry, electrochemistry, waveguide-based techniques, electrochemistry, AFM's simultaneous combination with electrochemistry, and electrochemical quartz crystal microbalance. A detailed description of each of these methods will be provided in the next chapter.

1.5 Characteristics of an electrochemical biosensor

Biosensor performance and efficiency are evaluated in terms of well-defined technical and functional characteristics according to the guidelines laid down by IUPAC (Theâvenot et al., 1999). Although the type of transducer and sample characterization are used as factors determining the efficiency and applicability of a sensor in different fields, its performance depends on such varied parameters as linear dynamic range, sensitivity, detection limit, selectivity, repeatability and reproducibility, accuracy, response time, recovery time, storage conditions, and ruggedness.

1.5.1 Linearity

The linearity of a sensor is determined based on how close its calibration curve is to a given straight line; in other words, system linearity is measured by the degree its calibration curve resembles a straight line.

1.5.2 Linear dynamic range

Determination of the analyte concentration in a system relies on the knowledge of the maximum and minimum values that the biosensor can measure in a test sample. A calibration graph is then plotted based on the results thus obtained, from which the analyte concentrations in test samples may be determined by interpolation.

1.5.3 Sensitivity

Sensitivity may be defined in analytical techniques as the slope of the calibration graph, whereas analytical sensitivity is the slope of the calibration graph divided by the standard deviation (Skoog et al., 2014). Put simply, sensitivity is the slightest difference in quantity read by an instrument. The linear portion of the calibration curve is often used to calculate more accurate sensitivity results.

1.5.4 Detection limit

The threshold quantity distinguishing the presence of a substance from its absence (a *blank value*) stated at a confidence level is referred to as the detection limit. Put differently, detection limit is the lowest concentration of an analyte that can be detected accurately by a sensor that enjoys a low enough signal to noise (S/N) ratio (Skoog et al., 2014). This parameter is typically calculated as three times the standard deviation of the baseline signal divided by sensitivity. Although a low signal to noise ratio puts a limit on the lower detection limit, it might enhance sensitivity.

1.5.5 Selectivity

The specificity of a sensor toward a given analyte to avoid false results owing to potentially interfering species is referred to as selectivity. It may be alternatively defined as a reagent's potential to discriminate between two or more substrates or two or more positions in the same substrate. Biosensor selectivity is influenced by the three important factors: the analyte tested, the transducer used, and the solution pH. For example, antibody and enzyme transducers are generally specific toward single analytes, whereas such biomolecules as microorganisms are highly nonspecific (Skoog et al., 2014); hence, the former has the widespread use as compared with the comparatively less application of the latter. Moreover, biosensor response might be either enhanced or diminished by the presence of interfering compounds. Interfering species that are structurally and/or chemically similar to the analyte typically enhance biosensor signal. In contrast, those that exhibit inhibitive effects toward the immobilized biocatalyst generally give rise to drastic reductions in the response of the sensing electrode.

1.5.6 Response and recovery time

Response time is a key parameter determining whether a biosensor can be scaled up from the lab to the industrial level. The factors involved in response time include sample temperature and concentration; bioreceptor thickness, geometry, and permeability; and agitation rate of the analyzing mixture. For practical reasons, two types of steady-state and transient response times are distinguished from each other, with the former being defined as the time required for a response signal to reach 95% of its steady state value and the latter defined as the time required for the response signal to reach its derivative.

Recovery time is defined as the minimum time required between two successive measurements, which depends on such factors as type, thickness, and permeability of the bioreceptor as well as analyte concentration.

1.5.7 Ruggedness

By ruggedness, it is meant the ability of a biosensor to show no calibration drifts due to minor physical or electrical variations.

1.5.8 Reproducibility and repeatability

The ability of a sensing system to produce identical responses under changing measurement conditions is referred to as its reproducibility, whereas repeatability refers to its ability to yield the same output for equal inputs applied over some period of time. Clearly, a biosensor with higher reproducibility and repeatability is one with higher reliability (Miller and Miller, 2000).

1.5.9 Accuracy

A sensing system is deemed accurate if it yields results whose correctness can be verified against the measured value of a measurand. Accuracy may be verified by either benchmarking the biosensor against a standard measurand or by comparing its output with a measurement produced by a system of superior accuracy (Miller and Miller, 2000).

1.5.10 Storage and operational stability

Stability is a measure representing changes in the biosensor baseline or sensitivity over a given period of time. Obviously, biosensors of commercial value are those that enjoy acceptable stability over time along with high sensitivity or selectivity. Biosensor stability may be expressed as operational or storage stability. Operational stability is influenced by such operational conditions as temperature, sample solution pH, the presence of organic solvents, and bioreceptor immobilization method. Storage stability, however, depends on such factors as storage conditions (dry or wet) as well as the composition, pH, and temperature of the buffer used for biosensor storage.

It may be concluded from the above that an ideal biosensor is one that is characterized by high selectivity, sensitivity, reproducibility, reliability, and stability;

low sensitivity to humidity and temperature; ease of calibration; ease of application; robustness and durability; small dimensions (portability); and short reaction and recovery times.

1.6 Biosensor applications

Biosensors have found wide application including point-of-care diagnosis, genetic problems (such as cancers, diabetes), evaluation and determination of analytes in biological samples, environmental monitoring, drug discovery, soil quality monitoring, water management, and food quality control (Karunakaran et al., 2015) (Fig. 1.3).

1.7 Electrochemical techniques

Many research problems that demand high degrees of accuracy, precision, sensitivity, and selectivity and that involve electroactive analytes with the potential to be detected via electrochemical methods warrant the application of such easy solutions collectively called electroanalytical techniques. These are indeed quantitative analytical methods based on the electrical properties of the solution of

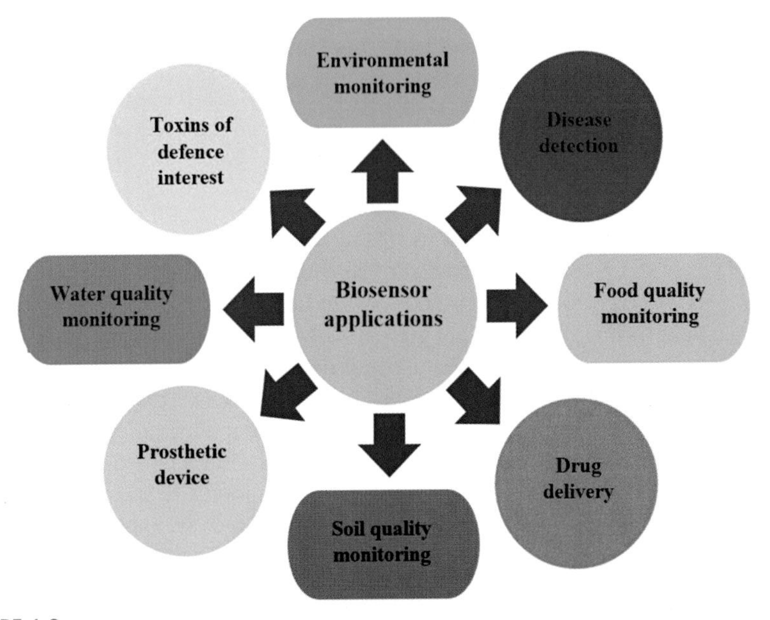

FIGURE 1.3

Application of biosensors in different areas.

the analyte. In responding to the requirements for the study of electrochemical reactions and determination of electrochemical properties of electroactive species, the field has witnessed the development of novel techniques for measuring the potential or current in an electrochemical cell (Bard and Faulkner, 2001). These methods are classified according to the aspects of the cells being controlled or measured.

Electrochemical-based transduction systems are typically characterized by their robustness, ease of application, portability, and low cost. The materials used in the electrodes in electrochemical biosensors include glassy carbon, carbon paste, graphite composites, carbon/graphite formulations, carbon nanotubes, graphene, and gold, among others. Thanks to their easy and reproducible fabrication at both laboratory and commercial scales, screen-printed electrodes (SPEs) have been widely used as the measuring element (Taleat et al., 2014). Several types of SPEs, functionalized or not, are now commercially available (e.g., Gwent Group Ltd), whereas many laboratories have their facilities for in-house production. However, not only the configuration of the electrode, and the materials used crucial, but the immobilization of the bioreceptor on the electrode surface is of vital importance as well.

1.7.1 Challenges facing biosensor research

Although biosensors have been around for the past 50 years, it is not more than decade research in the field has made its greatest contributions. This is perhaps the reason why very few, except for the lateral flow pregnancy tests and electrochemical glucose biosensors, have found their ways into the global retail markets. This failure to gain commercial success may be explained by difficulties faced with in commercializing academic research into viable prototypes and marketable products by the industry; stringent requirements of clinical application; and the almost nonavailability of researchers trained in biosensor technology or those with a commitment to teamwork and interdisciplinary interests. The situation is even more complicated by the fact that academic research lives in peer-reviewed journals that have conflicting interests or funded by institutions that might have connections with circles of power having their interests and conflicts. Funding agencies have their specific priorities, and there are legislators who approve funds only under their regulations. These considerations naturally drive researchers toward areas that are more fanciful attractive to funders. Despite all this, biosensor technology has been fortunate enough to win priority because of its potential applications. Hence, biosensors as practical devices can buy support. It is taken for granted that biosensor research depends on basic science; there is, however, no justification why it should not be "curiosity-driven" devoid of any practical or commercial application. Considering the commercial achievement and successful application of such biosensors as glucose sensors, it is reasonable to view biosensor research as a very lucrative activity for the industry. The only concern that seems to be looming large at this point is the risk the industry is reluctant to take as, in most case; the birth of a commercially viable device from an academic concept does not occur spontaneously but needs time and effort.

Acknowledgment

The author would like to acknowledge the help received from Mr. Arjun Ajith Mohan and Dr. Ramani T. in the preparation of figures presented in this chapter.

References

Bard, A.J., Faulkner, L.R., 2001. Electrochemical Methods, Fundamentals and Applications, second ed. John Wiley & Sons.

Buerk, D.G., 1993. Biosensors, Theory & Applications. Technomic Publishing Company.

Buerk, D.G., 1995. Biosensors: Theory and Applications. CRC press.

Clark Jr., L., Lyons, C., 1962. Electrode systems for continuous monitoring in cardiovascular surgery. Annals of the New York Academy of Sciences 102, 29–45.

Cosnier, S. (Ed.), 2013. Electrochemical Biosensors. CRC Press.

Ensafi, A.A., Kazemzadeh, A., 1999. Optical pH sensor based on chemical modification of polymer film. Microchemical Journal 63, 381–389.

Ensafi, A.A., Karimi-Maleh, H., Ghiaci, M., Arshadi, M., 2011. Characterization of Mn-nanoparticles decorated organo-functionalized $SiO_2{-}Al_2O_3$ mixed-oxide as a novel electrochemical sensor: application for voltammetric determination of captopril. Journal of Materials Chemistry 21, 15022–15030.

Ensafi, A.A., Amini, M., Rezaei, b., 2014. Impedimetric DNA-biosensor for the study of anti-cancer action of mitomycin C: comparison between acid and electroreductive activation. Biosensors and Bioelectronics 59, 282–288.

Fraden, J., 2004. Handbook of Modern Sensors, Physics, Designs, and Applications, third ed. Springer.

Grandke, T., Ko, W.H. (Eds.), 2008. Sensors: Fundamentals, vol. 1. Wiley.

Gründler, P., 2007. Chemical Sensors, an Introduction for Scientists and Engineers. Springer.

Janata, J., 2009. Principles of Chemical Sensors. Springer.

Karunakaran, C., Bhargava, K., Benjamin, R., 2015. Biosensors and Bioelectronics, first ed. Elsevier.

Ligler, L., Taitt, C. (Eds.), 2008. Optical Biosensors, second ed. Elsevier Science.

Miller, J.C., Miller, J.N., 2000. Statistics for Analytical Chemistry, fourth ed. Prentice Hall.

Rechnitz, G.A., Kobos, R.K., Riechel, S.J., Gebauer, C.R., 1977. A bio-selective membrane electrode prepared with living bacterial cells. Analytica Chimica Acta 94 (2), 357–365.

Scott, A.O., 1998. Biosensors for Food Analysis. Woodhead Publishing Series.

Skoog, A.A., West, D.M., Holler, F.J., Crouch, S.R., 2014. Fundamentals of Analytical Chemistry, ninth ed.

Steinem, C., Janshoff, A. (Eds.), 2007. Piezoelectric Sensors. Springer.

Taleat, Z., Khoshroo, A.R., Mazloum-Ardakani, M., 2014. Screen-printed electrodes for biosensing: a review, 181 (9–10), 865–891.

Theâvenot, D.R., Toth, K., Durst, R.A., Wilson, G.S., 1999. Electrochemical biosensors: recommended definitions and classification. Pure and Applied Chemistry 71 (12), 2333–2348.

Vetelino, J., Reghu, A., 2010. Introduction to Sensors. CRC Press.

Zhou, S., Zhao, Y., Mecklenburg, M., Yang, D., Xie, B., 2013. A novel thermometric biosensor for fast surveillance of β-lactamase activity in milk. Biosensors and Bioelectronics 49 (15), 99–104.

Yang, T., Zhao, X., He, Y., Zhu, H., 2018. Graphene-based sensors (Chapter 6). In: Zhu, H. (Ed.), Graphene, Fabrication, Characterizations, Properties and Applications. Academic Press, pp. 157–174.

Further reading

Zhang, X., Ju, H., Wang, J. (Eds.), 2008, Electrochemical Sensors, Biosensors and their Biomedical Applications, Academic press. Available from: https://www.bookdepository.com/Biosensors-Biodetection-Avraham-Rasooly/9781603275682?ref=grid-view.

Cosnier, S., 2015. Electrochemical Biosensors, 1st Edition, Jenny Stanford Publishing. Available from: https://www.crcpress.com/Electrochemical-Biosensors/Cosnier/p/book/9789814411462.

Yoon, J-Y., 2016. Introduction to Biosensors: From Electric Circuits to Immunosensors, 2nd ed., Available from: https://www.springer.com/gp/book/9783319274119.

Lalauze, R. (Ed.), 2012. Chemical Sensors and Biosensors, Wiley. Available from: https://www.wiley.com/en-us/Chemical+Sensors+and+Biosensors-p-9781848214033.

CHAPTER

Electrochemical detection techniques in biosensor applications

2

Behzad Rezaei, PhD, Neda Irannejad

Professor, Department of Chemistry, Isfahan University of Technology, Isfahan, Iran

2.1 Electrochemical detection techniques

Electrochemical biosensors transduce biological element-target detection events into detectable electrochemical signals. In biosensing measurements, the inherent electrochemical properties of the biological system are used to find access to valuable information. In this way, the active section of the bioelectrochemical species acts as the main transduction element (Grieshaber et al., 2008; Torrinha et al., 2018). In biosensors, among the various types of biorecognition components (cells, nucleic acids, antibodies, and microorganisms) (Chaubey and Malhotra, 2002; Eggins, 2002; Ronkainen et al., 2010), enzymes (D'Orazio, 2011; Martinkova, 2017) have found a special place owing to their special binding abilities and biocatalytic activities. In bioelectrochemical reactions, antibody, antibody fragments, or antigens are used as immunosensors to monitor the binding events. In (bio)electrochemistry, the investigation of reactions can be followed either by generating a detectable current (amperometric) (Rezaei et al., 2015), potential, or charge gathering (potentiometric) or by inducing a significant variation in the conductivity of the medium (conductometric) (Pohanka and Republic, 2008; Su et al., 2011; Thévenot et al., 2001a). Moreover, such other electrochemical measurements as electrochemical impedance spectroscopy (impedimetric methods), which measures the impedance of the sample (Daniels and Pourmand, 2007; Ensafi et al., 2016a; Franks et al., 2005; Katz and Willner, 2003) and field effects (which involve a transistor technology for current measurement as a result of potentiometric effects), have also been reported in the literature (Mir et al., 2009; Ohno et al., 2010; Thévenot et al., 2001b).

Given that reactions in electrochemical-based biosensors are commonly monitored near the surface of the electrode, the electrodes used play the essential role in the performance of the system. Based on the specific performance of an electrode, such features as the electrode materials, type of modification, and its geometry greatly affect its detection capacity (Yogeswaran and Chen, 2008). In an electrochemical sensing system, three or two chemically stable electrodes are commonly used. A typical three-electrode electrochemical cell consists of a working (or indicator), a counter, and a reference electrode. Fig. 2.1 shows a schematic diagram

Electrochemical Biosensors. https://doi.org/10.1016/B978-0-12-816491-4.00002-4

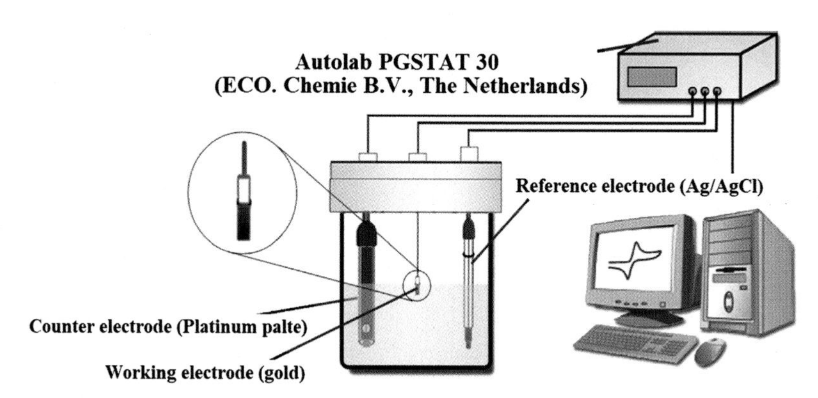

FIGURE 2.1

Schematic diagram of a three-electrode electrochemical cell system (Olad and Gharekhani, 2015).

of a three-electrode electrochemical cell system. The working electrode serves as the transducer in the bioelectrochemical or biochemical reaction, whereas the counter-electrode immersed into the electrolyte solution controls the possibility of applying currents to the working electrode. The reference electrode is usually set at a given distance from the reaction site and near the working electrode to give a known and stable potential. Among the reference electrodes available, including standard hydrogen, calomel (Hg_2Cl_2/Hg), and silver/silver chloride (Ag/AgCl) electrodes, the latter is the preferred one in bioelectrochemical systems as it does not necessarily need temperature control. Hydrogen electrode is not used in routine applications because of the difficulties associated with its preparation and the setup requirements. In an electrochemical cell based on three electrodes, the charge from the electrolysis process passes through the counterelectrode, so that the half-cell potential of the reference electrode remains constant (Bard and Faulkner, 2001). An electrochemical cell based on two electrodes consists of reference and working electrodes. In this system, the reference electrode conveys the charge with no adverse effects under very low current densities (Bartlett, 2008). The two- and three-electrode systems have been largely used effectively in many biosensing applications.

The advances made in nanotechnology and bioelectronics have made it possible to miniaturize the presently microscale sensor devices down to nanoscale ones. A significant advantage of these nanoscale systems is their ability to minimize the electrodes to such small sizes on the micrometer, or even as reported in some works, to the nanometer scale (Heinze, 1993; Wang, 1994). In the case of very small sample volumes (as low as a few microliters or less), this kind of miniaturized electrochemical biosensors of high sensitivity offers a great advantage (Goral et al., 2006). They owe their enhanced electrical properties to their greater surface-to-volume ratio. Moreover, they enjoy a higher detection sensitivity because of their nanoscale size, which matches that of the target biomolecules (Yang et al., 2010).

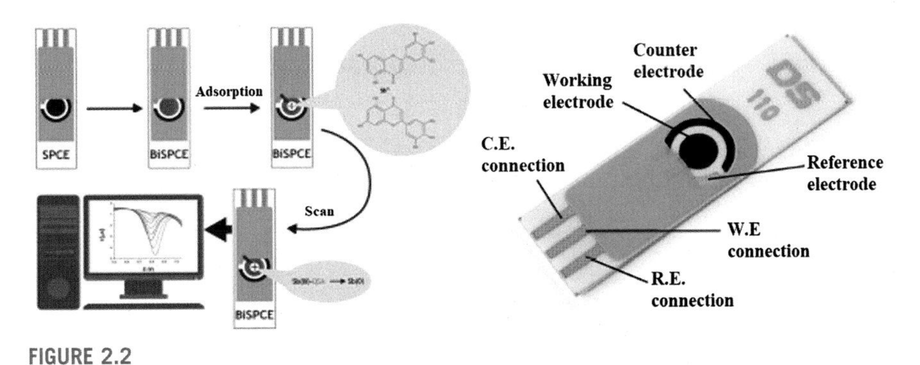

FIGURE 2.2

Screen-printed electrodes structure (Rojas-Romo et al., 2016).

Above all, what makes this kind of electrochemical systems a great choice for biosensing is the possibility they offer for making portable instruments at relatively low costs. Fig. 2.2 depicts one of the most popular electrochemical biosensors that uses screen-printed electrodes (SPEs) decorated with minielectrode (working, counter, and reference) structures, which offer the advantages of low production costs, high operation speeds, and ease of mass production (Li et al., 2012; Taleat et al., 2014; Wongkaew et al., 2018). These versatile SPEs of miniaturized size have been designed for highly specific on-site determination of target analytes by providing surface modification and the ability to connect to portable instruments. Finally, SPEs have become superior to classical solid electrodes, as they allow memory effects and tedious cleaning processes to be avoided (Mistry et al., 2014). It is the combination of these advantages that has led to a wide variety of applications for SPEs in the electrochemical immunosensors, enzyme-based biosensors, and DNA and aptasensors constructed (Rojas-Romo et al., 2016).

2.1.1 Amperometric method

Amperometric systems are electrochemical devices that determine continuously the current resulting from the reduction or oxidation of electroactive species at the surface of electrodes in a biochemical reaction (Emr, 1995; Ensafi et al., 2017c; Wang, 1999). The term "amperometry" refers to the process of tracing variations in current (due to electrochemical oxidation or reduction) through time while potential is kept constant in the cell or between the working and reference electrodes (Davis, 1985; Kawagoe and Wightman, 1994). This method measures current by directly stepping the potential to the favorite value or by maintaining it at the desired value. In this regard, peak current value measured in a linear potential range is directly related to the bulk concentration of the analyte present in the solution (Ensafi et al., 2017a). Amperometric biosensors enjoy excess selectivity because the potential resulting from the reduction or oxidation reaction used for detection is a distinctive property of the analyte species (Banica, n.d.; Eggins, 2002).

Under the conditions where the current is measured using a controlled variable potential, the corresponding method is called voltammetry (Ensafi et al., 2016b). In this method, the current response appearing as a peak or a plateau corresponds to the analyte concentration. Efficient voltammetric methods with a wide linear dynamic range that are suitable for low-level quantities include direct current voltammetry and polarography, linear sweep voltammetry (LSV), normal pulse voltammetry (NPV), differential pulse voltammetry (DPV) (Ensafi et al., 2017b, 2013), cyclic voltammetry (CV) (Ensafi et al., 2015b, 2015a), square-wave voltammetry (SWV), hydrodynamic methods, ac voltammetry, and stripping voltammetry (Chen and Shah, 2013; Rezaei et al., 2013a).

Using the dynamic chronoamperometry method, Ghiaci et al. (2016) evaluated and analyzed glucose (Gl) levels in biological samples by designing two new electrochemical sensors based on Ag nanoparticles (AgNPs) and decorated with anchored-type ligands based on the composition of the two amine compounds attached to a silica support (Ghiaci et al., 2016). Based on amperometric studies (Fig. 2.3), the superior glucose sensor showed a stable signal with a detection range of 28.6 μmol L^{-1} to 9.8 mmol L^{-1} glucose.

2.1.2 Potentiometric devices

In potentiometric devices, the accumulation charge potential is measured at the indicator electrode and compared with that at the reference electrode to obtain useful data on ion activity in the electrochemical reaction (under zero or no significant flow

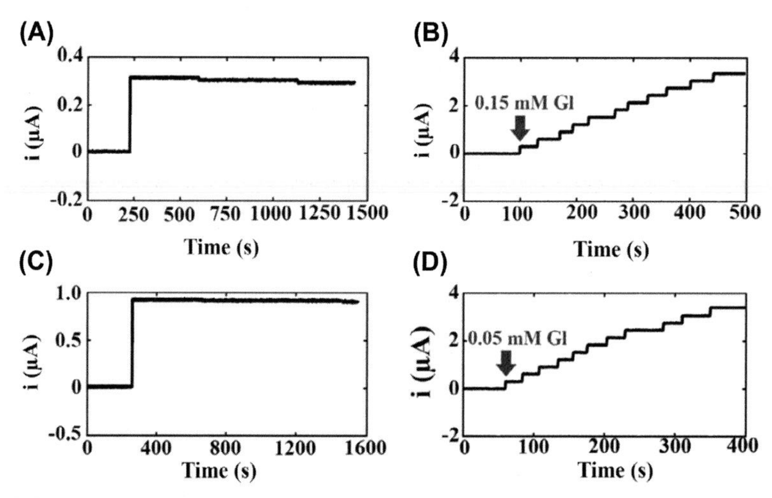

FIGURE 2.3

A and C show stability of the amperometric signal versus glucose concentration for AgNPs-decorated SiO_2-pro-NH/CPE and AgNPs-decorated SiO_2-pro-NH-cyanuric-NH_2/CPE, respectively, and B and D show their response to various glucose injections into the cell (Ghiaci et al., 2016).

currents through the indicator and reference electrodes) (Bakker, 2014). In such conditions, the measured potential is attributed to the number of electroactive species present in the sample. Using the Walther Nernst equation, the reduction potential is related to the concentrations of the analytes (Janata and Josowicz, 1997).

$$E = E^{\circ} - (RT/nF) \times \ln(\alpha\mathrm{Red}/\alpha\mathrm{Ox}) \quad (2.1)$$

where E° represents the standard-state potential, R is the gas constant, T is the constant temperature of the cell in Kelvin degrees, n is the number of electrons exchanged in the redox reaction, F is Faraday's constant, and αRed/αOx denotes the activity ratio of the reductant and oxidant species (Janata and Josowicz, 1997). Direct potentiometry is the direct measurement of analyte concentration with the application of the Nernst equation. Potentiometric devices currently have detection limits in the range of 10^{-8} to 10^{-11} M. The best and lowest detection limit has been recorded for potentiometric devices based on ion-selective electrodes (ISEs) (Bard and Faulkner, 2001). Considering that potentiometric sensors do not chemically affect the sample, they are suitable for measurement at very low concentrations and in samples with extremely low volumes. Detailed information on potentiometry and its detection limit may be found in Bakker and Pretsch (2005).

The potentiometric method is an appropriate alternative to electrical determination of the endpoint in a biochemical titration reaction. Potentiometric titration belongs to the chemical methods in which the endpoint of the titration is determined using a working electrode. In this method, variation in potential is identified as a function of a well-defined quantity (usually volume) of the added titrant of a known concentration (Bard and Faulkner, 2001; Buck, 1974). Titration devices have nowadays found a special place owing to the availability of suitable indicator electrodes of low cost and high reliability for the titrimetric studies of almost all chemical reactions. Potentiometric titration curves provide useful information on the endpoint position of the titration, whereas the shape and position of the curves also provide appropriate data about the processes accompanying the titration reaction (Buck, 1974).

2.1.3 Cyclic voltammetry

The voltammetric method is categorized in the class of electroanalytical techniques through which analytic evidence is acquired by changing the potential and determining the resultant current (Ensafi et al., 2017d). The vast variety of ways in which potential may be changed leads to numerous types of voltammetric methods, including polarography (DC voltage) (Bocarsly, 2012), LSV, differential staircase, NPV, reverse pulse, DPV, and SWV, among others (Rezaei et al., 2011c, 2015a, 2016, 2015c, 2013b). CV is one of the most common methods used to obtain useful information on redox potentials and to investigate the mechanisms and kinetic parameters (e.g., electrochemical rate constant) involved in the reactions of electroactive analyte solutions (Ensafi et al., 2016, 2017f). In this method, voltage is scanned from a given value (V_1) to a predetermined one (V_2) before it returns to

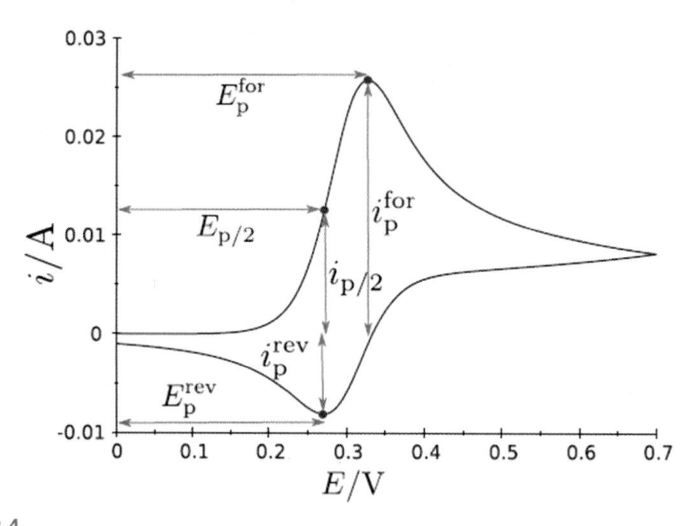

FIGURE 2.4

Example cyclic voltammogram (Bogdan et al., 2014).

its initial value (Fig. 2.4). Given that a meaningful chemical reaction needs enough time to proceed, the sweep rate $(V_2 - V_1)/(t_2 - t_1)$ will be a critical and decisive parameter in this technique so that varying the scan rate will lead to different results (Steudel, 1975).

Although the current between the working and counterelectrodes is traced, changes in the potential of the working electrode due to the reference electrode are controlled. The voltammograms attained from these measurements are, in fact, derived by plotting the measurements of current versus voltage. The current generated by the electrochemical reaction increases with increasing voltage toward the potential of the analyte's electrochemical reduction. With subsequent increases in potential toward V_2 and decreases in the current, a peak appears on the voltammogram indicating a decrease in analyte concentration near the electrode surface due to an excessive oxidation potential. When the voltage returns to its initial value of V_1 (to complete the CV), the reaction begins to reoxidize the product from the primary reaction. In this case, the current increases in the opposite polarity of the forward scan before it diminishes again as the voltage scanning toward V_1 continues (Bard and Faulkner, 2001). Thus, a reverse scan can provide useful information on the reaction reversibility at a specified scan rate (Bocarsly, 2012; Patolsky et al., 1999). The profile of the voltammogram for a known species depends strongly on the sweep rate and the changes in the electrode surface after each deposition step. It should be noted that catalyst concentration also plays a great role in determining the shape of the voltammogram so that the current increases with increasing concentration of a specific enzyme at a particular scan rate compared with the conditions in which the reaction is not catalyzed (Elgrishi et al., 2018).

Rezaei et al. (2008) used a film of multiwalled carbon nanotubes (MWCNTs) in the presence of a metal impurity serving as an electrocatalyst to construct a modified electrode for the effective measurement of noscapine levels in biological and pharmaceutical samples.

2.1.4 Chronoamperometry and chronocoulometry

Chronoamperometry is a well-established amperometric technique. In this method, a pulse potential is applied to a working electrode and the current passing through the cell is determined versus time (Kamat et al., 2010). Changes in the current appear in response to rises or decreases in the diffuse layers of the analyte at the surface of the working electrode. According to the definition by the IUPAC, the diffuse layer is equal to the surrounding region of an electrode in which analyte concentrations are different from those in the bulk solution. By applying an appropriate potential to the system, the local concentration of the analyte drops to zero. Under these conditions, a concentration gradient is generated that provides analyte transfer through diffusion from a higher concentration section (bulk solution) to the electrode surface (Lefrou and Fabry, 2012). The chronoamperometry processes are shown in Fig. 2.5.

Chronoamperometric experiments are performed in either of two general forms: a single potential step (applying a forward potential step and recording the resulting current) or a double potential step (applying a forward potential and returning the potential to the initial value during a given period of time). Cottrell's equation captures the current obtained at each point in time after a high single potential step (or large overpotential) is applied in a reversible redox reaction (Evans and Kelly, 1982):

$$\text{it} = \frac{nFAC_0D_0^{1/2}}{\pi^{1/2}t^{1/2}} \tag{2.2}$$

where n is the stoichiometric number of the electrons transferred in the redox reaction, F is the Faraday's constant (96,485 C/equivalent), A denotes the electrode

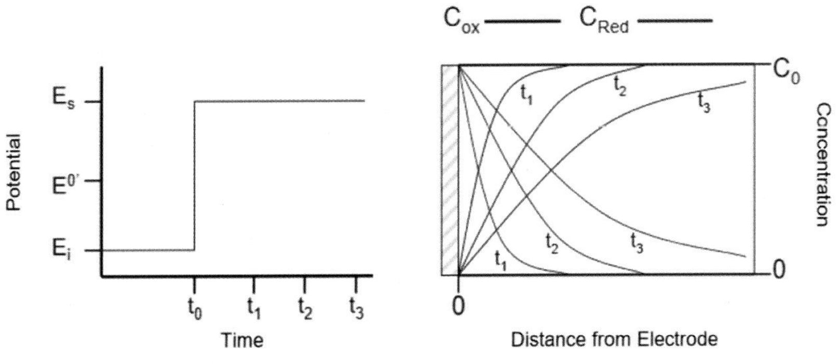

FIGURE 2.5

The chronoamperometry processes (Honeychurch, 2012).

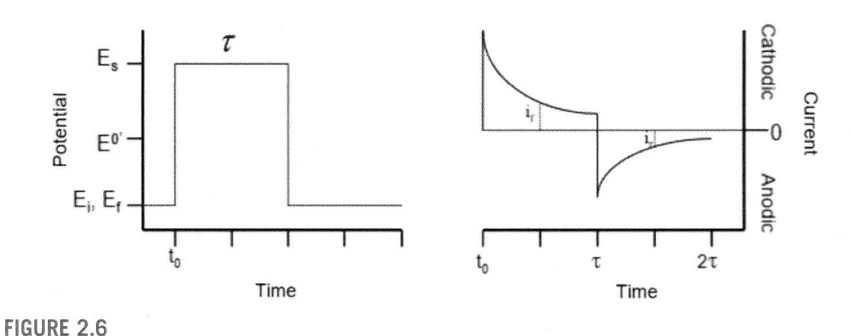

FIGURE 2.6

Double potential step chronoamperometry (Yu and Zhang, 1997).

surface area (cm), and C and D are the concentration (mol cm^{-1}) and diffusion constant, respectively, of the electroactive species (cm s^{-1}) in the solution (Evans and Kelly, 1982). Fig. 2.6 shows a double-step chronoamperometry. As can be seen, the forward potential step (E_i) in the potential step program on the left reaches a secondary value (E_s) during the interval time of τ. In the reverse potential step, the potential either returns back to the initial value (E_i) or changes to other values of E (depending on the experimental and system conditions). Generally, the electrical current for the opposite scanning potential conditions is also recorded in the interval τ (Evans and Kelly, 1982; Yu and Zhang, 1997).

It is deduced from Cottrell's equation that the observed current in the forward step decays as a function of τ. In general, the chronoamperometric method is able to measure the electrode surface area (A) based on the well-known redox reaction (known as n, C, and D). If an electrode area is known, it is then easy to measure the electroactive species in a solution. The double potential step methods provide a very comprehensive view of the rate constants of the chemical reaction, especially in product absorption systems (Yu and Zhang, 1997).

Ensafi et al. (2016c) were able to design an efficient, fast, and stable nonenzymatic glucose sensor by decorating silver nanoparticles on multiwalled carbon nanotubes (AgNPs/F-MWCNTs). The modified electrode was used as a nonenzymatic glucose sensor to determine glucose concentrations based on the hydrodynamic chronoamperometry method. Their results verified a low detection limit of 0.03 μM, a high sensitivity of 1.06×10^3 μA mM^{-1}, and a linear dynamic range of 1.3–1000 μM for their biosensor.

2.1.5 Chronocoulometry

In the chronocoulometric technique, the entire charge (Q) passing during the application of a step potential is recorded versus time. By integrating the current through the applied potential step, Q is calculated. Using Cottrell's equation, the Anson equation may be derived for chronocoulometric systems (Electrochemistry, 2006; Inzelt, 2014).

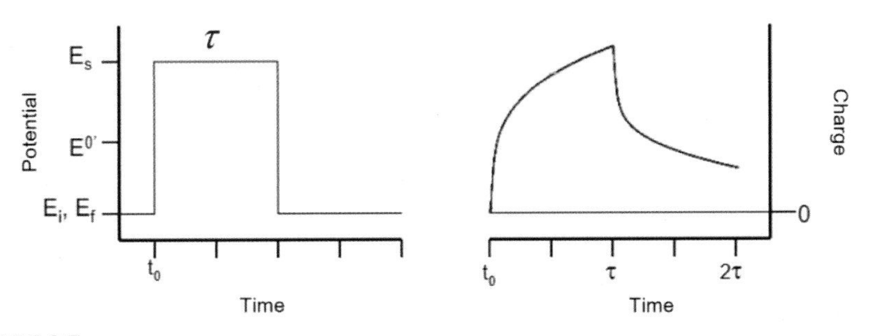

FIGURE 2.7

The double potential step chronocoulometry (Inzelt, 2014).

$$Q_d = \frac{2nFAC_0 D_0^{1/2} t^{1/2}}{\pi^{1/2}} \tag{2.3}$$

Fig. 2.7 shows the chronocoulometry experiment of a system controlled only by redox diffusion under a double potential step. The potential step programs for the chronocoulometry method comply with the procedures based on the chronoamperometry technique. The most prominent advantage of recording charges in current is that the most significant experimental information can be observed when Q increases during the potential step over time to reach its maximum value (due to the current decay to zero) at the termination of the step (El Harrad et al., 2018).

The chronocoulometric technique is preferable to the chronoamperometric one for measuring electrode area (A) and diffusion coefficients (D). On the other hand, it is the best method for measuring electrochemical reaction rates following an electron transfer because of the enhanced signals it provides. In the reverse potential step experiment, there is the possibility for the material to be reoxidized to the original material during the reverse potential step, as a reduced form of the redox reaction during the forward step remains in the district of the working electrode (El Harrad et al., 2018). According to the data obtained from the chronoamperometric technique, the signal recorded in the backward step is lower than that recorded in the forward step. The charge for the chemical and electrochemical reversible species in the backward step can be computed using the following equation (Anson and Osteryoung, 1983):

$$Q_r = \frac{2nFAC_0 D_0^{1/2}}{\pi^{1/2}} \left[\tau^{1/2} + (t - \tau)^{1/2} - t^{1/2} \right] \tag{2.4}$$

2.1.6 Field-effect transistor

Field-effect transistor (FET) is a kind of electrochemical device that uses an electric field to control the electrical behavior and the resistance of a channel between two electrodes in semiconductor materials. All FET systems make use of the three source, drain, and gate semiconductors, with no physical connection between the source and the gate ones. Under these conditions, a current path called the conduction channel is formed between the source and the drain. The device can act as an on/off switch by changing the gate-to-source voltage. The voltage applied to the gate determines the electric field power, acting as a control mechanism (Kaisti, 2017a). The current passed is measured by the actual charge motion (more precisely, electrons as the negative carrier for the n-type channel and the holes as the positive carrier for the p-type channel). Changes in the electric field potential at the gate electrode are used in conjunction with the source and drain electrodes to control conductivity (CHENG, 2015). Fig. 2.8 presents a schematic illustration of an application of FET for biological and chemical sensors.

Depending on the type of configuration and the dopant used in the semiconductor material, the charge carriers in the conducting channel are attracted or repelled because of the existence of negative or positive potentials at the gate electrodes. This can either fill or empty the depletion area of the charge carriers, which results in the operative electrical sizes of the conducting channel being formed or deformed. In this situation, the conductivity between the source and drain electrodes is controlled. FET operates in either of two general modes: linear and saturation

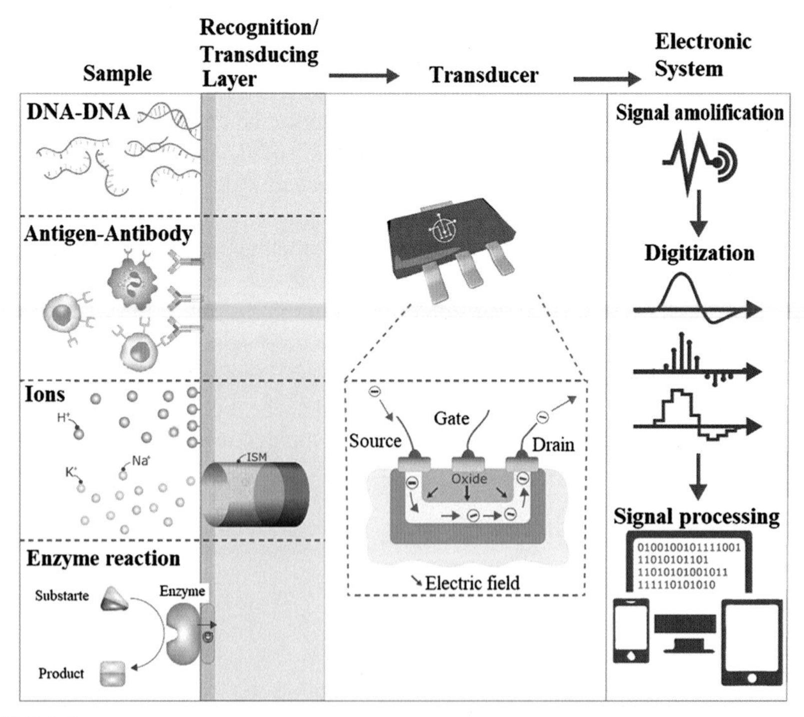

FIGURE 2.8

Illustration of a biological and chemical FET sensor (Kaisti, 2017b).

modes. Because of the lower drain-to-source voltage relative to the gate-to-source one, FET in the linear mode is more similar to a variable resistor that can be switched between the two conductive and nonconductive states.

In the saturation mode, the FET serves as a voltage amplifier in addition to its function as a constant current. In this mode, the constant current level is measured based on the gate-to-source voltage. FET devices have found vast applications in electrochemical biosensing systems because of their high operational capability at low signals and/or high impedance applications (Rashid, 2010). The types of FET so far known include metal-oxide-semiconductor field-effect transistor (MOSFET), junction gate field-effect transistor (JFET), metal-semiconductor field-effect transistor (MESFET), chemically sensitive field-effect transistor (CHEMFET), and ion-selective field-effect transistor (ISFET) (Luo et al., 2004; Schöning and Poghossian, 2006). One of the most extensively used FET systems is the MOSFET. As shown in Fig. 2.9, it is made of a metal-insulator-semiconductor in which a metal gate electrode is mounted on an oxide insulating layer. In operational FET conditions, use is made of a modulated electric field with the dimension and form of the source-drain determined by the channel distance modulation and channel shape modulation, respectively. In response to a target species, the current generated through the flow of carriers (electrons or holes) in the channel between the source and the drain is controlled by the gate electrode, thereby leading to a change in the drain current (I_d).

More specifically, use is made in bio-based FET (bio-FET) systems of a transistor device with a sensitive biolayer especially designed to detect biological molecules such as nucleic acids or proteins. The performance of the system is based on the semiconducting FET and an insulating layer (e.g., of SiO_2) serving as a transducer that separates the FET from the biological recognition element (e.g., a receptor or a probe molecule) that is selective to the analyte. Once the analyte is bound to the recognition element, the charge distribution at the surface changes in proportion to the change in the electrostatic semiconductor surface potential, leading to changes in current or conductivity to provide an effective means of detecting analyte concentration (Lowe et al., 2017).

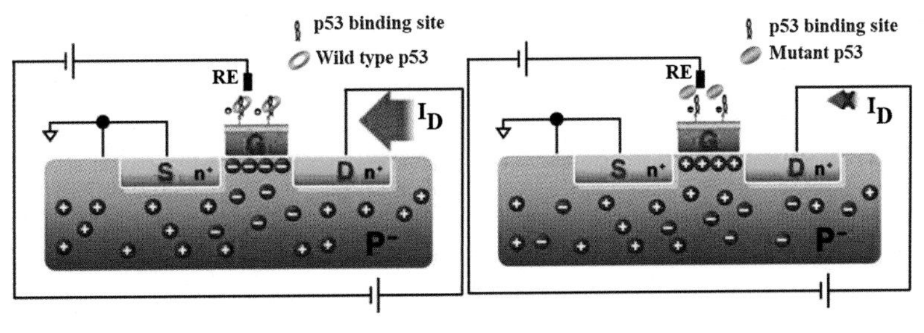

FIGURE 2.9

Representation illustration of the MOSFET-based detection of p53 binding to cognate DNA (Han et al., 2010).

Han et al. (2010) used an n-type MOSFET device to detect the ability of DNA to bind to a p53 tumor suppressor (found in more than 50% of human tumors) (Han et al., 2010). In biomolecules, the surface charge density (Vbio) is directly and strongly affected by the drain current (I_d), which decreases (or increases) in n-type FET devices in response to negatively (or positively) charged molecules on the electrode.

ISFET used as a transducer is one of the most promising tools in biological applications. In an ISFET machine, the metal electrode of the gate is eliminated, as it is replaced with an ISE (i.e., an electrolyte solution and a reference electrode). Hence, there is a great structural resemblance between an ISFET and a MOSFET. In the ISFET device, however, the charge density of the analyte molecules on the gate surface determines the magnitude of the current density (Schöning and Poghossian, 2002). To measure conformational changes in proteins, Park et al. (?) introduced the first functional description of an ISFET-based biosensor. As shown in Fig. 2.10, a modified FET receptor was used to examine the changes in the surface charge following structural changes in the maltose-binding protein (MBP) in response to maltose. Clearly, the configuration for the maltose-bound MBP changed as the N-terminus and C-terminus of the MBP drew closer to each other. Furthermore, immobilization of the maltose-free MBP on the oxide layer led to a downshift in drain current (I_d), which reduced to 1.175 μA owing to the structural changes occurring in the MBP as a result of treatment with maltose (Fehr et al., 2004). Investigation of the possible interpretations for this revealed that the reduced field effect could be partly attributed to the removal of a positive MBP from the oxide layer of the ISFET device (Schöning and Poghossian, 2002).

One of the most prominent uses of FETs is their use as enzymatic sensors. A typical example of an enzyme FET sensor is the glucose-sensitive one that contains a glucose oxidase membrane–modified gate surface and uses a glucose oxidase–based enzyme electrode to determine blood glucose levels. Caras et al. (1985)

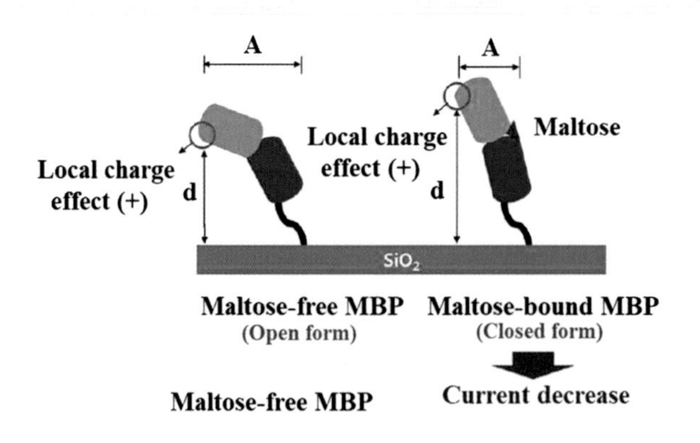

FIGURE 2.10

The research design for monitoring the conformational change in MPB by an ISFET-based biosensor (Fehr et al., 2004).

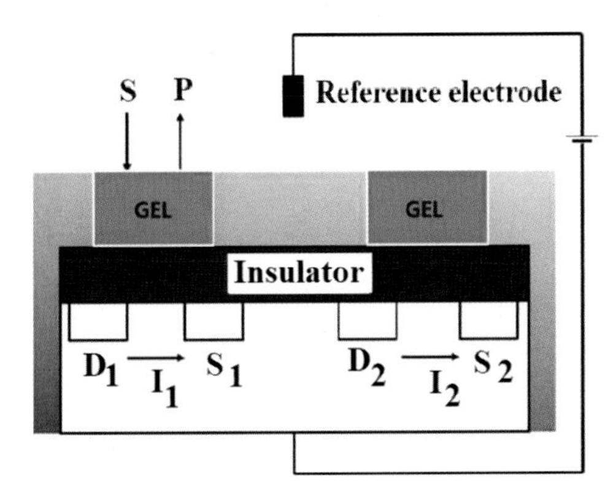

FIGURE 2.11

Scheme of a cross section of a glucose oxidase–based enzyme ISFET (Caras et al., 1985). P and S show the product and the substrate.

developed a glucose oxidase–based enzyme ISFET as a glucose sensor in which the enzyme is immobilized in a gel matrix of the chemically inactive and electrically neutral polyacrylamide (Fig. 2.11). In this sensor, enzyme loading is well controlled by the enzyme covalent immobilization in the uncharged polyacrylamide gel so that changes in the matrix structure associated with swelling or shrinkage due to ionic concentration alterations inside the gel are minimized (Caras et al., 1985). Generally speaking, in a FET-ion sensitive-based system, changes in ionic concentration in the solution result in changes in the current passing through the transistor.

FET sensors based on such nanostructures as semiconducting nanowires and carbon nanotubes have found a superior position in biosensing applications (Fig. 2.12). These systems exhibit a progressive sensitivity because of the confinement effect of the nanoscale channel (Park et al., 2014; Stern et al., 2007). Those even with nanowires are quite similar in operation to the typical FET systems.

The remarkable increase in sensitivity achieved by FET sensors containing nanostructures stems from the drastic increase in surface-to-volume (S/V) ratio as a result of wire diameters reduced to the nanometer scale. Thus, nanowires with a greater diameter offer a lower S/V ratio. Fig. 2.12 depicts a conception of conductivity along the nanowires under the influence of surface interactions (Grieshaber et al., 2008). Conductivity in these FET devices is strongly influenced by both surface and inner conductions.

Patolsky et al. (2004) used the *p*-type silicon nanowires in an FET to measure and detect the A-type influenza virus (Patolsky et al., 2004). Fig. 2.13 shows a general view of the antibody-modified silicone nanowire FET device used to measure the virus of influenza. Based on their studies, changes in pH are responsible for the variation in the conductivity related to the binding (unbinding) of virus particles. After optimization, an approximate value of 6.8 was selected for pH and the isoelectric point (IEP) of the target virus ranged from a pH value of 6.5–7.0. It is clear from the results that the FET nanowires served as a very useful tool for determining IEPs.

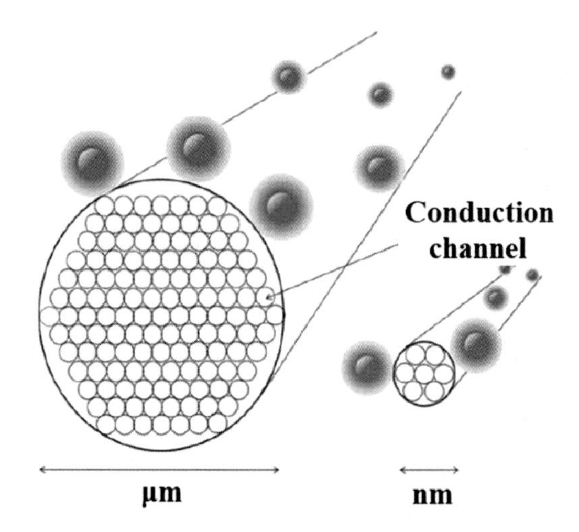

FIGURE 2.12

Diagram for explaining the relationship between nanowire surface interactions and its inner conduction (Elfström et al., 2007; Park et al., 2014).

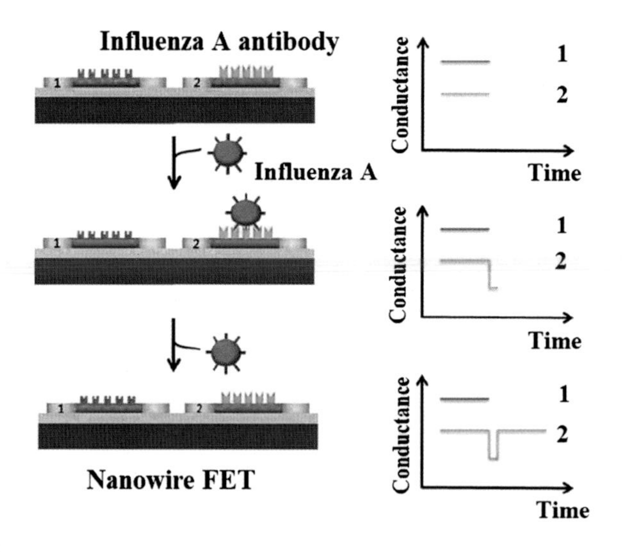

FIGURE 2.13

Representation of a virus-specific antibody-modified nanowire FET for determination of the influenza virus (Park et al., 2014; Patolsky et al., 2004). 1 and 2 show the modified nanowires signals of a specific and nonspecific antibody, respectively.

2.1.7 Electrochemical impedance spectroscopy

The electrochemical impedance technique is a simple, inexpensive, fast, and noninvasive one for measurements in biological systems. "Impedance" refers to a physical variable that examines the characteristics of the resistance of an electrical circuit in the presence of an alternating current applied between the electrodes (Huang et al., 2011; Rezaei et al., 2015b, 2011b, 2011a). In this system, the current flow response is measured by applying a small sinusoidal potential and detecting changes in frequency (*f*) from the applied potential over a wide frequency range. Examination of the mathematical relations obtained by varying the excitation frequency, the impedance response appears as a complex number as a function of frequency, which consists of both real and imaginary components (Tuorkey, 2012). EIS is capable of assessing the intrinsic material properties and investigating the particular processes involved in the conductivity/resistivity or the capacitivity of the electrochemical system (Fig. 2.14). These systems serve as very useful tools for the characterization and analysis of materials and biosensor transductions (Ensafi et al., 2017e, 2014a; 2014b).

The four elements of ohmic resistance, capacitance, constant phase element, and Warburg impedance (W) are used in the EIS analysis of electrolyte-based systems. To investigate the impedance behavior of each particular system, it is necessary to select an appropriate equivalent circuit (based on the elements defined in Table 2.1). Equivalent circuits are useful tools for approximating experimental impedance data, as they provide good descriptions of impedance components in parallel and/or in series. Randles circuit is most commonly used for electrodes immersed in electrolytes; it includes solution resistance (R_S), charge (electron) transfer resistance (R_{ct}), capacitance of double layer (C_{dl}), and mass transfer element, shown as Warburg impedance (W) (Bard and Faulkner, 2001).

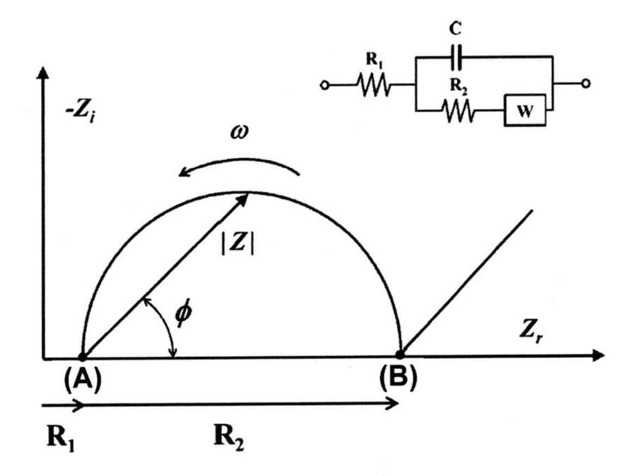

FIGURE 2.14

The typical Nyquist plot of an electrochemical cell containing working electrode into the electrolyte containing electroactive species and inset is the Randles' equivalent circuit.

Table 2.1 Definitions, frequency dependence, and phase shifts of the impedance components mostly used to explain the (bio)electrochemical behavior of systems.

Impedance component	Definition	Phase angle	Frequency dependence
R	$Z=R$	0°	No
C	$Z_C=1/j\omega C$	90 degrees	Yes
CPE	$Z_{CPE}=A^{-1}(j\omega)^a$	0–90 degrees	Yes
W (infinite)[a]	$Z_W=\sigma/\omega(1-j)$ $\sigma=RT/n^2F^2 2(1/D_O c_O+1/D_R c_R)$	45 degrees	Yes

[a] *With a limited diffusion layer, a different behavior is observed. An explanation for the two borderline cases with blocked and nonblocked transportations at the end of the diffusion region is given in Ref. [3].*

As seen in Fig. 2.14, in a typical Nyquist plot, the equation $\omega = 2\pi f = 1/R_{ct}C_{dl}$ is used to calculate the double-layer capacitance based on the frequency on the maximum point of the semicircle. A line at an angle of 45 degrees represents the Warburg limited behavior, which can be extrapolated from the real axis. Also, the intercept at this section is equal to $R_s + R_{ct} - 2\sigma C_{dl}$, in which σ represents the diffusion coefficient whose value can be subsequently calculated. In this method, R_s and R_{ct} can be straightforwardly obtained from the points (a) and (b), respectively.

EIS has been widely used to investigate the procedures for biosensor fabrication. It also plays an essential role in the study of such different kinds of bioanalytical species as whole cells, proteins, microorganisms, nucleic acids, antigens, and antibodies (Bonanni and del Valle, 2010; GUAN et al., 2004; Shipway et al., 2000). Ismail et al. (2011) used the EIS technique to examine clustered mineral ions. In the pathological calcification of the cardiovascular system, this agent is known to be a cause of high mortality. Therefore, electrical impedance changes are recorded by an aggregation of mineral ions in the supersaturated serum (Fig. 2.15). Inhibition of ectopic calcification is based on two modes: (1) formation of various acidic serum proteins, calciprotein particles, nanospherical complexes of calcium phosphate mineral, and fetuin-A and (2) stability of calcium phosphate prenucleation clusters using a fetuin-A monomer (Ismail et al., 2011).

Rezaei et al. (2011b) designed a monoclonal antibody immunosensor based on AgNPs to determine the presence of doxorubicin, an important anticancer drug, in human serum samples using the EIS-based technique. The results shown in Fig. 2.16 indicate that the immunosensor served as a useful tool in a clinical laboratory for screening picogram quantities of doxorubicin (Rezaei et al., 2011b).

2.1.8 Waveguide-based techniques and electrochemistry

Waveguide-sensing systems, mainly classified into single-mode and multimode ones, have been the subject of many research efforts in recent decades. The multimode waveguide systems are low cost and easy to manufacture owing to the use

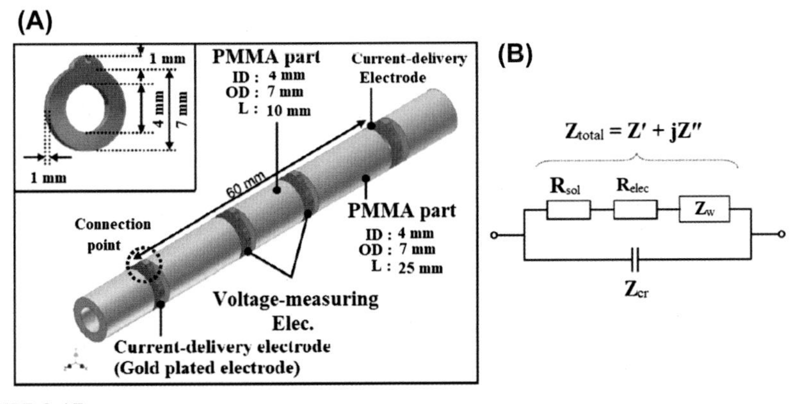

FIGURE 2.15

(A) EIS-determination cell (~1 mL volume) consists of four electrodes: the sensing (inner pair) and the stimulating (outer pair) electrodes. (B) The electrical equivalent circuit for modeling the EIS recorded data (Ismail et al., 2011).

of glass, polymer, or silica materials. In these systems, the thickness (in the order of several microns) is much more important than the wavelength of the light used for excitation, as it allows a simple edge coupling of the excitation light in the waveguide structure. Given the low signal intensity levels, fluorescence-based detection techniques are concentrated on the Peltier-cooled charge-coupled device cameras.

On the other hand, waveguides of the single mode are typically prepared using thin layer deposition methods or such alternatives as the sol-gel or ion deposition (Iii et al., 2006). These systems contain a very thin dielectric index film (<wavelength of excitation) deposited on a low index substrate (Mukundan et al., 2009). The single-mode waveguide has attracted more attention than the multimode or fiber waveguides thanks to its two orders of magnitude in supporting several thousands of reflections per centimeter of beam propagation for the wavelength of the visible region (Guo Qing Luo et al., 2005).

2.1.9 Electrochemical-surface plasmon resonance

Surface plasmon resonance (SPR) optical sensor is known as a robust technique for characterizing dielectric properties of surfaces, interfaces, and thin films. In the surface plasmons (charge density oscillations), the wavevector of the incoming light is matched with the surface plasmon through controlling the angle of the incident light beam, which leads to optical excitation at the interface between the dielectric material and the metal. The surface between the two media with different refractive indexes is covered with a tiny layer of a conductive material (often noble metals such as Au, Ag, or Cu). When the conductive layer is irradiated with an SPR angle (light of a different wavelength at a particular incidence angle), an evanescent wave is formed from the total internal reflection to excite an electron wave (termed as

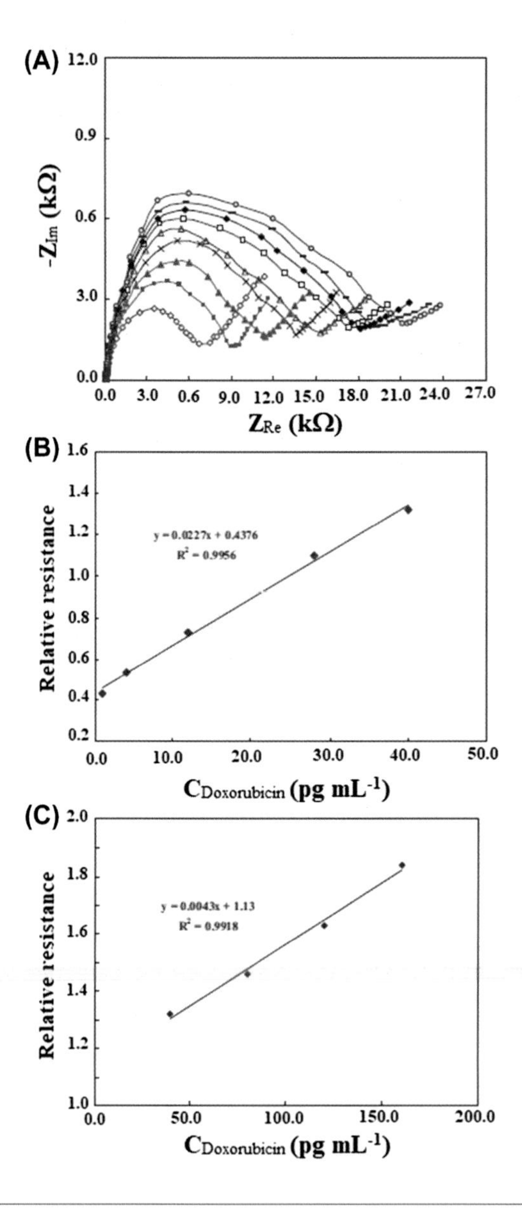

FIGURE 2.16

(A) Nyquist diagrams of the developed modified antibody/AgNP/HDT/gold electrode sensor after action with the various amounts ($pgmL^{-1}$) of doxorubicin. (B) Calibration curves in two concentration ranges (Behzad Rezaei et al., 2011b).

"the surface plasmon") that propagates through the metal surface (Division et al., 2010; Mcdonnell, 2001). Broadly defined, the SPR mechanism is based on measuring minor variations in refractive index in the vicinity of the surface of a conductive thin film (Campagnolo et al., 2004; Liedberg et al., 1995). In biosensor-based systems, SPR can be used to check biomolecular absorption/desorption events at surfaces to provide in situ, time-dependent, label-free, and quantifiable measurements of surface coverage.

In the meantime, an efficient electrochemical-SPR (EC-SPR) method has been developed for both the study of local electrochemical reactions on the surface of the electrode and investigation of the structure and activity of redox reactions to explore exceptionally minor changes in electrostatic fields (Boussaad et al., 2000; Hanken and Corn, 1997). In a typical EC-SPR system, a three-electrode assembly with SPR was used to investigate the interactions between the electric field and the chemical changes (Fig. 2.17). In a three-electrode system, the working electrode (the gold film in which the surface plasmon is excited) controls both the potential and the interfacial electric field. In general, the electrode potential sweeps between the two potentials at a constant sweep rate to give rise to potential alterations between the working and the counter (typically a platinum wire)-electrodes. To apply a known and constant potential to the working electrode, the potential is supplied to a reference electrode and the current is simultaneously recorded using a counter. The current generated is based on both the nonfaradaic (as a result of charging and discharging of the double layer capacitance) and the faradaic (due to the electrochemical redox reaction) components. Application of the EC-SPR technique has led to the substantially growing use of SPR to the extent that this method is now widely used in the study of enzymatic processes (Juan-Colás et al., 2017), molecular adsorption with controlled potential (Zhai et al., 2007), charge transfer reaction (Yao et al., 2004), and detection of DNA using electrochemical sensors (Juan-Colás et al., 2017).

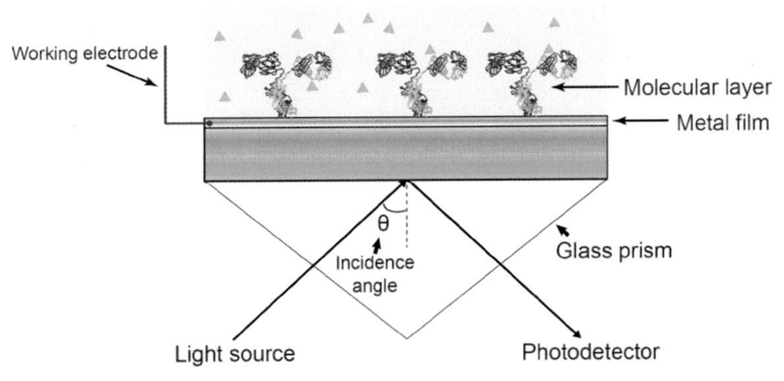

FIGURE 2.17

Diagram of electrochemical-SPR (EC-SPR) system for electrochemical studies (Juan-Colás et al., 2017).

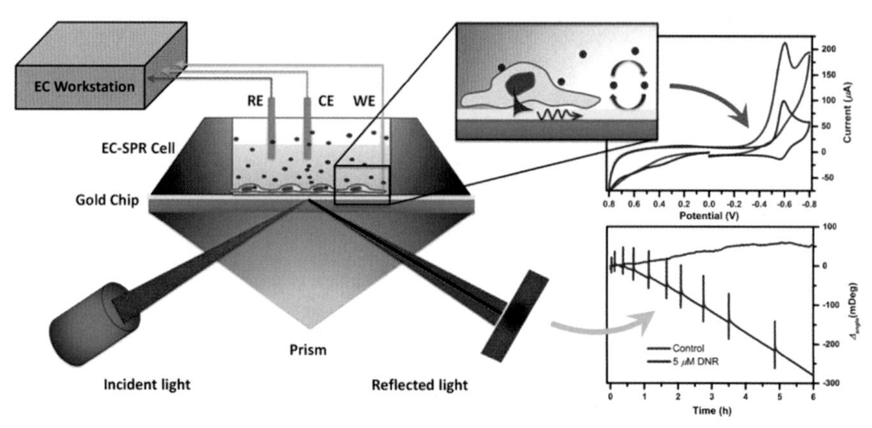

FIGURE 2.18

Schematic diagrams of a combination of electrochemical methods and SPR in the real-time estimation of the treatment of live cancer cells using DNR (Wu et al., 2015).

Wu et al. (2015) used a novel strategy based on the EC-SPR technique for estimating live cancer cell treatment with daunorubicin (DNR) at the interface of chips. Their results indicated that SPR signal variations were strongly influenced not only by changes in the mass and morphology of the adsorbed cancer cells but also by those in the refractive index of the medium solution (Fig. 2.18). They further used the electrochemical technique to monitor residual DNR concentrations outside the cells in the medium. Based on their observations, the combined SPR and electrochemical technique provided a powerful tool (with label-free and real-time features) for estimating the efficiency of the bioactive agent treatment in the cell. The proposed technique seems to have good prospects for inspecting clinical treatment procedures and pharmaceutical analyses (Wu et al., 2015).

2.1.10 Ellipsometry and electrochemistry

Ellipsometry is a powerful experimental method that measures changes in the polarization (elliptical polarization in general) of light. The measurements provide useful information on materials, especially on those in the phase boundary region. In this technique, a polarized light beam is reflected from a flat surface (which may be covered with a thin film) at a nonnormal angle and the resulting change in light polarization is measured (Fig. 2.19). Its capacity for use for both thin film and bulk material investigations has won this technique's applications in a wide variety of electrochemical systems such as films on metals made by passivation, absorbed ionic and molecular species, conducting polymers, and electrodeposited materials (Covington, 2010).

The unique features that make ellipsometry a beneficial research tool in electrochemical systems are (1) nondestructiveness, the exposed materials do not alter owing to the fact that the light used in this technique is in the visible or

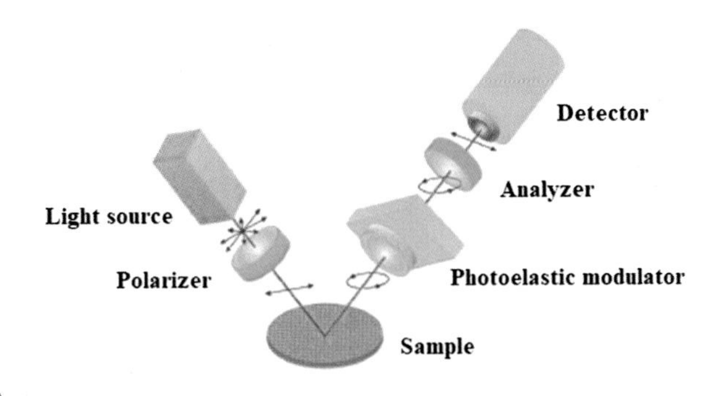

FIGURE 2.19

Schematic setup of an ellipsometry experiment (Dressel et al., 2008).

near-visible range; (2) in situ measurement, the measurements can be carried out simultaneously on a sample in an electrochemical environment without being taken out of the cell; (3) extremely sensitive measurements, quantitative measurements performed with this method have sufficient sensitivity to detect small ions or small layers of molecules; and (4) unambiguous determination of the thickness of anodic or other films of nanometer or angstrom sizes on electrode surfaces (Greef, 1970). Using the ellipsometry technique, Blaffart et al. (2013) studied electrostriction in thin films of anodic and thermal silicon dioxide. The authors showed that the electrical fields applied to their thin silica films resulted in the formation of tensile or compressive electrostrictive stresses (Fig. 2.20). By combining the curvature and ellipsometry measurements in-line, the electrostrictive stresses may be recorded

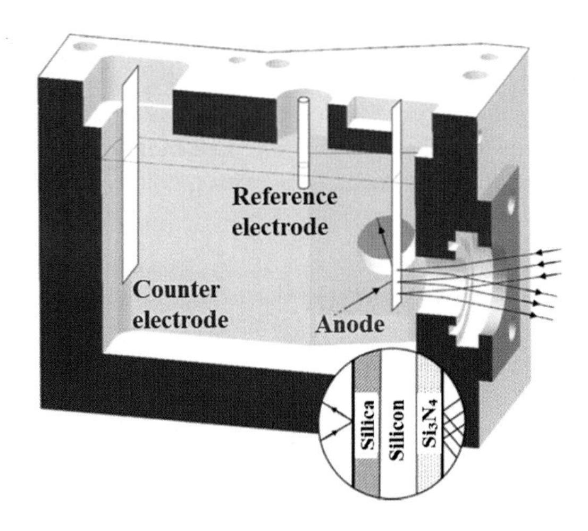

FIGURE 2.20

A cross-sectional observation of the used electrochemical cell displaying the optical beams and electrodes. Silica covered on the silicon sample (anodic or thermal) on the ellipsometer side and silicon nitride on the stress sensor side (Blaffart et al., 2013).

through both the cyclic polarization of the silicon thin films and the growth of the anodic silica (Blaffart et al., 2013).

2.1.11 Electrochemical atomic force microscopy

Scanning probe microscopy (SPM) has been designed to image and manipulate individual atoms or molecules at the nanoscale level (Boussaad et al., 2000). One of the most commonly known SPM techniques is the atomic force microscope (AFM), which is equipped with powerful and versatile tools for inspecting and imaging molecular surfaces (Marti et al., 1988). In the AFM technique, the interaction force between a probe (e.g., a sharp tip connected to a force-sensing cantilever) and the sample surface is used to obtain images. The details and the theoretical setup of the AFM techniques may be found in the literature (e.g., Dong and Shannon, 2000; Immoos et al., 2004; Muguruma and Kase, 2006; Reggente et al., 2017a).

Although the AFM technique is often used for individual characterization before and/or after an electrochemical analysis, what has nowadays become the focus of more interest is the combination of this technique and electrochemical ones. Electrochemical AFM (or, EC-AFM for short) is a special type of SPM in which a classical AFM is combined with electrochemical measurements. The combined technique is capable of in situ AFM measurements in an electrochemical cell, while simultaneously playing an important role in the examination of morphological changes at the electrode surface during the chemical reaction in question. This allows for the close examination of the solid-liquid interface (Fig. 2.21) (Toma et al., 2016). In this method, use is made of a standard AFM instrument and an electrochemical cell with the following three electrodes: a sample surface (where the process takes place) acting as a working electrode, AgCl wires used as a reference electrode, and platinum wires serving as the counterelectrode (Valtiner et al., 2011).

Depending on the polarization tip, two configurations are possible. The first is a "passive" probe configuration, in which the potential is applied to the electrochemical cell and the AFM probe, as a "passive" element, monitors the surface changes as a function of time (Gewirth and Niece, 1997). Morphology is monitored by current

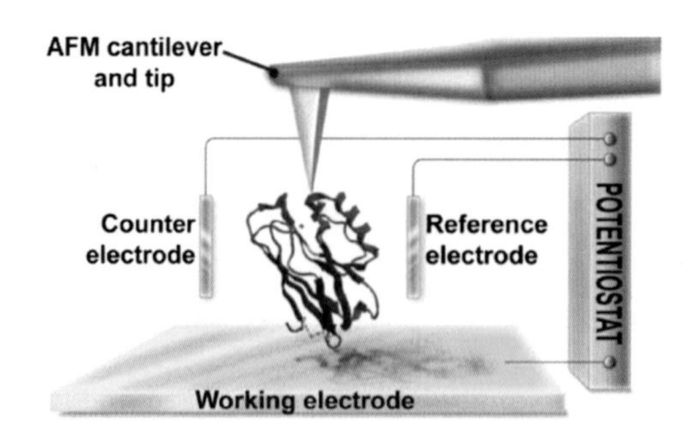

FIGURE 2.21

Electrochemical AFM system (Wu et al., 2016).

flows through the sample during the potential sweeping that can be accomplished in a variety of ways such as CV or pulse voltammetry (Reggente et al., 2017a).

The second configuration is an "active" probe one, in which surface electroactivity is investigated using a particular potential applied to the tip fixed to the substrate space during the scanning process (Gardner and Macpherson, 2002; Wain et al., 2011). Under these conditions, the electrolyte species have the potential to move to the electrode surface, where they might be oxidized or reduced (Fig. 2.22.A). In the conductive substrate (Fig. 2.22.B), the electrochemically oxidized species at the tip surface diffuse toward the substrate where they might reduce (R). In cases where an insulator substrate is used, the R species already oxidized under the influence of the electrochemical reaction cannot react with the surface at the beginning of the reaction. Under these conditions, the current identified at the tip decays exponentially at a very short distance from the substrate Fig. 2.22.C).

If a laser beam is to be used for morphological examinations, it is necessary to choose a transparent electrolyte in the electrochemical cell because it is only in this way that the laser beam can reach the sample and be deflected. For this purpose, the electrolyte solution must be diluted with regard to the solute concentration. Another important parameter in electrolyte selection is the noncorrosiveness of the electrolyte in contact with the AFM scanner and cantilever.

EC-AFM has found extensive applications in battery, electrode corrosion, and electrodepositing studies in which different electrode surfaces (metals such as copper, or polymers such as polyaniline) are used (Koinuma and Uosaki, 1995; Li et al., 2001; Singh et al., 2009). Khalakhan et al. (2018) used the EC-AFM technique to study the dependence of the thin film platinum catalyst on potentials applied during an electrochemical cycle (Khalakhan et al., 2018). Using quantitative analyses of the AFM results, these authors were for the first time able to obtain useful information regarding the surface coarsening rates through the electrochemical degradation of

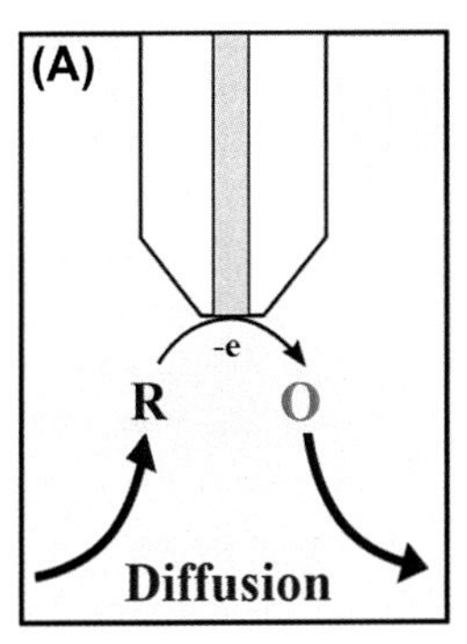

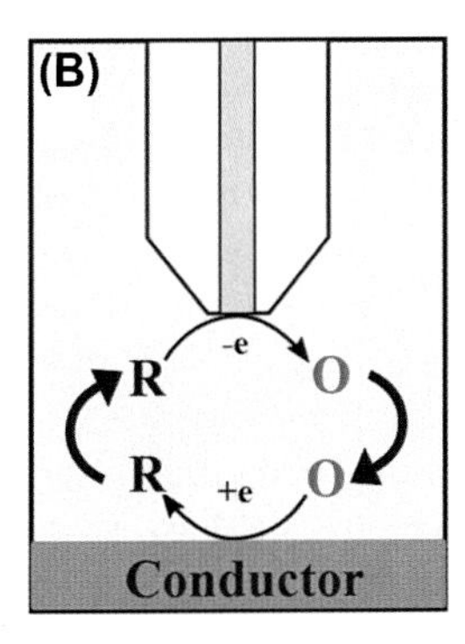

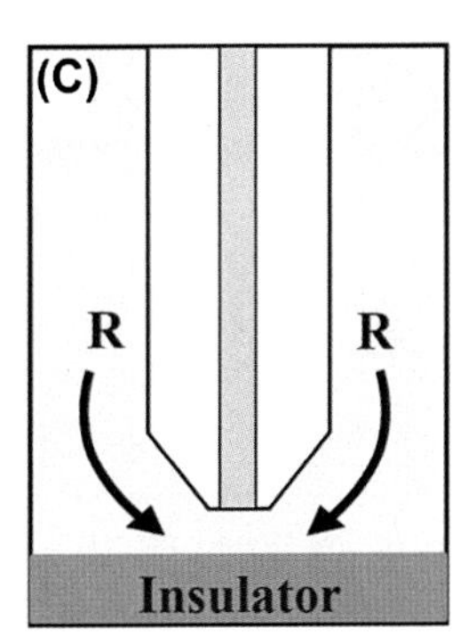

FIGURE 2.22

Scheme of the EC-AFM arrangement using an active probe setup (A) and of the reaction happening at the tip electrode if it is swept over a surface of the conductor (B) or an insulator substrate (C).

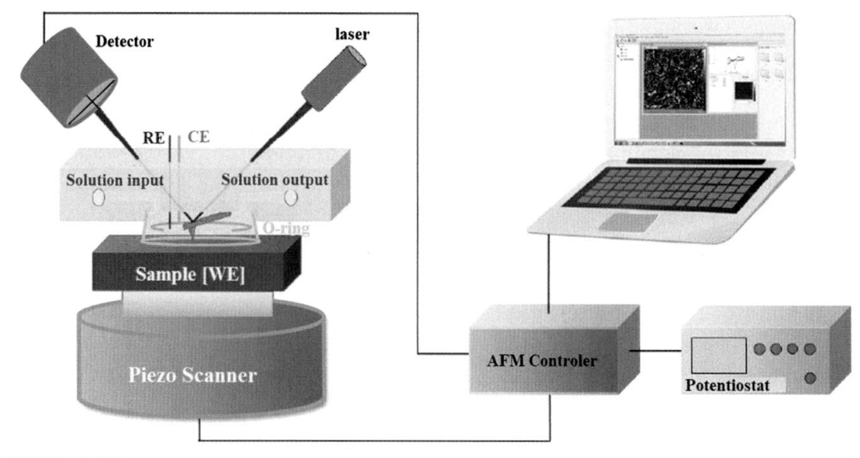

FIGURE 2.23

Diagram of EC-AFM method.

platinum. Fig. 2.23 demonstrates a schematic illustration of this device (Khalakhan et al., 2018).

One attractive application of the EC-AFM involves the study of special electrodes used in systems based on biosensors and chemosensors (Badia et al., 1999; Wain et al., 2011). In this regard, Eliaz et al. (2007) used the EC-AFM technique for both ex situ and in situ imaging as well as potentiodynamic and potentiostatic measurements to help promote the osseointegration formation of titanium-based alloys in orthopedic devices and implants (Reggente et al., 2017b).

2.1.12 Electrochemical quartz crystal microbalance using dissipation monitoring

Because of the development of new tools for investigating interface characteristics, new electrochemical methods have been developed in recent decades to identify and characterize the structure and composition of interfaces. Moreover, many electrochemical researchers have been focused on developing more sophisticated techniques for analyzing the electrode surface through coupling new tools to the already established electrochemical systems (Buttry and Ward, 1992). One of the most prominent techniques is the microcrystal quartz crystal. In quartz crystal microbalance (QCM), a tinny quartz crystal is sandwiched between two metal electrodes to create an alternating electric field across the crystalline structure. This, in turn, gives rise to the vibration of the crystal at its resonant frequency. Changes in the mass of the crystal and its electrode (as well as in other parameters) can be strongly influenced by the resonance frequency. Nowadays, the QCM has become a very powerful tool for examining electrochemical processes based on thin films, including monolayer and submonolayer films, through simultaneous measurement

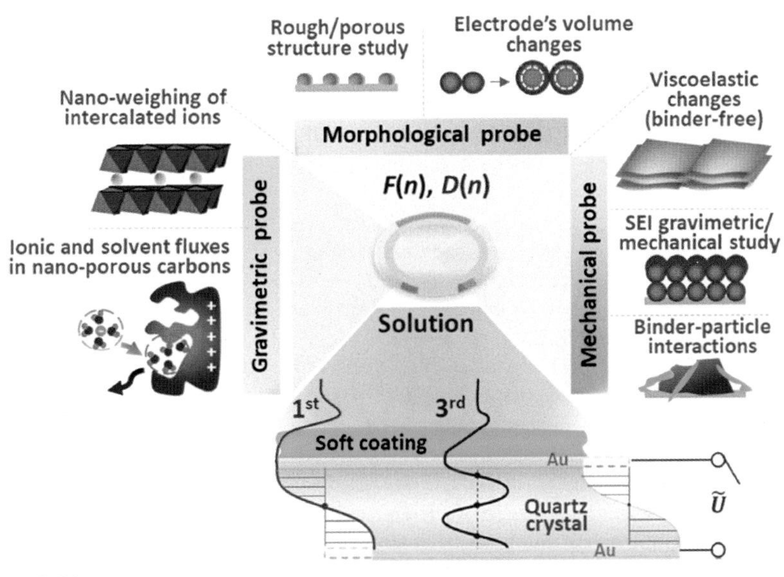

FIGURE 2.24

Applications of gravimetric and nongravimetric EQCM-D for a description of using electrodes in the energy-storage system. Bottom panel: The fundamental frequency of acoustic waves and their third overtone.

of mass changes in an electrochemical cell. This technique is known as the electrochemical quartz crystal microbalance (EQCM) (Buttry and Ward, 1992). Fig. 2.24 illustrates some of the EQCM applications as well as those of the gravimetric and nongravimetric EQCM-D using electrodes for deposition in energy storage systems (Shpigel et al., 2018).

Sibo Liu et al. (?) used a new strategy based on the QCM immunosensor to measurement and sensitive detection of C-reactive protein (CRP). Albumin and CRP at serum levels are known as practical and reliable systems used for predicting outcomes in patients with inflammation. The authors showed that QRM immunosensor sensitivity is much higher than the biomarker determination (Gao, 2018).

References

Anson, F.C., Osteryoung, R.A., 1983. Chronocoulometry: a convenient, rapid and reliable technique for detection and determination of adsorbed reactants. Journal of Chemical Education 60, 293.

Badia, A., Arnold, S., Scheumann, V., Zizlsperger, M., Mack, J., Jung, G., Knoll, W., 1999. Probing the electrochemical deposition and/or desorption of self-assembled and electropolymerizable organic thin films by surface plasmon spectroscopy and atomic force microscopy. Sensors and Actuators B: Chemical 54, 145–165.

Bakker, E., Pretsch, E., 2005. Potentiometric sensors for trace-level analysis. Trends in Analytical Chemistry 24, 199–207.
Bakker, E., 2014. Electrochemical sensors. Analytical Chemistry 76, 3285–3298.
Banica, F.-G., n.d. Chemical Sensors and Biosensors: Fundamentals and Applications.
Bard, A.J., Faulkner, L.R., 2001. Electrochemical Methods: Fundamentals and Applications.
Bartlett, P.N., 2008. Bioelectrochemistry Fundamentals, Experimental Techniques and Applications.
Blaffart, F., Van Overmeere, Q., Pardoen, T., Proost, J., 2013. In situ monitoring of electrostriction in anodic and thermal silicon dioxide thin films. Journal of Solid State Electrochemistry 17, 1945–1954.
Bocarsly, A.B., 2012. Cyclic voltammetry. In: Characterization of Materials. John Wiley & Sons, Inc., Hoboken, NJ, USA, pp. 837–850.
Bogdan, M., Brugger, D., Rosenstiel, W., Speiser, B., 2014. Estimation of diffusion coefficients from voltammetric signals by support vector and gaussian process regression. Journal of Cheminformatics 6–30.
Bonanni, A., del Valle, M., 2010. Use of nanomaterials for impedimetric DNA sensors: a review. Analytica Chimica Acta 678, 7–17.
Boussaad, S., Pean, J., Tao, N.J., 2000. High-resolution multiwavelength surface plasmon resonance spectroscopy for probing conformational and electronic changes in redox proteins. Analytical Chemistry 72, 222–226.
Buck, R.P., 1974. Ion selective electrodes, potentiometry, and potentiometric titrations. Analytical Chemistry 46, 28–51.
Buttry, D.A., Ward, M.D., 1992. Measurement of interfacial processes at electrode surfaces with the electrochemical quartz crystal microbalance. Chemistry Review 92, 1355–1379.
Campagnolo, C., Meyers, K.J., Ryan, T., Atkinson, R.C., Chen, Y.T., Scanlan, M.J., Ritter, G., Old, L.J., Batt, C.A., 2004. Real-Time, label-free monitoring of tumor antigen and serum antibody interactions. Journal of Biochemical and Biophysical Methods 61, 283–298.
Caras, S.D., Janata, J., Saupe, D., Schmitt, K., 1985. pH-based enzyme potentiometric sensors. Part 1. Theory. Anal. Chem. 57, 1917–1920.
Chaubey, A., Malhotra, B.D., 2002. Mediated biosensors. Biosensors and Bioelectronics 17, 441–456.
Chen, A., Shah, B., 2013. Electrochemical sensing and biosensing based on square wave voltammetry. Anal. Methods 5, 2158.
Cheng, S., 2015. Shanshan CHENG 程 姗姗.
Covington, A., 2010. Electrodeposition, Modern Aspects of Electrochemistry. Springer, New York, NY.
Daniels, J.S., Pourmand, N., 2007. Label-free impedance biosensors: opportunities and challenges. Electroanalysis 19, 1239–1257.
Davis, G., 1985. Electrochemical techniques for the development of amperometric biosensors. Biosensors 1, 161–178.
Division, V., Establishment, D., Road, J., 2010. Surface plasmon resonance, 627, 79–92.
Dong, Y., Shannon, C., 2000. Heterogeneous immunosensing using antigen and antibody monolayers on gold surfaces with electrochemical and scanning probe detection. Analytical Chemistry 72, 2371–2376.
D'Orazio, P., 2011. Biosensors in clinical chemistry — 2011 update. Clinica Chimica Acta 412, 1749–1761.

Dressel, M., Gompf, B., Faltermeier, D., Tripathi, A.K., Pflaum, J., Schubert, M., 2008. Kramers-Kronig-consistent optical functions of anisotropic crystals: generalized spectroscopic ellipsometry on pentacene. Optics Express 16, 19770–19778.

Eggins, B.R., 2002. Chemical Sensors and Biosensors, Analytical Techniques in the Sciences.

El Harrad, L., Bourais, I., Mohammadi, H., Amine, A., 2018. Recent advances in electrochemical biosensors based on enzyme inhibition for clinical and pharmaceutical applications. Sensors 18, 164.

Electrochemistry, A., 2006. Controlled-potential techniques. In: Analytical Electrochemistry. John Wiley & Sons, Inc., Hoboken, NJ, USA, pp. 67–114.

Elfstrom, N., Juhasz, R., Sychugov, I., Engfeldt, T., Karlstrom, A.E., Linnros, J., 2007. Surface Charge Sensitivity of Silicon Nanowires: Size Dependence. Nano Lett 7, 2608–2612.

Elgrishi, N., Rountree, K.J., McCarthy, B.D., Rountree, E.S., Eisenhart, T.T., Dempsey, J.L., 2018. A practical beginner's guide to cyclic voltammetry. Journal of Chemical Education 95, 197–206.

Eliaz, N., Eliyahu, M., 2007. Electrochemical processes of nucleation and growth of hydroxyapatite on titanium supported by real-time electrochemical atomic force microscopy. Biomedical materials research 80A, pp. 621–634.

Emr, S.A., 1995. Use of Polymer Films in Amperometric Biosensors and Chemically Modified Electrodes, p. 357.

Ensafi, A.A., Khoddami, E., Rezaei, B., 2013. A combined liquid three-phase microextraction and differential pulse voltammetric method for preconcentration and detection of ultra-trace amounts of buprenorphine using a modified pencil electrode. Talanta 116, 1113–1120.

Ensafi, A.A., Amini, M., Rezaei, B., 2014a. Impedimetric DNA-biosensor for the study of anti-cancer action of mitomycin C: comparison between acid and electroreductive activation. Biosensors and Bioelectronics 59, 282–288.

Ensafi, A.A., Amini, M., Rezaei, B., 2014b. Assessment of genotoxicity of catecholics using impedimetric DNA-biosensor. Biosensors and Bioelectronics 53, 43–50.

Ensafi, A.A., Lesani, S., Amini, M., Rezaei, B., 2015a. Electrochemical ds-DNA-based biosensor decorated with chitosan modified multiwall carbon nanotubes for phenazopyridine biodetection. Journal of the Taiwan Institute of Chemical Engineers 54, 165–169.

Ensafi, A.A., Sohrabi, M., Jafari-Asl, M., Rezaei, B., 2015b. Selective and sensitive furazolidone biosensor based on DNA-modified TiO_2 -reduced graphene oxide. Applied Surface Science 356, 301–307.

Ensafi, A.A., Khoddami, E., Rezaei, B., 2016. Development of a cleanup and electrochemical determination of flutamide using silica thin film pencil graphite electrode functionalized with thiol groups. Journal of the Iranian Chemical Society 13, 1683–1690.

Ensafi, A.A., Amini, M., Rezaei, B., Talebi, M., 2016a. A novel diagnostic biosensor for distinguishing immunoglobulin mutated and unmutated types of chronic lymphocytic leukemia. Biosensors and Bioelectronics 77, 409–415.

Ensafi, A.A., Karbalaei, S., Heydari-Bafrooei, E., Rezaei, B., 2016b. Biosensing of naringin in marketed fruits and juices based on its interaction with DNA. Journal of the Iranian Chemical Society 13, 19–27.

Ensafi, A.A., Zandi-Atashbar, N., Rezaei, B., Ghiaci, M., Esmaeili Chermahini, M., Moshiri, P., 2016c. Non-enzymatic glucose electrochemical sensor based on silver nanoparticle decorated organic functionalized multiwall carbon nanotubes. RSC Adv 6, 60926–60932.

Ensafi, A.A., Ahmadi, N., Rezaei, B., 2017a. Nickel nanoparticles supported on porous silicon flour, application as a non-enzymatic electrochemical glucose sensor. Sensors and Actuators B: Chemical 239, 807–815.
Ensafi, A.A., Farfani, N.K., Amini, M., Rezaei, B., 2017b. Developing a sensitive DNA biosensor for the detection of flutamide using electrochemical method. Journal of the Iranian Chemical Society 14, 1325–1334.
Ensafi, A.A., Fattahi-Sedeh, S., Jafari-Asl, M., Rezaei, B., 2017c. Thionine-functionalized graphene oxide, new electrocatalyst for determination of nitrite. Journal of the Iranian Chemical Society 14, 1069–1078.
Ensafi, A.A., Gorgabi-Khorzoughi, M., Rezaei, B., Jafari-Asl, M., 2017d. Electrochemical behavior of polyoxometalates decorated on poly diallyl dimethyl ammonium chloride-MWCNTs: a highly selective electrochemical sensor for determination of guanine and adenine. Journal of the Taiwan Institute of Chemical Engineers 78, 56–64.
Ensafi, A.A., Khoddami, E., Rezaei, B., 2017e. Aptamer@Au-o-phenylenediamine modified pencil graphite electrode: a new selective electrochemical impedance biosensor for the determination of insulin. Colloids and Surfaces B: Biointerfaces 159, 47–53.
Ensafi, A.A., Rezaloo, F., Rezaei, B., 2017f. CoFe2O4/reduced graphene oxide/ionic liquid modified glassy carbon electrode, a selective and sensitive electrochemical sensor for determination of methotrexate. Journal of the Taiwan Institute of Chemical Engineers 78, 45–50.
Evans, D.H., Kelly, M.J., 1982. Theory for double potential step chronoamperometry, chronocoulometry, and chronoabsorptometry with a quasi-reversible electrode reaction. Analytical Chemistry 54, 1727–1729.
Fehr, M., Ehrhardt, D.W., Lalonde, S., Frommer, W.B., 2004. Minimally invasive dynamic imaging of ions and metabolites in living cells. Current Opinion in Plant Biology 7, 345–351.
Franks, W., Schenker, I., Schmutz, P., Hierlemann, A., 2005. Impedance characterization and modeling of electrodes for biomedical applications. IEEE Transactions on Biomedical Engineering 52, 1295–1302.
Gao, K., 2018. Development of an electrochemical quartz crystal microbalance-based immunosensor for C-reactive protein determination. International Journal of Electrochemical Science 13, 812–821.
Gardner, C.E., Macpherson, J.V., 2002. Peer reviewed: atomic force microscopy probes go electrochemical. Analytical Chemistry 74, 576 A–584 A.
Gewirth, A.A., Niece, B.K., 1997. Electrochemical applications of in situ scanning probe microscopy. Chemistry Review 97, 1129–1162.
Ghiaci, M., Tghizadeh, M., Ensafi, A.A., Zandi-Atashbar, N., Rezaei, B., 2016. Silver nanoparticles decorated anchored type ligands as new electrochemical sensors for glucose detection. Journal of the Taiwan Institute of Chemical Engineers 63, 39–45.
Goral, V.N., Zaytseva, N.V., Baeumner, A.J., 2006. Electrochemical microfluidic biosensor for the detection of nucleic acid sequences. Lab on a Chip 6, 414.
Greef, R., 1970. An automatic ellipsometer for use in electrochemical investigations. Review of Scientific Instruments 41, 532–538.
Grieshaber, D., MacKenzie, R., Vörös, J., Reimhult, E., 2008. Electrochemical biosensors - sensor principles and architectures. Sensors 8, 1400–1458.
Guan, J.-G., Miao, Y.-Q., Zhang, Q.-J., 2004. Impedimetric biosensors. Journal of Bioscience and Bioengineering 97, 219–226.

Han, S.H., Kim, S.K., Park, K., Yi, S.Y., Park, H.-J., Lyu, H.-K., Kim, M., Chung, B.H., 2010. Detection of mutant p53 using field-effect transistor biosensor. Analytica Chimica Acta 665, 79–83.

Hanken, D.G., Corn, R.M., 1997. Electric field measurements inside self-assembled multilayer films at electrode surfaces by electrochemically modulated surface plasmon resonance experiments. Israel Journal of Chemistry 37, 165–172.

Heinze, J., 1993. Ultramicroelectrodes in electrochemistry. Angewandte Chemie International Edition in English 32, 1268–1288.

Honeychurch, K.C., 2012. 13 printed thick-film biosensors, Printed Films, pp. 366–409.

Huang, V.M., Wu, S.-L., Orazem, M.E., Pébère, N., Tribollet, B., Vivier, V., 2011. Local electrochemical impedance spectroscopy: a review and some recent developments. Electrochimica Acta 56, 8048–8057.

Iii, W.J.D., Wysocki, R.J., Armstrong, N.R., Saavedra, S.S., 2006. Electrochemical Copolymerization and Spectroelectrochemical Characterization of 3 , 4-Ethylenedioxythiophene and 3 , 4-Ethylenedioxythiophene-Methanol Copolymers on Indium-Tin Oxide, pp. 4418–4424.

Immoos, C.E., Lee, S.J., Grinstaff, M.W., 2004. Conformationally gated electrochemical gene detection. ChemBioChem 5, 1100–1103.

Inzelt, G., 2014. Chronoamperometry, chronocoulometry, and chronopotentiometry. In: Kreysa, G., Ota, K., Savinell, R.F. (Eds.), Encyclopedia of Applied Electrochemistry. Springer New York, New York, NY, pp. 207–214.

Ismail, A.H., Schäfer, C., Heiss, A., Walter, M., Jahnen-Dechent, W., Leonhardt, S., 2011. An electrochemical impedance spectroscopy (EIS) assay measuring the calcification inhibition capacity in biological fluids. Biosensors and Bioelectronics 26, 4702–4707.

Tuorkey, M.J., 2012. Bioelectrical impedance as a diagnostic factor in the clinical practice and prognostic factor for survival in cancer patients: prediction, accuracy and reliability. Journal of Biosensors & Bioelectronics 03.

Janata, J., Josowicz, M., 1997. Peer reviewed: a fresh look at some old principles: the Kelvin probe and the Nernst equation. Analytical Chemistry 69, 293A–296A.

Juan-Colás, J., Johnson, S., Krauss, T.F., 2017. Dual-mode electro-optical techniques for biosensing applications: a review. Sensors 17, 1–15.

Kaisti, M., 2017. Detection principles of biological and chemical FET sensors. Biosensors and Bioelectronics 98, 437–448.

Kamat, A., Huth, A., Klein, O., Scholl, S., 2010. Chronoamperometric investigations of the electrode-electrolyte interface of a commercial high temperature PEM fuel cell. Fuel Cells 10, 983–992.

Katz, E., Willner, I., 2003. Probing biomolecular interactions at conductive and semiconductive surfaces by impedance spectroscopy: routes to impedimetric immunosensors, DNA-sensors, and enzyme biosensors. Electroanalysis 15, 913–947.

Kawagoe, K.T., Wightman, R.M., 1994. Characterization of amperometry for in vivo measurement of dopamine dynamics in the rat brain. Talanta 41, 865–874.

Khalakhan, I., Choukourov, A., Vorokhta, M., Kúš, P., Matolínová, I., Matolín, V., 2018. In situ electrochemical AFM monitoring of the potential-dependent deterioration of platinum catalyst during potentiodynamic cycling. Ultramicroscopy 187, 64–70.

Koinuma, M., Uosaki, K., 1995. An electrochemical AFM study on electrodeposition of copper on p-GaAs(100) surface in HCl solution. Electrochimica Acta 40, 1345–1351.

Lefrou, C., Pierre Fabry, J.-C.P., 2012. Electrochemistry: The Basics, with Examples.

Li, Y., Maynor, B.W., Liu, J., 2001. Electrochemical AFM "Dip-Pen" nanolithography. Journal of the American Chemical Society 123, 2105–2106.

Li, M., Li, Y.-T., Li, D.-W., Long, Y.-T., 2012. Recent developments and applications of screen-printed electrodes in environmental assays—a review. Analytica Chimica Acta 734, 31–44.

Liedberg, B., Nylander, C., Lundstrom, I., 1995. Biosensing with surfacing plasmon resonance - how it all started. Biosensors and Bioelectronics 10, i–ix.

Lowe, B.M., Sun, K., Green, N.G., 2017. Field-effect sensors – from pH sensing to biosensing: sensitivity enhancement using streptavidin–biotin as a model system†. Analyst 142, 4173–4200.

Luo, X.-L., Xu, J.-J., Zhao, W., Chen, H.-Y., 2004. Glucose biosensor based on ENFET doped with SiO2 nanoparticles. Sensors and Actuators B: Chemical 97, 249–255.

Luo, G.Q., Hong, W., Hao, Z.-C., Liu, B., Li, W.D., Chen, J.X., Zhou, H.X., Wu, K., 2005. Theory and experiment of novel frequency selective surface based on substrate integrated waveguide technology. IEEE Transactions on Antennas and Propagation 53, 4035–4043.

Marti, O., Elings, V., Haugan, M., Bracker, C.E., Schneir, J., Drake, B., Gould, S.A.C., Gurley, J., Hellemans, L., Shaw, K., Weisenhorn, A.L., Zasadzinski, J., Hansma, P.K., 1988. Scanning probe microscopy of biological samples and other surfaces. Journal of Microscopy 152, 803–809.

Martinkova, P., 2017. Main streams in the construction of biosensors and their applications. International Journal of Electrochemical Science 12, 7386–7403.

Mcdonnell, J.M., 2001. Surface Plasmon Resonance : Towards an Understanding of the, pp. 572–577.

Mir, M., Homs, A., Samitier, J., 2009. Integrated electrochemical DNA biosensors for lab-on-a-chip devices. Electrophoresis 30, 3386–3397.

Mistry, K.K., Layek, K., Mahapatra, A., RoyChaudhuri, C., Saha, H., 2014. A review on amperometric-type immunosensors based on screen-printed electrodes. Analyst 139, 2289.

Muguruma, H., Kase, Y., 2006. Structure and biosensor characteristics of complex between glucose oxidase and plasma-polymerized nanothin film. Biosensors and Bioelectronics 22, 737–743.

Mukundan, H., Anderson, A., Grace, W.K., Grace, K., Hartman, N., Martinez, J., Swanson, B., 2009. Waveguide-based biosensors for pathogen detection. Sensors 9, 5783–5809.

Ohno, Y., Maehashi, K., Matsumoto, K., 2010. Label-Free Biosensors Based on Aptamer-Modified Graphene Field-Effect, pp. 18012–18013.

Olad, A., Gharekhani, H., 2015. Preparation and electrochemical investigation of the polyaniline/activated carbon nanocomposite for supercapacitor applications. Progress in Organic Coatings 81, 19–26.

Park, J., Hiep, H., Woubit, A., Kim, M., 2014. Applications of field-effect transistor (FET) -type biosensors, 23, 61–71.

Patolsky, F., Zayats, M., Katz, E., Willner, I., 1999. Precipitation of an insoluble product on enzyme monolayer electrodes for biosensor applications: characterization by faradaic impedance spectroscopy, cyclic voltammetry, and microgravimetric quartz crystal microbalance analyses. Analytical Chemistry 71, 3171–3180.

Patolsky, F., Zheng, G., Hayden, O., Lakadamyali, M., Zhuang, X., Lieber, C.M., 2004. Electrical detection of single viruses. Proceedings of the National Academy of Sciences 101, 14017–14022.

Pohanka, M., Republic, C., 2008. Electrochemical biosensors – principles and applications. Methods 6, 57–64.

Rashid, M.H., 2010. Microelectronic Circuits: Analysis & Design.

Reggente, M., Passeri, D., Rossi, M., Tamburri, E., Terranova, M.L., 2017. Electrochemical atomic force microscopy: in situ monitoring of electrochemical processes. In: AIP Conf. Proc, vol. 1873, p. 020009.

Rezaei, B., Mirahmadi Zare, S.Z., 2008. Modified glassy carbon electrode with multiwall carbon nanotubes as a voltammetric sensor for determination of noscapine in biological and pharmaceutical samples. Sensors and Actuators B 134, 292–299.

Rezaei, B., Majidi, N., Rahmani, H., Khayamian, T., 2011b. Electrochemical impedimetric immunosensor for insulin like growth factor-1 using specific monoclonal antibody-nanogold modified electrode. Biosensors and Bioelectronics 26, 2130–2134.

Rezaei, B., Saghebdoust, M., Sorkhe, A.M., Majidi, N., 2011c. Generation of a doxorubicin immunosensor based on a specific monoclonal antibody-nanogold-modified electrode. Electrochimica Acta 56, 5702–5706.

Rezaei, B., Zare, S.Z.M., Ensafi, A.A., 2011c. Square wave voltammetric determination of dexamethasone on a multiwalled carbon nanotube modified pencil electrode. Journal of the Brazilian Chemical Society 22, 897–904.

Rezaei, B., Askarpour, N., Ensafi, A.A., 2013. Adsorptive stripping voltammetry determination of methyldopa on the surface of a carboxylated multiwall carbon nanotubes modified glassy carbon electrode in biological and pharmaceutical samples. Colloids and Surfaces B: Biointerfaces 109, 253–258.

Rezaei, B., Askarpour, N., Ghiaci, M., Niyazian, F., Ensafi, A.A., 2015. Synthesis of functionalized MWCNTs decorated with copper nanoparticles and its application as a sensitive sensor for amperometric detection of H 2 O 2. Electroanalysis 27, 1457–1465.

Rezaei, B., Boroujeni, M.K., Ensafi, A.A., 2015a. Fabrication of DNA, o-phenylenediamine, and gold nanoparticle bioimprinted polymer electrochemical sensor for the determination of dopamine. Biosensors and Bioelectronics 66, 490–496.

Rezaei, B., Khosropour, H., Ensafi, A.A., Dinari, M., Nabiyan, A., 2015b. A new electrochemical sensor for the simultaneous determination of guanine and adenine: using a NiAl-layered double hydroxide/graphene oxide-multi wall carbon nanotube modified glassy carbon electrode. RSC Advances 5, 75756–75765.

Rezaei, B., Khosropour, H., Ensafi, A.A., Hadadzadeh, H., Farrokhpour, H., 2015c. A differential pulse voltammetric sensor for determination of glutathione in real samples using a trichloro(terpyridine)ruthenium(III)/Multiwall carbon nanotubes modified paste electrode. IEEE Sensors Journal 15, 483–490.

Rezaei, B., Boroujeni, M.K., Ensafi, A.A., 2016. Development of Sudan II sensor based on modified treated pencil graphite electrode with DNA, o-phenylenediamine, and gold nanoparticle bioimprinted polymer. Sensors and Actuators B: Chemical 222, 849–856.

Rojas-Romo, C., Serrano, N., Ariño, C., Arancibia, V., Díaz-Cruz, J.M., Esteban, M., 2016. Determination of Sb(III) using an ex-situ bismuth screen-printed carbon electrode by adsorptive stripping voltammetry. Talanta 155, 21–27.

Ronkainen, N.J., Halsall, H.B., Heineman, W.R., 2010. Electrochemical biosensors. Chemical Society Reviews 39, 1747.

Schöning, M.J., Poghossian, A., 2002. Recent advances in biologically sensitive field-effect transistors (BioFETs). Analyst 127, 1137–1151.

Schöning, M.J., Poghossian, A., 2006. Bio FEDs (Field-Effect devices): state-of-the-art and new directions. Electroanalysis 18, 1893–1900.

Shipway, A.N., Katz, E., Willner, I., 2000. Nanoparticle arrays on surfaces for electronic, optical, and sensor applications. ChemPhysChem 1, 18–52.

Shpigel, N., Levi, M.D., Sigalov, S., Daikhin, L., Aurbach, D., 2018. In situ real-time mechanical and morphological characterization of electrodes for electrochemical energy storage and conversion by electrochemical quartz crystal microbalance with dissipation monitoring. Accounts of Chemical Research 51, 69–79.

Singh, P.R., Mahajan, S., Rajwade, S., Contractor, A.Q., 2009. EC-AFM investigation of reversible volume changes with electrode potential in polyaniline. Journal of Electroanalytical Chemistry 625, 16–26.

Stern, E., Wagner, R., Sigworth, F.J., Breaker, R., Fahmy, T.M., Reed, M.A., 2007. Importance of the debye screening length on nanowire field effect transistor sensors. Nano Letters 7, 3405–3409.

Steudel, R., 1975. Properties of sulfur-sulfur bonds. Angewandte Chemie International Edition in English 14, 655–664.

Su, L., Jia, W., Hou, C., Lei, Y., 2011. Microbial biosensors: a review. Biosensors and Bioelectronics 26, 1788–1799.

Taleat, Z., Khoshroo, A., Mazloum-Ardakani, M., 2014. Screen-printed electrodes for biosensing: a review (2008–2013). Microchim. Acta 181, 865–891.

Thévenot, D.R., Toth, K., Durst, R.A., Wilson, G.S., 2001a. Electrochemical biosensors: recommended definitions and classification *. Analytical Letters 34, 635–659.

Thévenot, D.R., Toth, K., Durst, R.A., Wilson, G.S., 2001b. Electrochemical biosensors: recommended definitions and classification1International union of pure and applied chemistry: physical chemistry division, commission I.7 (biophysical chemistry); analytical chemistry division, commission V.5 (Electroanalytical). Biosensors and Bioelectronics 16, 121–131.

Toma, F.M., Cooper, J.K., Kunzelmann, V., McDowell, M.T., Yu, J., Larson, D.M., Borys, N.J., Abelyan, C., Beeman, J.W., Yu, K.M., Yang, J., Chen, L., Shaner, M.R., Spurgeon, J., Houle, F.A., Persson, K.A., Sharp, I.D., 2016. Mechanistic insights into chemical and photochemical transformations of bismuth vanadate photoanodes. Nature Communications 7, 12012.

Torrinha, Á., Amorim, C.G., Montenegro, M.C.B.S.M., Araújo, A.N., 2018. Biosensing based on pencil graphite electrodes. Talanta 190, 235–247.

Valtiner, M., Ankah, G.N., Bashir, A., Renner, F.U., 2011. Atomic force microscope imaging and force measurements at electrified and actively corroding interfaces: challenges and novel cell design. Review of Scientific Instruments 82, 023703.

Wain, A.J., Cox, D., Zhou, S., Turnbull, A., 2011. High-aspect ratio needle probes for combined scanning electrochemical microscopy — atomic force microscopy. Electrochemistry Communications 13, 78–81.

Wang, J., 1994. Analytical Electrochemistry.

Wang, J., 1999. Amperometric biosensors for clinical and therapeutic drug monitoring: a review. Journal of Pharmaceutical and Biomedical Analysis 19, 47–53.

Wongkaew, N., Simsek, M., Griesche, C., Baeumner, A.J., 2018. Functional nanomaterials and nanostructures enhancing electrochemical biosensors and lab-on-a-chip performances: recent progress, applications, and future perspective. Chemical Reviews, 8b00172.

Wu, C., Rehman, F. ur, Li, J., Ye, J., Zhang, Y., Su, M., Jiang, H., Wang, X., 2015. Real-time evaluation of live cancer cells by an in situ surface plasmon resonance and electrochemical study. ACS Applied Materials and Interfaces 7, 24848–24854.

Wu, H., Feng, X., Kieviet, B.D., Zhang, K., Zandvliet, H.J.W., Canters, G.W., Schön, P.M., Vancso, G.J., 2016. Electrochemical atomic force microscopy reveals potential stimulated height changes of redox responsive Cu-azurin on gold. European Polymer Journal 83, 529–537.

Yang, W., Ratinac, K.R., Ringer, S.P., Thordarson, P., Gooding, J.J., Braet, F., 2010. Carbon nanomaterials in biosensors: should you use nanotubes or graphene? Angewandte Chemie International Edition 49, 2114–2138.

Yao, X., Wang, J., Zhou, F., Wang, J., Tao, N., 2004. Quantification of redox-induced thickness changes of 11-ferrocenylundecanethiol self-assembled monolayers by electrochemical surface plasmon resonance. Journal of Physical Chemistry B 108, 7206–7212.

Yogeswaran, U., Chen, S., 2008. A review on the electrochemical sensors and biosensors composed of nanowires as sensing material. Sensors 8, 290–313.

Yu, J., Zhang, Z., 1997. Double potential-step chronoamperometry and chronocoulometry at an ultramicrodisk electrode: theory and experiment. Journal of Electroanalytical Chemistry 439, 73–80.

Zhai, P., Guo, J., Xiang, J., Zhou, F., 2007. Electrochemical surface plasmon resonance spectroscopy at bilayered silver/gold films. Journal of Physical Chemistry C 111, 981–986.

CHAPTER

3 Surface modification methods for electrochemical biosensors

N. Sandhyarani, PhD

Professor, School of Materials Science and Engineering, National Institute of Technology Calicut, Kozhikode, Kerala, India

3.1 Introduction

Quantitative and qualitative determination of biological species in biological or biochemical processes is of supreme importance in medical, pharmaceutical, food, environmental, and biotechnological applications. Hence the development of biosensors is always an interesting and hot topic of research. Generally, biosensors consist of a selective analyte interface or a bioreceptor, which is connected to a transducer. Bioreceptor is a biological or biomimetic material that binds or recognizes specifically to the target analyte molecule. The transducer converts the interaction event to a useful analytical signal. When a target analyte is present, the bioreceptor interacts with the target and triggers the signal generation and amplification through the transducer. Wide ranges of biorecognition elements have been used in biosensor development such as proteins (enzymes, antibodies), nucleic acids (DNA, RNA), whole cells, or cellular components, and based on the biorecognition elements, the biosensors are categorized into enzyme-based biosensors, antibodies-based biosensors, affinity-based biosensors, nucleic acid biosensors, and whole cell biosensors, respectively. Depending on the transducer used, biosensors are classified into optical biosensors, piezoelectric biosensors, ion-sensitive biosensors, calorimetric biosensors, and electrochemical biosensors. Among these different classifications, electrochemical biosensors are promising owing to their low production cost, high sensitivity, and ease of operation. It combines the analytical power of electrochemical technique with the specificity of bioreceptor. The detection mechanism is based on the fact that during the recognition element-analyte interaction event, electroactive species such as electrons are consumed or produced, generating an electrochemical signal that is detected by the electrochemical transducer. Electrochemical biosensors can be functioned in turbid media, have high sensitivity, are more acquiescent to miniaturization, and are the most widely used class of biosensors. In electrochemical biosensors, qualitative and quantitative information of the target molecules in the analyte is obtained in the form of either voltage or current, which

Electrochemical Biosensors. https://doi.org/10.1016/B978-0-12-816491-4.00003-6

are known as potentiometric biosensor or amperometric biosensor, respectively. The analyte concentration is obtained by calibrating a biosensor with the measured current with respect to the solutions of known concentrations.

3.2 The need for surface modification

All electrochemical biosensors rely on the detection of a biological interaction on the surface of the electrode. The sensitivity, stability, reusability, specificity, and cost of the sensor depend on the biorecognition element and its accessibility/reactivity. Given the functional mechanism of electrochemical biosensors, appropriate surface modification of the electrode is crucial in its performance. Surface modifications not only help in the proper immobilization of bioreceptors but also play a central role in noise control and enhanced sensitivity. More importantly, functionalization of the surface helps to abate the undesired nonspecific binding of the analyte or other components on the surface. Moreover, the conformation of biomolecules on the surface has a substantial role in defining the sensitivity and selectivity of a biosensor. In other words, the biosensor performance prominently depends on the materials used for the modification of the surfaces and also the surface chemistry or the functionalization method used in the immobilization of the bioreceptors. Various electrode surfaces used for the immobilization of biomolecules in the electrochemical biosensors are silicon, glass, gold, carbon nanotube (CNT), conducting polymers, graphene, etc. Several surface modification strategies are available now in literature, and different modification methods yield distinct biosensor performance. The choice of an immobilization method depends on the substrate where it is applied. In this chapter, several materials and functionalization methods used for the immobilization of biomolecules for the construction of sensitive biosensor are described.

3.3 Specific versus nonspecific binding

One important parameter in biomedical sensor is the specificity of a sensor to an analyte. Binding to the receptor/analyte of interest is called specific binding, and binding to the other sites/molecules is called nonspecific binding (Bongrand, 1998). Specific interaction in the biological molecules is the one-to-one interaction such as binding between an antigen and its antibody in the affinity immunosensors or the specific complementary strand binding in nucleic acid sensors. A specific interaction occurs when the orientation of the analyte fits the complementary physical properties on the receptor surface. Most times, biological samples contain complex mixtures of compounds such as proteins, biological fluids, and some interfering molecules, leading to unwanted adsorption on the electrode surface nonspecifically, and produce false signals. Nonspecific interactions are noncovalent binding of the biological molecules to the surface without any specific receptor. This is mainly due to the hydrophobic interaction and the electrostatic interaction of a biomolecule to the surface. Every nonpolar molecule and material is hydrophobic, and hence the hydrophobic interaction needs not to be a specific group. This leads to nonspecific binding on the surface. Similarly, electrostatic interaction relies on the electrical charge on the analyte and the ionic strength of the medium. The interaction can be either attractive or repulsive with respect to the signs of charge and is irrespective of the ionic species. Hence this leads to nonspecific adsorption (NSA). NSA of proteins on an artificial surface is commonly referred to as biofouling. NSA can limit the performance of biosensors by increasing the background signals as a result of binding in the absence of a recognition element. Hence, NSA of biomolecules is an obstinate encounter that always needs to be taken care of in the biosensor research especially in the label-free sensors (Sadana and Chen, 1996). An inert electrode surface can lessen the NSA and thereby decrease the background signal of the sensor. Suitable blocking agents such as bovine serum albumin have been used to alleviate the nonspecific binding. This is done by dipping the modified electrode in a solution of this blocking agent for a period followed by washing with buffer. Washing of electrodes with detergents such as Tween-20 in steps also helps to minimize nonspecific interaction. The impact of NSA in biosensor performance has inspired the biosensor research on understanding the mechanism of NSA at sensor surfaces and developing efficient strategies of surface modification to reduce the surface fouling. One of the most challenging aspects in biosensor research is the modification of sensor surfaces such that it immobilizes the biomolecules strongly without varying their original conformations and biochemical activities. In the following sections, few surface modification strategies will be discussed followed by the commonly used immobilization techniques for bioreceptors.

3.4 Surface modification strategies

Surface modification of an electrode surface is vital in biosensor research, as it provides a way to augment the biocompatibility and electron conductivity and also to reduce the surface fouling. Frequently used methods are functionalization of an electrode surface with self-assembled monolayers (SAMs), electrodeposition of polymers/nanomaterials, conducting polymers, various nanomaterials, and metal-organic frameworks (MOFs).

3.4.1 Self-assembled monolayers

SAMs are ordered assemblies of molecules formed at the solid surface by strong chemisorption of the head group of a molecule. This is one of the most efficient methods for making ultrathin organic films of controlled thickness on the surface with desired properties. The simple way of getting organized molecular assemblies on the substrate is by immersing the substrate into a dilute solution of the adsorbate

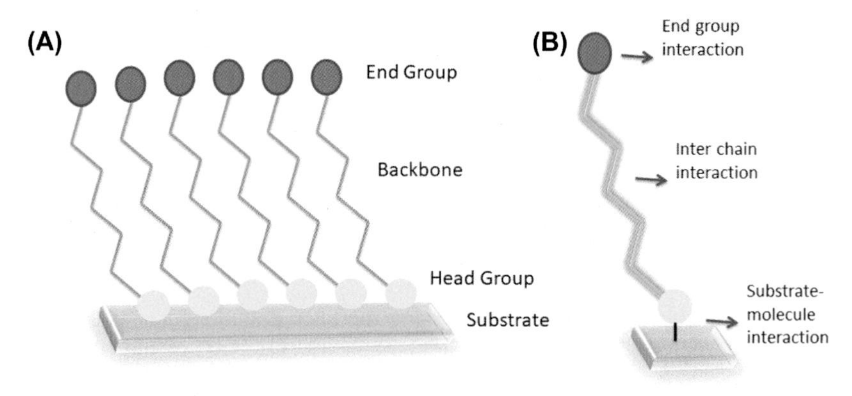

FIGURE 3.1

(A) Schematic of a self-assembled monolayer of alkane thiol on gold, and (B) the forces acting on the self-assembled monolayer.

at room temperature. The properties of SAMs, such as improved electrocatalysis, freedom from surface fouling, easy procedure of formation, and long stability make them an attractive tool for biosensor development (Chaki and Vijayamohanan, 2002). There are different kinds of SAMs reported and extensively studied, which involve carboxylate SAMs on the oxide surfaces, silane SAMs on glass/silicon surfaces, and alkane thiol SAMs on noble metals. The sensor research particularly emphasizes on alkane thiol monolayers on gold. These are known to form well-ordered assemblies that can be used to immobilize different electroactive moieties, which can function as an electrical wire between the electroactive site of a biomolecule and the electrode surface. Excellent SAM-forming ability of gold, easy availability, relatively inert nature, and the biocompatible nature make it a promising platform for sensor development. A schematic diagram of a SAM of alkane thiol is depicted in Fig. 3.1A.

Chemisorption of alkane thiols on a clean gold takes place through a simple oxidative addition as follows:

$$R - SH + Au \leftrightarrow RS - Au + e^{-} + H^{+} \tag{3.1}$$

The Au–S bond strength is 44 kcal mol^{-1}, and the van der Waals force between the neighboring methylene groups is 1.4–1.8 kcal mol^{-1} (Ulman, 1996). In addition to the above chemical bond, a lateral van der Waals interaction among the aliphatic chains of neighboring alkane thiol molecule allows the molecule to stand up vertically with a slight tilt of ~30 degrees. When the end group contains any functional groups, the interaction between the end groups is possible through van der Waals attraction, or hydrogen bonding depends on the nature of the functional groups (Ulman, 1996). The forces acting on SAMs of a functional alkane thiol are represented in Fig. 3.1B.

Although Au (111) is the ideal solid substrate for the closely packed SAMs of alkane thiol, polycrystalline gold is also reported as a stable substrate for SAM formation with a lower packing density of chains (Biebuyck et al., 1994). Biomolecules can easily be conjugated on the functional group of the SAMs using appropriate bioconjugation techniques, which are explained later in this chapter. The distance dependence of electron transfer between the electroactive proteins and metal electrodes has been the subject of extensive theoretical and experimental research in the recent past (Protsailo and Fawcett, 2000). Studies on the dependence of the electron transfer rate on tunneling distance (Song et al., 1993; Mathew and Sandhyarani, 2014) indicate that the rate of electron transfer (k_{ET}) varies exponentially with the distance between the electroactive center and the electrode surface. Fermin and coworkers carried out a systematic study (Zhao et al., 2005) on the distance dependence of electronic communication between metal nanoparticles and the metal electrode. They observed that when the chain length of thiol is greater than 13 Å, the apparent charge transfer resistance is independent of thiol length on the nanoparticles-terminated electrodes. They have suggested a mechanism for the electron transfer as the hot electron transport. The uniform distribution of the monolayer and the ease of varying the terminal functional groups make SAM an appropriate platform in biosensor development. Modification of the tailoring functional groups with receptor molecules is used in the development of sensitive biosensor surfaces (Mark et al., 2004). Various functionalities are necessary for biomolecule immobilization. Therefore, the formation of mixed SAMs is more appropriate than single-component SAMs in the biosensor research.

3.4.2 Electrodeposition

Electrodeposition is another well-established surface modification technique. This method is used for the growth of thin films on a surface. Change in the electrochemical potential of an electrode by the application of an external electric field leads to either reduction or oxidation reaction on the electrode surface. These reactions may either lead to a deposition or dissolution on the electrode surface depending on the change in potential. Electrodeposition takes place through two steps: (1) diffusion of charged species from the solution to the electrode and (2) reduction of charged species at the electrode surface and growth of particles or film on the surface. The reaction will continue till a new equilibrium state is reached. Morphology and nature of the film deposited on the electrode surface are determined by various parameters such as the interaction of the solute ion with solvent, ionic strength of the solution, rate of electron transfer reaction, solution viscosity, and diffusion coefficient. Electrodeposition can be accomplished at a constant current, at a constant potential, at pulsed current, or at pulsed voltage. Recently this technique has widely been used for the deposition of nanomaterials and hence found significant in the modification of electrode surface in electrochemical sensors. Electrodeposition technique offers advantages such as cost-effective equipment and materials, easy scaling up to industry level, high purity of deposited materials, uniform deposition on the substrates, and rigid control on the composition and structure of the deposits. Nanomaterials such as gold nanoparticles, gold nanorods, gold nanowires, gold nanoclusters, platinum nanoparticles, nickel nanoparticles, nickel nanowires, silver nanoparticles, copper nanowires, cobalt nanowires, and zinc oxide nanowires have been successfully fabricated by electrodeposition.

3.4.3 Conducting polymers

Polymers can interact physically or chemically with biomolecules (or cells) to produce biologically functional systems. For the development of biosensors, commonly used bioreceptors such as enzymes, antibodies, and cells were immobilized on or within polymeric systems (Hoffman, 1992). Conducting polymers have, in particular, opened a new scope for electrochemical biosensor by providing a suitable platform for bioreceptor immobilization and prompt electron transfer through them (Teles and Fonseca, 2008). Polymeric coatings can be used for modification of a wide range of electrode materials commonly used in electrochemical biosensors, which are realized either by electropolymerization or by solution casting of preformed polymers. Electropolymerization is a highly reproducible method, which is done at room temperature, and polymeric coating is made by electrochemical route. Electropolymerization can be done in situ with biomolecule entrapment or can be done as a two-step process. In situ electropolymerization is done by applying a suitable potential to the working electrode, which is immersed in an aqueous solution of a monomer and the biomolecule. During this process, the biomolecules get homogenously integrated into the polymer matrix and deposited on the surface. In two-step process, the polymer is first electropolymerized on the surface, and then the bioreceptor is immobilized.

3.4.4 Nanomaterials

Nanoscale materials exhibit exciting optical, electronic, and mechanical properties, which makes them suitable for a variety of applications (Jianrong et al., 2004). Various nanostructures such as metal nanoparticles, metal oxides, semiconductors, and nanotubes have extensively been used for constructing electrochemical biosensors. The biocompatibility, high surface area, biochemical and mechanical stability, nontoxicity, excellent electrocatalytic activity, and high conductivity are the contributing factors, which make these nanomaterials interesting surface modification materials (Chen and Chatterjee, 2013). Nanostructured materials play important roles such as immobilization platforms for biomolecules, catalysts of electrochemical reactions, and conducting centers, which allow efficient electron transfer between the bioreceptor and electrodes in different sensing systems (Grabar et al., 1996). Because of the attractive characteristics of gold nanoparticles such as their ability to provide a firm surface for the immobilization of biomolecules and the ability to act as an electron-conducting track between the bioreceptor and metal electrode surface, they are widely used in augmenting the efficiency of electrochemical biosensors (Daniel and Astruc, 2004; Liu et al., 2003). Moreover, the biocompatible nature of gold nanoparticles retains the activity of the bioreceptor, and the high surface area paves the way for the immobilization of a large number of biomolecules on the electrode surface. Improving the biocompatibility of gold nanoparticles and their use in the modification of electrodes is a subject of extensive research.

CNT has attracted enormous interest in biosensor research owing to its excellent conductivity and unique geometrical structure. It has potential application in the construction of mediator-free glucose sensors. Liu et al. (2005) entrapped GOx into a composite of chitosan and CNT and achieved direct electron transfer to the electrode. The microenvironment enabled conformational changes in GOx, which leads to an enhancement in the electron transfer rate, leading to a high sensitivity of 0.52 $\mu A\ mM^{-1}$.

Graphene and its derivatives such as graphene oxide (GO), reduced graphene oxide (rGO), graphene nanoribbon, and graphene nanosheet have attracted significant attention due to their excellent electronic, mechanical, and thermal properties. The use of graphene-related nanomaterials is explored extensively as electrode modification materials to construct various electrochemical biosensors. Compared with other carbon nanomaterials, graphene possesses advantages such as biocompatibility, high surface area, and rich anchor sites. Graphene can be functionalized with nanomaterials, biopolymers, polymers, or any organic or inorganic materials, and by incorporating other functional species, the synergic effect of the graphene and the materials used for functionalization can be achieved, leading to an enhanced sensitivity of electrochemical biosensor (Xu et al., 2017).

3.4.5 Metal-organic frameworks

MOFs are a prominent class of high surface area, porous crystalline solids, fabricated from inorganic ions and organic molecules linked with coordinate bonds. Recently, MOFs are emerged as an attractive immobilization matrix for biomolecules in biosensor applications because of the inherent physicochemical properties such as large and tunable pore size, pore volume, pore shape, and hydrophilic/hydrophobic balance. MOFs immobilize the biomolecules through the host-guest interactions and the possibility for tuning the properties of pores, and the polar/apolar nature of the MOFs enables it to modify the interaction with the guest molecule, which is the bioreceptor. MOFs are most widely used for enzyme immobilization, which is done by adsorption, covalent binding, cavity inclusion, or in situ synthesis. The tunable structural and chemical diversity and functionality together with the high surface area makes the MOFs a promising bioelectrode. For example, Chen et al. (2013) successfully used MOF as the fluorescent sensing platform to detect double-stranded DNA (dsDNA) with a detection limit of 1.3 nM and good selectivity.

A Cu-MOF with a high surface area and 3d interconnected channels was synthesized and used to construct a tyrosinase biosensor for the detection of bisphenol A. Here, Cu-MOF acted as a sorbent, which enhanced the available BPA concentration through π-π stacking interactions between BPA and Cu-MOF, resulting in an increased sensitivity of the sensor (Wang et al., 2015). An Au nanoparticle-encapsulated Cu-based MOF was effectively utilized for the detection of lipopolysaccharide (Shen et al., 2015). A 3D cerium-MOF was developed using Ce^{3+} ions and amino-functionalized ligand and applied for the detection of ATP diagnosis (Shi et al., 2017). In this biosensor construction, first, the Au electrode was restructured with the Ce-MOF by electrostatic interactions and weak covalent bonding, followed by the immobilization of aptamer of ATP onto the MOF through hydrogen bonding, π-π stacking interactions, and electrostatic interactions. A schematic illustration of the label-free electrochemical detection of ATP using Ce-MOF is presented in Fig. 3.2. The conformational change ensuing from the specific binding between the ATP and the aptamer resulted in a change in impedance, which was monitored using electrochemical impedance spectroscopy. The method was successfully utilized to detect the nanomolar concentration of ATP and was used for the detection of ATP in the serum of cancer patients that revealed the impending clinical application of this sensor.

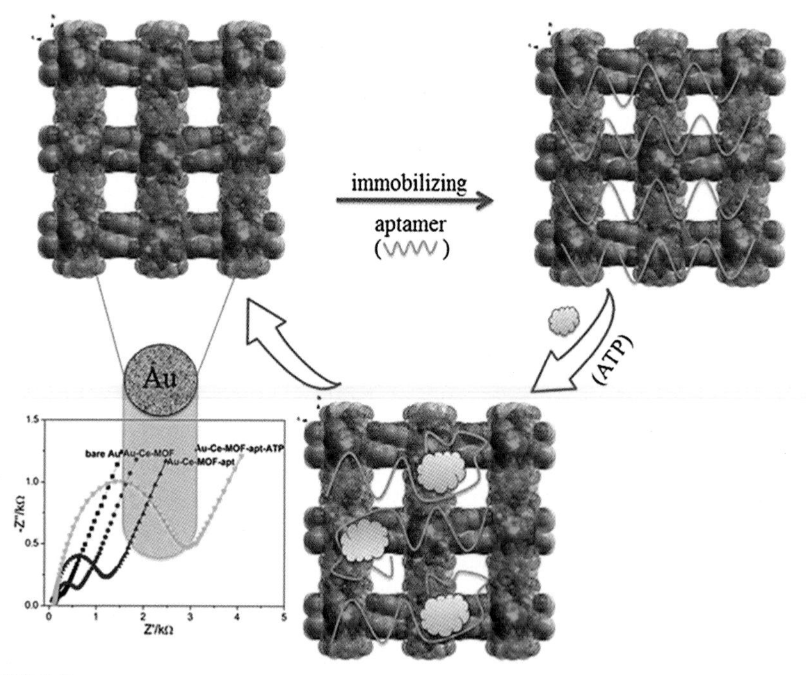

FIGURE 3.2

Schematic illustration of the electrochemical detection of ATP using Ce-MOF.

Reprinted with permission from reference Shi, P., Zhang, Y., Yu, Z., Zhang, S., 2017. Label-free electrochemical detection of ATP based on amino-functionalized metal-organic framework. Scientific Reports 7(1), 6500. Copyright from Nature Publishing Group, https://creativecommons.org/licenses/by/4.0/.

Xie et al. encapsulated hemin into Au nanoparticle-functionalized Fe-MIL-88 MOFs (hemin@MOFs) and applied for the detection of thrombin (TB). By using a signal amplification strategy with the aid of an enzyme, the authors reported a very low detection limit of 0.068 pM for TB. Here, the Au/hemin@MOF acted as the redox mediator and a solid catalyst together with the immobilization platform for conjugating a large number of biomolecules (Xie et al., 2015).

Thus, MOFs now emerge as a potential platform for the biosensor immobilization, leading to sensitive sensor development. Precise chemical and functional modifications of the MOFs offer the possibility of designing highly sensitive and selective sensors. The design of functional MOFs and the signal transduction strategies are detailed in the review article by Lei et al. (2014).

3.5 Immobilization techniques for biomolecules

In aqueous solution, biomolecules such as the enzyme may lose their activity because of the oxidation reactions or the loss of ternary structure at the air-water interface. The issue of loss of activity or stability can be resolved by integrating the biomolecules to an inert (chemically) support material, thereby extending their activity and stability. A variety of biomolecule immobilization techniques such as adsorption, covalent attachment, entrapment, cross-linking, immobilization of nanomaterials, and affinity immobilization have been developed and used effectively in biosensors. Selection of the immobilization method depends on the biomolecule to be immobilized, the nature of the surface on which the biomolecules are immobilized, and the transducing mechanism involved in the sensor. Various immobilization techniques are introduced here, and then a detailed account from the literature will be provided based on the type of biosensor/the biomolecule to be immobilized. Immobilization techniques are classified into two categories: reversible and irreversible methods. Reversible methods include adsorption, and irreversible methods include entrapment, covalent coupling, cross-linking, and affinity immobilization. Each of these is described below. The schematic illustrations of these immobilization methods are depicted in Fig. 3.3.

3.5.1 Physisorption

Physical adsorption is the simplest immobilization method, in which the biomolecules are attached to the surface through weak bonds such as van der Waals forces, hydrogen bonding, or hydrophobic interactions. Physisorption of biomolecules on an electrode surface is usually done by immersing the electrode surface into the biomolecule solution followed by a fixed incubation period. Then the electrode is washed with the buffer solution. Advantages of physisorption are that it is a simple and cost-effective method, which causes only little conformational change of the biomolecules, and it does not need surface functionalization. Immobilization through physisorption suffers from the drawbacks such as desorption of the

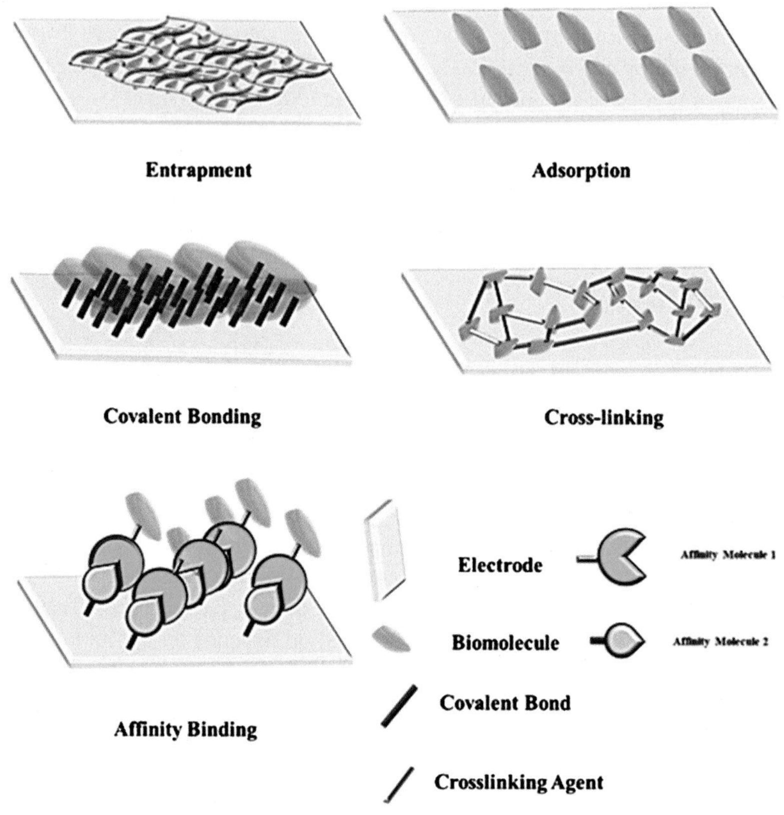

FIGURE 3.3

Schematic illustrations of the commonly used immobilization methods for biomolecules on electrode surfaces.

bioreceptor from the surface during the measurements and NSA of interfering molecules on the surface. Biomolecules can be immobilized onto the conducting polymers without any loss of activity. The excellent electron transfer property along with the biocompatibility of conducting polymers makes it an attractive platform for immobilization of biomolecules.

3.5.2 Entrapment

The sensitivity of a sensor can be increased by increasing the surface area of the immobilization matrix, which is generally attained by immobilizing the biomolecule in a high surface area, 3D porous matrix. The immobilization of biomolecule by the entrapment method is centered on the occlusion of a biomolecule within a sol-gel matrix or a polymeric network. It is based on the growth of the gel network or polymer chains around the biomolecule, and the entrapped biomolecules remain available for biorecognition because of the porous nature of the matrix used for entrapment (Gupta and Chaudhury, 2007). The efficiency of the method and sensitivity of the sensor depends on the diffusion of the analyte to the entrapped receptors, physicochemical properties of the surrounding medium, and stability of the entrapped bioreceptor. Although the entrapment method found to increase the sensitivity of the sensor, it suffers from nonspecific binding and low stability. This is overcome by the covalent immobilization of biomolecules inside the porous structures. The covalent immobilization improves the stability of the bioreceptor inside the matrix and hence can be used after repeated washing and measurements.

Bessueille et al. (2005) synthesized the oligonucleotide on a porous silicon substrate with a 10-fold enhancement of surface density and proved the biomolecule to be stable up to 25 cycles of hybridization/stripping. Conducting polymer hydrogels, which are the network of polymers containing water, resemble the biological environment and hence are widely used to entrap the biomolecules for biosensing applications. Because of their high hydrophilicity, good biocompatibility, and large surface area, these hydrogels contribute to the stability and extended shelf life of molecules and are explored as a promising immobilization matrix for biomolecules (Li et al., 2015). Recently, a novel polymer matrix was developed for the enzyme immobilization based on poly(3,4-ethylenedioxy-thiophene) (PEDOT), graphene oxide nano-sheets (GONs), and laccase (Lac) by electropolymerizing the EDOT along with GONs and Lac as dopants on glassy carbon electrode (GCE) (Maleki et al., 2017). An effective platform for enzyme entrapment was reported using a cationic polymer, poly(2-(dimethylamino)ethyl methacrylate) (MADQUAT)-modified single-walled carbon nanotube and reduced graphene oxide (SWCNT−rGO) nanocomposite. The matrix was found to be effective for the immobilization of a model enzyme alcohol dehydrogenase (ADH) and was used for the electrochemical detection of ethanol. The entrapped enzyme exhibited a fast exchange of electrons, and the SWCNT-rGO nanocomposite enhanced the biocatalytic activity of the immobilized ADH significantly (Adhikari et al., 2017).

3.5.3 Chemisorption

Chemisorption or chemical adsorption of biomolecules involved direct covalent attachment between biomolecules and the electrode surface via the functional group of the molecules present on the surface. Covalent attachment may be a non−site-directed attachment or site-directed attachment. In former, external agents such as organic molecules, metal bridges, or polymers are used, and unregulated covalent binding of the biomolecule to a surface can potentially restrict the active site for the analyte or denature it. In latter, the biomolecules bind via distinctive attachment sites directly to the molecule on the surface. The widely used method for the chemisorption of biomolecules is to use SAM interfacial layers with the terminal functional groups such as amine or carboxylic acid, which are normally activated using N-hydroxysuccinimide (NHS) ester. One of the most widely used methods of covalent immobilization is activated ester chemistry. The activated ester chemistry—mediated immobilization of biomolecules is discussed below.

3.5.4 Active ester chemistry

Immobilization by cross-linking is the method of chemically joining the molecules by a covalent bond via cross-linkers, which contain reactive ends that react to specific functional groups on biomolecules. Because of the stable nature of the covalent bonds formed between the biomolecule and the surface, the bioreceptor is not released into the solution during measurement, and hence cross-linking or covalent method is one of the most explored immobilization methods in the biosensor research. As explained earlier, the covalent bonds form between the biomolecule and the solid/electrode surface through the cross-linkers. This is mainly achieved through preformed functional groups on the surface. Depending on the functional groups available on the target and the surface, various reaction strategies and different cross-linkers have been developed.

Cross-linkers are selected on the basis of their reactivity or specificity for individual functional groups. The chemical specificity, spacer arm length (the distance between the conjugated groups), solubility in water, and cell membrane permeability of the cross-linker are the major factors to be considered for the selection of cross-linkers. Based on the chemical specificity, cross-linkers are classified mainly into two: homo-bifunctional having identical reactive groups at either end of a spacer arm and hetero-bifunctional having different reactive groups at either end. Homo-bifunctional cross-linkers are used for the single-step reactions, and hetero-bifunctional cross-linkers are used in multistep reactions. Selection of these cross-linkers depends on the nature of functional groups available on the protein surface. Reaction conditions used for the cross-linking also are important. To maintain the native structure and activity of the bioreceptor, cross-linking is generally performed in physiological conditions.

The major functional groups used for practical bioconjugation methods are primary amines, carboxy groups, and thiols (Liebana and Drago, 2016). The method of coupling these groups are explained below, and the most extensively used method for each of these groups is shown in Fig. 3.4.

Primary amines ($-NH_2$): N-terminus of the polypeptide chain and the side chain of lysine residues contain the primary amine group, which is positively charged under physiological condition. On the surface of proteins, these amine groups are normally facing outward and hence are available for conjugation without denaturing the biomolecule. The most important method used for the conjugation of the primary amine is that by using the N-hydroxysuccinimidyl ester (NHS ester) reactive group. Carbodiimides bind between the carboxyl groups of the electrode surface and the amino functionality of the biomolecule. The water-soluble 1-ethyl-3-(3-dimethylaminopropyl) carbodiimide hydrochloride (EDC) is the most extensively used carbodiimide for

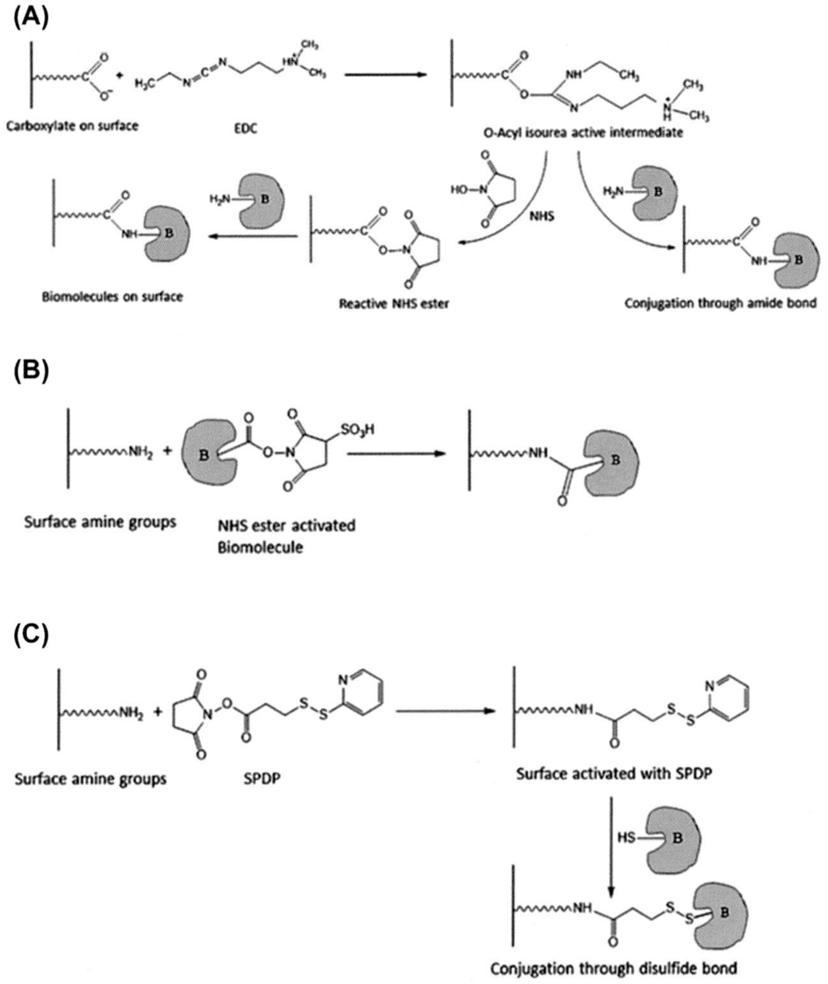

FIGURE 3.4

Scheme showing the immobilization of (A) NH_2-terminated biomolecules on carboxylate surface using EDC/NHS linker; (B) carboxylate-terminated biomolecules on amine surface using EDC/NHS linker; (C) thiol-terminated biomolecules on amine-terminated surface using SPDP linker.

conjugation of biomolecule that contains carboxyl or amino group. N-hydroxysuccinimide is used along with carbodiimide to improve the immobilization efficiency. Here, the immobilization surface containing carboxylate is first treated with EDC to form the acyl derivative, which then reacts with NHS to form the active ester. This reactive ester then reacts with the amino group on the bioreceptor (Fig. 3.4A).

Carboxy groups (−COOH): C-terminus of the polypeptide chain and the side chain of aspartic acid and glutamic acid residues contain the carboxy groups, which are negatively charged under physiological condition. The most commonly used conjugation method is by EDC-NHS coupling to the NH_2-modified surface. The conjugation is through the amine surface group and biomolecule carboxylate group. In this method, the biomolecule is first activated with EDC and sulfo-NHS, and then the activated intermediate is added to the amine surface, leading to the conjugation (Fig. 3.4B).

Thiols (−SH): Thio group exists in the side chain of cysteine and can be immobilized with N-succinimidyl 3-(2-pyridyldithio) propionate (SPDP) coupling on an amine surface as shown in Fig. 3.4C.

3.5.5 Affinity immobilization

In affinity immobilization, the principle of affinity between complementary biomolecules is utilized for biomolecule immobilization (Liu and Yu, 2016). Major advantage of the method is the significant selectivity of the interaction. Affinity immobilization methods depend on affinity tags, the biotin-avidin system, etc. Glycoproteins are immobilized via their carbohydrate moieties to surface-bound lectins, and antibody immobilization is facilitated by protein A or G. The well-known procedure for the affinity immobilization is the (strept)avidin-biotin interaction where one (strept)avidin binds to four molecules of biotin noncovalently. It is one of the strongest noncovalent bonds that exists with $K_D = 10^{-15}$ M. The strength of the (strept)avidin-biotin interaction along with its stability makes it particularly useful in the bioconjugate chemistry of any biomolecule.

3.6 Types of immobilization strategies used in various categories of sensors

In the following sections, some typical examples citing the literature reports of immobilization strategy used for the bioreceptor immobilization in three major categories of enzymatic, nucleic acid, and whole-cell biosensors are elaborated.

3.6.1 Enzymatic sensors

Electrochemical enzymatic sensors have attracted considerable attention owing to their specificity and sensitivity (Das et al., 2016). Enzymes are the biological catalysts containing an active site for a specific substrate and promote chemical reactions in biological systems. Enzymes can catalyze the reactions either as individual molecules in solution or as aggregates with other moieties or as immobilized on a surface. Electrochemical enzymatic sensors utilize an electrode modified with an enzyme as a bioreceptor for the signal transduction. The function of an enzyme in an electrode surface is generating an electroactive species in a stochiometric relationship to the substrate concentration (Davis, 1985). The electroactive species produced as a result of the enzymatic reaction are then detected by electrochemical oxidation or reduction.

An enzyme has a hydrophobic core in its native form, which is protected from contact with water in the physiological surroundings. With changes in the conditions, the enzyme may unfold, the binding site may be disrupted, and it loses its binding ability and catalytic activity. Hence, an appropriate immobilization strategy needs to be adapted for the biosensor development. Enzymes get denatured or undergo a reduction in bioactivity upon direct attachment of enzymes on a bare metal surface. Surface modification is required to make the metal electrode biocompatible with redox enzymes and to promote electron transfer from the active site to the electrode surface. Enzyme immobilization strategy used for sensor fabrication plays a significant role in the retention of the activity as well as stability of the enzyme. Different strategies have been used for enzyme immobilization, including entrapment, adsorption, cross-linking, and affinity immobilization. The support, which is used to immobilize enzyme, is of significant importance in defining the effectiveness of the enzyme. The support should possess characteristics such as resistance to compression, hydrophilicity, inertness to enzymes, biocompatibility, and ease of functionalization.

In porous entrapment method, a porous matrix is formed around the recognition element, which binds it to the sensor surface. Development of mesoporous electrodes has received considerable interest, as they entrap enzymes on the electrode surface and allow direct electron transfer from enzyme to electrode (Walcarius, 2015). Mesoporous silica is a widely used entrapment matrix. Entrapment of enzymes in conducting polymers has also received considerable interest in the direct electron communication between the active site and electrode (Naveen et al., 2017). Electropolymerization of polymer on the surface of electrode is an attractive and viable method of surface modification. Enzymes can be immobilized on the electropolymerized electrode surface by a single-step or a two-step method. In single step method, the electrode is dipped in a solution of monomer and enzyme, and by applying a suitable potential, the polymer is deposited on the surface. In this process, the enzyme also gets entrapped in the polymer matrix and gets deposited on the electrode surface. Njagi and Andreescu (2007) immobilized glucose oxidase (GOx) into a gold-polypyrrole nanocomposite through entrapment to fabricate a glucose sensor. There is a possibility of slight denaturation of an enzyme during electropolymerization, which can contribute to the exposure of active site outward making the enzyme active. In two-step process, the polymer is deposited on the electrode surface initially using electropolymerization and then, in the second step, enzyme is conjugated to the surface using either electrostatic interaction or covalent binding.

Recently, MOFs are used for the entrapment of enzymes (Zhang et al., 2017). The small cavities of MOFs lead to a decrease in the enzymatic activity and the substrate affinity. However, the cavities can be modified to alter the properties and to achieve efficient entrapment of enzyme, leading to better performance of the sensor (Chen et al., 2017). MOF-derived nanomaterials in porous carbon are demonstrated to possess high conductivity, thereby accelerating the electron transfer and distinctive microstructure to load huge amounts of enzymes. MOFs were also used as templates to prepare carbon-supported metal oxide nanocomposites, which retained the original structure of MOFs. These MOF-derived materials were used for effective immobilization of enzyme, which facilitated the availability of more active sites and electron transfer, thereby improving the performance of the sensor (Dong et al., 2018).

Physical adsorption of an enzyme is based on weak forces such as van der Waals force, ionic force, hydrophobic force, and hydrogen bond. Physical adsorption is the easiest method of enzyme adsorption, which is done by dipping the precleaned electrode in a solution of the enzyme for a fixed period. The enzyme gets adsorbed on the surface, and the unadsorbed enzyme is washed by the buffer. However, this method experiences drawbacks such as weak binding of the enzyme, which results in desorption of enzyme during the measurement. Wang et al. (2009) used physical adsorption of horseradish peroxidase (HRP) onto gold nanoparticles and then electrodeposited to ITO electrode for the development of hydrogen peroxide (H_2O_2) sensor (Wang et al., 2009). One efficient method of enzyme immobilization on the modified electrode surface by adsorption is using ionic interaction. For this purpose, carboxylate-functionalized electrode surface is treated with cationic solutions such as Ca^{2+}, and then the negatively charged enzyme molecules are immobilized on the surface through the ionic interaction between the negative charge of the enzyme and the surface-bound Ca^{2+}. This method helps to retain the bioactivity enzyme.

In covalent bonding, the reactive group on the sensor surface binds to the recognition element through covalent bonds. In this method, enzymes are immobilized on the surface through the functional groups, which are not required for their catalytic activity. This is usually carried out by the activation of the surface by reactive groups, using reagents such as EDC/NHS and glutaraldehyde followed by the coupling of an enzyme to the activated support. In the amperometric H_2O_2 sensor developed by Yu et al. (2003), a covalent bonding between carboxyl-terminated functional groups of SWCNT and lysine residue of the enzyme was used through a linker.

Another extensively used approach for the development of biosensors is immobilization of enzymes by cross-linking with glutaraldehyde, glyoxal, or hexamethylenediamine. A glucose sensor for the blood glucose monitoring in fish was realized by cross-linking between GOx and glutaraldehyde (Yonemori et al., 2009). In affinity-based immobilization strategy, affinity bonds between activated support and specific group of the protein are constructed. Impedimetric H_2O_2 sensor developed by Esseghaier et al. (2008) explored the streptavidin-biotin affinity for the effective immobilization of HRP on a gold electrode.

The advent of nanotechnology has opened up new avenues for development of highly sensitive biosensors. Properties of nanomaterials including high electrical conductivity, catalytic properties, stability, high surface area, biocompatibility, and effective loading of the biorecognition element are widely used for the improvement of sensor performance. They have been used to achieve direct contact of enzyme to the electrode surface to promote the electrochemical reaction. A variety of nanomaterials including nanoparticles, nanowires, nanotubes, and nanorods have been reported, which enable direct electron transfer between the enzyme and electrode. Guiseppi-Elie et al. (2002) achieved direct electron communication between an enzyme and electrode by adsorbing GOx onto SWCNTs drop casted on GCE. A study by Cai and Chen (2004) revealed that the biocatalytic activity of GOx remains stable after the immobilization in a CNT matrix, and hence the modified CNT promotes the electron transfer with a higher heterogeneous electron transfer rate constant of 1.53 s^{-1}. Liu et al. (2005) modified gold electrode with a monolayer of SWCNTs and subsequently immobilized GOx onto the monolayer and therefore, enabled an electrical connection to the electrode and achieved a higher rate constant of 9 s^{-1}. A combination of semiconductor quantum dots and the CNT is reported as an excellent immobilization platform of GOx allowing direct electron transfer (Liu et al., 2007). Improved electron transfer through silver nanoparticle monolayer is also reported. Lin et al. (2009) incorporated silver nanoparticles into a matrix of chitosan and CNTs to enable the electron communication to the electrode and achieved a lower detection limit of 100 nM for glucose (Lin et al., 2009). Oxide nanoparticles and semiconductor nanoparticles are also used as an electrical bridge, which wires the electron transfer between the active site of the enzyme to the electrode.

Even though different nanostructures find application in the development of mediator-free enzymatic biosensors, application of gold nanostructures is more attractive owing to their electron tunneling properties. Gold nanoparticles assembled on the electrode surface are reported as an excellent electron transfer mediator. The reports revealed that presence of gold nanoparticles on the electrode surface could enhance the sensitivity owing to its electron tunneling property. Numerous works have been reported in glucose sensors based on gold nanoparticles. Zhang et al. (2005a) developed a mediator-free glucose sensor by modifying a gold electrode with gold nanoparticles and GOx. They used self-assembled gold nanoparticles onto a gold electrode, which was modified with a dithiol via the Au−S bond. In another work, they modified gold electrode with a three-dimensional network of silica gel and immobilized glucose oxidase onto the modified electrode (Zhang et al., 2005a,b). They used gold nanoparticles to achieve direct electron transfer to the electrode and attained a high sensitivity of 8.3 $\mu A\ mM^{-1}\ cm^{-2}$.

Yan et al. (2008) used gold nanoparticles of average diameter 5 nm onto a GCE modified with silver chloride and polyaniline, to allow efficient electron transfer, which leads to a lowest detection limit of 4 pM. German et al. investigated the role of gold nanoparticles on the electron transfer and observed that the presence of gold nanoparticles between GOx and carbon rod electrode acts as an electrical wire enabling the direct electron transfer. The developed sensor surface exhibited an excellent storage stability of 49 days (German et al., 2010). We performed meticulous research on the features affecting the performance of an electrochemical-enzymatic biosensor (Mathew and Sandhyarani, 2014). This study has been executed using an enzyme/insulator-metal/insulator-metal surface based on enzyme HRP, mercaptoalkanoic acid−stabilized gold nanoparticle, and mercaptoalkanoic acid SAM on the gold electrode surface. The sensor performance has been evaluated with respect to the distance between the conducting centers and the hydrophilicity of the biosensor surface. Distance-dependent electron transfer was examined by changing the length of the mercaptoalkanoic acid molecule attached to the gold electrode surfaces. Hydrophilicity of the sensor surface was altered by using citrate, and mercaptoalkanoic acid protected gold nanoparticles. Electrochemical behavior of the modified electrodes was studied using cyclic voltammetry. The current response was observed to be decreasing exponentially with increasing the distance between metal nanoparticles and metal electrode. The hydrophilicity of the gold nanoparticles exhibited a significant role in determining the sensitivity. Good electron communication through the electrode was observed when the electrode surface was modified with mercaptopropionic acid (MPA), and hydrophilic citrate stabilized gold nanoparticles.

Development of a highly sensitive H_2O_2 sensor using biocompatible cystine-based cyclic bisureas stabilized gold nanoparticles, and HRP has been reported from our research group (Mathew and Sandhyarani, 2011). The sensor was realized by the sequential assembly of MPA, cystine-based cyclic bisureas (CBU)-gold nanoparticle (AuNP) composite, and HRP on a polycrystalline gold electrode. A schematic representation of the fabrication process of the sensor surface is illustrated in Fig. 3.5. Presence of copious hydrogen bonds, from the chemical structure of the CBU on the surface of the immobilized Au nanoparticle, ensured the proper immobilization of the enzyme in a favorable orientation, thereby preserving its enzymatic activity. The biosensor surface altered with gold nanoparticles and CBU revealed a lower detection limit of 50 nM for H_2O_2 with a Michaelis-Menten constant K_M^{app} value 4.5 μM.

The same electrode platform was then used for the fabrication of a glucose sensor with improved performance by coimmobilization of HRP and glucose oxidase (GOx) (Mathew and Sandhyarani, 2014). The enzymes HRP and GOx retained their catalytic activity for an extended time indicated by the low K_M^{app} value. The sensor exhibited a detection limit of 100 nM with a linear range of 100 nM to 1 mM and a high sensitivity of 217.5 μA mM^{-1} cm^{-2} at a low potential of −0.3 V.

3.6.2 Affinity-based biosensors

Affinity-based biosensors are analytical devices that use biorecognition elements such as DNA or RNA, antibody, a receptor protein, or biomimetic material along with the transducer (Hahn et al., 2005). This class of sensor gains high sensitivity and selectivity from the strong affinity between the complementary pairs. The selectivity, stability, and sensitivity of the sensor depend on the bioaffinity elements present. Antibodies are the most widely used bioreceptors in affinity-based sensors because of their high specificity, versatility, and availability. Sometimes, antibody-based biosensors are termed as immunosensors in which an antibody or antigen is immobilized on the sensor surface, and the binding event of antigen or antibody in the analyte solution is measured. Immobilization of antibodies and their

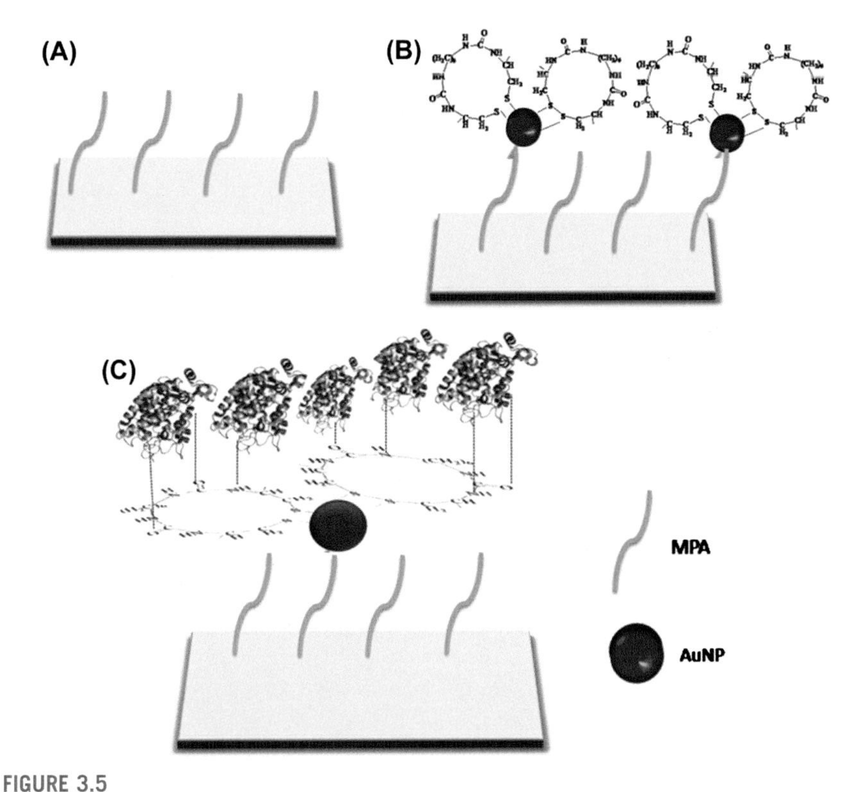

FIGURE 3.5

Schematic representation of (A) Au/MPA, (B) Au/MPA/CBU-AuNP, and (C) Au/MPA/CBU-AuNP/HRP surfaces.

orientation on the transducer surface is very important in the development of affinity-based antibody sensor because of the stoichiometric relationship between the antibody and antigen. The commonly used methods for the immobilization of antibodies are adsorption, entrapment, covalent bonding (directive and indirective), and cross-linking, by using proteins such as protein A or protein G or by a biotin-avidin system. Another major type of affinity-based sensor is DNA biosensor where nucleic acids are used as the bioreceptor. Here the single-stranded DNA (ssDNA) is immobilized on the surface of sensor, and the hybridization of the complementary sequence is monitored as a transducer signal. Electrochemical methods help in distinguishing the prehybridized ssDNA and the hybridized dsDNA.

3.6.2.1 Antibodies-based biosensors

Antibodies are globular serum proteins, made up of hundreds of amino acids, which are arranged in a highly ordered sequence. An antibody consists of two fragments: an antigen binding (Fab) region located at the ends of the two arms and a fragment crystallizable region (Fc), which is less variant. Each antibody comprises four

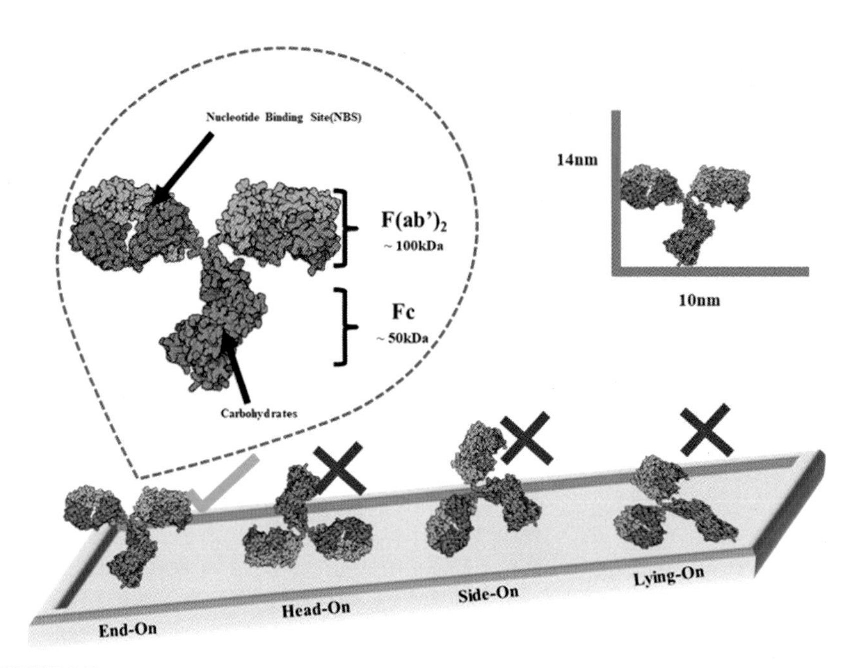

FIGURE 3.6

Antibody dimensions, binding sites, and orientations on the surface. "*End-on*" alignment is the most favorable orientation for antigen binding.

polypeptides, two heavy chains and two light chains joined to form a "Y"-shaped molecule, as shown in Fig. 3.6. The amino acid sequence in the tips of heavy and light chains varies greatly among different antibodies constituting the variable region. This variation in the amino acid sequence contributes to the antigen specificity and binding affinity of each antibody. Based on the constant region structure and immune function, the antibodies are mainly categorized into five major classes: immunoglobulin M (IgM), immunoglobulin G (IgG), immunoglobulin A (IgA), immunoglobulin D (IgD), and immunoglobulin E (IgE).

Monoclonal antibodies are derived from identical immune cells that are clones of a unique parent cell, whereas polyclonal antibodies are made by multiple plasma cells. Both monoclonal and polyclonal antibodies are effectively exploited in many biosensors. An antibody can specifically bind to an antigen with high specificity and high affinity (North, 1985). These high specificities and strong affinities of antibodies for specific antigens make them an ideal biorecognition element. Monoclonal antibodies recognize more specific region of antigens than polyclonal antibodies. The sensitivity, selectivity, stability, immobilization, labeling, and antibody size are the key parameters that determine the efficiency of antibodies-based biosensors. Immobilization of antibodies at an electrode surface plays the most crucial role in determining the performance of the biosensor system as the orientation of antibody has paramount importance in the specific binding to antigen. One of the major limitations of the antibody immobilization methods is the arbitrary orientation of the antibody molecule (see Fig. 3.6). Owing to the random orientation, the antigen binding domains on a substantial percentage of the immobilized antibody become unreachable to the antigen. Proper orientation of antibody is to be taken care of during the immobilization step in the development of antibody biosensors. Immobilization methodologies used for antibodies encompass adsorption, entrapment, cross-linking, covalent binding, and the use of protein A or protein G.

One of the regularly used antibody immobilization methods is physical adsorption. Although the method is simple and is satisfactory for a limited number of assays, this cannot be used for multiple assays owing to desorption of antibody during analysis. Antibodies are also reported to be entrapped in polymers, such as polyacrylamide (PAM), polyvinyl alcohol (PVA), epoxy, or inorganic sol-gels. Although entrapment approaches permit for optimization of variables such as bioreceptor loading, major concerns are the antibody leakage, its orientation, and maintenance of the biological activity.

Protein A or protein G binding method is used to maintain the biological activity and orientation of antibodies on the sensor surface. First, these proteins are conjugated on the sensor surface either covalently or noncovalently as described earlier for the immobilization of proteins/enzymes. These proteins bind to the Fc region of antibodies, and the antigen binding site will be free of steric interference to augment the binding of antigen. It is demonstrated that the application of this method considerably increases the antigen binding ability of IgG immobilized to the sensor. Another commonly used method to regulate antibody orientation is by avidin/biotin system, where avidin is a protein and biotin is a low-molecular-weight cofactor, which binds to avidin with high affinity. Here avidin is immobilized on the sensor surface, and biotin binds to avidin, which then binds to the Fc region of an antibody. SAMs are also used to maintain the orientation of antibody and to increase its available antigen-binding sites. A preferred orientation with a higher antigen/antibody ratio of mouse IgG1 adsorbed to an NH_2-terminated SAM over a COOH-terminated SAM is demonstrated by exploiting the antibody dipole and the charged SAM surfaces (Chen et al., 2003).

Covalent coupling of capture antibodies is considered as the most appropriate orientation-controlled immobilization method for an antibody on the surface. However, site-directed attachment to the antibody is essential for its oriented immobilization. Various conjugation chemistries are used for the covalent attachment of an antibody depending on the substrate functionality and antibody target group. The conjugation chemistries are selected based on the available functional groups for immobilization as explained in the section on active ester chemistry (Section 3.5.4).

3.6.2.2 Nucleic acid biosensors

Electrochemical DNA sensors integrate a sequence-specific probe and an electrochemical signal transducer. DNA sensors can detect pathogenic or genetic diseases by monitoring the target DNA strands related to the disease. The target DNA binds to the immobilized complementary single strand and produces a signal that indicates the amount of DNA present (Homs, 2002). Instantaneous and sensitive detection of a trace amount of the specific DNA sequences has become a vital topic of research, and highly sensitive DNA microarrays or DNA sensors were developed (Sassolas et al., 2008). Generally, the surface of a DNA sensor entails an immobilized ssDNA, which hybridizes to its complementary target DNA sequence. The hybridization process is monitored by appropriate signal transducers such as electrochemical (Wang, 2006), optical (Ermini et al., 2013), or mass-sensitive elements (Rasheed and Sandhyarani, 2016), which produce readable signals such as current, light, or frequency, respectively. DNA sensors are categorized based on their transducer type such as (1) optical DNA sensors, (2) piezoelectric DNA sensors, (3) electrical DNA sensors, and (4) electrochemical DNA sensors.

Electrochemical DNA sensor provides a simple, precise, and economical platform for DNA detection where the hybridization event of DNA is monitored with electrochemical transducers (He et al., 2005). The probe sequences can be effectively immobilized on a variety of electrode substrates. In an electrochemical DNA sensor, electrochemical signals based on the oxidation of DNA bases, the redox reactions of reporter probes present on the electrode by specific DNA-target interactions, or charge transport reactions facilitated through the π-stacked base pair are used (Drummond et al., 2003). Nanomaterials are used for amplification of the signals in DNA sensors providing very high sensitivity (femto molar to the zepto molar range), and it can be used for multiple DNA target detection with different nanoparticles (Jianrong et al., 2004).

Stable conjugation of DNA strands at the electrode surface has a crucial role in the efficiency of DNA electrochemical sensors. The accuracy and sensitivity of a DNA sensor significantly depend on the extent of the immobilized DNA strands and its orientation on the surface. DNA can be easily conjugated to nanomaterials without changing its biological activity. Various nanomaterials such as gold nanoparticles (AuNPs) (Rasheed and Sandhyarani, 2017a), carbon nanomaterials (Rasheed and Sandhyarani, 2017b), and polymeric nanoparticles (Saberi et al., 2013) are being used to modify the electrode materials in the construction of sensing platforms. Because of the unique properties such as enormous surface area, biocompatibility, and ample binding sites, nanomaterials can provide amplified electrochemical detection signals.

Electrostatic adsorption is one of the methods widely used for DNA immobilization on an electrode surface. Charged nanoparticles can electrostatically adsorb biomolecules with opposite charges through attractive forces. In addition to this, covalent interactions are also used for the immobilization of biomolecules on nanoparticles. To immobilize DNA onto the nanoparticle surface, the DNA strands are frequently altered with special functional groups, which can be conjugated with particular nanoparticles. For example, DNA is usually modified with a thiol group or amino group at one end for immobilization on gold nanoparticles, and DNA-immobilized gold nanoparticles are widely used as an electrode material (Sapsford et al., 2013).

A systematic investigation of the detection strategies is performed for highly sensitive and specific detection of BRCA1 gene using labeled gold nanoparticles. The sensor used a "sandwich" strategy, where one half of the target DNA (DNA-t) binds to the immobilized capture probe DNA (DNA-c) on the surface, and the other half hybridizes to a gold nanoparticle–conjugated reporter probe DNA (DNA-r.AuNP) (Fig. 3.7A). In the absence of target DNA, hybridization does not take place, leaving the surface-confined capture probe DNA unhybridized. The foremost advantage of the sandwich sensor is that no alteration or pretreatment is necessary for the target BRCA1 DNA sequence. Gold electrode was used as the substrate, and different strategies were studied to enhance the detection capability of the sensor. Functionalized gold nanoparticles have been used to make the surface more conductive. The

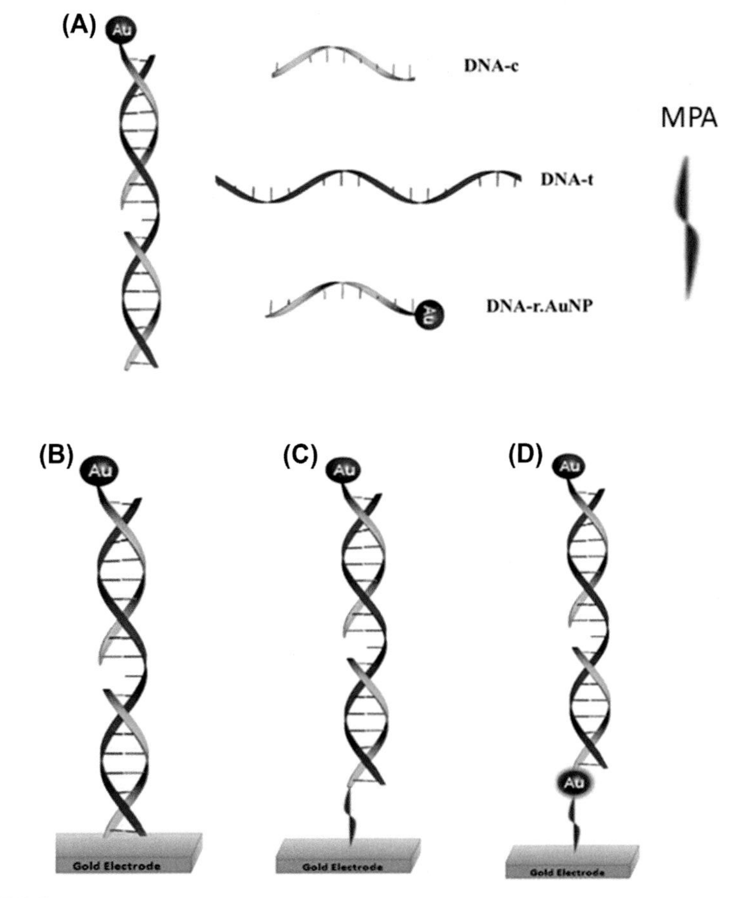

FIGURE 3.7

(A) The sandwich- type hybridization, (B) Au/DNA-c|DNA-t|DNA-r.AuNP, (C) Au/MPA/DNA-c|DNA-t|DNA-r.AuNP, and (D) Au/MPA/AuNP/DNA-c|DNA-t|DNA-r.AuNP.

optimized conditions for the best performing sensor have been investigated by the systematic modifications of the sensor, and functions of each modification elements were studied (Rasheed and Sandhyarani, 2015). In the first method, immobilization of DNA-c on the gold electrode was achieved by the self-assembly of DNA strands (Au/DNA-c|DNA-t|DNA-r.AuNP; Fig. 3.7B). No linkers were used in the experiments, which resulted in a limited number of immobilized DNA strands because of the lack of close-packed arrangements. To increase the number of immobilized DNA-c on the surface, in the next experiment, an SAM of mercaptopropionic acid (MPA) was used (Au/MPA/DNA-c|DNA-t|DNA-r.AuNP; Fig. 3.7C). MPA monolayer also helps to prevent the nonspecific adsorption of DNAs on the sensor surface. In the next experiment, functionalized gold nanoparticles were used to nullify the insulating nature of MPA monolayer and to increase the sensitivity of the sensor (Au/MPA/AuNP/DNA-c|DNA-t|DNA-r.AuNP; Fig. 3.7D). A detection limit of 100 aM was achieved by the use of gold nanoparticles on the MPA surface and gold nanoparticles as the label for DNA-r. The developed DNA sensor showed good reproducibility, stability, and reusability, and it demonstrated good selectivity against noncomplementary sequences and three-base mismatch complementary sequences. Thus by using appropriately functionalized gold nanoparticles and the electrode surface, it is possible to significantly enhance the sensitivity because of the high electron conduction through the gold nanoparticles.

Because of the unique electrochemical characteristics of carbon nanomaterials, they are extensively explored for faradaic and nonfaradaic processes. They transfer electrons from or to the adsorbed molecules, which can be detected easily. These characteristics make the carbon nanostructures favorable materials for electrode modification in the biosensor. The major disadvantage is being the same kind of charge-transfer effect produced by the occurrence of any charged ions within the Debye length around these carbon nanostructures. This interference of the charged molecules/ions disturbs the selectivity of the sensor.

Similarly, there is a possibility of NSA resulting from various interactions between DNA and the carbon nanostructures. This can be evaded using membranes or functional layers. CNTs are one of the widely explored materials for DNA immobilization, which provide a high surface area for immobilization of DNA molecules and significantly improve the electrochemical properties of the sensors (Wang, 2005). The use of graphene and graphene derivatives, as electrode materials, has been extensively investigated to construct various electrochemical DNA biosensors (Gan and Hu, 2011). DNA bases are immobilized on graphene or CNT surfaces through van der Waals interaction (π interaction), and the solvation energy arises from the solvent molecules. Covalent binding using the active ester chemistry can be used here also, where graphene or CNT surfaces are functionalized to attain more available sites for covalent binding of biomolecules, which also helps in improving its biocompatibility.

Conductive polymers are also used as electrode materials in DNA sensors (Hao et al., 2014). But the relatively low conductivity when compared with CNT and non-oriented nanofiber morphology leads to low sensitivity in these types of sensors. To get high sensitivity, conducting polymers are coupled with different nanomaterials (Park et al., 2014).

3.6.3 Whole-cell biosensors

These are intact biological structures such as a cell, microorganism, or a specific cellular component that are capable of specifically binding to certain species. Bacterial whole-cell biosensors are highly useful for the toxicity measurements and detection of biochemical oxygen demand. The method used for the immobilization of cells on the transducer surface is important in determining the geometry, structural, biological, and chemical stability of the cells. One of the earliest methods used for immobilization of cells is adsorption method, in which the cells are attracted to the solid matrix/porous matrix followed by the adsorption. The interaction between the matrix and cells is governed by van der Waals, electrostatic, hydrophobic, and hydration forces. The adsorption process is widely used for the immobilization of viable cells. The appropriately modified techniques used for immobilization of enzymes are used for immobilization of cells. From them, entrapment and surface attachment techniques are the commonly used procedures for cells. Entrapment of cells is done by the addition of gelling or cross-linking agent to the polymer or sol-gel system.

A microbial biosensor was developed by entrapping *Pseudomonas syringae* on a highly porous microcellular polymer and used for the detection of biological oxygen demand (Kara et al., 2009). The configuration of the sensor allowed continuous regeneration and growth of immobilized microorganisms in the disk, and thereby the sensor exhibited good stability and lifetime. Oriented immobilization of bacteriophage on the surface is important over random orientation to attain the selectivity and sensitivity in the detection of live bacteria. A polyethyleneimine-functionalized CNT was used for the immobilization of T2 bacteriophage (virus) through covalent binding for the detection of the live bacterial cell (Zhou et al., 2017). In this work, an electric field—induced immobilization is combined with charge-directed immobilization and covalent cross-linking for oriented phage immobilization. Initially, multiwalled CNTs were functionalized with polyethyleneimine to obtain positive surface charge, and then a positive potential was applied for phage deposition, which then linked to the surface covalently using 1-pyrenebutanoic acid succinimidyl ester (PBSE) (Fig. 3.8). This method helped in the oriented immobilization of phage on the surface and resulted in a high loading of bioreceptor on the surface, which, in turn, contributed to increased sensitivity and selectivity.

Bacteria-mediated bioimprinted films were also explored for the selective bacterial detection. In this method, reduced graphene sheets and chitosan were electrodeposited on indium tin oxide, which served as a matrix for bacterial attachment. Followed by the absorption of sulfate-reducing bacteria (SRB), a thin layer of chitosan was deposited around the bacteria. SRB was then washed off from the surface to get a bioimprint on the biosensor surface. The binding of target SRB was

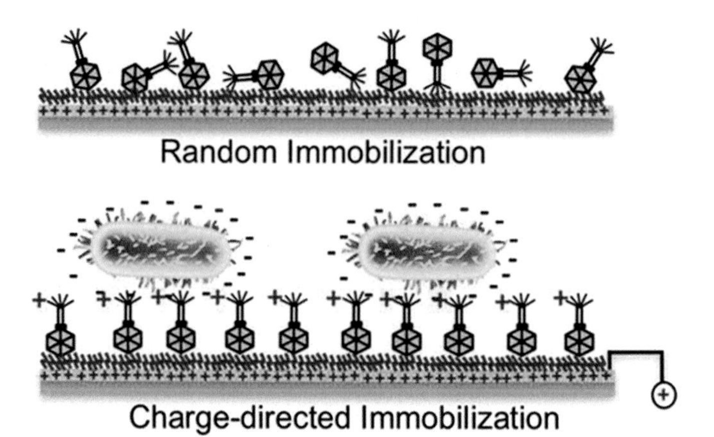

FIGURE 3.8

A random immobilization versus oriented immobilization of bacteriophage on the transducer surface. In this work, the oriented immobilization is achieved by a combination of electric field induced, charge directed, and covalent cross-linking immobilization methods.

Reprinted with permission from Zhou, Y., Marar, A., Kner, P., Ramasamy, R.P., 2017. Charge-directed immobilization of bacteriophage on nanostructured electrode for whole-cell electrochemical biosensors. Analytical Chemistry 89(11), 5734–5741. Copyright from American Chemical Society.

monitored by impedance change of the biosensor, and this imprint could capture target SRB in a range of 104–108 CFU mL^{-1}.

3.7 Conclusion

From the literature cited in the chapter, it is evident that surface modification of electrode contributes to a significant role in determining the efficiency and applicability of electrochemical biosensor. The materials used for surface modifications should be biocompatible and stable to ensure the stable immobilization of bioreceptors. Generally used materials for surface modification include SAMs, conducting polymers, and various nanomaterials. Important properties such as large surface area, high electron transfer rate, and good biocompatibility of the nanomaterials are efficiently utilized in the development of electrochemical biosensors. The immobilization strategy should be selected appropriately, considering the application and the nature of the bioreceptor. For example, in an affinity-based sensor, an antibody should be immobilized on the surface in the "end on" orientation to enable it to bind the antigen. Similarly, the upright or near upright orientation of DNA is important in the effective hybridization of the target. Covalent binding through active ester chemistry is one of the best suitable and stable methods of immobilization if the covalent bond is not affecting the reactivity. Selection of an appropriate surface modification approach and immobilization of bioreceptor is the focus point in defining the efficacy of the electrochemical biosensor.

Acknowledgment

The author would like to acknowledge the help received from Mr. Arjun Ajith Mohan and Dr. Ramani T. in the preparation of figures presented in this chapter.

References

Adhikari, B.R., Schraft, H., Chen, A., 2017. A high-performance enzyme entrapment platform facilitated by a cationic polymer for the efficient electrochemical sensing of ethanol. Analyst 142 (14), 2595–2602.

Bessueille, F., Dugas, V., Vikulov, V., Cloarec, J.P., Souteyrand, E., Martin, J.R., 2005. Assessment of porous silicon substrate for well-characterised sensitive DNA chip implement. Biosensors and Bioelectronics 21 (6), 908–916.

Biebuyck, H.A., Bain, C.D., Whitesides, G.M., 1994. Comparison of organic monolayers on polycrystalline gold spontaneously assembled from solutions containing dialkyl disulfides or alkanethiols. Langmuir 10 (6), 1825–1831.

Bongrand, P., 1998. Specific and nonspecific interactions in cell biology. Journal of Dispersion Science and Technology 19 (6–7), 963–978.

Cai, C., Chen, J., 2004. Direct electron transfer of glucose oxidase promoted by carbon nanotubes. Analytical Biochemistry 332 (1), 75–83.

Chaki, N.K., Vijayamohanan, K., 2002. Self-assembled monolayers as a tunable platform for biosensor applications. Biosensors and Bioelectronics 17 (1–2), 1–12.

Chen, A., Chatterjee, S., 2013. Nanomaterials based electrochemical sensors for biomedical applications. Chemical Society Reviews 42 (12), 5425–5438.

Chen, S., Liu, L., Zhou, J., Jiang, S., 2003. Controlling antibody orientation on charged self-assembled monolayers. Langmuir 19 (7), 2859–2864.

Chen, L., Zheng, H., Zhu, X., Lin, Z., Guo, L., Qiu, B., Chen, G., Chen, Z.N., 2013. Metal–organic frameworks-based biosensor for sequence-specific recognition of double-stranded DNA. Analyst 138 (12), 3490–3493.

Chen, W., Yang, W., Lu, Y., Zhu, W., Chen, X., 2017. Encapsulation of enzyme into mesoporous cages of metal–organic frameworks for the development of highly stable electrochemical biosensors. Analytical Methods 9 (21), 3213–3220.

Daniel, M.C., Astruc, D., 2004. Gold nanoparticles: assembly, supramolecular chemistry, quantum-size-related properties, and applications toward biology, catalysis, and nanotechnology. Chemical Reviews 104 (1), 293–346.

Das, P., Das, M., Chinnadayyala, S.R., Singha, I.M., Goswami, P., 2016. Recent advances on developing 3rd generation enzyme electrode for biosensor applications. Biosensors and Bioelectronics 79, 386–397.

Davis, G., 1985. Electrochemical techniques for the development of amperometric biosensors. Biosensors 1 (2), 161–178.

Dong, S., Peng, L., Wei, W., Huang, T., 2018. Three MOF-templated carbon nanocomposites for potential platforms of enzyme immobilization with improved electrochemical performance. ACS Applied Materials & Interfaces 10 (17), 14665–14672.

Drummond, T.G., Hill, M.G., Barton, J.K., 2003. Electrochemical DNA sensors. Nature Biotechnology 21 (10), 1192.

Ermini, M.L., Mariani, S., Scarano, S., Campa, D., Barale, R., Minunni, M., 2013. Single nucleotide polymorphism detection by optical DNA-based sensing coupled with whole genomic amplification. Analytical and Bioanalytical Chemistry 405 (2–3), 985–993.

Esseghaier, C., Bergaoui, Y., Ben Fredj, H., Tlili, A., Helali, S., Ameur, S., Abdelghani, A., 2008. Impedance spectroscopy on immobilized streptavidin horseradish peroxidase layer for biosensing. Sensors and Actuators B: Chemical 134 (1), 112–116.

Gan, T., Hu, S., 2011. Electrochemical sensors based on graphene materials. Microchimica Acta 175 (1–2), 1.

German, N., Ramanaviciene, A., Voronovic, J., Ramanavicius, A., 2010. Glucose biosensor based on graphite electrodes modified with glucose oxidase and colloidal gold nanoparticles. Microchimica Acta 168 (3–4), 221–229.

Grabar, K.C., Allison, K.J., Baker, B.E., Bright, R.M., Brown, K.R., Freeman, R.G., Fox, A.P., Keating, C.D., Musick, M.D., Natan, M.J., 1996. Two-dimensional arrays of colloidal gold particles: a flexible approach to macroscopic metal surfaces. Langmuir 12 (10), 2353–2361.

Guiseppi-Elie, A., Lei, C., Baughman, R.H., 2002. Direct electron transfer of glucose oxidase on carbon nanotubes. Nanotechnology 13 (5), 559.

Gupta, R., Chaudhury, N.K., 2007. Entrapment of biomolecules in sol–gel matrix for applications in biosensors: problems and future prospects. Biosensors and Bioelectronics 22 (11), 2387–2399.

Hahn, S., Mergenthaler, S., Zimmermann, B., Holzgreve, W., 2005. Nucleic acid based biosensors: the desires of the user. Bioelectrochemistry 67 (2), 151–154.

Hao, Y., Zhou, B., Wang, F., Li, J., Deng, L., Liu, Y.N., 2014. Construction of highly ordered polyaniline nanowires and their applications in DNA sensing. Biosensors and Bioelectronics 52, 422–426.

He, P., Xu, Y., Fang, Y., 2005. A review: electrochemical DNA biosensors for sequence recognition. Analytical Letters 38 (15), 2597–2623.

Hoffman, A.S., 1992. Immobilization of biomolecules and cells on and within polymeric biomaterials. In: Biologically Modified Polymeric Biomaterial Surfaces. Springer, Dordrecht, pp. 61–65.

Homs, M.C.I., 2002. DNA sensors. Analytical Letters 35 (12), 1875–1894.

Jianrong, C., Yuqing, M., Nongyue, H., Xiaohua, W., Sijiao, L., 2004. Nanotechnology and biosensors. Biotechnology Advances 22 (7), 505–518.

Kara, S., Keskinler, B., Erhan, E., 2009. A novel microbial BOD biosensor developed by the immobilization of P. Syringae in micro-cellular polymers. Journal of Chemical Technology & Biotechnology: International Research in Process, Environmental & Clean Technology 84 (4), 511–518.

Lei, J., Qian, R., Ling, P., Cui, L., Ju, H., 2014. Design and sensing applications of metal–organic framework composites. TRAC Trends in Analytical Chemistry 58, 71–78.

Li, L., Shi, Y., Pan, L., Shi, Y., Yu, G., 2015. Rational design and applications of conducting polymer hydrogels as electrochemical biosensors. Journal of Materials Chemistry B 3 (15), 2920–2930.

Liebana, S., Drago, G.A., 2016. Bioconjugation and stabilisation of biomolecules in biosensors. Essays in Biochemistry 60 (1), 59–68.

Lin, J., He, C., Zhao, Y., Zhang, S., 2009. One-step synthesis of silver nanoparticles/carbon nanotubes/chitosan film and its application in glucose biosensor. Sensors and Actuators B: Chemical 137 (2), 768–773.

Liu, Y., Yu, J., 2016. Oriented immobilization of proteins on solid supports for use in biosensors and biochips: a review. Microchimica Acta 183 (1), 1–19.

Liu, S., Leech, D., Ju, H., 2003. Application of colloidal gold in protein immobilization, electron transfer, and biosensing. Analytical Letters 36 (1), 1–19.

Liu, Y., Wang, M., Zhao, F., Xu, Z., Dong, S., 2005. The direct electron transfer of glucose oxidase and glucose biosensor based on carbon nanotubes/chitosan matrix. Biosensors and Bioelectronics 21 (6), 984–988.

Liu, J., Chou, A., Rahmat, W., Paddon-Row, M.N., Gooding, J.J., 2005. Achieving direct electrical connection to glucose oxidase using aligned single walled carbon nanotube arrays. Electroanalysis: An International Journal Devoted to Fundamental and Practical Aspects of Electroanalysis 17 (1), 38–46.

Liu, Q., Lu, X., Li, J., Yao, X., Li, J., 2007. Direct electrochemistry of glucose oxidase and electrochemical biosensing of glucose on quantum dots/carbon nanotubes electrodes. Biosensors and Bioelectronics 22 (12), 3203–3209.

Maleki, N., Kashanian, S., Maleki, E., Nazari, M., 2017. A novel enzyme based biosensor for catechol detection in water samples using artificial neural network. Biochemical Engineering Journal 128, 1–11.

Mark, S.S., Sandhyarani, N., Zhu, C., Campagnolo, C., Batt, C.A., 2004. Dendrimer-functionalized self-assembled monolayers as a surface plasmon resonance sensor surface. Langmuir 20 (16), 6808–6817.

Mathew, M., Sandhyarani, N., 2011. A novel electrochemical sensor surface for the detection of hydrogen peroxide using cyclic bisureas/gold nanoparticle composite. Biosensors and Bioelectronics 28 (1), 210–215.

Mathew, M., Sandhyarani, N., 2014a. Detection of glucose using immobilized bienzyme on cyclic bisureas–gold nanoparticle conjugate. Analytical Biochemistry 459, 31–38.

Mathew, M., Sandhyarani, N., 2014b. Distance dependent sensing capabilities of enzymatic biosensor surface constructed with gold nanoparticle immobilized on self assembled monolayer modified gold electrode. Sensor Letters 12 (8), 1286–1294.

Naveen, M.H., Gurudatt, N.G., Shim, Y.B., 2017. Applications of conducting polymer composites to electrochemical sensors: a review. Applied Materials Today 9, 419–433.

Njagi, J., Andreescu, S., 2007. Stable enzyme biosensors based on chemically synthesized Au–polypyrrole nanocomposites. Biosensors and Bioelectronics 23 (2), 168–175.

North, J.R., 1985. Immunosensors: antibody-based biosensors. Trends in Biotechnology 3 (7), 180–186.

Park, S., Kwon, O., Lee, J., Jang, J., Yoon, H., 2014. Conducting polymer-based nanohybrid transducers: a potential route to high sensitivity and selectivity sensors. Sensors 14 (2), 3604–3630.

Protsailo, L.V., Fawcett, W.R., 2000. Studies of electron transfer through self-assembled monolayers using impedance spectroscopy. Electrochimica Acta 45 (21), 3497–3505.

Rasheed, P.A., Sandhyarani, N., 2015. Attomolar detection of BRCA1 gene based on gold nanoparticle assisted signal amplification. Biosensors and Bioelectronics 65, 333–340.

Rasheed, P.A., Sandhyarani, N., 2016. Quartz crystal microbalance genosensor for sequence specific detection of attomolar DNA targets. Analytica Chimica Acta 905, 134–139.

Rasheed, P.A., Sandhyarani, N., 2017a. Electrochemical DNA sensors based on the use of gold nanoparticles: a review on recent developments. Microchimica Acta 184 (4), 981–1000.
Rasheed, P.A., Sandhyarani, N., 2017b. Carbon nanostructures as immobilization platform for DNA: a review on current progress in electrochemical DNA sensors. Biosensors and Bioelectronics 97, 226–237.
Saberi, R.S., Shahrokhian, S., Marrazza, G., 2013. Amplified electrochemical DNA sensor based on polyaniline film and gold nanoparticles. Electroanalysis 25 (6), 1373–1380.
Sadana, A., Chen, Z., 1996. Influence of non-specific binding on antigen-antibody binding kinetics for biosensor applications. Biosensors and Bioelectronics 11 (1–2), 17–33.
Sapsford, K.E., Algar, W.R., Berti, L., Gemmill, K.B., Casey, B.J., Oh, E., Stewart, M.H., Medintz, I.L., 2013. Functionalizing nanoparticles with biological molecules: developing chemistries that facilitate nanotechnology. Chemical Reviews 113 (3), 1904–2074.
Sassolas, A., Leca-Bouvier, B.D., Blum, L.J., 2008. DNA biosensors and microarrays. Chemical Reviews 108 (1), 109–139.
Shen, W.J., Zhuo, Y., Chai, Y.Q., Yuan, R., 2015. Cu-based metal–organic frameworks as a catalyst to construct a ratiometric electrochemical aptasensor for sensitive lipopolysaccharide detection. Analytical Chemistry 87 (22), 11345–11352.
Shi, P., Zhang, Y., Yu, Z., Zhang, S., 2017. Label-free electrochemical detection of ATP based on amino-functionalized metal-organic framework. Scientific Reports 7 (1), 6500.
Song, S., Clark, R.A., Bowden, E.F., Tarlov, M.J., 1993. Characterization of cytochrome c/alkanethiolate structures prepared by self-assembly on gold. The Journal of Physical Chemistry 97 (24), 6564–6572.
Teles, F.R.R., Fonseca, L.P., 2008. Applications of polymers for biomolecule immobilization in electrochemical biosensors. Materials Science and Engineering: C 28 (8), 1530–1543.
Ulman, A., 1996. Formation and structure of self-assembled monolayers. Chemical Reviews 96 (4), 1533–1554.
Walcarius, A., 2015. Mesoporous materials-based electrochemical sensors. Electroanalysis 27 (6), 1303–1340.
Wang, J., Wang, L., Di, J., Tu, Y., 2009. Electrodeposition of gold nanoparticles on indium/tin oxide electrode for fabrication of a disposable hydrogen peroxide biosensor. Talanta 77 (4), 1454–1459.
Wang, X., Lu, X., Wu, L., Chen, J., 2015. 3D metal-organic framework as highly efficient biosensing platform for ultrasensitive and rapid detection of bisphenol A. Biosensors and Bioelectronics 65, 295–301.
Wang, J., 2005. Carbon-nanotube based electrochemical biosensors: a review. Electroanalysis: An International Journal Devoted to Fundamental and Practical Aspects of Electroanalysis 17 (1), 7–14.
Wang, J., 2006. Electrochemical biosensors: towards point-of-care cancer diagnostics. Biosensors and Bioelectronics 21 (10), 1887–1892.
Xie, S., Ye, J., Yuan, Y., Chai, Y., Yuan, R., 2015. A multifunctional hemin@ metal–organic framework and its application to construct an electrochemical aptasensor for thrombin detection. Nanoscale 7 (43), 18232–18238.
Xu, J., Wang, Y., Hu, S., 2017. Nanocomposites of graphene and graphene oxides: synthesis, molecular functionalization and application in electrochemical sensors and biosensors. A review. Microchimica Acta 184 (1), 1–44.

Yan, W., Feng, X., Chen, X., Hou, W., Zhu, J.J., 2008. A super highly sensitive glucose biosensor based on Au nanoparticles—AgCl@ polyaniline hybrid material. Biosensors and Bioelectronics 23 (7), 925—931.

Yonemori, Y., Takahashi, E., Ren, H., Hayashi, T., Endo, H., 2009. Biosensor system for continuous glucose monitoring in fish. Analytica Chimica Acta 633 (1), 90—96.

Yu, X., Chattopadhyay, D., Galeska, I., Papadimitrakopoulos, F., Rusling, J.F., 2003. Peroxidase activity of enzymes bound to the ends of single-wall carbon nanotube forest electrodes. Electrochemistry Communications 5 (5), 408—411.

Zhang, S., Wang, N., Yu, H., Niu, Y., Sun, C., 2005a. Covalent attachment of glucose oxidase to an Au electrode modified with gold nanoparticles for use as glucose biosensor. Bioelectrochemistry 67 (1), 15—22.

Zhang, S., Wang, N., Niu, Y., Sun, C., 2005b. Immobilization of glucose oxidase on gold nanoparticles modified Au electrode for the construction of biosensor. Sensors and Actuators B: Chemical 109 (2), 367—374.

Zhang, C., Wang, X., Hou, M., Li, X., Wu, X., Ge, J., 2017. Immobilization on metal—organic framework engenders high sensitivity for enzymatic electrochemical detection. ACS Applied Materials & Interfaces 9 (16), 13831—13836.

Zhao, J., Bradbury, C.R., Huclova, S., Potapova, I., Carrara, M., Fermín, D.J., 2005. Nanoparticle-mediated electron transfer across ultrathin self-assembled films. The Journal of Physical Chemistry B 109 (48), 22985—22994.

Zhou, Y., Marar, A., Kner, P., Ramasamy, R.P., 2017. Charge-directed immobilization of bacteriophage on nanostructured electrode for whole-cell electrochemical biosensors. Analytical Chemistry 89 (11), 5734—5741.

CHAPTER

4 Typically used carbon-based nanomaterials in the fabrication of biosensors

Esmaeil Heydari-Bafrooei, PhD [1], Ali A. Ensafi, PhD [2]

Associate Professor, Department of Chemistry, Faculty of Science, Vali-e-Asr University of Rafsanjan, Rafsanjan, Iran[1]; Professor, Department of Chemistry, Isfahan University of Technology, Isfahan, Iran[2]

4.1 Introduction

The use of carbon nanomaterials (CNMs) such as fullerenes, carbon nanotubes (CNTs), carbon nanoparticles (CNPs), graphene, and derivatives obtained from them in many research fields is increasing so rapidly, and the breadth and depth of their applications are increasing continuously (EL-Seesy and Hassan, 2019; Zhang et al., 2018; Hebbar et al., 2017; Voronkova et al., 2017; Rasheed and Sandhyarani, 2017; Heydari-Bafrooei and Askari, 2017a; Heydari-Bafrooei and Shamszadeh, 2017; Heydari-Bafrooei and Askari, 2017b; Heydari-Bafrooei et al., 2016; Heydari-Bafrooei and Shamszadeh, 2016). The potential applications for CNM across many research fields therefore make a supreme topic for evaluation and assessment of their use in fabrication of biosensors (Ensafi et al., 2014a, 2014b; Heydari-Bafrooei et al., 2017; Zhao et al., 2018; Dong et al., 2018). This CNM only contains sp^2 carbon atoms, which make a perfect hexagonal array of atoms.

One of the important problems in the construction of all biosensors is to provide an appropriate biosensing surface for perfect immobilization of the biorecognition element on it, although the biosensing interface must have all the characteristics required to provide a suitable signal from the transducer. The biosensing surface should be such that the analyte interacts selectively with the biorecognition element, which in turn increases the selectivity of the biosensor. If the surface can be modifiable by increasing and/or decreasing of different functional groups or by doping of various materials on the surface, the biosensor signal can be increased, which will increase the sensitivity of the biosensor and improve the response time (Ensafi et al., 2013a, 2014b, 2014c, 2016a).

The more fundamental challenge in the fabrication of biosensors is to provide conditions that can be used from biosensors in complex biological samples and even in confined environments such as inside cells for in vivo research (Baranwal

Electrochemical Biosensors. https://doi.org/10.1016/B978-0-12-816491-4.00004-8

and Chandra, 2018; Wu et al., 2018a; Liu et al., 2017; Ensafi et al., 2013b). Furthermore, multiplexing biosensing systems, which make this possible to test several analytes simultaneously on one device, is an important challenge that researchers are facing (Bizzarri et al., 2018; Leirs et al., 2017; Lu et al., 2016; Ensafi et al., 2013c). Researchers have used different types of nanomaterials to solve these challenges. Examples of such materials include metal nanoparticles (Wu et al., 2018b; Gupta and Meek, 2018; Ensafi et al., 2016b), metal and nonmetal oxide nanoparticles (Picca et al., 2018), polymeric nanoparticles (Melnychuk and Klymchenko, 2018; Shrivastava et al., 2016), and different other nanomaterials (Ensafi et al., 2013d, 2014d, 2015a, 2015b, 2016c; Zhu et al., 2018; Yang et al., 2017). Therefore, novel biosensors based on nanomaterials have received significant attention, and much of the electrochemical sensing and biosensing research is devoted to the synthesis and modification of new nanomaterials.

Carbon-based materials can take on different structures with different physical properties because of the extreme flexibility of its bonding. One of the well-known molecules that consist only of carbon atoms is graphene. The graphene structure consists of a two-dimensional (2D) sheet of carbon atoms in a hexagonal configuration (honeycomb structure). In graphene, the carbon atoms are sp^2 hybridized, and one carbon atom is bonded with three other carbon atoms with a bonding angle of 120 degrees, all of which are placed in the same plane. For this reason, a network of regular hexagons is formed (Fig. 4.1). Different functional groups, as well as hydrogen atoms, can be located in free space (outside the plane) (Erol et al., 2018). Carbon atoms in fullerenes, another carbon allotrope, are arranged

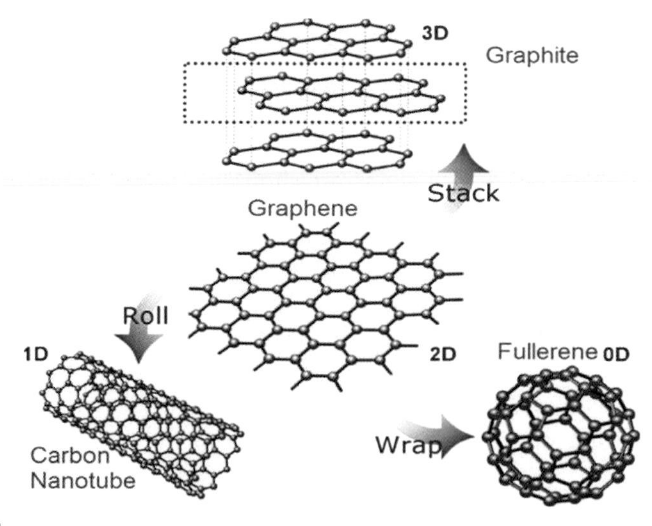

FIGURE 4.1

Schematic structure of graphene, graphite, CNTs, and fullerene.

(Reprinted with permission from Ghany, N.A.A, Elsherif, S.A., Handal, H.T., 2017. Revolution of graphene for different applications: state-of-the-art. Surfaces and Interfaces 9, 93–106)

in a spherical form (buckyball structure), and hence they do not have any dimensions. Fullerenes can form directly from graphene by the creation of pentagons on the edge of graphene, which is followed by the warping of the graphene plane into a bowl-like shape, that is why it is called wrapped-up graphene (Fig. 4.1) (Afreen et al., 2015). CNTs are formed by the rolling up of graphene sheet into a cylinder of nanometer-size diameter.

Given that the nanotubes only have hexagons, they are one-dimensional (1D) objects. Nanotubes can either be single-walled (SWCNT) or multi-walled (MWCNT). SWCNTs consist of only one graphene tube, whereas MWCNTs consist of several concentric graphene tubes (2–100) with the coaxial arrangement.

Graphene, CNTs, and fullerenes have several interesting advantages for fabricating electrochemical biosensors such as excellent chemical, thermal and mechanical stability, biocompatibility, and component diversity (Afreen et al., 2015). These properties make CNM unique to achieve the main objectives of a biosensor that is selective, low-cost, sensitive, and rapid detection of molecules or organisms. Furthermore, the simplicity of the synthesis of CNM-based patterns makes it feasible simultaneous detection of multiple molecules by electrochemical biosensors arrays (Ensafi et al., 2013c). Moreover, the unique mechanical flexibility of CNM allows us to build flexible and wearable biosensors to constantly screen the chemical signals transmitted from the human body (Liu et al., 2018). CNMs that include CNT, nanowires, NP, nanopores, nanoclusters, and graphene and its derivatives have the substantial character in the construction of selective and sensitive biosensors.

4.2 Properties of Carbon Nanomaterials

The facility to syntheses the CNMs with arbitrary functional groups and to form nanocomposite with other materials has opened new ways to design novel carbon-based supports with unique electrochemical properties. One of the most important differences between CNMs and other carbon materials is the metallic or semiconducting properties of CNMs, which increases their electrocatalytic properties, resulting in facilitation of the electron transport at the electrode/solution interface. The outstanding electrochemical properties have led CNMs to be used as catalysts for a different electrochemical process such as hydrogenation, sensors, and fuel cells (Afreen et al., 2015; Liu et al., 2018).

4.2.1 Properties of Carbon Nanotubes

The physical and chemical stability of CNTs is very high, and they conduct electricity. The CNT has a seamless and flexible structure, which is the same as soft graphite, yet one of the hardest materials with high strength. CNT is at least 100 times stronger than steel, but only one-sixth as heavy (Gupta et al., 2017). Depending on their chirality, CNT can have metallic or semiconducting characteristics.

Therefore, some nanotubes are expected to have higher conductive than copper, whereas others behave more like silicon.

CNTs have robust optical absorption in the near-infrared region, and scientists use this property for drug delivery applications. By functionalization of CNT with a tumor cell, near-infrared ray destructed disease cells selectively using specific binding entities. Thermal properties of CNT are also one of the interesting topics for researchers, as many of their applications come from this unique feature. Nanotubes can conduct heat by ballistic phonon propagation. Additional important matter in the properties of CNT is the problem of toxicity. The clarification of toxicity elements of CNT is still incomplete, and studies are still ongoing.

Because of their one-dimensional nature, charge transfer in CNT was found to occur via quantum effects and only spread along the tube axis, while there is no scattering. These unique electrical and structural properties of CNTs make them ideal for use in biosensors so that any change in electron transfer can mean the occurrence of a biological phenomenon whose purpose is to detect it. The advantage of using CNTs over 0D fullerene and carbon dots in electrochemical biosensors is because of its extraordinary electrical conductivity, which greatly increases the biosensor's sensitivity. Also, by connecting to each other, the CNTs create an interconnected superconductive network, which is not observed in other carbon allotropes. The surface of CNTs can be modified with diverse chemical groups, or biomolecules can be regulated the surface area, dispersibility, mechanical, conductivity, optical absorption, and strength capacities to achieve improved performance of electrochemical biosensors. For these reasons, CNT is extensively used as a surface modifier to design the electrochemical biosensors.

4.2.2 Properties of graphene and graphene oxide

Overall, the derivatives of graphene-based materials contain graphene oxide (GO), single and multilayer graphene, graphene quantum dots, and reduced graphene oxide (rGO). GO has oxygen-based functional groups, which makes it easy to disperse in organic and aqueous solvents. This is an extraordinary advantage of graphene, which allows it to combine with other compounds, such as polymers and ceramics. However, GO is insulated regarding electrical conductivity. To restore the electrical conductivity, it is essential to retrieve the hexagonal honeycomb lattice of graphene again by reduction of GO to rGO. The rGO would not be easily dispersed because it has a tendency to aggregation. Like CNTs, graphene can also be chemically functionalized, thereby improving its properties for use in numerous applications.

Each type of graphene-based materials has various and adjustable mechanical, physical, electrical, defect density, and chemical properties, which has made the researchers hope for their usage in the development of a different electrochemical (bio)sensors. Graphene has advantages over CNTs, the most important of which is the easy to disperse, plentiful surface functional groups, high surface area, and easy to pattern. The high surface area of graphene-based nanomaterials makes them suitable for adsorbing and binding to biomolecules, which is the key issue in making multifunctional biosensors. For example, the Raman spectroscopy reveals that graphene and its derivatives contain a larger density of structural defects that these defects are ethers, hydroxides, epoxides, alcohols, carboxylate, and carboxylic groups at the edges and basal plane of the graphene sheets (Mondal et al., 2013). The presence of these defects can allow the graphene to be conjugated to multifunctional biomolecules via different interactions such as covalent bonding, electrostatic interactions, hydrogen bonding, and π-π stacking.

4.2.3 Properties of fullerene

Like all other carbon allotropes, fullerene also has considerable physical, chemical, and electrical properties and is therefore thought to be an ideal material for fabricating nanoassemblies for numerous applications. The fullerene has a spherical structure, which makes it possible to accelerate the transfer of charge and electron in various applications. Fullerene can be converted into a multifunctional material, which is done by functionalization of it with covalent and noncovalent interactions. The fullerenes are mechanically overstrong and tolerate very high pressures.

4.3 Different synthesis methods of carbon nanomaterials for biosensing application

To realize the electrochemical biosensing applications, it is of key significance to synthesize CNMs with specific structures and a good synthesis yield. However, to overcome the lack of solubility and compatibility/bonding/inertness of CNMs toward other materials, surface modification through chemical functionalization is always required. The satisfactory structure and the surface properties are crucial for biosensing application of CNMs.

The synthesis of fullerenes is carried out easily and in ambient conditions by using initial laser ablation of graphite, then arc discharge, and finally thermal decomposition of aromatic organic compounds. Kroto et al. (1985) discovered fullerene for the first time by sublimation of graphite using pulsed laser beam when they were studying the formation of interstellar dust (Kroto et al., 1985). Solvent extraction, electrochemical methods, HPLC separation, and "SAFA" (stir and filter approach) were used for purification of synthesized fullerene (Xu et al., 2016). However, more potent synthetic procedures and more powerful separation methods are now desired to improve the accessibility of fullerene-based materials.

The extensively used approaches for synthesizing of CNTs are laser ablation, chemical vapor deposition (CVD), and arc discharge. The first SWCNT and MWCNT obtained by Iijima of the NEC Corporation were synthesized by arc discharge. Furthermore, since 1985, when fullerene was discovered using laser radiation, this technique (laser ablation) has been used to produce not only fullerenes but also CNT. By radiation of high-power laser beam on the high purity graphite, carbon is evaporated and converted to MWCNT at 1200°C in an Ar or a N_2 atmosphere. By impregnation of graphite with some metal catalysts such as nickel and cobalt, SWCNT bundles were synthesized (Xu et al., 2016). The facilities for the synthesis of CNT using laser ablation and arc discharge are very similar to that used to produce fullerenes. In the arc discharge, MWCNT is synthesized by applying arc between two ultrapure carbon rods under optimal conditions, but for SWCNT it is necessary to use appropriate metal catalysts such as Fe, Co, Ni, Pt, and Rh. CVD is a versatile and economic method for synthesis of a various types of CNTs with different morphologies and crystallinity. In the CVD process, carbon-containing sources (usually gaseous form) are heated in an elevated temperature furnace and break into atomic carbons and passes over a transition metal catalyst (usually Co, Ni, or Fe), initiating the growth of CNT. Although the yield of production of CNT by CVD is very high, but the percentage of the nanotubes that are structurally defective is higher than those produced by other methods. After synthesis, CNT is purified by refluxing in strong acid, sonication-assisted dispersion in aqueous solutions of the surfactant, or air oxidation procedure (Kroto et al., 1985). Purification is a key stage in the preparation of CNT to remove impurities such as transition metals used as catalysts throughout the synthesis, fullerenes, and amorphous carbon.

In extensive studies to date, seven different routes were used for the synthesis of graphene. These methods are mechanical exfoliation, chemical reduction of graphite oxide, chemical exfoliation, pyrolysis, epitaxial growth, CVD, and plasma synthesis (Xu et al., 2016). Mechanical exfoliation is the first accepted method for large production of graphene. Oxidation of graphite to graphite oxide using concentrated sulfuric acid, potassium permanganate, and nitric acid and reduction of it to graphene is also extensively used. Hydroxylamine, $NABH_4$, ascorbic acid, phenyl hydrazine, hydroquinone, glucose, pyrrole, and alkaline solutions are some of the reducing agents that can be utilized for the synthesis of graphene. Electrochemical reduction is also a robust technique for a large amount of production of graphene. In chemical exfoliation, heating or ultrasonication is used to exfoliate graphene. There are several methods for this purpose, but most of them are based on Hummer methods. The major differences between these methods lie in the type of oxidants used and the toxicity. During the pyrolysis method, chemical synthesis of graphene is performed by solvothermal method. For example, pyrolization of sodium ethoxide was used for the enhancement of detachment of graphene sheets. Epitaxial growth is based on heat and cools down of an electrically insulating surface such as silicon carbide (SiC) and growth of graphene on it. Pressure and heating rate are very important factors that determined the size and structure of synthesized graphene. CVD is a simple method for producing of continuous, large-area, and uniform graphene films. However, graphene based on CVD approach displays moderate charge transferability. Generally, providing a high-temperature growth condition ($>1000°C$) is essential for the CVD method. This high temperature restricts the choice of the substrate as not all substrates can endure the high temperatures. In this case, the joining of plasma sources in the CVD scheme supports the production of graphene at the lower growth temperature on the chosen substrate with the help of a metal catalyst. Plasma synthesis of graphene contains both plasma-enhanced CVD and plasma doping to produce graphene. In these methods, a quartz tube that surrounded by a furnace for temperature control creates the rf plasma source.

4.4 Modification of carbon nanomaterials

Functionalization of CNMs is one of the most key fields in carbon research, which increases their electrochemical application, especially in the field of (bio)sensors. By functionalization of CNMs, a new and interesting class of nanomaterials is generated, which has both unique properties of CNMs and special properties of a functional moiety that provides the opportunity to design new electrochemical sensors with exceptional capabilities. In fact, it can be said that most sensing and biosensing applications of CNMs depend heavily on their modification. Therefore, to increase the properties of CNMs, their functionalization is inevitable. The two main objectives for functionalization of CNMs are to (1) increase their solubility in water and other solvents to easy the manufacturing process of carbon-based applications and (2) increase compatibility with other materials. Their modification is generally separated into two groups: covalent modifications and noncovalent modifications. The covalent functionalization allows us to modulate the structural and electrical parameters.

In the case of CNTs, the covalent modification is often performed through chemical reactions such as halogenation, oxidation, electrochemical reactions, and cycloaddition because of the presence of oxygen functionalities that simplifies the procedure of chemical modification by joining other useful (bio)molecules. These modifications alter the shape, structure, and even the length of CNTs or generate some specific chemical groups on the CNT edges. Covalent modifications typically improve the biocompatibility and hydrophilicity of CNTs and are therefore extensively utilized in biosensor research. Esterification, amidation, and ionic interaction are the nature of bonding between the CNTs and the functional molecules or groups, in the presence or absence of coupling agents. Although covalent modification functionalizes CNTs more effectively, this modification process has a negative effect on the mechanical and electronic properties of carbon. Noncovalent modification not only increases the solubility of carbon in water, but also this kind of process is completely nondestructive and does not affect the microstructure and intrinsic properties of CNTs. The possibility of noncovalent functionalization of CNTs is owed to the presence of their benzene ring structures, which allows a noncovalent interaction between (bio)molecules and CNTs. This makes it possible to control the biological performances of CNTs for in vivo electrochemical biosensing application.

Graphene lacks any functional groups and therefore is not suitable for electrochemical applications. Therefore, it is necessary to modify it for new electrochemical biosensing use. For graphene, carboxylic and hydroxyl groups have been added covalently onto the surface, which is often made possible by oxidants and highly strong acids. These functional groups on the surface of graphene act as chemical handles that provide conditions for binding molecules such as carbohydrates,

DNA, proteins, aptamers, and polymers, therefore improving the specificity, sensitivity, and biocompatibility of it. The highly negatively charge of graphene and its derivatives facilitate the electrostatic absorption of positive charge molecules. Other noncovalent interactions such as π-π stacking, van der Waals interactions, and hydrophobic also contribute to this process of physical absorption. DNA and aptamers are biomolecules that can easily be attached to graphene using this process.

4.5 Application of carbon-based nanomaterials as biosensing

In fact, rapid advances in nanotechnology have released a promising perspective for its electrochemical applications. Apart from the intrinsic miniaturization and high sensitivity of electrochemical-based detections, easy and rapid modification of the electrode surface using new nanomaterials is another advantage of these methods. Undoubtedly, carbon-based nanomaterials are one of the most significant nanostructures in the analytical scene. As mentioned, carbon-based nanomaterials are the nanostructures with exceptional chemical, geometrical, electronic, and mechanical properties, which in particular provide desirable properties resulting from enhanced charge transfer and their robust absorption aptitude. The main analytical benefits predictable and obtainable from the use of CNM as modifiers in electrochemical biosensors are as follows:

1. *Lower detection potentials:* The higher surface-to-volume ratio of the CNM-modified electrodes reduces the current densities and actually decreases the "overpotentials." This electrochemical phenomenon is called the electrocatalytic effect, which has a direct impact on electrocatalysis of analyte and reduces the determination potentials and increases the sensitivity of the detection (Vilela et al., 2014).
2. *Higher currents:* When the electrode surface area was increased by carbon-based nanomaterials, the scale of the redox exchange is increased and subsequently sensitivity increases. It should be kept in mind that the increase of the signal always does not indicate an increase in sensitivity, because the increase of the signal may be accompanied by increased background noise. In fact, in analytical measurements, the signal to noise ratio (S/N) level is more important than the signal alone. As long as the increase in the noise is greater than the signal, the detection limit is not improved.
3. *Higher stability and resistance to passivation:* The root of this advantage is the CNM-based (bio)sensors with the high active surface area and the mass transport rate, prevents the electrode fouling, and reduces the accumulation of oxidation products. The outcome of this antifouling effect is the better reproducibility.
4. *High compatibility and functionality with biomolecules:* Biofunctionalization is required in all of the biosensors, and as mentioned, selective functionalization

strategy for each CNM developed. Briefly, CNM characteristically proposes us enhanced sensitivity, reproducibility, selectivity, and separation efficiency.

4.6 Biosensor using carbon nanotubes

Among CNMs the CNT is a particularly favorable modifier for biosensor construction because of high surface areas, high mechanical strength, outstanding chemical and thermal stability, and high electronic conductivity for the electron transfer reactions. CNT is a great choice for use in multiplexed biosensing because of its ultra-large surface area, which permits immobilization of several molecules, simultaneously. The blend of excellent electrochemical properties, nanometer dimensions, and exceptional conductivity has caused the CNT to be directly attached to different redox enzymes for improved detection. With the alignment of CNT on the electrode, it is possible to create biosensors that show higher conductivity for electron transfer compared with arbitrarily spread CNT. This result is due to the fact that the nanotube tips often facilitate the electron transfer more than the side-walls, and it is only necessary for the electrons to pass through one tube, rather than having to jump from tube to tube, to be transferred to the bulk electrodes. Today, many research groups are working on using CNT to construct new biosensors.

4.7 Carbon nanomaterial–based aptamer and DNA biosensors

One of the interesting developments in the construction of new electrochemical biosensors is the use of aptamers. Aptamers are single-stranded sequences that can bind specifically to the target molecules (e.g., proteins and drugs) that can be considered oligonucleotide analogs of antibodies as biorecognition layer in biosensors. The values of the dissociation constant for complexes of target-aptamer are in the range of nanomolar to picomolar that this shows high affinity and specificity of this bioreceptors. By immobilization of the different aptamers on the electrode surfaces, a special kind of electrochemical sensors can be designed to selective sense a varied range of molecular targets. Moreover, aptamers can be immobilized on the surface of the electrodes modified with CNT via π stacking between the bases of DNA and the CNT walls. This has led to a lot of research into the use of CNT in the construction of electrochemical aptasensors for detection of clinical markers, drugs and other pharmaceuticals, toxic molecules, pesticides, and other environmental pollutants, pathogenic bacteria, and viruses and viral antigens. For instance, an aptasensor, which detects potentiometrically the thrombin change, was developed by covalently linking the 15-mer thrombin DNA aptamer to the carboxylic groups of the SWCNT surface (Düzgün et al., 2010). The potentiometric signal following aptamer-target interaction was significantly changed that can be directly detected thrombin within 15 s. The sensor showed a low detection limit (80 nM) in the broad linear range of 5−1000 ppt. The linear range is 0.1−1 μM. To improve the performance of the sensor, Liu et al. (2010) described an ultrasensitive voltammetric sensor based on amine-modified capture probe (12-mer) immobilized on the CNT-modified GCE (Fig. 4.2). Hybridization reaction on the electrode surface occurred between the target aptamer probe (21-mer) contains thrombin-binding aptamer (15-mer) labeled with ferrocene (Fc) and capture probe (12-mer). The aptamer probe (15-mer) labeled with Fc dissociates from the biosensor, upon the introduction of the analyte; subsequently, the peak current amplitude declines considerably. Afterward, the linear curve ranging from 1.0 pM to 0.5 nM with a detection limit of 0.5 pM was observed. The authors have emphasized that MWCNTs not only work as a carrier for large amount immobilization of electrochemical capture probe molecules but also enhance the electroanalytical signal and sensitivity of the electrochemical apta-assay. In a separate report, an amperometric biosensor for sensing of thrombin was introduced by our group using a platform consisting a nanocomposite of TiO_2NPs and MWCNTs that provide a good substrate for immobilization of aptamer (Heydari-Bafrooei et al., 2016). The aptasensor exhibited low detection limit (in fM range) and specificity in CSF or blood of MS, epilepsy, polyneuropathy, and Parkinson's disease patients. High sensitivity, stability, and selectivity of the sensor are due

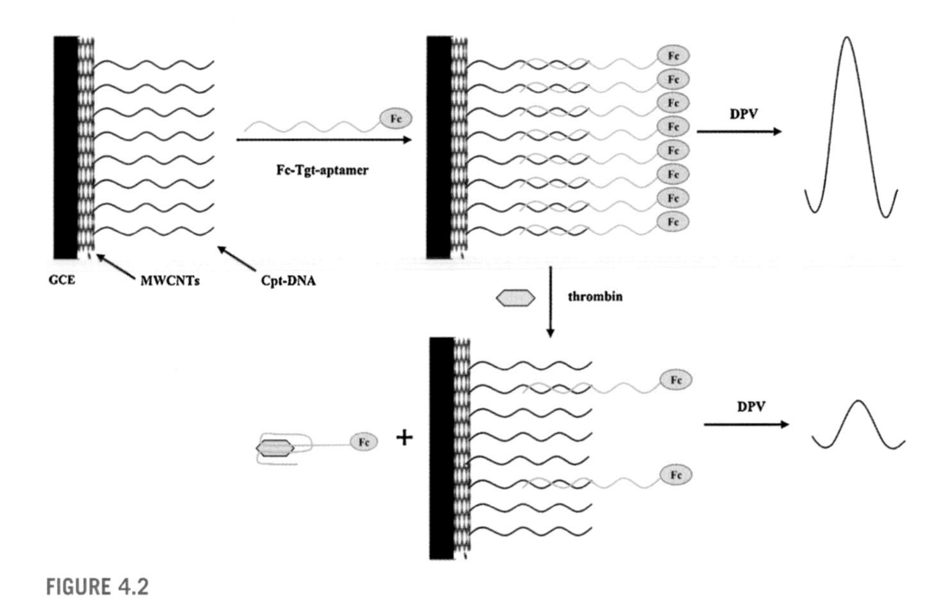

FIGURE 4.2

CNT-based aptasensor design and the procedure of thrombin detection.

(Reprinted with permission from Liu, X.R., Li, Y., Zheng, J.B., Zhang, J.C., Sheng, Q.L., 2010. Carbon nanotube-enhanced electrochemical aptasensor for the detection of thrombin. Talanta 81, 1619–1624)

to good in-plane conductivity, fast electron-transfer kinetics, and the high surface-to-volume ratio of MWCNTs and a porous morphology of TiO_2 nanocrystals.

Jiang et al. (2015) constructed an ultrasensitive and low-cost electrochemical apta-assay for the quantitative detection of acetamiprid. Immobilization of the aptamer was done on the GCE modified with the nanocomposite, and after soaking in a droplet of the acetamiprid and creation of a complex between aptamer and target, an increased impedimetric signal was captured on the electrode, the variation of EIS response being proportioned to the acetamiprid concentration in the solution. The amplified signal is due to the existence of the silver nanoparticles and nanocomposites on the surface of the electrode that simplifies the electron transfer. Furthermore, the modified electrode makes available a good platform for stabilization of a great quantity of aptamer. The proposed aptamer-based biosensor displayed a low limit of detection (LOD) of 33 fM in spiked cucumber and tomatoes with good reproducibility and stability.

Kumar et al. (2015) fabricated a label-free electrochemical apart-assay for the determination of myoglobin (Mb). The procedure is based on the immobilization of the individual aptamer on the rGO/MWCNT-modified screen-printed electrode (SPE). Synergistic effects of the hybrid of rGO and MWCNT improve the electrical conductivity of the electrode and facilitate chemical functionality, which leads to their enhanced sensitivity of the biosensor. Furthermore, because of selective affinity interaction of the surface-confined aptamer with protein, the myoglobin immobilization on the surface of the rGO/MWCNT was efficiently performed. The direct electrochemical reduction of Fe(III) to Fe(II) was applied for detection of myoglobin (Fig. 4.3). The aptasensor provides a dynamic linearity range from 1 to 4000 ng mL^{-1}.

Chen et al. (2015) designed an ultrasensitive sandwich-type electrochemical aptasensor for the MUC1 determination (Chen et al., 2015). The assay was developed based on electropolymerization of o-phenylenediamine (PoPD) on the gold electrode and then deposition of AuNPs on it. Actually, the role of PoPD-AuNP film is a capture probe for the immobilization of thiolated primary aptamers (APT1). In the following, another nanohybrid as tracing tag was developed via deposition of AuNPs, thionine (Thi), and thiolated aptamers (APT2) on the silica/MWCNT core-shell nanocomposites (AuNP/SiO_2@MWCNTs). The AuNP/SiO2@MWCNTs enhance the surface area of the electrode for immobilization of the secondary aptamer and amount of loading of the electrochemical probe thionine. Exposure of the sensor to a solution containing MUC1 leads to the construction of a complex compound between the Thi-AuNP/SiO_2@MWCNT-APT2 and the PoPD-AuNP-APT1. Formation of such complex changes DPV peak values as an electrochemical signal. The detection limit of the aptasensor was reported 1 pM. Despite thrombin, Mb, and MUC1, other protein biomarkers such as epidermal growth factor receptor (EGFR), human epidermal growth factor receptor 2 (HER2), angiogenin, and epithelial cell adhesion molecule (EpCAM) have also been sensed by CNM-based electrochemical aptasensors.

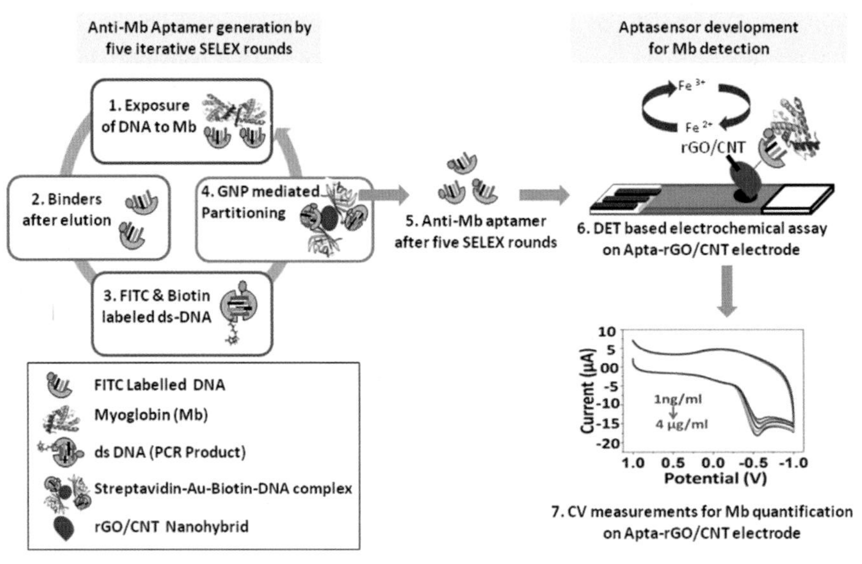

FIGURE 4.3

(Left) Schematic showing the production of a specific antimyoglobin aptamer using SELEX method and (right) the construction of rGO/CNT-modified aptamer-based sensor for the detection of myoglobin.

(Reprinted with permission from Kumar, V., Shorie, M., Ganguli, A.K., Sabherwal, P., 2015. Graphene-CNT nanohybrid aptasensor for label free detection of cardiac biomarker myoglobin. Biosensors and Bioelectronics 72, 56–60)

The characteristics of tumor cells that differentiate them from proteins and nucleic acid biomarkers are more binding sites and large surface area. The targets of some nucleic acid aptamers are diverse cell types and key molecules that exist on the cells surface, which makes possible to detect the cancer cells. Electrochemical aptasensors that use CNTs for signal amplification have been widely considered for the detection of cancer cells. For instance, Liu et al. (2013) developed a super sandwich assay with probes of aptamer-DNA concatamer-CdTeQDs and conductive composites of MWCNTs/polydopamine/AuNP (MWCNTs@PDA@AuNPs) for capturing of cancer cells. For signal amplification, the hybridization is realized by ADNA and CDNA. As a result, linear range was found in the wide range of 10^2–10^6 cells mL^{-1}, and the limit of detection was obtained to be 50 cells mL^{-1}. An excellent feature of this sensor is discrimination between cancer cells and normal cells. The nanohybrid of MWCNTs@PDA@AuNPs plays a key role in this biosensor through enhancement of electrical conductivity and surface area, which makes it possible to immobilize concanavalin A (ConA) for selective and sensitive detection of the cell.

Different CNT-based aptasensors were reported for determination of analytes related to food quality and safety. For example, an apta-assay based on amperometric detection for the ultrasensitive determination of chlorpyrifos was improved based on an easy method with double-assisted signal amplification strategy (Jiao et al., 2016). The increasing use of ferrocene in the construction of electrochemical biosensors is due to the unique redox characteristics, good biocompatibility, aromaticity, and sandwich structure, which made it have an extensive request for use in electrochemical sensors. Chitosan with rich amino groups displays a tough film-forming aptitude. The porous matrix of chitosan films affords an ideal location for immobilization of the biorecognition elements and facilitates diffusion of substrate and/or inhibitors. Also, chitosan shows the advantages of inexpensiveness, nontoxicity, susceptibility to chemical modification, and biocompatibility. Because of its desired characteristics, chitosan has extensively been utilized as an ideal matrix for immobilization of biomolecules in biosensor devices. Ferrocene-branched chitosan composite and MWCNTs amplified the electrochemical signal. Also, it organized the effective immobilization of aptamers on the transducer. Furthermore, mesoporous carbon functionalized by chitosan has also a high permeability, perfect dispersibility, and a large specific surface area, which was utilized to powerful immobilization of greater quantities of material. Although the aptasensor concept has some degree of difficulty (preparation of modified electrode), the sensor showed good stability and reproducibility, and its applicability for the very sensitive determination of chlorpyrifos in fruits and vegetables confirmed that it is satisfactory for real sample analysis. Despite chlorpyrifos, other important analytes in food analysis such as salmonella, bisphenol A, aldicarb, streptomycin, saxitoxin, tetracycline, and tanamycin have also been sensed by CNM-based electrochemical aptasensors.

Over the past years, research in our group was focused on the developing electrochemical biosensors for detecting DNA damage and genotoxicity (Heydari-Bafrooei and Shamszadeh, 2016, 2017; Heydari-Bafrooei and Askari, 2017b; Heydari-Bafrooei et al., 2016, 2017; Ensafi et al., 2013a, 2013b, 2013d, 2014a, 2014b, 2014c, 2014d, 2015a, 2015b, 2016a, 2016b, 2016c; Lu et al., 2016). A few years ago, our team developed an MWCNT-based electrochemical DNA biosensor for detection of DNA damage by immobilization of $\{MWCNTs/PDDA/ds\text{-}DNA\}_2$ film on the graphite electrode and methylene blue (MB) as an indicator (Fig. 4.4) (Ensafi et al., 2013b). Detection of DNA damage was performed by immersion of the modified electrode into an aqueous solution containing MB for loading of the MB into the ds-DNA existing on the modified electrode surface. The loaded MB on the surface displayed a sharp DPV peak at −230 mV versus Ag/AgCl. After immersion of the MB-loaded modified electrode in a blank solution, the MB was steadily released from the film, but by dipping the electrode into the MB solution again, the complete reloading of MB into the films was got, representing the incorporation of the MB is reversible. However, after incubation of the $\{MWCNTs/PDDA/ds\text{-}DNA\}_2$-modified graphite electrode in the solution containing glutathione, chromium (VI), and hydrogen peroxide as genotoxic agent, the damaged MWCNTs/PDDA/ds-DNA films were not reappeared to their original form and completely loaded state by reloading MB, thus displaying reduced DPV peak currents.

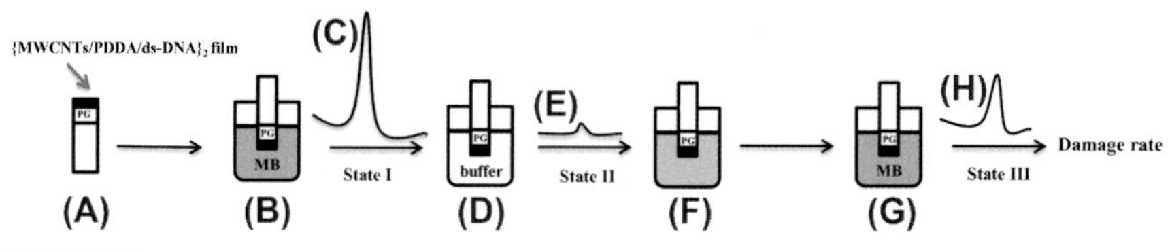

FIGURE 4.4

Schematic diagram of the procedure presented by Ensafi et al. (2013b). for detection of DNA damage: (A) immobilization of {MWCNTs/PDDA/ds-DNA}2 film on the graphite electrode; (B) loading of MB into the ds-DNA existing on the modified electrode surface; (C) DPV scan of the MB-loaded modified electrode in the blank solution; (D) releasing of MB in the buffer solution; (E) DPV scan again in the blank solution; (F) immersion of the MB-released modified electrode in the solution containing damaging agents; (G) dipping the electrode into the MB solution again and reloading of MB; and (H) DPV scan in the blank solution.

With permission from the publisher.

In another work, an electrochemical procedure was developed for the perfect checking of genotoxicity of catecholics (Ensafi et al., 2014b). As a novel work, catechol was encapsulated onto the MWCNTs (CA@MWCNTs) through potential cycling treatment on the surface of the MWCNT/PG electrode between −400 and 1000 mV in a phosphate buffer solution (pH 7.0) containing 1.0 mmol L^{-1} CA at a scan rate of 50 mV s^{-1} ($n = 30$). Voltammograms of the electrode during cycling (Fig. 4.5A&B) showed two pairs of redox peaks located at 23 and 220 mV versus Ag/AgCl, which is directed to the encapsulation of the CA onto the MWCNTs.

Furthermore, appearing an obvious quantity of black spots in the TEM image of MWCNTs after consecutive cycling treatment of MWCNTs approves that CA was encapsulated effectively onto the MWCNTs (Fig. 4.5D&E). Then, DNA was absorbed physically on the surface of the CA@MWCNT/PGE for the detection of DNA damage induced by radicals produced by catecholics. The impedance changes of the bare and modified electrodes in different stages of surface modification are indicative of a successful encapsulation of CA and immobilization of DNA (Fig. 4.5C). CA, as the damaging agent, was immobilized strongly on carbon surfaces by this procedure. Electrochemical impedance spectroscopy was used to explore the effect of metal ions on the strength of the oxidative DNA damage induced by CA.

4.8 Carbon nanomaterial–based immunosensors

Despite the many efforts to substitute them with other recognition elements (artificial receptors, biomimetic polymers, and aptamers), antibodies are still the most common bioreceptor molecules in research and commercial affinity assays. Enough loading of enzyme and finest activity are two main factors for a good

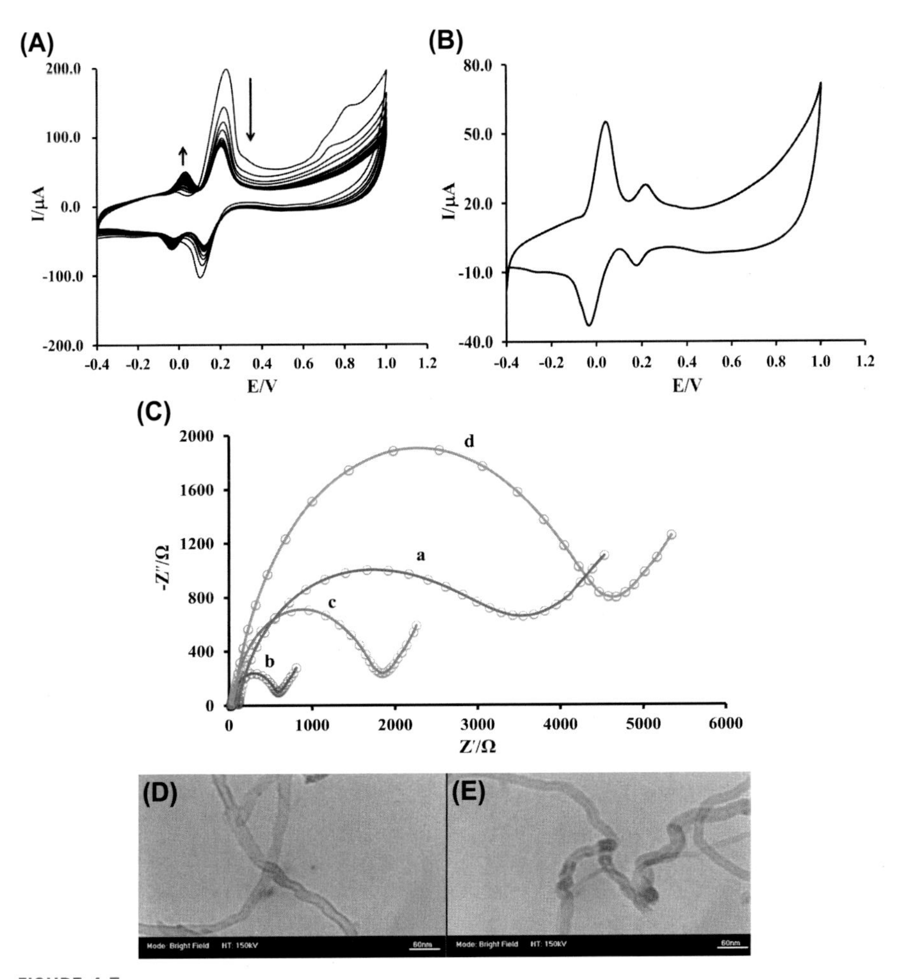

FIGURE 4.5

Cyclic voltammograms of the MWCNT/PG electrode during cycling (A) and CA@MWCNT/PG electrode (B) in the phosphate buffer (pH 7.0) containing 1.0 mmol L^{-1} CA at a scan rate, 50 mV s^{-1}. (C) Impedance spectra of (a) bare PGE; (b) MWCNT/PGE; (c) CA@MWCNT/PGE; and (d) DNA/CA@MWCNT/PGE in 0.1 mol L^{-1} PBS (pH 4.0) and 5.0 mmol L^{-1} $Fe(CN)_6^{3-/4-}$ containing 0.1 mol L^{-1} KCl; and TEM images of (D) MWCNTs and (E) CA@MWCNTs.

(Reprinted with permission from Ensafi, A.A., Amini, M., Rezaei, B., 2014. Assessment of genotoxicity of catecholics using impedimetric DNA-biosensor. Biosensors and Bioelectronics 53, 43–50)

immunosensors. In preparing immunosensors, it should be noted that the enzymes are softly treated because exposure to high temperature and chemical reactions lead to missing their activity. As stated above, unique optical and electronic properties of CNMs for use as transducer makes it as a unique surface for immobilization of enzymes for making biosensors. There are many methods for enzymes immobilization on the surface of CNM-based substrates offered in the literature. The most used enzyme immobilization methods are (1) physical methods that include weak interactions such as van der Waals, hydrophobic, hydrophilic, hydrogen bonds, or ionic or electrostatic, (2) chemical methods, where stronger interactions are applied (formation of covalent bonds between enzymes and CNTs), and (3) entrapment that includes polymerization of enzymes to gel after mixing with monomer solution. Various CNT-based biosensors conjugated to enzymes are developed for the detection of environmental pollutants, biomarkers, drugs and other pharmaceuticals, pathogenic bacteria, toxic molecules and viruses, and viral antigens. For instance, an electrochemical immunosensor was developed based on a film containing the nanocomposite of MWCNTs, chitosan, and thionine for quantitative sensing of insecticide chlorpyrifos, which belong to organophosphate family, in fruits and vegetables such as cabbage or lettuce (Sun et al., 2012). Chitosan has exceptional film-making and adhesion aptitude and can dissolve MWCNTs adequately without assembling into bundles. Association of chitosan with MWCNTs provides a suitable microenvironment for immobilizing molecules and enhancing the electrocatalytic activity of MWCNTs. Thionine easily interacts with MWCNTs via π-π stacking noncovalent interactions because of aromaticity and the presence of great volume of amino functional groups with hydrophilic properties. Subsequently, electroactivity of the MWCNTs was enhanced by thionine. As discussed in the previous section of this chapter, covalent bonding is the most used approach for biomacromolecules immobilization on the surface of the electrode. In this work, the authors used the glutaraldehyde as a cross-linking agent for covalently immobilization of antichlorpyrifos monoclonal antibody on the surface of electrode modified with MWCNTs, chitosan, and thionine.

A potentiometric immunosensor for detection of naphthalene acetic acid was developed with an experimental thin lipid film biosensing setup developed on graphene nanosheets, incorporated with auxin-binding protein 1 receptor (Bratakou et al., 2016). The data suggest a high affinity toward hormone and noninterference of indole-3-acetic acid and biologically inactive 2-NAA. The response time of the biosensor is satisfactorily fast (order of min) and LOD for phytohormone is in the nanomolar range. According to the authors, a portable and miniaturized version of the sensor can be established for the fast on-site monitoring of vegetables and fruits without the need for a skilled technician.

In another study, SPE was modified with graphene sheets and, after electrochemical treatment with the 2-aminobenzyl amine, was used for detection of parathion (Mehta et al., 2016). In this label-free method, for occurring the biorecognition event, no cross-linkers have been used. Producing of a large amount of the amino functional groups on the surface of electrode upon functionalization with 2-aminobenzyl amine allows simple immobilization of antiparathion antibodies. The immunosensor showed a very low LOD of 52 pg L^{-1} within the concentration range of 0.1–1000 ng L^{-1} by impedimetric experiments. For the investigation of the capacity of the sensor for the real sample analysis, carrot and tomato were selected.

Cross-validation against standard method (HPLC) and the fabricated immunosensor showed specificity to parathion against other pesticides.

Besides pesticides, there are a number of electrochemical immunosensors for detection of other biological elements such as fungi in fruits and vegetable samples using electrochemical immunosensors (Fernández-Baldo et al., 2009). A microfluidic biosensor and a continuous-flow biosensor system based on CNT-modified SPE were developed for quantification of *Botrytis cinerea.* The *B. cinerea* is a plant-pathogenic fungus, which can cause a disease known as gray mold. The biosensor was constructed by immobilization of purified antigens of *B. cinerea* on an SPE, and the detection of *B. cinerea* is based on a competitive immunoassay method. The immunologically reaction between *B. cinerea*–specific monoclonal antibody with the *B. cinerea* antigens is sensed by HRP enzyme and 4-tertbutyl catechol as an enzymatic mediator. Then the reduction of produced 4-tertbutyl o-benzoquinone to 4-tertbutyl catechol is monitored on CNT-modified SPE. The signal is the oxidation peak current of the product of the enzymatic reaction, which it is proportional to the enzyme activity and subsequently, to the number of attached antibodies to the modified electrode surface. The response time is 30 min, and the LOD for an electrochemical method and the ELISA procedure are 0.02 and 10 $\mu g\ mL^{-1}$, respectively.

Xu et al. (2004) constructed electrochemical urease and acetylcholinesterase biosensors based on the immobilization of the enzymes on the SWCNT-based platform by a sol-gel hybrid material for detection of urea and acetylthiocholine (Xu et al., 2004). These immunosensors are utilized. A highly sensitive organophosphorus pesticides biosensor is fabricated based on amino-functionalized CNTs decorated on glassy carbon electrode and immobilization of acetylcholinesterase as recognition layer (Yu et al., 2015). Lenihan et al. (2004) offered a new method for bioconjugation of CNTs for the construction of novel immunosensors in which a protein layer is created on the surface of CNT by incubation with streptavidin. Li et al. (2015) constructed an electrochemical immunosensor for the detection of CEA by immobilizing anti-CEA primary antibody (Ab1) on the GCE modified with Pb(II) @AuNP@MWCNT-Fe_3O_4. The presence of the different components on the surface of the biosensor has numerous benefits such as increase in the surface area, which induced the higher amounts absorption of Ab1, and a synergetic effect on the electrocatalysis of the H_2O_2 reduction reaction, which made possible a significant improvement in the analytical signal. The immunosensor achieved a detection limit of 1.7 $fg\ mL^{-1}$, and it showed a good linear relationship between H_2O_2 reduction current and logged C_{CEA} over the range from 0.0005 $pg\ mL^{-1}$ to 50 $ng\ mL^{-1}$. Effect of potential interferents was evaluated on the biosensor response, and no interferences from ascorbic acid, glucose, and bovine serum albumin were observed. In another study, carboxylated SWCNTs and anti-AFP primary antibody (Ab1) were covalently attached inside the channels of aminopropylethoxysilane-grafted mesoporous silica (MPS) for label-free immunosensing of AFP (Fig. 4.6). The anti-AFP/cSWCNT/TMCS–MPS (where the presence of trimethylchlorosilane (TMCS) is for blocking of the external silanol groups of MPS) and graphene were immobilized on the surface of the electrode via an LBL self-assembly. The

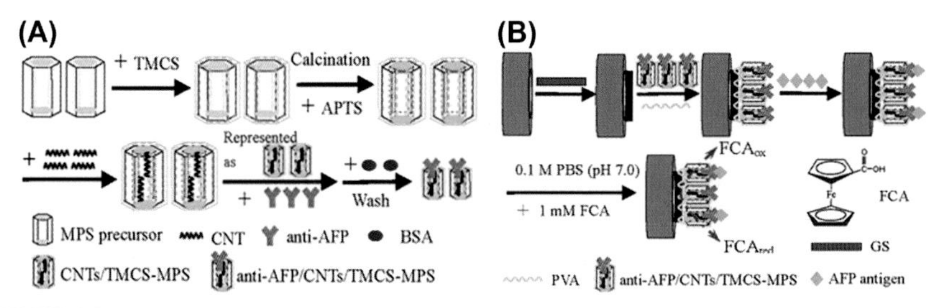

FIGURE 4.6

(A) Preparation of a label-free immunoprobe and its usage in the design of the electrochemical label-free immunosensing of the AFP presented by Lin et al. (2013).

With permission from the publisher.

analytical signal is the electrochemical oxidation signal of the redox probe FcCOOH. The presence of CNT placed inside the MPS and the graphene has enhanced the electrochemical signal (Lin et al., 2013).

4.9 Conclusions

The use of CNMs with different size, shape, and compositions in the manufacture of biosensors is due to their unique properties. To specify the biological interactions in biosensors, covalent and noncovalent interactions are used for functionalization of CNMs with biomolecules. The biomolecules-functionalized CNMs have a synergetic effect on the conductivity, catalytic activity, and biocompatibility. These three factors are key elements for the design of novel electrochemical biosensing devices. The plenty of binding sites for functionalization, good biocompatibility, ability in the multiple analyses, and so on are issues that need to be considered in the development of electrochemical bioassays based on the biofunctional CNM systems. This chapter has generally focused on the introduction of CNM, synthesis, modifications, as well as electrochemical biosensing of them. The modifications of CNMs are extensively used for improved applications in electrochemical biosensors. Particularly, the use of CNMs in biosensing technology improved selectivity, sensitivity, and linear dynamic range. The biocompatible properties of CNMs make possible use of them for in situ detection of different analytes in living cells.

References

Afreen, S., Muthoosamy, K., Manickam, S., Hashim, U., 2015. Functionalized fullerene (C60) as a potential nanomediator in the fabrication of highly sensitive biosensors. Biosensors and Bioelectronics 63, 354–364.

Baranwal, A., Chandra, P., 2018. Clinical implications and electrochemical biosensing of monoamine neurotransmitters in body fluids, in vitro, in vivo, and ex vivo models. Biosensors and Bioelectronics 121, 137–152.

Bizzarri, A.R., Moscetti, I., Cannistraro, S., 2018. Surface enhanced Raman spectroscopy based immunosensor for ultrasensitive and selective detection of wild type p53 and mutant p53R175H. Analytica Chimimica Acta 1029, 86–96.

Bratakou, S., Nikoleli, G.-P., Siontorou, C.G., Karapetis, S., Nikolelis, D.P., Tzamtzis, N., 2016. Electrochemical biosensor for naphthalene acetic acid in fruits and vegetables based on lipid films with incorporated auxin-binding protein receptor using graphene electrodes. Electroanalysis 28, 2171–2177.

Chen, X., Zhang, Q., Qian, C., Hao, N., Xu, L., Yao, C., 2015. Electrochemical aptasensor for mucin 1 based on dual signal amplification of poly(o-phenylenediamine) carrier and functionalized carbon nanotubes tracing tag. Biosensors and Bioelectronics 64, 485–492.

Dong, S., Peng, L., Wei, W., Huang, T., 2018. Three MOF-templated carbon nanocomposites for potential platforms of enzyme immobilization with improved electrochemical performance. ACS Applied Materials & Interfaces 10, 14665–14672 [15].

Düzgün, A., Maroto, A., Mairal, T., O'Sullivan, C., Rius, F.X., 2010. Solid-contact potentiometric aptasensor based on nucleic acid aptamer functionalized carbon nanotubes for the direct determination of proteins. Analyst 135, 1037–1041.

EL-Seesy, A.I., Hassan, H., 2019. Investigation of the effect of adding graphene oxide, graphene nanoplatelet, and multiwalled carbon nanotube additives with n-butanol-Jatropha methyl ester on a diesel engine performance. Renewable Energy 132, 558–574.

Ensafi, A.A., Amini, M., Rezaei, B., 2013. Biosensor based on ds-DNA decorated chitosan modified multiwall carbon nanotubes for voltammetric biodetection of herbicide amitrole. Colloids and Surfaces B: Biointerfaces 109, 45–51.

Ensafi, A.A., Amini, M., Rezaei, B., 2013. Detection of DNA damage induced by chromium/glutathione/H_2O_2 system at MWCNTs–poly(diallyldimethylammonium chloride) modified pencil graphite electrode using methylene blue as an electroactive probe. Sensors and Actuators B: Chemical 177, 862–870.

Ensafi, A.A., Heydari-Bafrooei, E., Rezaei, B., 2013. Different interaction of codeine and morphine with DNA: a concept for simultaneous determination. Biosensors and Bioelectronics 41, 627–633.

Ensafi, A.A., Heydari-Bafrooei, E., Rezaei, B., 2013. DNA-based biosensor for comparative study of catalytic effect of transition metals on autoxidation of sulfite. Analytical Chemistry 85, 991–997.

Ensafi, A.A., Nasr-Esfahani, P., Heydari-Bafrooei, E., Rezaei, B., 2014. Redox targeting of DNA anchored to MWCNT and TiO_2 nanoparticles dispersed in poly dialyldimethylammonium chloride and chitosan. Colloids and Surfaces B: Biointerfaces 121, 99–105.

Ensafi, A.A., Amini, M., Rezaei, B., 2014. Assessment of genotoxicity of catecholics using impedimetric DNA-biosensor. Biosensors and Bioelectronics 53, 43–50.

Ensafi, A.A., Amini, M., Rezaei, B., 2014. Impedimetric DNA-biosensor for the study of anticancer action of mitomycin C: comparison between acid and electroreductive activation. Biosensors and Bioelectronics 59, 282–288.

Ensafi, A.A., Jamei, H.R., Heydari-Bafrooei, E., Rezaei, B., 2014. Development of a voltammetric procedure based on DNA interaction for sensitive monitoring of chrysoidine, a banned dye, in foods and textile effluents. Sensors and Actuators B: Chemical 202, 224–231.

Ensafi, A.A., Nasr-Esfahani, P., Heydari-Bafrooei, E., Rezaei, B., 2015. Determination of atropine sulfate using a novel sensitive DNA–biosensor based on its interaction on a modified pencil graphite electrode. Talanta 131, 149–155.

Ensafi, A.A., Kazemnadi, N., Amini, M., Rezaei, B., 2015. Impedimetric DNA-biosensor for the study of dopamine induces DNA damage and investigation of inhibitory and repair effects of some antioxidants. Bioelectrochemistry 104, 71–78.

Ensafi, A.A., Lesani, S., Amini, M., Rezaei, B., 2016. Electrochemical ds-DNA-based biosensor decorated with chitosan modified multiwall carbon nanotubes for phenazopyridine biodetection. Journal of the Taiwan Institute of Chemical Engineers 54, 165–169.

Ensafi, A.A., Amini, M., Rezaei, B., Talebi, M., 2016. A novel diagnostic biosensor for distinguishing immunoglobulin mutated and unmutated types of chronic lymphocytic leukemia. Biosensors and Bioelectronics 77, 409–415.

Ensafi, A.A., Jamei, H.R., Heydari-Bafrooei, E., Rezaei, B., 2016. Electrochemical study of quinone redox cycling: a novel application of DNA-based biosensors for monitoring biochemical reactions. Bioelectrochemistry 111, 15–22.

Erol, O., Uyan, I., Hatip, M., Yilmaz, C., Tekinay, A.B., Guler, M.O., 2018. Recent advances in bioactive 1D and 2D carbon nanomaterials for biomedical applications. Nanomedicine 14, 2433–2451.

Fernández-Baldo, M.A., Messina, G.A., Sanz, M.I., Raba, J., 2009. Screen-printed immunosensor modified with carbon nanotubes in a continuous-flow system for the Botrytis cinerea determination in apple tissues. Talanta 79, 681–686.

Ghany, N.A.A., Elsherif, S.A., Handal, H.T., 2017. Revolution of graphene for different applications: state-of-the-art. Surfaces and Interfaces 9, 93–106.

Gupta, S., Meek, R., 2018. Metal nanoparticles-grafted functionalized graphene coated with nanostructured polyaniline 'hybrid' nanocomposites as high-performance biosensors. Sensors and Actuators B 274, 85–101.

Gupta, U.S., Kumar, K., Sohale, K., Bansal, M., 2017. A review of structures and properties of carbon nanotubes. International Journal of Advance Research in Science and Engineering 6, 863–873.

Hebbar, R.S., Isloor, A.M., Inamuddin, Asiri, A.M., 2017. Carbon nanotube- and graphene-based advanced membrane materials for desalination. Environmental Chemistry Letters 15, 643–671.

Heydari-Bafrooei, E., Askari, S., 2017. Ultrasensitive aptasensing of lysozyme by exploiting the synergistic effect of gold nanoparticle-modified reduced graphene oxide and MWCNT in a chitosan matrix. Microchimica Acta 184, 3405–3413.

Heydari-Bafrooei, E., Askari, S., 2017. Electrocatalytic activity of MWCNT supported Pd nanoparticles and MoS_2 nanoflowers for hydrogen evolution from acidic media. International Journal of Hydrogen Energy 42, 2961–2969.

Heydari-Bafrooei, E., Shamszadeh, N.S., 2016. Synergetic effect of CoNPs and graphene as cocatalysts for enhanced electrocatalytic hydrogen evolution activity of MoS2. RSC Advances 6, 95979–95986.

Heydari-Bafrooei, E., Shamszadeh, N.S., 2017. Electrochemical bioassay development for ultrasensitive aptasensing of prostate specific antigen. Biosensors and Bioelectronics 91, 284–292.

Heydari-Bafrooei, E., Amini, M., Ardakani, M.H., 2016. An electrochemical aptasensor based on TiO_2/MWCNT and a novel synthesized Schiff base nanocomposite for the ultrasensitive detection of thrombin. Biosensors and Bioelectronics 85, 828–836.

Heydari-Bafrooei, E., Amini, M., Saeednia, S., 2017. Electrochemical detection of DNA damage induced by Bleomycin in the presence of metal ions. Journal of Electroanalytical Chemistry 803, 104–110.

Jiang, D., Du, X., Liu, Q., Zhou, L., Dai, L., Qian, J., Wang, K., 2015. Silver nanoparticles anchored on nitrogen-doped graphene as a novel electrochemical biosensing platform with enhanced sensitivity for aptamer-based pesticide assay. Analyst 140, 6404–6411.

Jiao, Y., Jia, H., Guo, Y., Zhang, H., Wang, Z., Sun, X., Zhao, J., 2016. An ultrasensitive aptasensor for chlorpyrifos based on ordered mesoporous carbon/ferrocene hybrid multiwalled carbon nanotubes. RSC Advances 6, 58541–58548.

Kroto, H., Heath, J., O'Brien, S., Curl, R.F., Smalley, R.E., 1985. C60: buckminsterfullerene. Nature 318, 162–163.

Kumar, V., Shorie, M., Ganguli, A.K., Sabherwal, P., 2015. Graphene-CNT nanohybrid aptasensor for label free detection of cardiac biomarker myoglobin. Biosensors and Bioelectronics 72, 56–60.

Leirs, K., Leblebici, P., Lammertyn, J., Spasic, D., 2017. Fast multiplex analysis of antibodies in complex sample matrix using the microfluidic Evalution™ platform. Analytica Chimica Acta 982, 193–199.

Lenihan, J.S., Gavalas, V.G., Wang, J., Andrews, R., Bachas, L.G., 2004. Protein immobilization on carbon nanotubes through a molecular adapter. Journal of Nanoscience and Nanotechnology 4, 600–604.

Li, F., Jiang, L., Han, J., Liu, Q., Dong, Y., Li, Y., Wei, Q., 2015. A label-free amperometric immunosensor for the detection of carcinoembryonic antigen based on novel magnetic carbon and gold nanocomposites. RSC Advances 5, 19961–19969.

Lin, J., Wei, Z., Zhang, H., Shao, M., 2013. Sensitive immunosensor for the label-free determination of tumor marker based on carbon nanotubes/mesoporous silica and graphene modified electrode. Biosensors and Bioelectronics 41, 342–347.

Liu, X.R., Li, Y., Zheng, J.B., Zhang, J.C., Sheng, Q.L., 2010. Carbon nanotube-enhanced electrochemical aptasensor for the detection of thrombin. Talanta 81, 1619–1624.

Liu, H., Xu, S., He, Z., Deng, A., Zhu, J.-J., 2013. Supersandwich cytosensor for selective and ultrasensitive detection of cancer cells using aptamer-DNA concatamer-quantum dots probes. Analytical Chemistry 85, 3385–3392.

Liu, W., Dong, H., Zhang, L., Tian, Y., 2017. Development of an efficient biosensor for the in vivo monitoring of Cu+and pH in the brain: rational design and synthesis of recognition molecules. Angewandte Chemie International Edition 56, 16328–16332.

Liu, X., Shi, L., Gu, J., 2018. Microbial electrocatalysis: redox mediators responsible for extracellular electron transfer. Biotechnology Advances 36, 1815–1827.

Lu, X., Liang, X., Dong, J., Fang, Z., Zeng, L., 2016. Lateral flow biosensor for multiplex detection of nitrofuran metabolites based on functionalized magnetic beads. Analytical and Bioanalytical Chemistry 408, 6703–6709.

Mehta, J., Vinayak, P., Tuteja, S.K., Chhabra, V.A., Bhardwaj, N., Paul, A.K., Kim, K.-H., Deep, A., 2016. Graphene modified screen printed immunosensor for highly sensitive detection of parathion. Biosensors and Bioelectronics 83, 339–346.

Melnychuk, N., Klymchenko, A.S., 2018. DNA-functionalized dye-loaded polymeric nanoparticles: ultrabright FRET platform for amplified detection of nucleic acids. Journal of the American Chemical Society 140, 10856–10865.

Mondal, P., Sinha, A., Salam, N., Roy, A.S., Jana, N.R., Islam, S.M., 2013. Enhanced catalytic performance by copper nanoparticle e graphene based composite. RSC Advances 3, 5615–5623.

Picca, R.A., Manoli, K., Luciano, A., Sportelli, M.C., Palazzo, G., Torsi, L., Cioffi, N., 2018. Enhanced stability of organic field-effect transistor biosensors bearing electrosynthesized ZnO nanoparticles. Sensors and Actuators B 274, 210–217.

Rasheed, P.A., Sandhyarani, N., 2017. Carbon nanostructures as immobilization platform for DNA: a review on current progress in electrochemical DNA sensors. Biosensors and Bioelectronics 97, 226–237.

Shrivastava, S., Jadon, N., Jain, R., 2016. Next-generation polymer nanocomposite-based electrochemical sensors and biosensors: a review. TRAC Trends in Analytical Chemistry 82, 55–67.

Sun, X., Cao, Y., Gong, Z., Wang, X., Zhang, Y., Gao, J., 2012. An amperometric immunosensor based on multi-walled carbon nanotubes-thionine-chitosan nanocomposite film for chlorpyrifos detection. Sensors 12, 17247–17261.

Vilela, D., Martín, A., González, M.C., Escarpa, A., 2014. Carbon nanotube electrochemical detectors in microfluidics. In: Pumera, M. (Ed.), Nanomaterials for Electrochemical Sensing and Biosensing. CRC Press, New York, pp. 138–168.

Voronkova, M.A., Luanpitpong, S., Rojanasakul, L.W., Castranova, V., Dinu, C.Z., Riedel, H., Rojanasakul, Y., 2017. SOX9 regulates cancer stem-like properties and metastatic potential of single-walled carbon nanotube-exposed cells. Scientific Reports 7, 11653.

Wu, Q., Wei, X., Pan, Y., Zou, Y., Hu, N., Wang, P., 2018. Bionic 3D spheroids biosensor chips for high-throughput and dynamic drug screening. Biomedical Microdevices 20, 82.

Wu, Z., Sun, L., Zhou, Z., Li, Q., Huo, L., Zhao, H., 2018. Efficient nonenzymatic H_2O_2 biosensor based on ZIF-67 MOF derived Co nanoparticles embedded N-doped mesoporous carbon composites. Sensors and Actuators B 276, 142–149.

Xu, Z., Chen, X., Qu, X., Jia, J., Dong, S., 2004. Single-wall carbon nanotube-based voltammetric sensor and biosensor. Biosensors and Bioelectronics 20, 579–584.

Xu, J., Lu, X., Li, B., 2016. Synthesis, functionalization, and characterization. In: Chen, C., Wang, H. (Eds.), Biomedical Applications and Toxicology of Carbon Nanomaterials. Wiley-VCH Verlag GmbH & Co. KGaA, Weinheim, Germany, pp. 1–28.

Yang, Z.-H., Ren, S., Zhuo, Y., Yuan, R., Chai, Y.-Q., 2017. Cu/Mn double-doped CeO_2 nanocomposites as signal tags and signal amplifiers for sensitive electrochemical detection of procalcitonin. Analytical Chemistry 89, 13349–13356.

Yu, G., Wu, W., Zhao, Q., Wei, X., Lu, Q., 2015. Efficient immobilization of acetylcholinesterase onto amino functionalized carbon nanotubes for the fabrication of high sensitive organophosphorus pesticides biosensors. Biosensors and Bioelectronics 68, 288–294.

Zhang, W., Zhang, J., Lu, Y., 2018. Stimulation of carbon nanomaterials on syntrophic oxidation of butyrate in sediment enrichments and a defined coculture. Scientific Reports 8, 12185.

Zhao, F., Wu, J., Ying, Y., She, Y., Wang, J., Ping, J., 2018. Carbon nanomaterial-enabled pesticide biosensors: design strategy, biosensing mechanism, and practical application. TRAC Trends in Analytical Chemistry 106, 62–83.

Zhu, J., Wen, M., Wen, W., Du, D., Zhang, X., Wang, S., Lin, Y., 2018. Recent progress in biosensors based on organic-inorganic hybrid nanoflowe. Biosensors and Bioelectronics 120, 175–189.

CHAPTER

Typically used nanomaterials-based noncarbon materials in the fabrication of biosensors

5

Fatemeh Farbod, Mohammad Mazloum-Ardakani, PhD

Department of Chemistry, Faculty of Sciences, Yazd University, Yazd, Iran

5.1 Introduction

The world of nanomaterials, in particular, noncarbon nanomaterials, is infinitely vast and fascinating enough for scientists to keep creating new nanomaterials with interesting and efficient properties. In nanotechnology, through dimensional control on the 1−100 nm scale, functional systems are designed and constructed. The focus of the research in this regard is shifting from carbon, as just one of about a hundred elements in the periodic table, to the rest of elements. Some of these elements have certain properties to make them desirable candidates for forming nanomaterials.

During the past decade, the use of noncarbon nanomaterials has been remarkable for biosensors and always associated with innovation and upgrading. These nanomaterials, with improved performance and efficiency, have been used in the production of biosensors, particularly electrochemical biosensors, with the advantages of portability for decentralized analysis, reasonable cost, instrumental simplicity, great sensitivity, and selectivity.

Noncarbon nanomaterials generally have interesting electronic, physical, and chemical traits such as great surface energy, large surface-to-volume ratio, and adjustable surface properties. The special chemical traits of such nanomaterials make surface modification easy with a diversity of functional groups to connect molecules of different natures. Also, their electronic properties have decreased the volume of electronic devices to a large extent. These intrinsic and distinctive properties have made noncarbon nanomaterials adequate for application in manufacturing biosensor systems.

A wide range of nanomaterials including oxide nanoparticles, metal nanoparticles, nanowires, quantum dots (QDs), nanocore-shells, and even composite nanomaterials have come to be of noteworthy applications in different kinds of electrochemical biosensors, such as amperometric, potentiometric, and impedimetric sensors, for broad analytical aspects. Noncarbon nanomaterials, endowed with different natures, figures, sizes, and compositions, have been applied to serve broad

Electrochemical Biosensors. https://doi.org/10.1016/B978-0-12-816491-4.00005-X

functions in the manufacture of electrochemical biosensors. Their major functions can be classified as follows:

1. modifying electrotransducer surface and substrate for immobilization of biomolecules;
2. amplifying electrochemical signals;
3. serving as signal-producing probe (as a label);
4. mimicking the enzymatic behavior.

5.1.1 Modifying electrotransducer surface and substrate for immobilization of biomolecules

One of the attractions of nanomaterials is providing the possibility of design and construction of diverse nanoscaffolds to immobilize biological or analyte species. Immobilization of biomolecules leads to an increase in their physiological stability. This, in turn, results in their advantage of being applicable in commercial electrochemical biosensors. Modification of transducer surfaces by using nanomaterials is commonly done to reduce the diffusion limitations, speed up the transfer of electrons across the electrode surface, and enhance the effective surface area to load significant amounts of biomolecules.

In recent years, several kinds of nanomaterials have been manufactured in different nanostructured forms, such as nanotubes, nanoflowers, nanowires, nanofibers, and nanospheres for immobilization purposes. Nanomaterials with a powerful propensity to absorb biomolecules have a major role in the design of biosensors by providing those sensors a good ability for immobilization of biomolecules.

There are several nanoparticles that carry an electrical charge and, thus, provide good electrostatic attraction to attach biomolecules with opposite charges; it is of great importance to attach biomolecules onto nanomaterials. It is noteworthy that, if biomolecules are immobilized in a bulk material medium, their denaturation and loss of bioavailability pose serious problems. However, if they are absorbed on nanomaterials, their bioavailability is maintained.

5.1.2 Electrochemical signal amplification

It is possible to detect ultralow analyte concentrations by the signal amplification strategy. Signal amplification methods that are based on nanomaterials have gained great significance for their immense sensitivity and selectivity in analyte detection. Certain noncarbon nanomaterials such as semiconductor nanoparticles, metal nanoparticles, and multimetal nanostructures have appeared as amplifiers of electrochemical signals in the fabrication of electrochemical biosensors (Ding et al., 2013). There are two conceivable ways of nanomaterial signal amplification as follows:

1. Catalytic reaction-based signal amplification: This procedure combines the unique features of biosensors and the catalytic properties of nanomaterials, which leads to extremely sensitive sensors. Biofunctionalized nanoparticles are

capable of producing synergistic effects and playing catalytic roles to accelerate electrochemical signals. This leads to the two advantages of low detection limits and wide linear correlation ranges.

2. Nanomaterials as mediators: A thick layer of proteins surrounds the active sites on biomolecules. This blocks the electrical contact between redox biomolecules and the electrode surface. To cope with this problem, the conductivity properties of metal nanoparticles are used to improve the rate of electron transfer between electrodes and molecules. Metal nanoparticles can serve as electrical wires or mediators of electron transfer. Hence, they are suitable substitutes for common intermediates in electrochemical biosensors.

In addition to metals, some semiconductor nanoparticles and oxide nanoparticles contribute to the improvement of the electron transfer rate between the protein and the electrode. This improvement in the electron flow is possible owing to the specific arrangement of biomolecules and nanoparticles as well as the conductivity of nanoparticles.

5.1.3 Nanomaterials as signal-producing probes (i.e., as labels)

The advent of nanotechnology paved the way for getting electrochemical signals in biosensors by means of nanomaterial labels. The scope and power of nanomaterials can be illustrated in the modification of target analytes or biomolecules. Nanomaterial-labeled biomolecules can maintain their bioactive property and identify counterpart particles. The ultrasensitivity of affinity-based electrochemical biosensors results from the great signal enhancement achieved by the use of nanomaterial labels.

Nanoparticles consist of numerous sites that are electrochemically active. By labeling biomolecules or analytes with nanoparticles, more electrical species are loaded upon the surface of electrodes and, consequently, the sensor sensitivity is enhanced. In this respect, gold, PbS, CdS, ZnS, silver, and Pt nanoparticles have been investigated.

Many nanoparticles have proved to be inactive in the potential range of aqueous electrolytes or possess a redox potential close to solvent barriers. In the former case, signal interference with the background current is possible. This problem lies there with the intrinsic stability of nanoparticles as well as the existence of stabilizing agents on nanosurfaces. The problem actually hardens the direct electrochemical detection of labels. To remove this drawback, first nanoparticles are dissolved in a solution by means of a powerful oxidizing agent, and, then, electrochemical measurements are conducted (Zhou et al., 2013). There is a correlation between the concentration of metal ions and the number of nanoparticles, which can be attributed to the concentration of biomolecules in the solution.

For the process of detecting nanoparticle labels, to skip the step of dissolving those labels, nanoparticles may be explored for their catalytic property, and their number may be monitored on the electrode surface. The monitoring is to be done indirectly through the catalyzation of the process. After the labeling process is carried out, a solution of nanoparticle electrochemically active substrates is used to incubate the sensing platforms. In such conditions, the biosensor behaves as an ultramicroelectrode array. This procedure is applicable to such nanoparticle labels as Au, Pd, and Pt, which have strong catalytic properties.

5.1.4 Mimicking the enzymatic behavior

Owing to their exclusive characteristics, such as remarkable surface-area-to-volume ratio, optical properties, amazing catalytic activity, and the presence of a plenty of reactive groups on their surface, noncarbon nanomaterials have induced a great deal of research on their ability to mimick enzymatic behavior.

Nowadays, peroxidase-, catalase-, and oxidase-like activities are generally known as the capabilities of various kinds of nanomaterials (Mazloum-Ardakani et al., 2016). Some of the commonly studied noncarbonic nanomaterials in this regard are Fe_3O_4 nanoparticles, gold nanoparticles, nickel oxide nanoparticles, nanoceria, and metallic nanoparticle composites (Xie et al., 2012).

Gaining such benefits as high stability, low cost, and stable catalytic activities is the result of using nanomaterial-based enzyme mimetics in biosensors. Biosensors of this type are widely applied for many purposes in biotechnology and biomedicine. For instance, numerous electrochemical biosensors that function on the basis of the enzyme-like activity of gold, Fe_3O_4, and silver are in use to determine glucose and hydrogen peroxide electrochemically.

However, a limitation of enzyme-like nanomaterials, which discourages their use in amperometric biosensors, is the difficulty of regenerating them for subsequent measurements.

5.2 Classification of noncarbon nanomaterials and their application in electrochemical biosensors

Noncarbon nanomaterials that are fabricated from metals, polymeric substances, or semiconductors have been vastly examined for their ability to promote the efficiency of electrochemical biosensors. In recent decades, an explosion has occurred in useful applications of nanomaterials particularly in electrochemical biosensors. Noncarbon nanomaterials have diversity, but they may be divided into five main groups as follows:

1. Metal and metal oxide
2. Core-shell
3. QDs
4. Wires, fibers, and sheets
5. Composites

Hereon, the applications of these types of biosensors will be discussed.

5.2.1 Metal and metal oxide

Metal and metal oxide materials as the basis for electrochemical biosensors can be typically divided into two types: noble metal nanoparticles and transition metal oxides. Noble metals include Au, Pt, and Pd, whereas transition metal oxides for electrochemical biosensors develop from monometallic oxides (such as NiO, Co_3O_4, and Fe_3O_4) into single-phase bimetallic oxides such as $NiCo_2O_4$, $MnCo_2O_4$ (Huang et al., 2018), $NiMoO_4$ (Liao et al., 2016), $NiFe_2O_4$ (Luo et al., 2010), and $CoMn_2O_4$.

Noble metal nanoparticles were first found exhibiting large surface-to-volume ratios, extraordinary conductivity, size-related electronic properties, remarkable electrochemical performances, low cytotoxicity, and biocompatibility. They have been widely used to fabricate novel electrochemical sensing devices and improve their efficiencies. There are several distinct strategies, such as layer by layer, self-assembly, and hybridization, to modifying surfaces with noble metal nanoparticles. Through these techniques, noble metal nanoparticles are able to control the microenvironment of biological molecules very well. This control is implemented by retaining the activity of biomolecules and facilitating the electron transfer between their redox centers and the electrode surface. In addition, noble metallic nanoparticles participate in biorecognition events by joining chemical labels, biomolecules, and nanomaterials of various types. This leads to signal transduction and amplification. Another attractive capability of metal nanoparticles in the development of biosensors is their catalytic properties in biochemical reactions.

Catalysis is one of the most important widely investigated and extensively utilized chemical applications of metallic nanoparticles. Transition metals, especially valuable metals, have an excellent catalytic aptitude to take part in various organic reactions. In a medium where a reaction takes place, nanoparticles serve as conventional homogeneous catalysts, but, after the reaction, it is easy to regenerate them. The existence of NPs on the surface of the electrode, as an immobilization platform for biomolecules, raises the rate of electron transfer and improves the electrochemical output signals (Jianrong et al., 2004).

Furthermore, some nanomaterials provide an environment similar to the native system, in which the bioactivity of proteins is maintained, more freedom in the orientation is provided, and denaturation of biomolecules is prevented. However, there exist deeply buried redox centers that protein shells insulate, and the redox active sites may be located too far from the electrode surface. In these conditions, electron transfer cannot occur directly, and most of the proteins and redox enzymes do not have a direct electrical connection to the surface of the electrode.

Metallic nanoparticles, especially noble metal nanoparticles, have been proposed to solve this problem. In this case, using nanoparticles as connectors is recommended to cater a pathway for electron relay from the redox center area to the electrode surface. Furthermore, noble metal nanoparticles provide a microenvironment in which proteins are freer in their displacement. This leads to the reduction of the insulating impact of the protein shells (Wang, 2012).

For the first time, Noorbakhsh and Salimi (2011) used a thin oxide layer of NiOx nanoparticles to immobilize DNA probe. The immobilization occurred directly through the formation of phosphate-metal oxide bonds. There was only a simple technique used in this regard, and no chemical cross-linkers or a complex surface modification was needed. What they immobilized was the ssDNA probe sequences belonging to the taxon: 32630 tumor necrosis factor. The process took place on the surface of a NiOx NPs-modified GCE. In their study, the NiOx NPs film significantly enhanced the amount of the loaded DNA probe, thereby raising the sensitivity for the detection of the target DNA. They confirmed the formation of the NiOx NPs layer on the surface of the GCE by the cyclic voltammogram responses of oxide formation and nickel dissolution in an alkaline environment, which revealed an anodic peak of 0.48 V and a cathodic peak of 0.41 V. These two peaks were due to the oxidation of $Ni(OH)_2$ to NiO(OH) and the reduction of NiO(OH) to $Ni(OH)_2$, respectively. The analytical performance of the developed DNA sensor was good for the detection of specific DNA sequences. The sensor could also make a distinction between a completely complementary target sequence and noncomplementary ones, which are characterized as three-base and single-base mismatched DNA sequences. According to differential pulse voltammetry (DPV) investigations, this sensor showed the linear range of 4×10^{6} M to 1×10^{-8} M, the limit of detection of 68 pM (S/N = 3), and the limit of quantification of 230 pM (S/N = 10) for a shorter concentration range.

In another study, Salimi (2008) performed the electrodeposition of CoOx nanoparticles on the surface of a GCE to fabricate a sensitive and selective nitride sensor. They also applied cyclic voltammetry for immobilization of flavin adenine dinucleotide (FAD) on cobalt oxide nanoparticles. Because of their excellent biocompatibility, the CoOx nanoparticles exhibited a very good rate of adsorption for FAD, resulting in highly promoted FAD loading and improved redox reversibility. In the sensing interface, CoOx nanoparticles not only provided a friendly medium for the immobilization of FAD molecules but also facilitated it for the electron to flow between the analyte and the electrode surface. They also investigated the voltammetric properties of FAD/CoOx/GCE in a 0.1 M PBS. A pair of reduction-oxidation peaks were observed at the formal potential of −0.42 V (Fig. 5.1, curve I). The anodic-to-cathodic peak potentials separation was found to be 30 mV, and the ratio of the anodic-to-cathodic peak currents was almost one. These findings indicated that the process in question was a quasi-reversible one, which was practiced for FAD at the surface of a GCE modified with CoOx nanoparticles. As a result, it can be claimed that CoOx films have a great effect on electrode reaction kinetics and supply an appropriate medium for FAD to transfer electrons with a GCE. It is also implied that the restricted orientations of CoOx nanoparticles may favor the direct relay of electrons between FAD molecules and the electrode that underlies.

The surface coverage and the heterogeneous electron transfer rate constant (k_s) belonging to the FAD immobilized on the CoOx film glassy carbon electrode were found to be 1.47×10^{-9} mol Cm^{-2} and 0.85 ± 0.1 s^{-1}, denoting the great

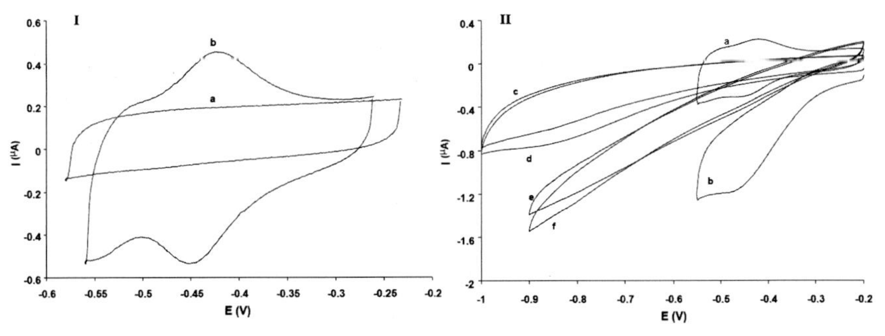

FIGURE 5.1

(I) (a) CVs of CoOx/GCE and (b) FAD/CoOx/GCE. The electrolyte is a PBS (pH 7) at the scan rate of 100 mV s^{-1}. (II) (a) CVs of FAD/CoOx/GCE in a buffer solution (pH 7) and (b) in the presence of nitrite. (c) and (d) are the same results as (a) and (b) for CoOx/GCE. (e) and (f) are as (a) and (b) for FAD/unmodified GCE.

Reprinted with permission from Salimi, A., Hallaj, R., Mamkhezri, H., Mohamad, S., Hosaini, T., 2008. Electrochemical properties and electrocatalytic activity of FAD immobilized onto cobalt oxide nanoparticles: application to nitrite detection. Journal of Electroanalytical Chemistry 619, 31–38.

loading capability of CoOx nanoparticles and the highly facile electron transfer between FAD and the CoOx nanoparticles. The cyclic voltammograms of CoOx/GCE and FAD/CoOx/GCE in the presence and absence of nitrite are provided in Fig. 5.1, curve II. At the CoOx/GCE, an irreversible peak of −0.85 V emerged for nitrite reduction. However, at the FAD/CoOx/GCE, at −0.45 V, a noticeable increase was observed in the reduction current of the FAD redox film owing to the catalytic reduction of nitrite. This is in contrast with the oxidation peak that had largely faded. The decline in the overvoltage (400 mV) and the rise in the peak current of nitrite reduction demonstrate that the FAD immobilized on CoOx nanoparticles possesses a great catalytic capability for nitrite reduction. The rotating modified electrode used at the applied potential of −0.4 V for the amperometric measurement of nitrite showed the detection limit (S/N = 3) and the sensitivity of 0.2 μM and 10.5 nA μM^{-1}, respectively. What may be concluded is that, owing to their great adsorption ability, high biocompatibility, and little detriment to the biological activity, CoOx nanoparticles promise formidable usages in bioelectrochemistry and biosensors.

Nanostructured thin metal oxide films have gained a lot of attention owing to their electrical and magnetic features, porosity, and unique morphology. These characteristics have made those films appropriate for application in diverse fields such as energy storage and generation, catalysis, gas sensors designing, and biology. Kumar et al. (2013) made a hydrogen peroxide amperometric biosensor on the basis of the direct electrochemical exertion of myoglobin (Mb) on a nanostructured cerium dioxide (CeO_2) film with high porosity (Fig. 5.2). The CeO_2 nanostructured film with a large surface area underwent an electrodeposition process on an ITO substrate to provide a large substrate to immobilize myoglobin on the electrode.

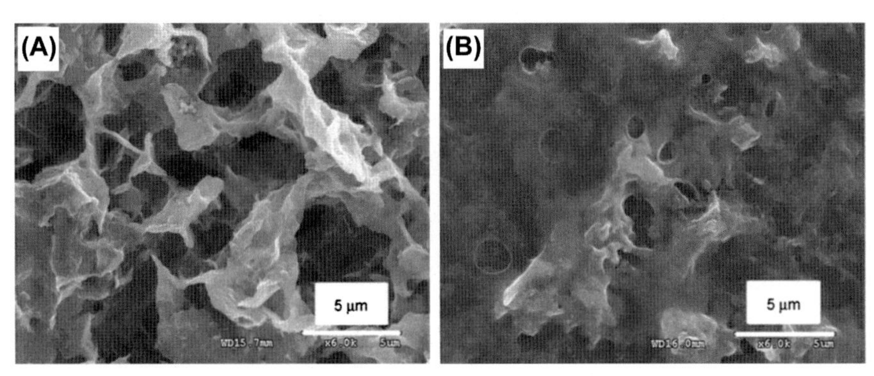

FIGURE 5.2

(A) The cerium film formed on the ITO electrode surface by an electrodeposition method and (B) myoglobin immobilized on the cerium oxide film ITO surface.

Reprinted with permission from Kumar, A., Lee, T., Min, J., Choi, J., 2013. An enzymatic biosensor for hydrogen peroxide based on CeO_2 nanostructure electrodeposited on ITO surface. Biosensors and Bioelectronics 47, 385–390.

CeO_2 also provided good conductivity as well as capability of transferring electrons between the Mb and the ITO surface that underlay. These properties were of benefit for the electrochemical detection of H_2O_2 with a very low limit of detection (0.6 μM). This platform offers the benefits of high efficiency, facility, and fast measurement of targets, which is vital for fabricating user-friendly sensors with sensitive and selective responses.

Metal nanoparticles have proved to be of benefit as catalytic labels in biorecognition events for the enlargement of particles. These nanoparticles label biological species without making alterations in their biological activities. Assays of affinity can afterward be carried out through monitoring of the electrochemical signals of the metal nanoparticles. Pt NPs with different shapes, sizes, and morphologies show a distinctive capability of catalyzing the hydrogenation, oxidation, and dehydrogenation of a variety of molecules. Polsky et al. (2006) utilized nucleic acid–functionalized Pt NPs as catalysts for aptamer/protein detection and the amplified recognition of DNA hybridization through electrochemical procedures. Initially, they exhibited the electrocatalytic ability of Pt-NPs to reduce H_2O_2. They observed an intense cathodic current in the region of −0.1V to −0.6V in cyclic voltammograms. This is in correspondence with the Pt-NP-catalyzed reduction of H_2O_2; hence, confirmation was made of Pt-NP to be applicable as a catalytic label. Then, they adopted two analytical procedures to use Pt-NPs in the analysis of DNA and thrombin. Fig. 5.3A exhibits the amplified electrochemical procedure for the analysis of the DNA analyte.

In the first amplified electrochemical process of the DNA analyte analysis, the thiolated primer, considered as complementary to a segment of the analyte, is assembled upon an Au electrode. A tricomponent Pt-NP-labeled structure is what results

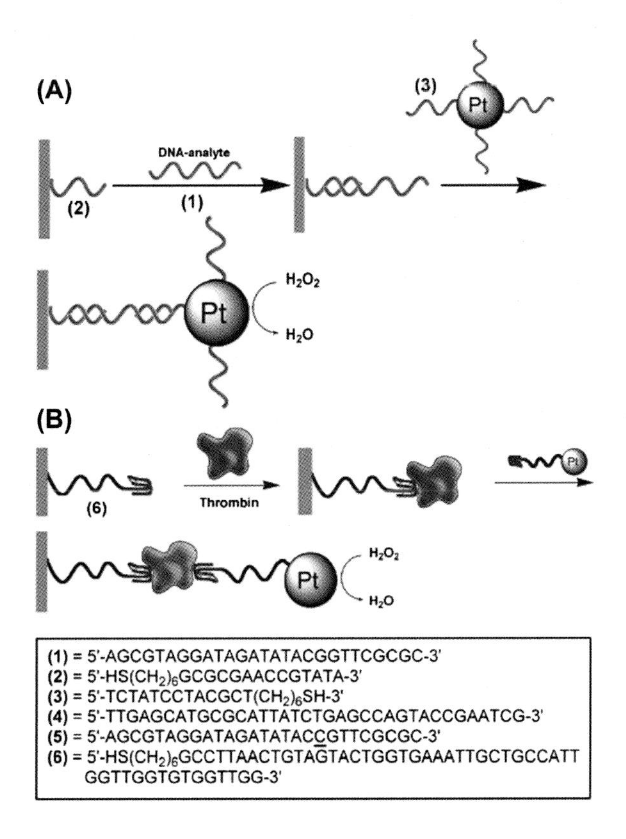

FIGURE 5.3

Analytical procedure for the amplified electrochemical analysis of the DNA analyte.

Reprinted with permission from Polsky, R., Gill, R., Kaganovsky, L., Willner, I., 2006. Nucleic acid-functionalized Pt Nanoparticles: catalytic labels for the amplified electrochemical detection of biomolecules. Analytical Chemistry 78, 2268–2271.

from the hybridization of the produced duplex DNA with nucleic acid and functionalized Pt-NPs. Then, as an amplifying reaction to detect DNA, H_2O_2 is reduced in a Pt-NP-electrocatalyzed manner. In the second procedure (**Fig. 5.3.B**), alteration of Au electrodes occurs with thiolated nucleic acid, containing a thrombin aptamer segment. The Au electrodes are exposed to solutions containing various thrombin concentrations, and then the modified electrodes react with the Pt-NPs functionalized with thiolated nucleic acid. In thrombin, there are two binding sites for the aptamer; therefore, the nucleic acid–modified Pt-NPs get attached to the thrombin complex present on the surface. The Pt-NP labels attached to thrombin then serve as electrocatalytic sites for H_2O_2 reduction. Polsky et al. (2006) reported the sensitivity limit of 1×10^{-11} M for DNA detection and 1×10^{-9} M for the detection of thrombin. The advantage of this sensor lies in the use of Pt-NPs catalytic labels rather than antibodies or enzymes that are sometimes used in the amplified detection of DNA or antigens.

The major task in biosensor fabrication is to immobilize a bioactivator on the surface of a transducer. The more active an immobilized material, the larger its loading, the less resistant the electron transport. A contributing factor in these conditions is the possibility of access to a regenerable sensor surface. To immobilize bioactive materials, the use of magnetic nanoparticles as substrates is considered to be a very effective platform technique. Those nanoparticles can be separated by means of an external magnetic field. Li et al. (2010) developed an amperometric immunosensor for CEA recognition on the basis of the carcinoembryonic antibody (anti-CEA) covalently immobilized on core/shell-structured Au/Fe_3O_4 nanocomposites as well as CEA and HRP-labeled CEA, making a competitive immune reaction with anti-CEA (Fig. 5.4I). The CVs of an SPCE with nano-Au on the surface in a PBS are presented in Fig. 5.4B. A characteristic peak (curve a) is observed for gold oxidation, as Au/Fe_3O_4 nanoparticles attach to the SPCE surface by a magnetic field. The peak, however, disappears as the Au/Fe_3O_4 nanoparticles react with thiourea and anti-CEA (curve b), denoting the saturated assembly of thiourea through a covalent coupling.

As Li et al. (ibid) observed, if an extra assembly operation was carried out for the Au/Fe_3O_4-modified electrode in a gold colloid, the formation of gold oxide would be enhanced in the CV scans of the nano-Au/Au/Fe_3O_4/SPCE in a PBS (curve c), implying that more of the Au nanoparticles were loaded. On the other hand, the gold oxide peak would decrease as the nano-Au/Au/Fe_3O_4/SPCE had a reaction with a 20 ng mL^{-1} anti-CEA solution. For detection of CEA, the immunosensor already incubated was put in a PBS containing hydroquinone and H_2O_2. Recordings were made of the DPV curves, so were quantitative measurements of the CEA content in the solution. The measurements were based on the rates of peak current decline.

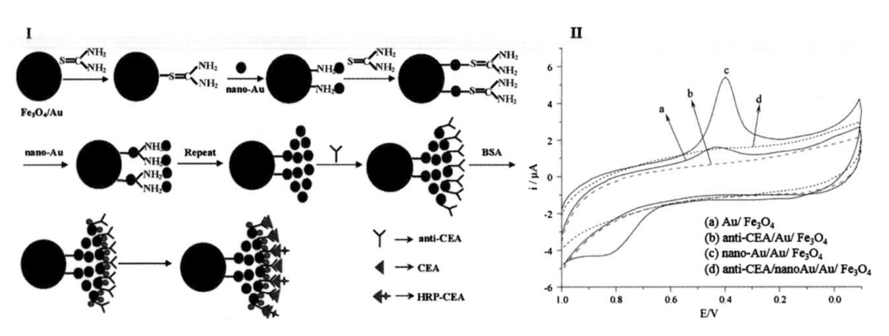

FIGURE 5.4

(I) Schematic display of the process for preparation of anti-CEA/nano-Au/Au/Fe_3O_4 NPs. (II) CVs of the modified SPCEs in a 0.02 mol L^{-1} PBS (pH 7.4), scan rate = 100 mV s^{-1}.

Reprinted with permission from Li, J., Gao, H., Chen, Z., Wei, X., Yang, C.F., 2010. An electrochemical immunosensor for carcinoembryonic antigen enhanced by self-assembled nanogold coatings on magnetic particles. Analytica Chimica Acta 665, 98–104.

This nanomaterial increased the specific surface area, which, in turn, raised the loading rate of immune reagents, such as CEA antibodies. Thus, the transfer of electrons was enhanced. The sensitivity and the response properties of the immunosensor were greatly impacted by the biological compatibility and the surface area of the multilayered nano-Au. The linear range of CEA detection was 0.005–50 ng mL^{-1}, and LOD was 0.001 ng mL^{-1}. Basically, LOD is almost 500 times as sensitive as the traditional enzyme-linked immunosorbent assay used for CEA detection.

It should also be noted that the large surface area along with ease of bioconjugation makes NPs very good carriers for other electroactive labels in bioassays. Some researchers have reported that, if NPs are loaded with other NPs, enzymes, aptamers, etc., labels with enhanced signals can be obtained, which would be a remarkable alternative for electrochemical systems. Zhao et al. (2015) introduced an affordable and highly sensitive and specific process to detect proteins. In this process, copper nanoparticles are formed through hybridization chain reactions (HCRs), which involve small molecules such as folate-linked DNA as a probe. As schematically presented in Fig. 5.5, for this purpose, they prepared a gold electrode modified with capture DNA and then formed copper nanoparticles on the electrode

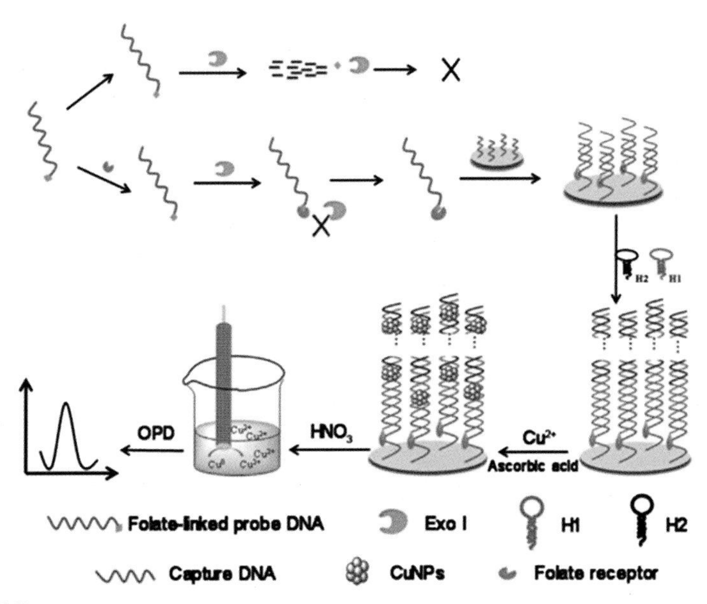

FIGURE 5.5

Schematic display of the electrochemical detection of protein based on hybridization chain reaction-assisted formation of copper nanoparticles.

Reprinted with permission from Zhao, J., Hu, S., Cao, Y., Zhang, B., Li, G., 2015. Electrochemical detection of protein based on hybridization chain re-action-assisted formation of copper nanoparticles. Biosensors and Bioelectronics 66, 327–331.

surface through the HCR-assisted procedure. They selected the folate receptor (FR) as a model. In the primary and metastatic stages of many cancers, the receptor is found overexpressed on the surface of the cancer cells.

To perform electrochemical measurements, they first immersed a CuNPs-coated electrode in 1 mL 0.5 M HNO_3 to release the copper ions. The resulting solution was diluted with PBS and incubated with an o-phenylenediamine solution at 80°C in a water bath to oxidize OPD. Cyclic voltammetry was used so as to characterize the HCR-assisted formation of CuNPs on the surface of the electrode. This was done by recording the electrochemical responses of the DAP generated in the Cu^{+2}-catalyzed oxidation of OPD. Thus, for FR electrochemical detection, the electrochemical responses of the generated DAP increased in pace with the addition of FR. This method allows the sensitive detection of FR within the linear range of 0.01–100 ng mL^{-1} with the limit of detection of 3 pg mL^{-1}. By means of this electrochemical technique, it is possible to monitor large quantities of metal ions that are released through the dissolution of nanoparticles. The technique has also the advantage of ultrahigh sensitivity, allowing the obvious amplification of signals for the detection of various types of biomolecules.

5.2.2 Core-shell

Core-shell nanomaterials consist of a core made of one material but with a shell of a different material laid on the top of it. As compared with simple nanoparticles, core-shell nanoparticles are significantly advantageous for biological applications. For instance, they have better biocompatibility and cytocompatibility as well as lower cytotoxicity. They also conjugate with other bioactive molecules better. The supremacy of core-shell nanoparticles may be realized especially from the way they function when they are toxic, which would cause the host organs and tissues many troubles. In such a case, a kind of material can be coated on the core to reduce the toxicity of the nanoparticles and make them biocompatible. Owing to this property, core-shells can be relied on as a useful alternative to conventional sensors.

Sometimes, nontoxic shells improve the core material properties (Chatterjee et al., 2014). The ease of fabrication of core materials has also played an important part in attracting scientists to this kind of materials. In many biological applications, it is of major importance to conjugate biomolecules onto the surface of particles. As it occurs in certain cases, an intended material may not be easily conjugated with a specific type of biomolecules. To solve the problem, an appropriate biocompatible material can be used for coating. It is actually the shell material that accounts for such surface properties of core-shell nanoparticles as biocompatibility and conjugation with bioactive materials. This is due to the function of reactive moieties that exist on the surface. By adjusting the shell thickness, it is possible to achieve adequate contrasting properties, as contrast agents, and to bind biomolecules for the specific purpose of biosensing.

There are two types of materials widely used in electrochemical biosensors. They include noble metals, such as gold and silver, and magnetic materials, such as iron as well as its sulfides and oxides. As an instance of the applicability of these materials, one may refer to microchips manufactured with core-shell nanocomposites, which can be used particularly in in situ diagnostic kits. In these kits, the core functions as an electrode/transducer system, whereas the shell immobilized with sensing molecules serves as an analyte probe.

Asif et al. (2017) synthesized a multipurpose type of core-shell nanomaterial to fabricate (Fe_3O_4@CuAl NSs) hybrid materials (Fig. 5.6I). This was with CuAl layered double hydroxides (LDHs) controllably integrated onto the surface of iron oxide (Fe_3O_4) nanospheres. The LDH phase of the fabricated materials was interiorly tunable, which is an advantage for such materials. Their applicability was investigated for ultrasensitive amperometric detection of H_2O_2 and as a cancer diagnostic probe. Owing to the synergistic impact created through the combination of the p-type semiconductive channels of LDHs with multifunctional properties, numerous surface-active sites, and unique morphology, the Fe_3O_4@CuAl NSs proved to be of an amazing electrocatalytic potential for H_2O_2 reduction. Also, the biosensor displayed remarkable electrochemical sensing characteristics, including a wide linear range with eight orders of magnitude, the low detection limit of 1 nM (S/N = 3), long stability, and acceptable reproducibility. Because of its excellent efficiency, the biosensor could successfully be used to determine H_2O_2 concentrations in human serum and urine samples through electrochemical in vitro techniques.

According to Fig. 5.6II, in contrast to the CV diagrams of CuAl LDH and Fe_3O_4@CuAl NSs, the Fe_3O_4 NSs catalyst shows the redox peaks related to the Fe(II)/Fe(III) redox couple. As compared with Fe_3O_4 NSs, Fe_3O_4 NSs @CuAl NSs exhibit well-defined reduction peaks, which can be attributed to the charge

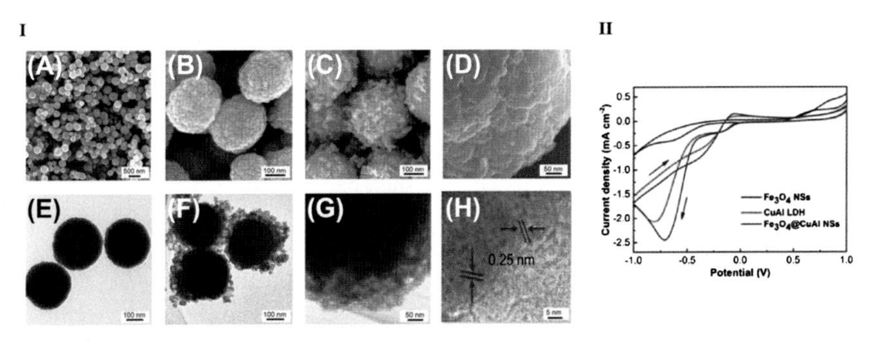

FIGURE 5.6

(I) (A) and (B) SEM images of Fe_3O_4 nanospheres; (C) and (D) SEM images of Fe_3O_4@ CuAl NSs with different rates of magnification; (E) TEM images of Fe_3O_4 nanospheres; (F) and (G) Fe_3O_4@CuAl NSs with different rates of magnification; (H) HRTEM image of the CuAl LDH shell. (II) CVs of various electrodes in the H_2O_2 solution.

Reprinted with permission from Asif, M., Liu, H., Aziz, A., Wang, H., Wang, Z., Xiao, F., Liu, H., 2017. Core-shell iron oxide-layered double hydroxide: high electrochemical sensing performance of H2O2 biomarker in live cancer cells with plasma therapeutics. Biosensors and Bioelectronics 97, 352–359.

transfer from Fe_3O_4 NSs to CuAl LDH layers. This transfer leads to a small modification in the electronic structure of CuAl LDH in Fe_3O_4 NSs@CuAl NSs and the significant improvement of electrocatalysis. Once electrons are partially transferred from Fe_3O_4 NSs to CuAl LDH at a nanoscale interface, there emerges a synergistic effect that makes as-prepared Fe_3O_4 NSs @CuAl NSs catalytically more active.

Nanospheres with Fe_3O_4 magnetic cores and Au shell structures have gained remarkable attention by virtue of the combined capacity of Fe_3O_4 and Au. Fe_3O_4 NPs have a highly paramagnetic nature, providing an appropriate technique to purify, separate, and isolate biological samples through an external magnetic field, whereas the functional reagents are bound to the surface of the particles. Indeed, Fe Au-NPs, as core-shell structures, possess the properties of both AuNPs and magnetism. They offer superb electrochemical signals for the increase of sensitivity, decrease of background, and separation. Owing to their biocompatibility and large active surface area, Fe_3O_4-Au-NPs are considered as great candidates for bioconjugation with cells, antigens, and antibodies. Ahmadi et al. (2014) reported a competitive electrochemical immunosensor developed on the basis of antigen-antibody reactions. They used antigen (Ag)-labeled Fe_3O_4-Au-NPs and the surface of a PVA-modified SPCE to detect the serum digoxin. In this approach, they combined a disposable chip with an Fe_3O_4-Au NPs tracing tag to improve magnetic separation, signal amplification, and background signal reduction. To prepare the disposable chip for digoxin detection, capture antibody, secondary antibody, and polyvinyl alcohol were immobilized on the surface of a modified SPCE. The electrochemical signals obtained from the binding of the antigen (Ag)-Fe_3O_4-Au-NPs complex was produced after $AuCl_4$ was electrooxidized in HCl and then reduced in a DPV mode. This immunosensor was able to detect digoxin in the linear range of 0.5–5 ng mL^{-1} with the limit of detection of 0.05 ng mL^{-1}. The magnetic core-shell Fe_3O_4-Au-NPs composite serves as an electronic tracing tag for appropriate acquisition of detection signals. The composite also provides a controllable procedure to fabricate a specific and sensitive platform through an ultrasensitive analysis.

Xu et al. (2015) applied an Au@Pd core-shell nanostructure to amplify the signal of an electrochemical thrombin aptasensor by using the synergetic catalysis of enzymes and porous Au@Pd core-shell nanostructures (Fig. 5.7I).

As shown in Fig. 5.7II, the electrode modified with Tb, GOx, and hemin/G-quadruplex multilabeled Au@Pd bioconjugates had much higher electrochemical signal variation than the other two bioconjugates. This observation provides a formidable proof for the practicality of the strategy devised for signal amplification. Amplified responses may arise from a porous Au@Pd core-shell nanocarrier that has a large surface area and a great catalytic function to reduce local H_2O_2. H_2O_2 is produced when glucose is oxidized through GOx. Such nanoalloys enhance the active sites for the better immobilization of the redox mediator Tb, hemin/G-quadruplex, and GOx. They also promote the transfer of electrons in Tb owing to their synergetic effect created through catalysis with GOx and hemin/G-quadruplex.

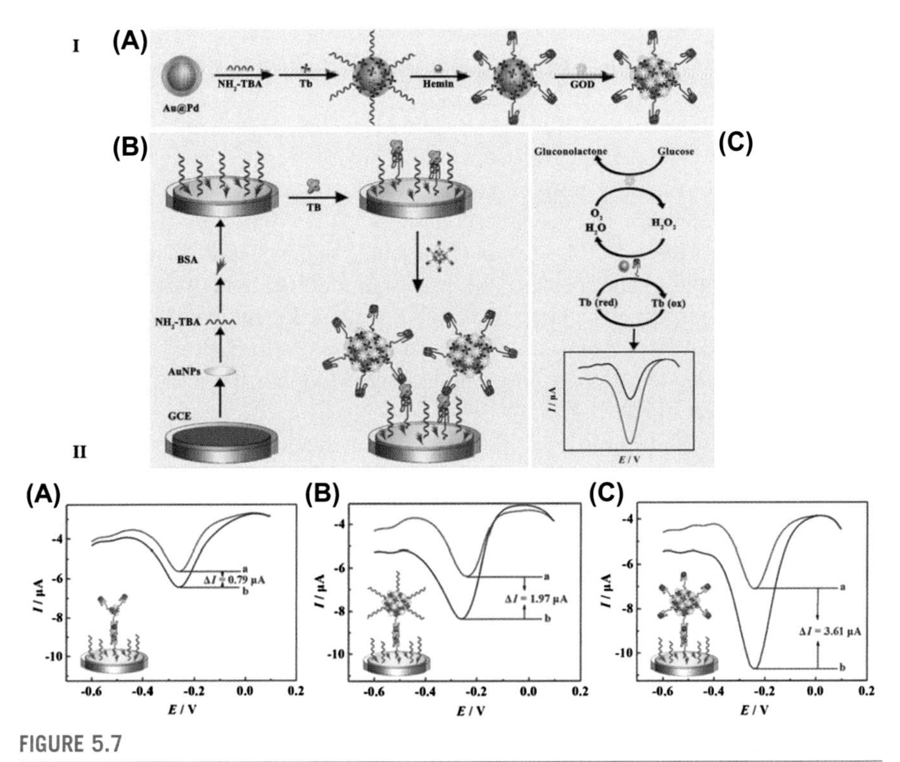

FIGURE 5.7

(I) The procedure of the secondary TB aptamer (A) and the stepwise process of manufacturing an electrode interface for the TB aptamer sensor and (B) with the signal amplification strategy (C). (II) DPV responses of different strategies.

Reprinted with permission from Xu, W., Yi, H., Yuan, Y., Jing, P., Chai, Y., Yuan, R., Wilson, G.S., 2015. An electrochemical aptasensor for thrombin using synergetic catalysis of enzyme and porous Au@Pd core – shell nanostructures for signal amplification. Biosensors and Bioelectronics 64, 423–428.

5.2.3 Quantum dots

QDs are semiconductive crystalline clusters with a size range of 2–10 nm. QDs have been of interest usually as labels for electrochemical biosensors. Compared with the existing labels, QDs are more stable and cheaper. They show high sensitivity, faster binding kinetics, and high reaction rates for various types of biosensors. For electrochemical detection of the QDs existing in bioassays, indirect electrical detection is the most widely applied procedure. The procedure consists of initial oxidative dissolving of materials in acidic media followed by the detecting of metal ions through such a powerful, sensitive, electroanalytical technique as anodic stripping voltammetry. In studies on QDs, very low limits of detection, in the order pM, have been achieved, which is owing to the release of a good number of metal ions from NPs. In addition, a major characteristic that makes good electroactive labels out of QDs is their capability of performing multidetection. This holds true especially about the QDs that are made of various inorganic crystals and have distinctive electrochemical responses. QDs barcode technology was developed for DNA or protein detection (Wen et al., 2017).

Mazloum-Ardakani et al. (2015) developed a CdS-QDs electrode system. It served as a transducing surface on which a specific designed synthetic ApoB-100 probe could be covalently immobilized and DNA was hybridized. The electrode was also used to monitor DNA synthesis in the sensitive detection of R3500Q mutation of apolipoprotein B-100 (ApoB-100) gene. As they reported, CdS-QDs are highly biocompatible and capable of probe immobilization. These properties make CdS-QDs sensitive, accurate, and stable enough for the detection of ApoB-100 genes. The developed biosensor was also reported to have provided an appropriate potential for detection of target oligonucleotides. In this case, the limit of detection was found to be 3.4×10^{-17} M.

Hansen et al. (2006) utilized CdS and PbS QDs nanocrystal tracers to fabricate double-analyte electrochemical aptasensors. For this purpose, extremely low LODs were coupled with high selectivity. This displacement protocol involved the coimmobilization of a number of thiolated aptamers, binding of their CdS-tagged thrombins and PbS-tagged lysozymes on an Au substrate, addition of certain protein samples, and, finally, displacement monitoring by detecting the remaining nanocrystals electrochemically. If the remaining captured nanocrystals are dissolved in 0.1M HNO_3, Pb^+ and Cd^+ are produced. These ions are then detected at a coated GCE by electrochemical stripping. Single-analyte or thrombin sensing was performed to test the fabricated device in terms of selectivity and the sensitivity. The researchers analytically proved the effective dependency of the results on concentration, expanding from 20 to 500 ng L^{-1}, with an LOD of 0.5 pM. The capability of the biosensor for simultaneous detection of the trace levels of various proteins was investigated. The size and the position of the metal peaks reflected the detection as well as the concentration of the target proteins. There was no obvious change in the second tag response, implying that the cross-interferences were negligible. In contrast, when the two proteins were simultaneously added, similar reductions occurred in sharp and well distinct Pb^+ and Cd^+ peaks. High selectivity, very low limits of detection, and ability to simultaneously determine up to five protein targets, in accordance with how many nonoverlapping metal peaks are involved in a single run, are the advantages of this novel biosensor (Fig. 5.8).

Zhang et al. (2014) proposed an electrochemical biosensor to deal with *Mycobacterium tuberculosis* (Mtb) DNA. They utilized CdSe QDs as a label in combination with MspI endonuclease and Au NPs for signal amplification and selective detection. MspI endonuclease makes the recognition of the duplex symmetrical sequence 5′-CCGG-3′ and performs the catalyzation of the double-strand DNA cleavage between the two cytosines, but it has no action on single-strand DNA. They immobilized 5′-SH-(CH2)62-GGTCTTCGTGGCCGGCGTTCA(CH2)6-biotin-3′, as the probe DNA, on the surface of a GCE modified with Au NPs and then connected the probe to the CdSe QDs-streptavidin (SA) label. After it was hybridized with the target DNA (5′-TGAACGCCGGCCACGAAGACC-3′), MspI endonuclease could identify the

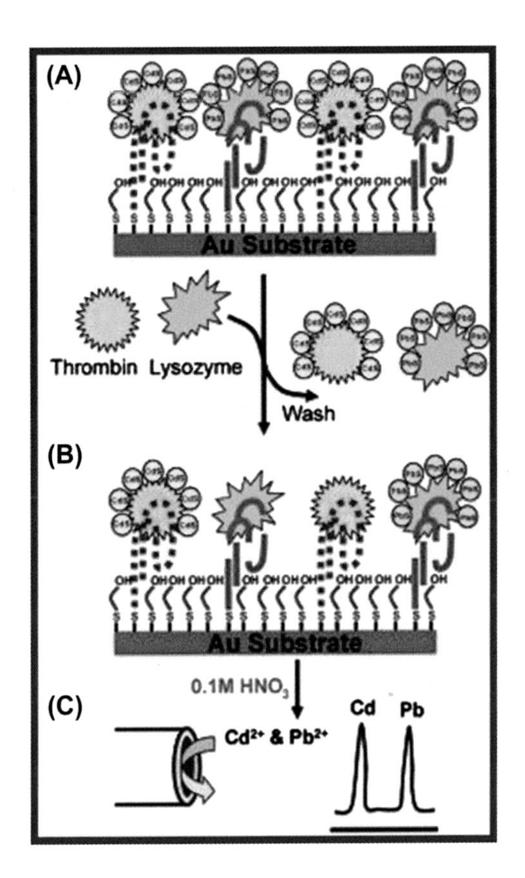

FIGURE 5.8

Fabrication process of double-analyte electrochemical aptasensors.

Reprinted with permission from Hansen, J.A., Wang, J., Kawde, A., Xiang, Y., Gothelf, K. V., Collins, G., 2006. Quantum-Dot/aptamer-based ultrasensitive multi-analyte electrochemical biosensor. Journal of the American Chemical Society 128, 2228–2229.

DNA-specific sequence in the duplex DNA and cleave the dsDNA fragments linked with CdSe QDs from the electrode. The CdSe QDs remaining on the biosensor were dissolved as Cd^{2+} through getting the biosensor dipped in an HNO_3 solution. This biosensor was able to determine Mtb DNA up to 8.7×10^{-15} M with a linear range of 1.0×10^{-14} to 1.0×10^{-9} M. The sensor also proved to have high selectivity in distinguishing the mismatched DNA.

The most usual genetic alterations are caused by single-nucleotide polymorphisms (SNPs), which are viewed as point mutations. Liu et al. (2005) displayed how distinct bioelectronic signs can be produced by inorganic nanocrystal tags for known two-base mutations as well as unknown individual SNPs in a single-stranded DNA target. To do so, they designed a miniaturized high-performance and low-cost electrochemical device for decentralized DNA diagnostics. In summary, an efficient nanocrystal-based bioelectronic procedure was introduced to codify individual SNPs using the aforementioned device. In their study, it was shown that the electrodiverse population of inorganic nanocrystal tags fulfills the significant task of yielding distinct four-potential voltammetric signatures for unknown SNPs. Through phosphoramidite chemistry and using a cysteamine linker, ZnS, CdS, PbS, and CuS were linked to adenosine, cytidine, guanosine, and thymidine mononucleotides, respectively. When there was a correspondence between a mutation site and a labeled chain terminator, the captured extension product would give an electrical signal. Thus, every mutation would capture various nanocrystal-mononucleotide conjugates. The T-G mutation is an example, in which T and G captured A-ZnS and C-CdS, respectively. This occurred after T-CuS and G-PbS were base-paired to the captured A-ZnS and C-CdS, respectively. What resulted was a characteristic four-potential voltammogram, the peak potentials of which confirmed the recognition of the mismatch. This method depends on multiple signals and results in distinct multipotential fingerprints for specific SNPs in a single voltammetric run (Fig. 5.9).

Hua et al. (2013) developed a selective procedure to detect MCF-7 breast cancer cells by means of CdTe QDs-based SiO_2NPs-bio-probes and aptamer-functionalized magnetic beads. First, they coupled MBs to an MUC-1 aptamer, which is a 25-base oligonucleotide (Apt1) with special linking features for the MUC1 peptide to link with MUC1-positive cells. Then, they added an MB/Apt1 suspension to a buffer that contained MCF-7 cells with a specific concentration to prepare MB/Apt1/MCF-7. Subsequently, they prepared nanobioprobes by coating CdTe QDs and Apt2 (AS1411) on SiO_2NPs. In the next place, they diluted MB/Apt1/MCF7 with a PBS and mixed it with the Apt_2/QDs/SiO_2 NPs suspension. The mixture was incubated for 45 min to obtain the conjugated forms of MB/Apt1/MCF_7/Apt_2/QDs/SiO_2. Finally, the CdTe QDs loaded on the cells underwent dissolution and converted to Cd^{2+}, the quantity of which could be determined by means of a bifilm-coated GCE through the SWV method. The oxidation of Cd showed a well-separated peak around -0.868 V (Fig. 5.10A). The voltammograms of MB/Apt1/MCF-7/Apt2/QDs/SiO_2 in diverse concentrations of MCF-7 cells are presented in Fig. 5.10B. According to the linear correlation equation, the correlation coefficient (R^2) was 0.9867, and the limit of detection was found to be 85 cells mL^{-1} at 3σ. This rate was much lower than those obtained in fluorescence detection assays.

5.2.4 Composites

It is possible to coat nanomaterials with biomolecules or polymers to make them more biocompatible as well as capable of exact targeting. Indeed, integration of nanomaterials to biosensors has made many desirable targeting and detection functions feasible. This has resulted in the widespread use of nanomaterials in the fabrication of electrochemical sensors. Conjugate nanomaterials that are confected from a combination of nanostructures exhibit the properties of those constituent nanostructures. For example, one can refer to the properties and the electronic structure

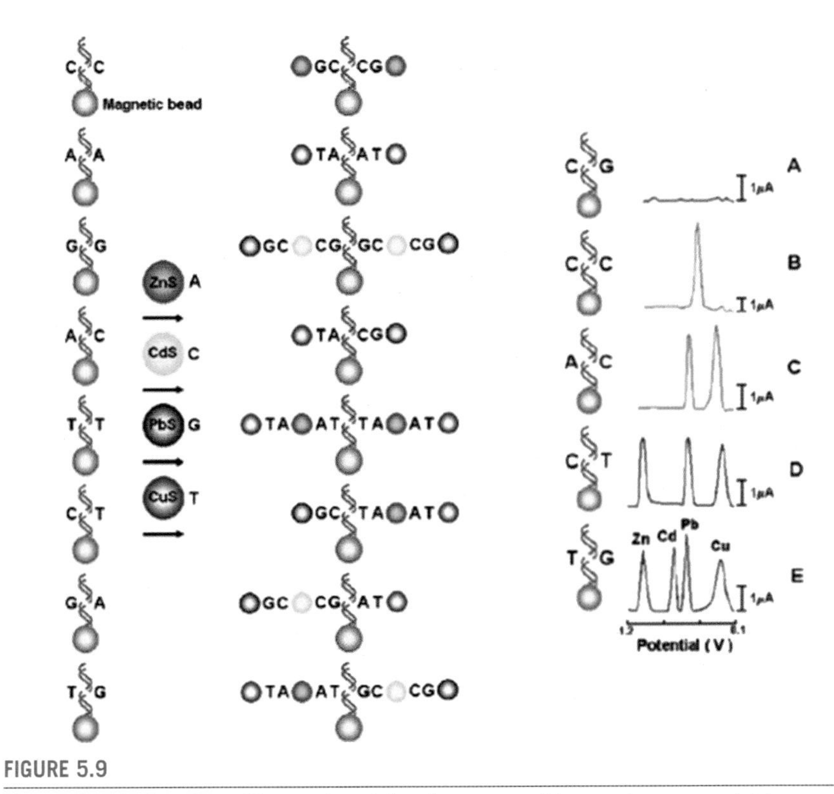

FIGURE 5.9

Electrochemical coding of base mismatches by means of inorganic nanocrystal tracers.

Reprinted with permission from Liu, G., Lee, T.M.H., Wang, J., 2005. Nanocrystal-based bioelectronic coding of single nucleotide polymorphisms, Journal of the American Chemical Society 127, 38–39.

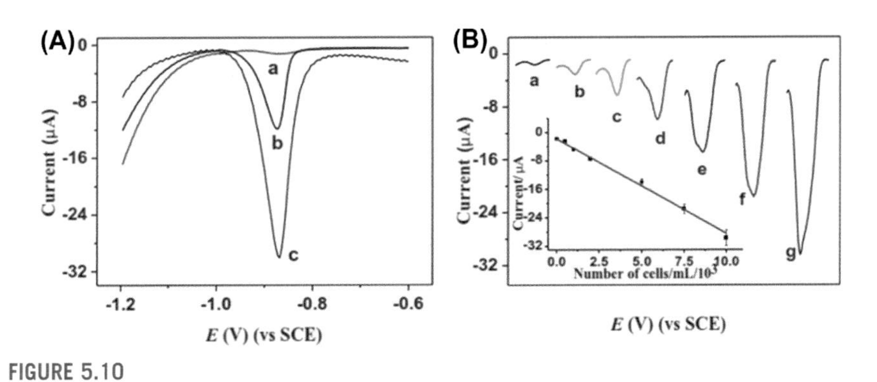

FIGURE 5.10

(A) SWV curves of various electrodes. (B) Calibration curve of the SWV peak current.

Reprinted with permission from Hua, X., Zhou, Z., Yuan, L., Liu, S., 2013. Selective collection and detection of MCF-7 breast cancer cells using aptamer-functionalized magnetic beads and quantum dots based, Analytica Chimica Acta 788, 135–140.

of the individual nanomaterials as well as the synergistic effect of the constituents. If appropriately designed, these hybrids can be reliably utilized for electrochemical biosensors. They can, indeed, be used as supports in immobilization supports, electrochemical signal-generating probes, and signal amplifiers. As an instance, iron-silica hybrids are used to immobilize biomolecules owing to their proper biocompatibility. Fe_3O_4-chitosan, magnetic nanoparticles-silica, and PANI/PVA-AgNP are some other examples of nanocomposites commonly used for immobilization purposes. Silver- and gold-based nanocomposites have been exploited for signal amplification and as probes that generate signals in affinity-based electrochemical sensors. Also, Pt and ZnO can possibly be of use when combined with different types of nanomaterials to improve the efficiency of amperometric sensors (Hayat et al., 2014).

5.2.4.1 Polymer-based nanocomposites

So far, a variety of polymers have been used effectively to design nanobiosensors. In the case of electrochemical nanobiosensors, polymers are applicable for coating of nanoparticles. They can also serve as anchors to adsorb biomolecules or other nanomaterials to make different nanocomposites.

The electroanalytical applications of polymer nanocomposites, as a bunch of nanostructures, in the manufacture of electrochemical biosensors have attracted a lot of attention. In comparison with single-component electrodes, polymeric nanocomposite-modified electrodes possess superior properties such as larger surface areas, higher electric conductivity, and faster electron transfer rates. These features have made it possible to construct electrochemical biosensors with higher selectivity and sensitivity, lower LODs, and better reproducibility and stability. Polymeric nanocomposites may be synthesized on nanoscales and in different forms, thus allowing diverse biological and chemical sensors of novelty to be constructed. So far, to achieve better rates of sensor efficiency, nanocomposites of diversified combinations have been tried. We continue to focus on polymer composites with noncarbon nanomaterials. Conductive polymers possess unique characteristics such as resistance to corrosion, flexibility, scalability, and lightweight. They may also be easily modified by chemists for specific requirements. These properties make conductive polymers favorable alternatives for the specific materials currently applied in sensors. As a matter of fact, a milestone in modern analytical science is the preference for conductive polymers. They have been broadly explored as materials that are able to boost signals for electroanalytical practices (Shirakawa et al., 1977). Recently, the properties of conducting polymer nanocomposites have been enhanced to remove the inherent shortcomings of pure CPs. Metal and metal-oxide nanoparticles have, since long, been widely investigated as materials with the ability of electrochemical sensing. This has been owing to their appealing characteristics, i.e., unique physical, chemical and electronic features, small size, and adaptability to construct novel and high-performance sensors. Integration of metal nanoparticles to conducting polymer matrices helps to enhance the electrocatalytic efficiency of many electrodes (Shrivastava et al., 2016).

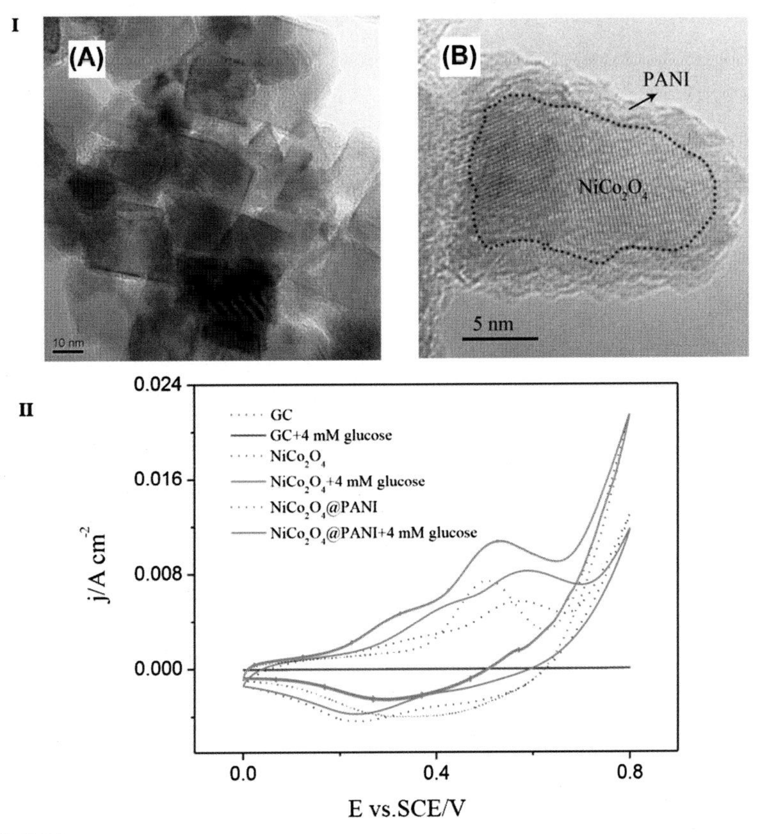

FIGURE 5.11

(I) (a) and (b) are HRTEM images of $NiCo_2O_4$ @PANI. (II) CV plots of various modified GC electrodes and $NiCo_2O_4$@PANI nanocomposite.

Reprinted with permission from Yu, Z., Li, H., Zhang, X., Tan, W., Zhang, X., Zhang, L., 2016. Facile synthesis of $NiCo_2O_4$@ Polyaniline core–shell nanocomposite for sensitive determination of glucose, Biosensors and Bioelectronics 75, 161–165.

Yu et al. (2016) investigated the biosensing properties of $NiCo_2O$@polyaniline core-shell nanocomposite and $NiCo_2O$ nanoparticles for glucose detection based on a GCE. They confirmed the morphology and the structure of the $NiCo_2O$@polyaniline core-shell by HRTEM images (Fig. 5.11I). The diameter of such $NiCo_2O_4$@ PANI nanoparticles is averagely 25 nm, which greatly shortens the ion-transport pathways. Also, $NiCo_2O_4$ nanoparticles have distinct lattice fringes and are very crystalline. The range of thickness for the PANI shell is from several nanometers to tens of nanometers. The structure of the PANI shell integrates the advantages of good contact and efficient diffusion distances for electron transfer, which elevates the sensor function.

The electrocatalytic activity of $NiCo_2O$@polyaniline for glucose oxidation is depicted in Fig. 5.11II. The dotted lines represent the CVs of different electrodes in a 0.1 M NaOH solution. The CV current of the $NiCo_2O_4$@PANI/GCE is much greater than those of the bare GCE and the $NiCo_2O_4$-modified electrode. This is ascribed to the greater surface area and the excellent conductivity of the $NiCo_2O_4$@PANI composite. Also, in comparison with the bare GCE and the $NiCo_2O_4$-modified electrode, the $NiCo_2O_4$@PANI composite-modified electrode exhibited two completely separate redox peaks at 0.35 and 0.50 V belonging to the transition of the Ni(II)/Ni(III) redox couple. Thus, the $NiCo_2O_4$@PANI composite is viewed as capable of creating a better electrical contact between the electrolyte and the redox-active centers. It makes the GCE modified with $NiCo_2O_4$@PANI nanocomposite highly sensitive for electrochemical detection. Noticeably, the oxidation peak potential on the $NiCo_2O_4$@PANI/GCE shifted negatively as compared with that of the $NiCo_2O_4$-modified GC electrode. Also, the peak separation for $NiCo_2O_4$@PANI was smaller than the one for $NiCo_2O_4$, illustrating that the conductivity of the $NiCo_2O_4$@PANI/GCE was raised. It was further observed that a nanocomposite is catalytically more active than $NiCo_2O_4$ nanoparticles for glucose oxidation. In optimal conditions, the electrochemical glucose biosensor exhibited superior performance, i.e., low detection limit (0.3833 μM), wide linear range (0.0150–4.7350 mM), and high sensitivity (4.55 mA mM^{-1} cm^{-2}). Therefore, the $NiCo_2O_4$ nanocomposite may be considered as a promising material with which to fabricate biosensors. Owing to the outstanding conductivity of polyaniline shell, the $NiCo_2O_4$@polyaniline core-shell composite allows fast electron transfer kinetics. This characteristic has made the core-shell composite highly sensitive for the detection of glucose.

Chitosan and its water-soluble derivative, i.e., N-trimethyl chitosan (TMC), are the polymers commonly used for construction of nanocomposites. Through electrostatic interactions, TMC often forms a nanocomplex with anionic compounds (e.g., Fe_3O_4 and Au NPs). In their study, Daneshpour et al. (2016) described a simple electrochemical nanobiosensor with which to analyze gene-specific methylation at a trace level. The nanosensor was made on the basis of a DNA probe labeled by Fe_3O_4/TMC/Au nanocomposites for the purpose of signal amplification. The nanosensor development was also based on a specific anti-5-methylcytosine monoclonal antibody entrapped in a thin polythiophene film for the purpose of capturing and separating methylated DNA from nonmethylated DNA. As observed, a great quantity of Au nanoparticles assembled on the Fe_3O_4/TMC complex would amplify the recognition signals successfully. This ability can be ascribed to the increase made in the surface area of the nanocomposite.

5.2.4.2 Multimetallic nanocomposites

On the purpose of enhancing the performance of electrochemical biosensors, various alloy nanomaterials have so far been synthesized and used as substrates. In fact, the synergistic effect of diverse nanomaterials helps to improve the performance of biosensors that are based on hybrid nanomaterial. Metal oxide nanomaterials, especially for nanomaterial arrays, have been explored widely as electrochemical biosensing substrates to replace high-cost noble metal nanomaterials. This is owing to their wonderful chemical stability, low cost, environment-friendliness, and good selectivity. As metal oxide nanomaterials such as ZnO, CeO_2, NiO, TiO_2, and Fe each have a distinct isoelectric point (IEP), they are of benefits for the immobilization of different enzymes. Cui et al. (2016) reported an easy hydrothermal procedure to fabricate porous NiO/CeO_2 hybrid nanoflake arrays that were later utilized to fabricate an electrochemical glucose biosensor. This amperometric glucose biosensor proved to have good features, such as low LOD, high sensitivity, and expanded linear range.

Here is the glucose-sensing mechanism on the basis of NiO/CeO_2 hybrid nanoflake arrays. Glucose is enzymatically catalyzed to gluconic acid and H_2O_2 when it reaches the active sites of glucose oxidase. Then, H_2O_2 is transferred to the surface of NiO/CeO_2 nanoflake arrays, on which it is oxidized to O_2 by NiO and CeO_2 and presents amperometric responses. It is to be noted that both NiO and CeO_2 have catalytic performance for H_2O_2, and their hybridization causes a reduction in the charge transfer resistance of as-prepared samples. This, in turn, induces electron transfer kinetics during reactions that are catalyzed enzymatically.

Yang et al. (2012) introduced the ZnO/Cu nanocomposite for biosensing applications and as a platform for direct electrochemical treatment of enzymes. It is known that conventional electrode substrates modified with ZnO are usually made by the postpasting of ZnO, but this procedure does not provide robust mechanical adhesion nor does it create electrical contacts between the substrate and the nanostructured ZnO. To overcome this problem, Xu et al. (2015) prepared ZnO nanostructures in an in situ manner and without using any organic reagent.

In their procedure, Xu et al. used two highly pure metallic targets of copper and zinc to deposit a Cu-–Zn alloy layer. They did it by reactive RF magnetron sputtering on ITO substrates. Then, the Cu–Zn-coated ITO was placed in a 50-mL Teflon-lined stainless steel autoclave with deionized water. After the heat treatment, a prickly kind of ZnO/Cu nanocomposite was formed in situ on the ITO sublayer. To immobilize GOx and HRP, they prepared a homogeneous mixture containing CS, AuNPs, Con A, HRP, and GOx. It served as an electrodeposition solution with which to make a biocomposite film. To do electrodeposition, they immersed the prickly ZnO/Cu nanocomposite in a CS-Au/Con A/HRP-GOx biocomposite solution. Then, they used it as a cathode by applying a voltage of −0.1 V for 5 min (Fig. 5.12I). Basically, the large active area of ZnO/Cu nanocomposites is an advantage for construction of biosensors through immobilization of biomolecules. According to the CV scans obtained at various sweep rates, Xu et al. showed that the redox peak current would increase when the scan rate rose. The peak current was linearly proportional to the scan rate in the range of 0.1–0.35 V. Their statistical results were obtained by considering the way the peak potential was related to the natural logarithm of the sweep rate for a CS-Au/Con A/HRP-GOx/ZnO/Cu

I

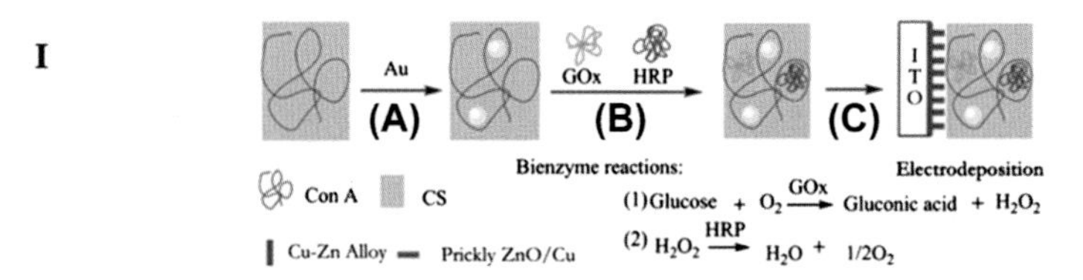

II

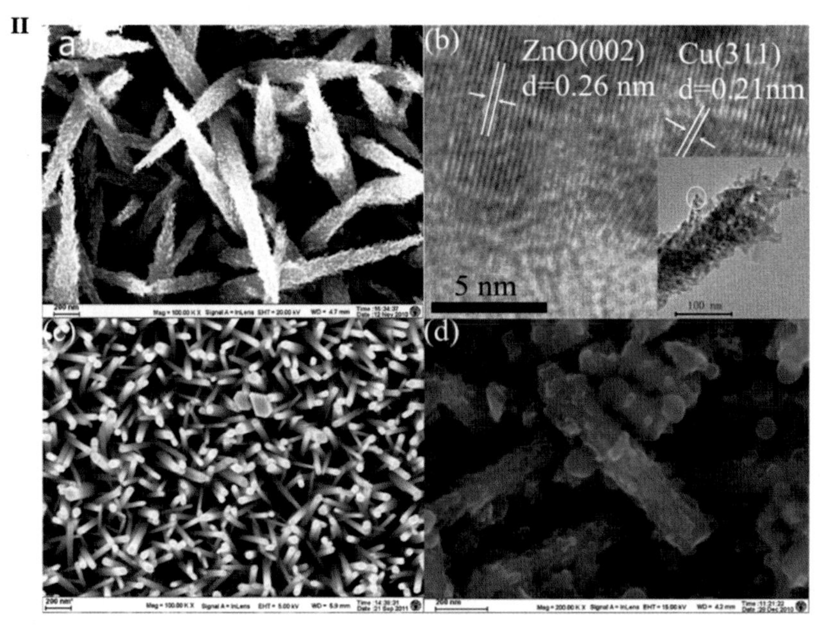

FIGURE 5.12

(I) Fabrication of a glucose biosensor with ZnO/Cu nanocomposite. (II) (a) FESEM image of the prickly ZnO/Cu, (b) HRTEM and TEM (inset) images of the nanocomposite, (c) FESEM image of the ZnO nanorods, and (d) FESEM image of the prickly ZnO/Cu after absorption of the CS-Au/Con A/HRP-GOx.

Reprinted with permission from Yang, C., Xu, C., Wang, X., 2012. ZnO/Cu nanocomposite: a platform for direct electrochemistry of enzymes and biosensing applications. Langmuir 28, 4580–4585.

electrode in a buffer solution. This relationship was calculated according to Laviron equation as follows:

$$E_p = E^{0'} + \frac{RT}{\alpha nF} - \frac{RT}{\alpha nF}(ln\nu) \tag{5.1}$$

where n and the charge transfer coefficient (α) were found to be 1.9 and 2.6, respectively. The redox reaction between ITO and the enzyme was, hence, a two-electron transfer process. According to the calculated parameters and Eq. (5.2), the constant (k_S) for the apparent heterogeneous electron transfer rate was $0.67 \pm 0.06\ s^{-1}$

$$
\begin{aligned}
logK_s = {} & \alpha \log(1-\alpha) + (1-\alpha)\log(1-\alpha) \\
& -\log\left(RT/_{nFv}\right) - \alpha(1-\alpha) \\
& -\log(nF\Delta E_p/2.3RT)
\end{aligned}
\tag{5.2}
$$

The linear calibration curves for the amperometric response of the biosensor to the glucose concentration presented a linear range of 1–15 mM, and the corresponding correlation coefficient was 0.993. Also, the biosensor had a sensitivity of 97 nA mM^{-1}, and the LOD turned out to be ~0.04 mM (S/N = 3). The remarkable point about the latter value is that it is close to the value obtained for nitrogen-doped carbon nanotubes.

As reported by Yang et al. (2010), silver NPs were first doped in TiO_2 and chitosan colloids to make homogeneous and porous Ag-TiO_2-Cs nanocomposites. This nanocomposite was electrochemically very active for redox reactions and excellently capable of forming films. To fabricate a biosensor, the researchers attached Pt hollow nanospheres to the free amino groups in chitosan, thereby, to immobilized human chorionic gonadotropin antibody (HGC). In a comparative inquiry, a bovine serum albumin (BSA)/anti-HCG/Ag-TiO_2-Cs/GCE was prepared along with a BSA/anti-HCG/hollow nano-Pt/Ag–Cs/GCE. The designed immunosensor proved to have good sensitivity and a wide calibration range because of the synergetic impact of Pt hollow nanospheres and Ag-TiO_2-Cs nanocomposites. Also, the Ag-TiO_2-Cs nanocomposites turned out with high redox activity and suitable stability. Those nanomaterials that had a high surface energy and a large surface area were expected to immobilize a large amount of antibody, as they really did.

5.2.4.3 Biomolecule-contained nanocomposite

Cao et al. (2015) used a simple biomimetic method to prepare three-dimensional porous Ag nanoflowers incorporated with protein-inorganic nanomaterial BSA. The Ag nanoflowers were placed on the surface of a GCE and linked to a targeting lectin molecule of *Sambucus nigra* agglutinin (SNA) to detect DLD-1 cancer cells in the human colon.

An impedance technique was applied for the quantitative monitoring of tumor cells. Fig. 5.13II presents the Nyquist diagram for the captured DLD-1 cells/SNA/BSA-incorporated Ag nanoflower electrode. As it is shown, there is an increase in the semicircle diameter or the electron transfer resistance, which is due to a rise in the number of DLD-1 cells in a range from 1.35×10^2 to 1.35×10^7cells mL^{-1}. The detection limit of 40 cells mL^{-1} (S/N = 3) was implemented for this biosensor (Fig.5.13II). The specific advantages of the biosensor were reproducibility and stability. According to the DPV results, after incubation of different cells including human astrocyte 1800 cells, human embryonic kidney 293 cells (HEK 293), and DLD-1 cells in the same concentration, no obvious change was detected in the peak current while DLD-1 cells provided a substantial DPV peak response. The

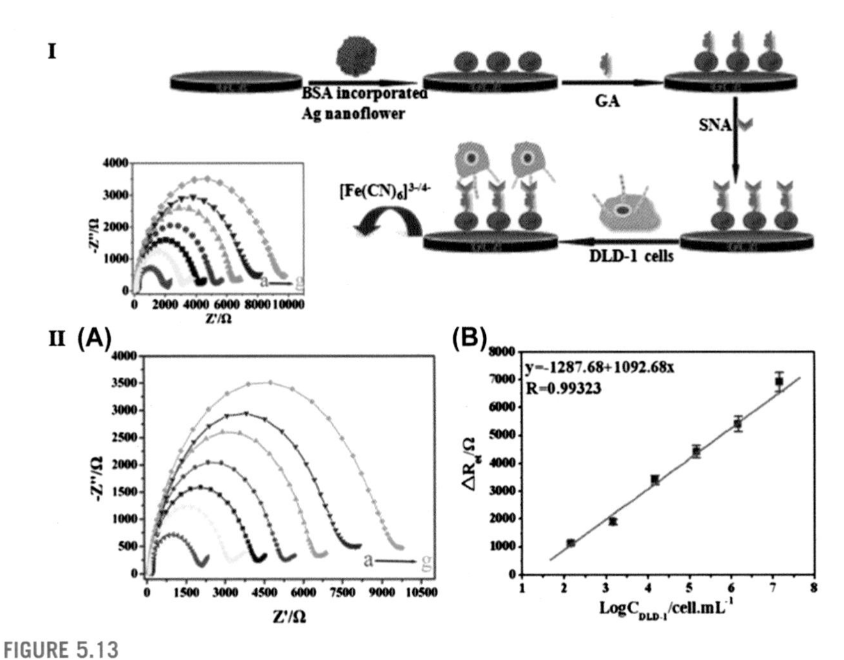

FIGURE 5.13

(I) Schematic illustration of an EIS cytosensor. (II) (A) Nyquist plot of the cytosensor after recognition with different cell concentrations. (B) Correlation curve of the change in electron transfer resistance versus the logarithm of DLD-1 cell concentration.

Reprinted with permission from Cao, H., Yang, D., Ye, D., Zhang, X., Fang, X., Zhang, S., 2015. Protein-inorganic hybrid nano fl owers as ultrasensitive electrochemical cytosensing Interfaces for evaluation of cell surface sialic acid. Biosensors and Bioelectronics 68, 329–335.

results suggest that Ag nanoflowers incorporated with BSA make an appropriate platform with good biocompatibility, enhanced cell-immobilization capacity, and retained activity of the immobilized cells. The nanoflowers are endowed with these merits because the porous structure has a large surface area and the natural BSA layer acts as a biocompatible support.

5.2.4.4 Semiconductor material–based nanocomposites

Wu et al. (2012) utilized silver-hybridized mesoporous silica nanoparticles (MSNs) to fabricate an electrochemical immunosensor to detect PSA tumor markers. MSNs increased the capacity of the primary antibody for immobilization, and Ag NPs enhanced the electron transfer rate. A large surface area was supplied by the Ag@MSNs supplied to fix the biomolecules, suggesting that a more flexible orientation would be adopted by the biomolecules. On the basis of the hindrance on the transfer of electrons, the immunosensor provided for the direct detection of PSA concentrations in incubation solutions. This label-free immunosensor showed acceptable reproducibility, storage stability, accuracy, and satisfactory precision.

Yuan et al. (2008) composed a metal nanostructure with SiO_2 NPs to design a novel nanomaterial for enzyme fixation and biosensor preparation. They could achieve high sensitivity and fast responses by preparing a homogeneous and dendritic nano-Ag/3-mercaptopropanesulfonic acid (MPS)-silicon dioxide (SiO_2) nanocomposite. To prepare the nanocomposite, they grew a silver nanostructure on SiO_2 nanoparticles by using MPS as a connective material. The reaction of the −SH end of MPS with nano-Ag resulted in the formation of a stable metal-sulfur bond. Also, an adhesion-promoting layer was introduced by MPS between the Ag and SiO_2 nanoparticles. As the GCE surface had a good affinity to the SiO_2 nanoparticles, the dendritic nano-Ag/MPS SiO_2 nanocomposite was appropriately immobilized on the electrode surface. As horseradish peroxidase was further immobilized on the film, a highly sensitive biosensor was produced for H_2O_2 at an approximately low cathodic potential with no need for an electron transfer mediator. The wide linear range of the designed biosensor indicated good analytical performance and practicability. The nanocomposites illustrated a dendritic but homogeneous open nanoarchitecture, which helped to improve the loading of enzymes or proteins on the electrode surface. This proposed method was more persuasive, more convenient, and less expensive than the electrochemical method for preparation of silver nanostructures through deposition of silver.

As commonly known, silica microspheres have some unique properties, such as large internal spaces inside the shells and structural rigidity. Substances can go in and out of the interior hollow space through a pathway provided by the mesopores inside the shells. Also, outstanding improvement is made in the products, resulting from the catalytic reaction as well as the diffusivity of the substrates.

To fabricate a novel artificial superoxide dismutase sensor, hollow mesoporous silica microspheres (HSMs) were used as a solid support. Owing to its advantages such as easy recovery, repeated utility, and resistance to extreme conditions, that novel sensor could solve many practical problems. Wu et al. (2012) explored HSMs as a new type of labels for amperometric immunosensors. Indeed, HSMs were used to serve as an immobilization matrix with the ability to upload a high concentration of enzyme HRP simultaneously and as a secondary antibody carrier. The immunosensor showed high sensitivity, good reproducibility and stability, a low detection limit (6.00 pg mL^{-1}), and linear responses in the range of 0.01–10 ng mL^{-1} PSA.

5.2.5 Nanowires, nanofibers, and nanosheets

5.2.5.1 Nanowires and nanofibers (1D nanomaterials)

Owing to their unique properties and capability of developing high-density nanoscale sensors, one-dimensional (1D) nanostructures, such as nanowires and nanofibers, have been paid a great deal of attention. 1D systems are, indeed, structures with the smallest dimension that can be used for the efficient transfer of electrons and hence crucial in the design and function of nanoscale devices. The electrical properties of such systems can be strongly impacted by minor perturbations. This is due to their tunable electron transport properties, quantum confinement effect, and high surface-to-volume ratio.

A conductive polymer nanowire with the role of sensing elements provides higher sensitivity than a polymer film, and this is due to its improved surface-to-volume ratio. Yang et al. (2014) reported a novel approach to fabricate enzyme-entrapped conducting polymer nanofibers with enhanced lifetime and higher sensitivity in comparison with conductive polymer film counterparts. They presented a strategy to develop a glucose enzymatic biosensor based on nanofibers. To exhibit the excellent role of nanofibers, they prepared GOx entrapped in PEDOT films and nanofibers. Then, each biosensor was evaluated in terms of sensitivity, durability, and electrical properties. Generally, nanostructured sensing elements promote sensitivity as a result of enhanced surface-to-volume ratios. The researchers regarded the benefits of suitable nanoscale matrices to entrap GOx, reduce the Pt microelectrodes impedance, and increase the degree of GOx entrapment in the PEDOT as the result of using nanofibers rather than polymer films in their designed biosensors. With regard to the reduction of the impedance of Pt microelectrodes, it is to be noted that the impedance underwent a reduction of 77% at the recording sites. This reduction was caused by PEDOT nanostructures rather than PEDOT films.

As found in the study above, entrapment directly during electropolymerization is a reproducible procedure that allows for spatially controlled deposition, maintains the activity of enzymes, and provides better stability than the covalent binding technique does. Also, entrapment of enzymes eliminates the need for harsh chemicals commonly used for covalent binding of enzymes to the surface of electrodes. As for GOx, it maintains its ability to function after entrapment in conductive polymers. Besides, compared with soluble GOx, it shows more resistance to denaturation from the changes that occur in pH or temperature.

Paul et al. (2016) developed an ultrasensitive diagnostic/triaging kit to detect malarial parasites early and rapidly by using copper-doped zinc oxide nanofibers. In their impedimetric biosensor, doping of Cu in ZnO not only enhanced the nanofibers conductivity but preconcentrated the target analyte on the nanofiber surface too. This could occur because an electrical field was intrinsically created at the CuO/ZnO heterojunction interface. The impedimetric detection signal of the electrode modified by Cu-doped ZnO nanofibers exhibited an admirable sensitivity of 28.5 KΩ/(g^{-1} mL)/cm^2 and a detection limit of 6.8 ag/mL.

Tang et al. (2008) reported an electrochemical sandwich-type enzyme immunosensor and applied horseradish peroxidase—encapsulated nano-Au hollow microspheres as labels (HRP-GHS). Although conjugated to the secondary anti-CEA antibody, the nano-Au-encapsulated HRP label was used for the improvement of the analytical sensitivity of immunoassays (Fig. 5.14I). The research focused on the double-amplified characteristics of HRP-GHS-antiCEA. The aim of the study was to achieve certain sophisticated HRP-GHS-based probes for molecular biology, genomics as well as diagnosis and treatment of cancer and infectious illnesses.

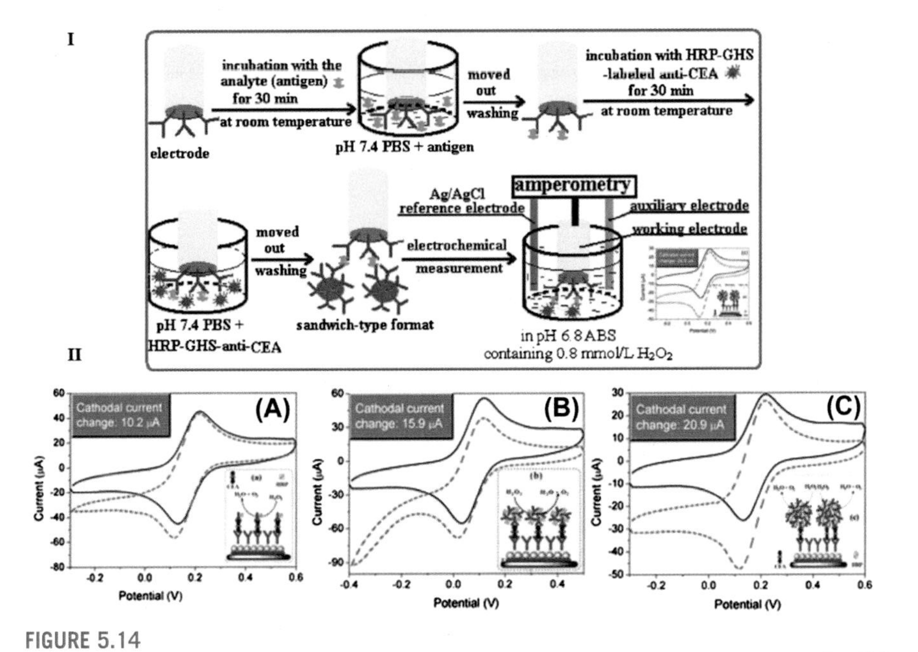

FIGURE 5.14

(I) Measurement procedure by the electrochemical sandwich-type enzyme immunosensor. (II) CVs of the immunosensors before (solid line) and after (dotted line) addition of H_2O_2 by utilizing various labels.

Reprinted with permission from Tang, D., Ren, J., 2008. In situ amplified electrochemical immunoassay for carcinoembryonic antigen using horseradish peroxidase-encapsulated nanogold hollow microspheres as labels. Analytical Chemistry 80, 8064–8070.

5.2.5.2 Nanosheets (a 2D nanomaterial)

Dichalcogenides, MoS_2, and hexagonal boron nitride (h-BN) nanosheets are examples of two-dimensional noncarbon nanomaterials that are structurally similar to graphene. These 2D nanosheets were used as a suitable supporting underlayer for loading organic and inorganic molecules or stabilizing nanoparticles. By virtue of their easy functionalization and large active surface area, the nanosheets could form hierarchical composites. Hence, these nanomaterials were recommended as probes and targets for biomarker detection. With regard to their appealing surface chemical and structural features, 2D nanosheets can be functionalized further with other active materials and/or biomolecules to develop biointerfaces of novelty. These biointerfaces are supposed to allow for the analysis of cancer biomarkers in a specific, sensitive, and multiplexed manner. The electrochemical procedures that may be reliably used in this regard are DPV, CV, EIS, SWV, and amperometry. If modified on specific electrode substrates, 2D nanomaterials can also be used to build up new kinds of electrode systems such as micro/nanoelectrodes and flexible electrodes (43).

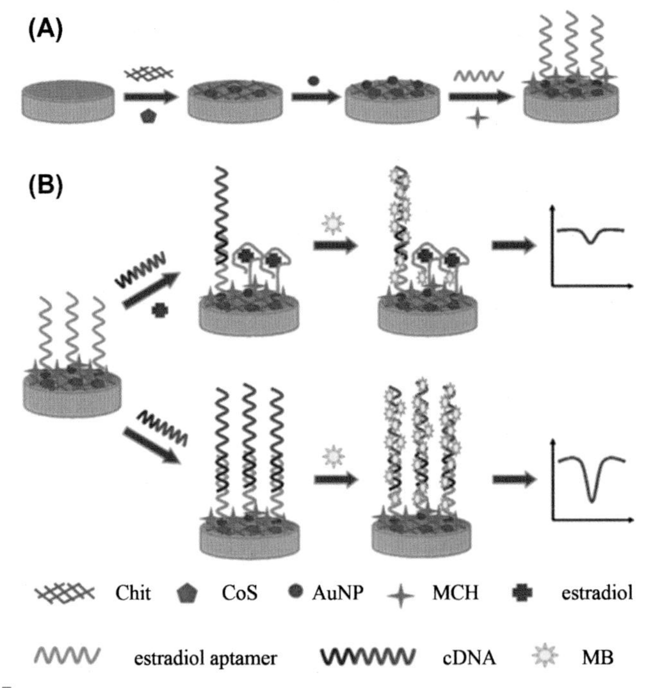

FIGURE 5.15

Schematic representation of the electrochemical aptasensor.

Reprinted with permission from Huang, K.J., Liu, Y.J., Zhang, J.Z., Cao, J.T., Liu, Y.M., 2015. Aptamer/Au nanoparticles/cobalt sulfide nanosheets biosensor for 17β-estradiol detection using a guanine-rich complementary DNA sequence for signal amplification. Biosensors and Bioelectronics 67, 184–191.

Huang et al. (2015) designed a biosensor for the electrochemical determination of 17β-estradiol. According to their design, first, an aptamer was assembled on AuNPs/CoS nanosheets/GCE. At this step, MB was used as an indicator, and cDNA (i.e., segmental complementary DNA with a rich G content) served as a signal amplifier. Then, through a hydrothermal method, 2D CoS nanosheets were synthesized. At this step, L-cysteine was used as a sulfur donor. In this procedure, the researchers obtained a CoS/GCE by coating the GCE with a CoS suspension. Then, they electrodeposited AuNPs on the CoS/GCE by using a constant potential. Afterward, through the Au-thiol chemistry, they immobilized a 17β-estradiol aptamer covalently onto the AuNPs/CoS/GCE. Subsequently, the electrode was incubated in MCH so that the nonspecific binding sites of AuNPs would be blocked. In this procedure, 17β-estradiol of various concentrations and a buffer solution of 2.0×10^{-7} M cDNA were applied on the aptamer/AuNPs/CoS/GCE. Eventually, an MB solution was drop-casted on the modified GCE (Fig. 5.15).

The analysis of the recommended method showed that the DPV peak currents would decrease if 17β-estradiol was added in the concentration range of 0.001−1 nM. The DPV peak currents also proved to be linearly correlated to the logarithm of the concentrations with the detection limit of 7.0×10^{-13} M. Low detection limit and good practical operation for urine samples were the advantages of this biosensor. These favorable results are attributed to the 2D CoS nanosheets, which highly promoted the effective specific surface of the electrode. Also, as an AuNPs film was formed on the CoS nanosheets, the electron transfer rate was boosted and the immobilized amount of aptamer was increased. All this occurred in low limits of detection.

Yang et al. (2016) used bulk MoS_2 to prepare a thin-layered molybdenum disulfide via an ultrasonic exfoliation procedure. Then, on the purpose of DNA sensing, they electrodeposited ZnO nanosheets on a MoS_2 scaffold. Serving as an excellent conductive skeleton, MoS_2 can provide a large accessible electrolytic surface area and make a direct pathway for fast electron transfer (Fig. 5.16).

Su et al. (2015) reported the fabrication of shape-controlled gold nanoparticles (AuNPs)-decorated thionine-MoS_2 nanocomposites. In their procedure, through electrostatic interactions as well as π-π, thionine was absorbed on the surface of MoS_2 sheets. With respect to its planar aromatic structure, thionine, indeed, served as an electrochemical indicator. Above all, thionine teamed up with MoS_2 and created a synergistic effect to tune the shape of the AuNPs. It is to be noted that, with the change of the ratio of thionine while HAuCl4 concentration was kept constant, AuNPs of various shapes, such as clover-like, flower-like, triangular, and spherical AuNPs, were produced. By means of the presented MoS_2-based

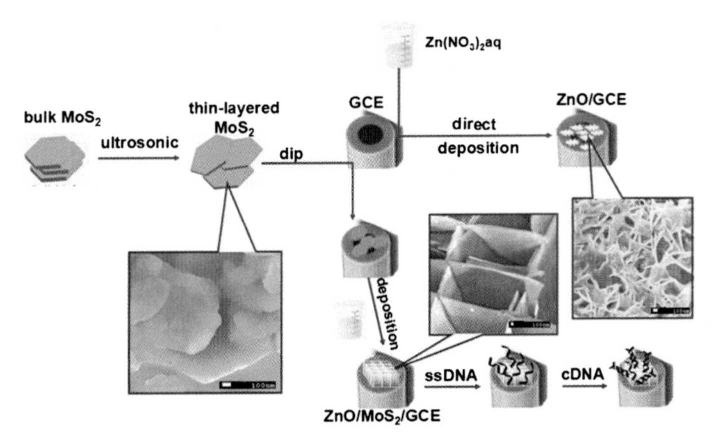

FIGURE 5.16

Fabrication scheme of ZnO/MoS_2 for DNA immobilization and hybridization.

Reprinted with permission from Yang, T., Chen, M., Kong, Q., Luo, X., Jiao, K., 2016. Toward DNA electrochemical sensing by free-standing ZnO nanosheets grown on 2D thin-layered MoS_2. Biosensors and Bioelectronics 89, 538–544.

electrochemical sensing platform, CEA could be detected label-free. The detection signal depended on the peak current of thionine, which declined as antigen and antibody reacted. Applied as a detector of CEA in serum samples, this SWV electrochemical sensor exhibited excellent selectivity, high sensitivity, and durable stability. Hence, AuNPs-Thi-MoS_2 nanocomposites were proposed as a promising sensing platform for fabrication of simple, ultrasensitive, and label-free electrochemical biosensors.

5.3 Conclusion

In this chapter, function of noncarbon nanomaterials in the fabrication of electrochemical biosensors was investigated. Applications of noncarbon nanomaterials are classified mainly as electrotransducer modifier surface and substrate for immobilization of biomolecules, electrochemical amplifier signals, as signal-producing probe and mimicking the enzymatic behavior in the designing of various advanced biosensors. In recent years, huge noncarbon-based nanostructures in several kinds of metal and metal oxide, core-shell, QDs, wires, fibers, and sheets and composites were registered in the electrochemical biosensor fabrication, owing to the unique properties of these nanomaterials such as large surface-to-volume ratios, conductivity, size-related electronic properties, remarkable electrochemical performances, ease of functionalization, low cytotoxicity, and biocompatibility. Hence create unique and applicable devices based on noncarbon nanomaterials, acting as advanced electrochemical biosensors, in the field of diagnosis of biomarkers and determination of biological species have a strong position which confirms with many scientific reported applications. Such electrochemical biosensors would exhibit wide linear response range and great sensitivity, selectivity, and promising advantage of practicability. Therefore, there is a bright and growing perspective for research on these nanomaterials for designing and building new biosensors to detect early illnesses and reduce the time consumption and cost of diagnosis.

References

Ahmadi, A., Shirazi, H., Pourbagher, N., 2014. An electrochemical immunosensor for digoxin using core − shell gold coated magnetic nanoparticles as labels. Molecular Biology Reports 41, 1659−1668.

Asif, M., Liu, H., Aziz, A., Wang, H., Wang, Z., Xiao, F., Liu, H., 2017. Core-shell iron oxide-layered double hydroxide: high electrochemical sensing performance of H_2O_2 biomarker in live cancer cells with plasma therapeutics. Biosensors and Bioelectronics 97, 352−359.

Cao, H., Yang, D., Ye, D., Zhang, X., Fang, X., Zhang, S., 2015. Protein-inorganic hybrid nano fl owers as ultrasensitive electrochemical cytosensing Interfaces for evaluation of cell surface sialic acid. Biosensors and Bioelectronics 68, 329−335.

Chatterjee, K., Sarkar, S., Rao, K.J., Paria, S., 2014. Core/shell nanoparticles in biomedical applications. Advances in Colloid and Interface Science 209, 8–39.

Cui, J., Luo, J., Peng, B., Zhang, X., Zhang, Y., Wang, Y., Qin, Y., 2016. Synthesis of porous NiO/CeO_2 hybrid nanoflake arrays as a platform for electrochemical biosensing. Nanoscale 8, 770–774.

Daneshpour, M., Syed, L., Izadi, P., Omidfar, K., 2016. Femtomolar level detection of RASSF1A tumor suppressor gene methylation by electrochemical nano-genosensor based on Fe_3O_4/TMC/Au nanocomposite and PT-modi fi ed electrode. Biosensors and Bioelectronics 77, 1095–1103.

Ding, L., Bond, A.M., Zhai, J., Zhang, J., 2013. Utilization of nanoparticle labels for signal amplification in ultrasensitive electrochemical affinity biosensors: a review. Analytica Chimica Acta 797, 1–12.

Hansen, J.A., Wang, J., Kawde, A., Xiang, Y., Gothelf, K.V., Collins, G., 2006. Quantum-Dot/aptamer-based ultrasensitive multi-analyte electrochemical biosensor. Journal of the American Chemical Society 128, 2228–2229.

Hayat, A., Catanante, G., Marty, J.L., 2014. Current trends in nanomaterial-based amperometric biosensors. Sensors 14, 23439–23461.

Hua, X., Zhou, Z., Yuan, L., Liu, S., 2013. Selective collection and detection of MCF-7 breast cancer cells using aptamer-functionalized magnetic beads and quantum dots based. Analytica Chimica Acta 788, 135–140.

Huang, K.J., Liu, Y.J., Zhang, J.Z., Cao, J.T., Liu, Y.M., 2015. Aptamer/Au nanoparticles/cobalt sulfide nanosheets biosensor for 17β-estradiol detection using a guanine-rich complementary DNA sequence for signal amplification. Biosensors and Bioelectronics 67, 184–191.

Huang, L., Chen, K., Zhang, W., Zhu, W., Liu, X., Wang, J., Wang, R., 2018. ssDNA-tailorable oxidase-mimicking activity of spinel $MnCo_2O_4$ for sensitive biomolecular detection in food sample. Sensors and Actuators B: Chemical 269, 79–87.

Jianrong, C., Yuqing, M., Nongyue, H., Xiaohua, W., 2004. Nanotechnology and biosensors. Biotechnology Advances 22, 505–518.

Kumar, A., Lee, T., Min, J., Choi, J., 2013. An enzymatic biosensor for hydrogen peroxide based on CeO_2 nanostructure electrodeposited on ITO surface. Biosensors and Bioelectronics 47, 385–390.

Li, J., Gao, H., Chen, Z., Wei, X., Yang, C.F., 2010. An electrochemical immunosensor for carcinoembryonic antigen enhanced by self-assembled nanogold coatings on magnetic particles. Analytica Chimica Acta 665, 98–104.

Liao, S.H., Lu, S.Y., Bao, S.J., Yu, Y.N., Wang, M.Q., 2016. $NiMoO_4$ nanofibres designed by electrospining technique for glucose electrocatalytic oxidation. Analytica Chimica Acta 905, 72–78.

Liu, G., Lee, T.M.H., Wang, J., 2005. Nanocrystal-based bioelectronic coding of single nucleotide polymorphisms. Journal of the American Chemical Society 127, 38–39.

Luo, L., Li, Q., Xu, Y., Ding, Y., Wang, X., Deng, D., Xu, Y., 2010. Amperometric glucose biosensor based on $NiFe_2O_4$ nanoparticles and chitosan. Sensors and Actuators B: Chemical 145, 293–298.

Mazloum-ardakani, M., Aghaei, R., Moaddeli, M., 2015. Quantum-dot biosensor for hybridization and detection of R3500Q mutation of apolipoprotein B-100 gene. Biosensors and Bioelectronics 72, 362–369.

Mazloum-Ardakani, M., Amin-Sadrabadi, E., Khoshroo, A., 2016. Enhanced activity for non-enzymatic glucose oxidation on nickel nanostructure supported on PEDOT:PSS. Journal of Electroanalytical Chemistry 775, 116–120.

Noorbakhsh, A., Salimi, A., 2011. Development of DNA electrochemical biosensor based on immobilization of ssDNA on the surface of nickel oxide nanoparticles modified glassy carbon electrode. Biosensors and Bioelectronics 30, 188–196.

Paul, K.B., Kumar, S., Tripathy, S., Rama, S., Vanjari, K., Singh, V., 2016. A highly sensitive self assembled monolayer modified copper doped zinc oxide nanofiber interface for detection of Plasmodium falciparum histidine-rich protein-2: targeted towards rapid, early diagnosis of malaria. Biosensors and Bioelectronics 80, 39–46.

Polsky, R., Gill, R., Kaganovsky, L., Willner, I., 2006. Nucleic acid-functionalized Pt Nanoparticles : catalytic labels for the amplified electrochemical detection of biomolecules. Analytical Chemistry 78, 2268–2271.

Salimi, A., Hallaj, R., Mamkhezri, H., Mohamad, S., Hosaini, T., 2008. Electrochemical properties and electrocatalytic activity of FAD immobilized onto cobalt oxide nanoparticles : application to nitrite detection. Journal of Electroanalytical Chemistry 619, 31–38.

Shirakawa, H., Louis, E.J., MacDiarmid, A.G., Chiang, C.K., Heeger, A.J., 1977. Synthesis of electrically conducting organic polymers: halogen derivatives of polyacetylene, (CH)x. Journal of the Chemical Society, Chemical Communications 16, 578–580.

Shrivastava, S., Jadon, N., Jain, R., 2016. Next generation polymer nanocomposites based electrochemical sensors and biosensors: a review. Trends in Analytical Chemistry 82, 55–67.

Su, S., Zou, M., Zhao, H., Yuan, C., Xu, Y., Zhang, C., Wang, L., 2015. Shape-controlled gold nanoparticles supported on MoS 2 nanosheets: synergistic effect of thionine and MoS_2 and their application for electrochemical label-free immunosensing. Nanoscale 7, 19129–19135.

Tang, D., Ren, J., 2008. In situ amplified electrochemical immunoassay for carcinoembryonic antigen using horseradish peroxidase-encapsulated nanogold hollow microspheres as labels. Analytical Chemistry 80, 8064–8070.

Wang, J., 2012. Electrochemical biosensing based on noble metal nanoparticles. Microchimica Acta 177, 245–270.

Wen, L., Qiu, L., Wu, Y., Hu, X., Zhang, X., 2017. Aptamer-modified semiconductor quantum dots for biosensing applications. Sensors 17, 1736.

Wu, D., Li, R., Wang, H., Liu, S., Wang, H., Wei, Q., Du, B., 2012. Hollow mesoporous silica microspheres as sensitive labels for immunoassay of prostate-specific antigen. Analyst 137, 608–613.

Xie, J., Zhang, X., Wang, H., Zheng, H., Huang, Y., 2012. Analytical and environmental applications of nanoparticles as enzyme mimetics. Trends in Analytical Chemistry 39, 114–129.

Xu, W., Yi, H., Yuan, Y., Jing, P., Chai, Y., Yuan, R., Wilson, G.S., 2015. An electrochemical aptasensor for thrombin using synergetic catalysis of enzyme and porous Au@Pd core – shell nanostructures for signal amplification. Biosensors and Bioelectronics 64, 423–428.

Yang, C., Xu, C., Wang, X., 2012. ZnO/Cu nanocomposite: a platform for direct electrochemistry of enzymes and biosensing applications. Langmuir 28, 4580–4585.

Yang, G., Kampstra, K.L., Abidian, M.R., 2014. High performance conducting polymer nanofiber biosensors for detection of biomolecules. Advanced Materials 26, 4954–4960.

Yang, H., Yuan, R., Chai, Y., Zhuo, Y., Su, H., 2010. Electrochemical immunoassay for human chorionic gonadotrophin based on Pt hollow nanospheres and silver/titanium dioxide nanocomposite matrix. Journal of Chemical Technology & Biotechnology 85, 577–582.
Yang, T., Chen, M., Kong, Q., Luo, X., Jiao, K., 2016. Toward DNA electrochemical sensing by free-standing ZnO nanosheets grown on 2D thin-layered MoS_2. Biosensors and Bioelectronics 89, 538–544.
Yu, Z., Li, H., Zhang, X., Tan, W., Zhang, X., Zhang, L., 2016. Facile synthesis of $NiCo_2O_4$@ Polyaniline core–shell nanocomposite for sensitive determination of glucose. Biosensors and Bioelectronics 75, 161–165.
Yuan, P., Zhuo, Y., Chai, Y., 2008. Dendritic silver/silicon dioxide nanocomposite modified electrodes for electrochemical sensing of hydrogen peroxide. An International Journal Devoted to Fundamental and Practical Aspects of Electroanalysis 20, 1839–1844.
Zhang, C., Lou, J., Tu, W., Dai, Z., 2014. Ultrasensitive electrochemical biosensing for DNA using quantum dots combined with restriction endonuclease. Analyst 140, 506–511.
Zhao, J., Hu, S., Cao, Y., Zhang, B., Li, G., 2015. Electrochemical detection of protein based on hybridization chain re-action-assisted formation of copper nanoparticles. Biosensors and Bioelectronics 66, 327–331.
Zhou, N., Li, J., Chen, H., Liao, C., Chen, L., 2013. A functional graphene oxide-ionic liquid composites–gold nanoparticle sensing platform for ultrasensitive electrochemical detection of $Hg2^{+}$. Analyst 138, 1091–1097.

CHAPTER

Types of monitoring biosensor signals

6

Hamid R. Zare[1], Zahra Shekari[2]

Professor, Department of Chemistry, Faculty of Science, Yazd University, Yazd, Iran[1]; Dr, Department of Chemistry, Faculty of Science, Yazd University, Yazd, Iran[2]

6.1 Introduction

A biosensor is an analytical device that works based on a specific interaction of an analyte (target) and a biological element. It produces a measurable response, which is proportional to the analyte concentration. As its name suggests, a biosensor is a combination of two parts: a biological sensing element (bioreceptor) and a sensor element (transducer). The target interacts with a specific bioreceptor and induces a change in the biomolecule. This change is converted into a signal detectable and quantifiable by a transducer. A transducer might be optical, electrochemical, thermometric, piezoelectric, or calorimetric. It can be coupled to an appropriate signal processor.

A bioreceptor is used to detect a specific analyte, and it does not detect other species. This feature makes biosensors very useful for measuring various analytes in different domains. Several types of bioreceptors, including enzymes, antibodies, nucleic acids, and cells, are used in the design of biosensors. Sometimes, a synthetic molecule is used to mimic the properties of biological receptors. Synthetic antibodies, molecularly imprinted polymers, and aptamers are some of these molecules, known as biomimetic receptors. Biosensors can be classified based on their bioreceptor type or signal transduction method. So far, various biosensors have been developed, but, according to the screening of signals, they are divided into direct and indirect detection biosensors. In the case of direct detection biosensors, biological interactions are directly measured, but indirect detection biosensors need secondary elements for detection.

6.2 Signal monitoring

Signal monitoring by biosensors is based on one of the two approaches: direct or indirect detection approach. Accordingly, biosensors can be divided into two categories: direct recognition biosensors and indirect recognition biosensors. In the former, biological interactions are directly measured in real time, whereas the latter

Electrochemical Biosensors. https://doi.org/10.1016/B978-0-12-816491-4.00006-1

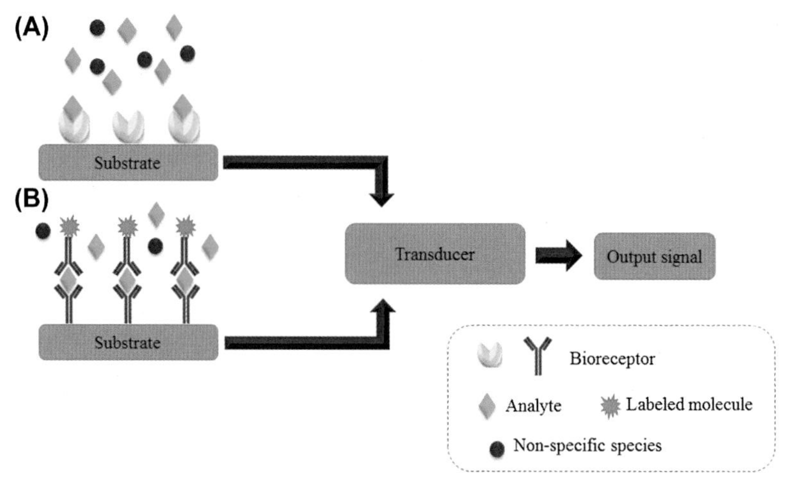

FIGURE 6.1

Schematic diagram of (A) a direct detection biosensor and (B) an indirect detection biosensor based on a sandwich method.

depend on secondary elements for the detection of targets. Fig. 6.1 is the schematic display of label-free (direct) and label-based (indirect) biosensors. An ideal biosensor is supposed to detect target molecules directly and without using labels. As compared with indirect detection biosensors, direct detection ones are faster and simpler, but they usually have more limitations in detecting.

6.3 Direct detection biosensors

In the direct detection method, measurements are performed based on the biological interaction of the components, and there is no need for labeled target molecules. In the direct detection mode (with label-free biosensors), however, biorecognition elements are immobilized on the sensor surface, target molecules bind to them, and the sensor response is measured. The response is proportional to the concentration of the captured target. Physical changes, such as the changes in the optical, mechanical, or electrical properties induced by biological interactions, are measured by a transducer. The advantages of the direct detection mode are simplicity, high speed, and ease of detection, thus keeping down the assay cost and the analysis time. Also, undesirable effects of labels, such as instability and steric hindrances, are eliminated in this mode. Direct detection is generally preferred when binding of an analyte produces an appropriate response in one or more concentration ranges. This method is suitable for the analytes of medium to large sizes. The main direct detection systems are optical-based biosensors such as surface plasmon resonance (SPR) biosensors, mechanical systems such as quartz crystal resonators, and electrochemical biosensors most commonly including electrochemical impedance spectroscopy (EIS) and field-effect transistors (FETs).

6.3.1 Direct optical biosensors

Optical detection is widely practiced by biosensing technologies, including both direct and indirect biosensors. In the case of a direct optical biosensor, the interaction of light with analyte leads to a change in the properties of light. This change in light properties, which is proportional to the concentration of the analyte, is the basis of the measurement. For example, in adsorption spectroscopy, analyte measurement is performed by measuring the light absorption at specific wavelengths that match the absorption bands of analytes. Also, optical methods based on the measurement of analyte-induced refractive index changes have attracted much attention, especially for affinity biosensing. By these methods, it is possible to measure the changes of the refractive index induced by the binding of analyte molecules to special biorecognition molecules (e.g., antibodies, nucleic acids, and peptides). These biorecognition molecules are immobilized on the surface of an optical transducer and are able to recognize and capture analyte molecules. The transducers that are able to measure analyte-induced refractive index changes are mainly based on (1) interferometry (as done by means of optical interferometers), (2) spectroscopy of optical waveguides (e.g., resonant mirror sensor, grating coupler sensor, and SPR sensor), and (3) optical resonators (e.g., microsphere resonator sensor and ring resonator sensor). Most of these transducers use an evanescent field to observe biomolecular interactions (Gauglitz and Homola, 2015). SPR is the most common technique in direct optical biosensors. It is known as an accepted technique in biosensing, especially to screen and characterize the kinetics of biomolecular interactions. SPR allows real-time and label-free detection of targets. In a direct SPR biosensor, the biorecognition element is immobilized on the prism surface, and then the target binds to the immobilized bioreceptor and causes a change in the plasmon resonance value. The change of the plasmon resonance value is proportional to the concentration of the captured target. SPR has attracted significant attention in the field of biosensors owing to its outstanding advantages including rapid and real-time analysis without labeling, being cost-effective, less reagent consumption, and possible automation and miniaturization. Kausaite-Minkstimiene et al. (2009) reported a direct, label-free method of detecting the human growth hormone antibodies (anti-HGH) by means of SPR biosensors. In their study, HGH antigen was immobilized on the surface of an SPR chip, which was modified by a self-assembled monolayer of 11-mercaptoundecanoic acid. Then, the monoclonal anti-HGH was directly bound to an HGH-modified surface. The concentration range of 0.25 nM−10 mM and the detection limit of 2.47 nM were reported for the detection of anti-HGH by the direct detection method. The disadvantage of direct detection, in comparison with indirect modes, is its high detection limit. The sandwich approach can improve the detection limits of direct SPR biosensors, and it also can be of help in complex matrices (Homola et al., 2002; Stern et al., 2016).

6.3.2 Direct electrochemical biosensors

EIS is the most common label-free electrochemical technique, which is successfully used to study biological interactions. A sensing device whose response is based on the impedance changes before and after binding a target on a modified electrode surface is known as impedimetric biosensor. This device is very capable of studying the interfacial properties associated with biorecognition processes occurring at the surfaces of different modified electrodes. Various bioreceptors such as enzymes, antibodies, nucleic acids, cells, and microorganisms can be immobilized on the surface of an electrode to develop an impedimetric biosensor. One of the major advantages of impedimetric biosensors is their low detection limit. Moreover, as these biosensors are able to do label-free detections, they are suitable for real-time monitoring. For impedimetric biosensors, the change in the impedance value is proportional both to the biochemical reaction rate and to the analyte concentration. Various methods have been developed to make these devices more functional and trustworthy. To name but a few, one can refer to designing of novel electrodes, functionalizing of transducer surfaces using nanomaterials, and tethering of bioreceptors. Nanostructure materials have been widely used in impedimetric biosensors as biomolecule immobilizing matrices/substrates to improve the electrochemical detection of the sensors. Modification of the electrode surface with nanomaterials increases the surface area, which leads to the enhanced immobilization of the bioreceptor on the modified electrode surface. It also facilitates the involved electroanalytical processes. In this regard, various nanomaterials have been applied, including carbon nanomaterials (Gupta et al., 2013; Gutés et al., 2013; Patil et al., 2015; Yagati et al., 2016), metal and metal oxide nanoparticles (Kaushik et al., 2010; Kuzin et al., 2015; Liu et al., 2009; Shiravand and Azadbakht, 2017), nanowires (Subramanian et al., 2014), nanopores (Wu et al., 2015), and nanocomposites (Shukla et al., 2012; Zhu et al., 2015). Also, an impedimetric aptasensor was developed to determine the carcinoembryonic antigen (CEA) biomarker (Shekari et al., 2017). This aptasensor was fabricated based on the covalent immobilization of an amine-modified CEA aptamer on the surface of a modified glassy carbon electrode (GCE) using gold nanoparticles (AuNPs) incorporated in amino-functionalized MCM-41 (Fig. 6.2). Then, the changes in the interfacial charge transfer resistance (R_{ct}) of the redox marker were used for the quantitative determination of CEA. The low detection limit of 9.8×10^{-4} ng mL^{-1} was reported through the EIS method.

In another study, Rushworth et al. (2014) reported a label-free biosensor to detect amyloid-beta oligomers (AβOs). These oligomers are the primary neurotoxic species in Alzheimer's disease. In their study, a fragment of the cellular prion protein (PrP^C, residues95−110) was used as a biorecognition element, a polymer-functionalized gold screen-printed electrode was used as a substrate, and the biotinylated PrP^C (95−110) was attached through a biotin/NeutrAvidin bridge to the substrate. This biosensor has some significant advantages such as being label-free, which allows detecting of natural oligomers, and the use of a synthetic

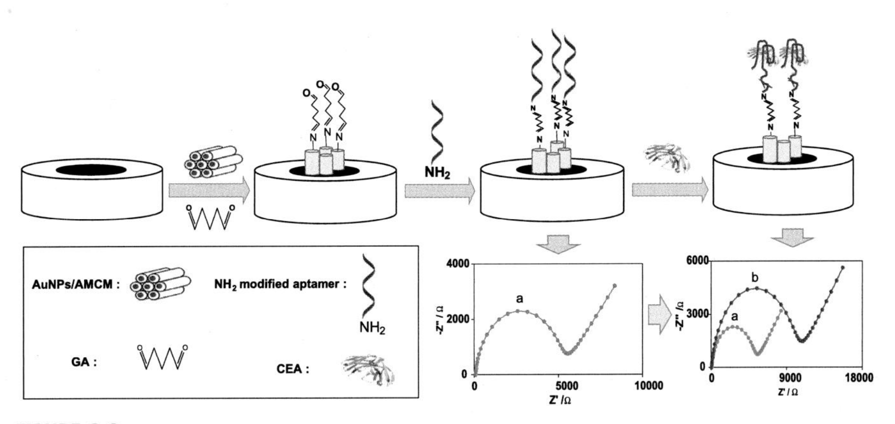

FIGURE 6.2

Schematic diagram of the different steps of constructing an impedimetric aptasensor for CEA determination.

Reproduced from Shekari, Z., Zare, H.R., Falahati, A., 2017. Developing an impedimetric aptasensor for selective label–free detection of CEA as a cancer biomarker based on gold nanoparticles loaded in functionalized mesoporous silica films. Journal of the Electrochemical Society 164(13), B739–B745 with permission.

PrP^C fragment as a bioreceptor instead of a pan-Aβ antibody, which makes it specific for AβO detection.

FETs make up another class of label-free electrochemical biosensors. They are the transistors that are gated by changes in the electric field. As most biomolecules transport electrostatic charges and biological activities involve electrical potential changes, FET-based biosensors are promising candidates for applications, which require high sensitivity and fast response time. FETs have a source, a drain, and a gate on the surface of which biomolecules are immobilized. To enhance the performance of FET biosensors, such nanomaterials as carbon nanotubes (Marques et al., 2017), nanowire (Gong et al., 2018; Jakob et al., 2017), nanoribbon (Lin et al., 2017b; Liu et al., 2018), and graphene (Tu et al., 2018) have been used in their fabrication. For example, Liu et al. (2018) used a highly sensitive In_2O_3 nanoribbon FET biosensor with a fully integrated on-chip gold side gate to determine glucose in various body fluids such as sweat and saliva. In their research, the functionalization of electrodes with glucose oxidase, chitosan, and single-walled carbon nanotubes (SWCNTs) led to a very wide concentration range and a low detection limit (10.0 nM) for the biosensor.

6.3.3 Direct piezoelectric (mechanical) biosensors

Piezoelectric transducers such as resonant crystals and cantilevers are applied as direct sensing transducers. Resonant crystal biosensors such as quartz crystal microbalance monitor changes at the resonating frequency of a quartz crystal. This ability is due to the mass changes produced by the biological processes occurring near the crystal surface.

Cantilever biosensors have been greatly favored as label-free detectors that can measure vibrational frequencies or bending stress changes. This ability is due to the mass change that occurs on the cantilever surface. Cantilever biosensors can be divided into two groups including bending or static cantilevers and dynamic or resonating cantilevers. In the static approach, the attachment of an analyte to the cantilever surface changes the cantilever surface stress, resulting in a deflection response that is proportional to the analyte concentration. In the dynamic approach, the measurement is based on the binding-induced changes that occur in the cantilever resonant frequency due to mass change or stiffness change. Butt (1996) reported the first application of cantilever sensors in biological systems. Since then, various biomolecules have been detected using cantilever platforms (Johansson et al., 2006; Savran et al., 2004; Xu and Mutharasan, 2009; Yue et al., 2008). Microcantilever arrays can be used for the simultaneous detection of multiple biomarkers (Joo et al., 2012). For example, in an attempt by Wang et al. (2016), a biosensor based on microcantilever arrays was fabricated to detect multiple liver cancer biomarkers. In that research, a microcavity was postulated at the free end of a cantilever for local reactions between antibody and antigen, and the microjet printing system was used to realize the local immobilization of different antibodies on the cantilever array.

6.4 Indirect detection biosensors

Indirect detection biosensors function on the basis of secondary elements or labels. Several secondary elements such as enzymes, fluorescent tags, and redox labels can be used for indirect detection. In this type of detection, the target must first be bound to a labeled secondary molecule. Indirect detection biosensors (i.e., label-based biosensors) are very sensitive and have lower detection limits than direct detection biosensors, but they typically present a slower response. This is due to the use of a labeled secondary molecule. Indirect transducers are of optical and electrochemical types.

6.4.1 Indirect optical biosensors

In optical biosensors, light is emitted or modified by label molecules. Fluorescence is a common method of designing indirect optical biosensors. Other methods including densitometry, colorimetry, and chemiluminescence can also be used to design indirect biosensors, depending on the type of the label used. In fluorescent sandwich assays, bioreceptor can be easily labeled with a variety of fluorescent tags (fluorophores) to detect the fluorescence. In sandwich-based assays, the target is captured between two biorecognition elements, one of which is usually immobilized on a substrate surface and the other is labeled with signal tags

(Grate et al., 2009; Kiening et al., 2005; Shojaei et al., 2014; Tennico et al., 2010). When a fluorophore is excited, the biosensor transducer can detect the light emission. Zhou et al. (2019) developed a fluorescent aptasensor for alpha-fetoprotein (AFP) determination. In this aptasensor, the AFP aptamer was covalently bound with the donor CdTe quantum dots (QDs), and the anti-AFP monoclonal antibody was conjugated with the acceptor AuNPs. After incubation of the QDs-labeled aptamer and the AuNPs-labeled antibody with their targets, the specific interaction among the labeled aptamer, the AFP, and the labeled antibody made the QDs and the AuNPs close together such that the fluorescence of CdTe QDs was quenched through the Förster resonance energy transfer (FRET) between the QDs and the AuNPs (Fig. 6.3). In another study, a sensitive colorimetric immunosensor for influenza virus H5N1 was developed based on enzyme-encapsulated liposomes (Lin et al., 2017a). The principles of this immunosensor are shown in Fig. 6.4. As it can be seen, the capture antibody was first immobilized in the wells of a microplate by physical adsorption, and then, H5N1 was interacted by the immobilized antibody. Afterward, a sandwich structure was formed by the interaction between H5N1 and the biotinylated detection antibody. Following that, the horseradish peroxidase (HRP)–encapsulated liposomes were connected to the biotinylated detection antibody via a biotin-avidin-biotin bond. The addition of a mixture of 3,3′,5,5′-tetramethylbenzidine (TMB) and H_2O_2 led to the direct lysing of the liposomes and the

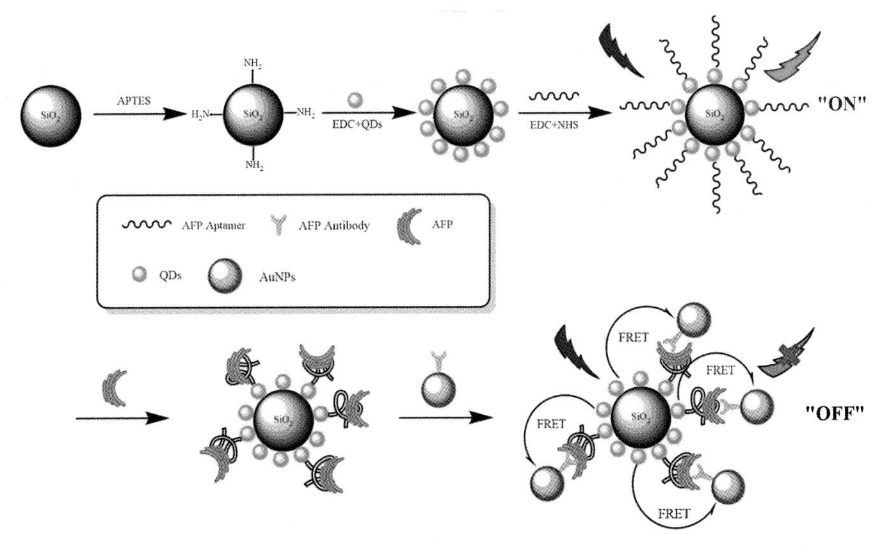

FIGURE 6.3

Schematic illustration of a fluorescent aptasensor for detection of AFP based on the FRET of sandwich-structured QDs-AFP-AuNPs.

Reproduced from Zhou, L., Ji, F., Zhang, T., Wang, F., Li, Y., Yu, Z., Jin, X., Ruan, B., 2019. An fluorescent aptasensor for sensitive detection of tumor marker based on the FRET of a sandwich structured QDs-AFP-AuNPs. Talanta 197, 444–450 with permission.

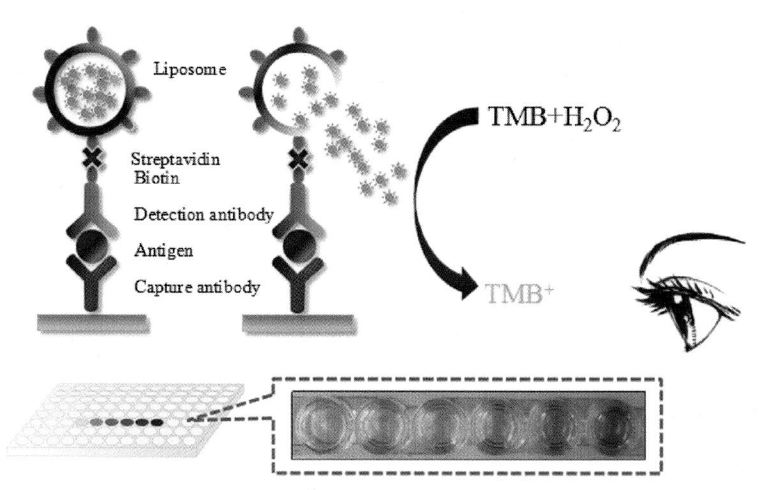

FIGURE 6.4

Schematic illustration of an immunosensor for H5N1 based on HRP-encapsulated liposome.

Reproduced from Lin, C., Guo, Y., Zhao, M., Sun, M., Luo, F., Guo, L., Qiu, B., Lin, Z., Chen, G., 2017a. Highly sensitive colorimetric immunosensor for influenza virus H5N1 based on enzyme-encapsulated liposome. Analytica Chimica Acta 963, 112–118 with permission.

release of a large amount of HRP. The released HRP catalyzed the H_2O_2-mediated oxidation of TMB and changed the color of the system. This color change was detectable by naked eyes and was measured by the UV-Vis method.

In addition, the electrogenerated chemiluminescence (ECL) technique was used for sandwich-based biosensors. For example, Xie et al. (2016) reported a sandwich ECL biosensor for the analysis of galactosyltransferase (Gal T) activity. The biosensor was based on a graphitic carbon nitride (g-C_3N_4) nanosheet–modified GCE as a substrate and polystyrene microsphere (PSM) as a probe. In that study, a GCE was modified by the g-C_3N_4 nanosheet, and then, bovine serum albumin conjugated with N-acetyl glucosamine (GlcNAc-BSA) was immobilized on the g-C_3N_4-modified GCE. In the next step, galactose was introduced onto the modified electrode surface and interacted with GlcNAc-BSA, resulting in decreased ECL signals of the GlcNAc-BSA/g-C_3N_4 nanosheet–modified GCE. Afterward, *Artocarpus integrifolia* lectin (AIA) conjugated with PSM nanoprobes was adsorbed on the surface of the modified electrode through a specific interaction between the captured galactose and AIA. The poor conductivity of the nanoprobes caused a significant decrease in the ECL signals and led to the high sensitivity of the Gal T activity analysis.

6.4.2 Indirect electrochemical biosensors

The most common electrochemical transducers are amperometric devices that detect the changes in a current at a constant potential, potentiometric devices that detect the changes in a potential at a constant current, conductometric systems that detect changes of the conductivity between two electrodes (at a constant potential), and impedimetric devices that measure the voltage-to-alternating current ratio. As discussed earlier, impedimetric biosensors are the main type of label-free electrochemical biosensors. Also, FET-based biosensors, as potentiometric biosensors, are another type of direct electrochemical biosensor. However, other electrochemical biosensors often need a label for analyte detection (Tang et al., 2011; Ting et al., 2009; Wang et al., 2014; Zhang et al., 2010). The labels used in electrochemical biosensors can be enzymes, nanomaterials, or redox labels. For example, platinum nanoparticles (PtNPs)–decorated carbon nanocages (CNCs) were used as signal tags for thrombin detection by a sandwich-type electrochemical aptasensor (Gao et al., 2016). The CNCs were applied as a carrier to load the PtNPs catalyst (PtNPs/CNC), and then, thrombin capture probe (TBA) was immobilized on the PtNPs/CNC. In this study, thrombin and TBA-PtNPs/CNCs were incubated on the TBA immobilized on the surface of a gold electrode. The electrocatalytic reduction current for H_2O_2 was used for the quantitative determination of thrombin through the differential pulse voltammetry (DPV) method.

In another study, Yin et al. (2015) reported an electrochemical biosensor to measure the activity of protein kinase A (PKA). Fig. 6.5 illustrates the preparation procedure and the response mechanism of the biosensor. First, the peptide was immobilized on the surface of a gold nanoparticles–modified GCE (AuNPs/GCE) and then phosphorylated through a PKA reaction. Afterward, a biotinylated phosphate binding reagent (Phos-tag-biotin) was specifically bound to the phosphorylated peptide immobilized on the electrode surface. Finally, avidin-functionalized

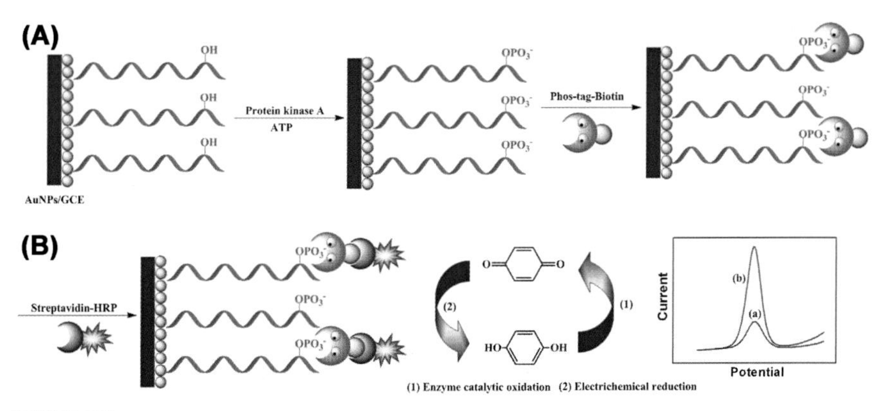

FIGURE 6.5

The preparation procedure of a Phos-tag-based electrochemical biosensor for a PKA activity assay.

Reprinted from Yin, H., Wang, M., Li, B., Yang, Z., Zhou, Y., Ai, S., 2015. A sensitive electrochemical biosensor for detection of protein kinase A activity and inhibitors based on Phos-tag and enzymatic signal amplification. Biosensors and Bioelectronics 63, 26–32 with permission.

horseradish peroxidase (avidin-HRP) was accumulated on the modified electrode surface via an immunoreaction between avidin and biotin. With regard to the electrocatalytic effect of HRP, a sensitive electrochemical signal was reported for benzoquinone. This signal was related to the PKA activity.

6.5 Signal monitoring in electrochemical DNA biosensors

Because nucleic acids have been widely used to fabricate various biosensors as well as bioanalytical assays, this section focuses on the direct and indirect monitoring of nucleic acid–based biosensors. Biorecognition elements in these biosensors are oligonucleotides, with a known sequence of bases or a fragment of DNA or RNA. Nucleic acid biosensors are based on the duplex formation between two strands of nucleic acid (DNA or RNA) and play the role of a biorecognition layer to detect biochemical or chemical species.

In a hybridization process, a single-strand DNA (ssDNA), known as a probe, can be hybridized with a complementary strand and form a hybrid duplex DNA (dsDNA) at the electrode surface. This process possesses high efficiency and good specificity; thus, the detection of the complementary strand DNA or RNA can be easily recognized. Artificial nucleic acids, such as aptamer or DNAzyme, can be used as a probe molecule to detect a wide variety of target molecules from small molecules, such as inorganic ions and organic molecules (e.g., potassium and adenosine triphosphate), to large biomolecules such as peptides and proteins as well as whole organisms such as bacteria and cells.

Electrochemical transducers are very useful for sequence-specific biosensing of DNA because they are sensitive, portable, low cost, and able to couple with modern microfabrication technologies. Moreover, electrochemical techniques serve as innovative ways to connect nucleic acid assays with signal-generating elements to amplifying electrical signals. In electrochemical DNA biosensors, the processes can also be followed by both direct and indirect detection methods.

6.5.1 Electrochemical DNA biosensors based on direct detection methods

The direct detection of DNA requires no labeling step. This methodology is simple, not time-consuming, quite sensitive, and applicable, but it cannot be used in multiplex detections. Some methods that are based on direct detection are presented below.

6.5.1.1 Electrochemical DNA biosensors based on inherent signals of DNA bases

The sensitive electrochemical detection of DNA can be done by monitoring the appearance of the oxidation peaks of DNA bases. It is possible to do this monitoring through an electrochemical technique because the potential window in this technique is wide in the positive direction. The electrochemical signals of DNA bases at mercury electrodes were monitored for the first time by Palecek (1980). Oxidation of guanine and adenine residues in polynucleotides at the surface of a carbon electrode was reported by Brabec and Dryhurst (1978) and Brabec (1981). Because the potential windows of solid electrodes, such as carbon electrodes, are more positive than those of mercury electrodes, solid ones are better for studying nucleic acids oxidation. In contrast, mercury electrodes are suitable for the investigation of nucleic acids reduction.

Electrochemical oxidation of purine and pyrimidine DNA bases on carbon electrodes indicated that all bases (guanine, adenine, thymine, and cytosine) can be oxidized with a pH-dependent mechanism, but purine bases (guanine and adenine) are more sensitive and oxidize at less positive potentials (Abbaspour and Noori, 2008; Oliveira-Brett et al., 2002a, 2002b; Yin et al., 2010). The direct strategy is based on the signal transduction induced directly from the oxidation of guanine or adenine moieties in DNA strands. The electrochemical activities of guanine and adenine were studied at the surfaces of different electrodes including carbon, gold, indium tin oxide, and polymer-coated electrodes and by various methods such as square-wave voltammetry and chronopotentiometry (Pournaghi-Azar et al., 2007; Singhal and Kuhr, 1997; Steenken and Jovanovic, 1997; Wang et al., 1995, 1996). In addition, the details of the guanine oxidation mechanism were studied by Steenken and Jovanovic (1997). The oxidation signal of guanine was used to determine the low level of RNA (Wang et al., 1995) and the single-stranded DNA at a carbon paste electrode (Wang et al., 1996). These studies were conducted through the adsorptive potentiometric stripping analysis.

Pournaghi-Azar et al. (2007) reported a label-free DNA biosensor to determine human interleukine-2 gene by guanine moiety oxidation signals before and after the hybridization of the probe with a target. Wang and Kawde (2001) developed a pencil-based renewable biosensor for label-free electrochemical detection of DNA hybridization. Also, DNA hybridization was monitored by guanine and adenine signals (Meric et al., 2002). Guanine and adenine oxidation signals were used to monitor the amplified real samples in a polymerase chain reaction (PCR) (Lucarelli et al., 2002; Masarik et al., 2003; Schülein et al., 2002). Direct electrochemical oxidation of DNA on a gold electrode is also reported in the literature (Ferapontova and Domínguez, 2003; Pang et al., 1995).

A label-free electrochemical genosensor was developed by Ozkan-Ariksoysal et al. (2008) to detect catechol-*o*-methyltransferase Val108/158Met polymorphism based on guanine inherent signals. In this research, the electrochemical detection of DNA hybridization was done by two different methods (Fig. 6.6). In method 1, the target DNA containing guanine was immobilized on the surface of a GCE and then hybridized with inosine-substituted probes. The hybridization process caused a decrease in the magnitude of the guanine oxidation peak current. In method 2, a GCE was modified by an inosine-substituted probe, and then, hybridization occurred between the probe and the DNA target. The hybridization process was confirmed by the appearance of the guanine signal. In another study, guanine and adenine

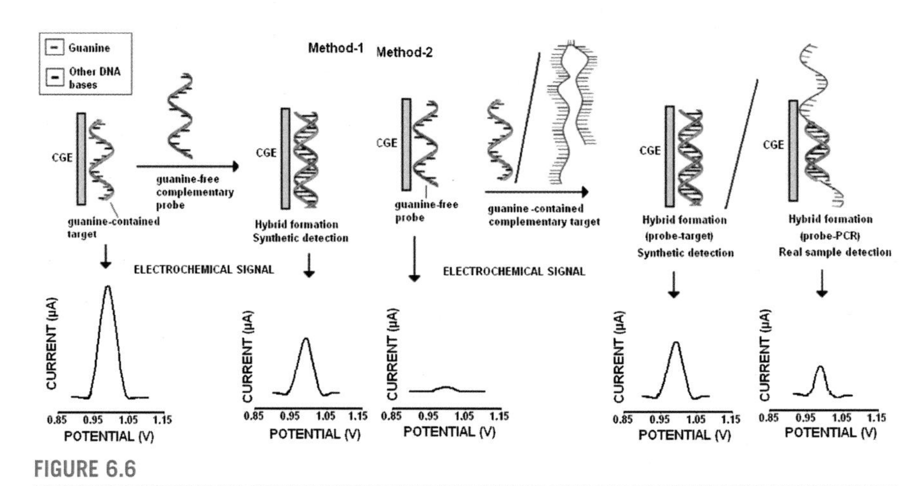

FIGURE 6.6

The schematic representation of a DNA hybridization biosensor using guanine-containing a target-modified GCE (method 1) and a guanine-free probe-modified GCE (method 2).

Reproduced from Ozkan-Ariksoysal, D., Tezcanli, B., Kosova, B., Ozsoz, M., 2008. Design of electrochemical biosensor systems for the detection of specific DNA sequences in PCR-amplified nucleic acids related to the catechol-O-methyltransferase Val108/158Met polymorphism based on intrinsic guanine signal. Analytical Chemistry 80(3), 588–596 with permission.

oxidation signals were used to detect the interaction between cisplatin and DNA on a reduced graphene oxide–modified GCE (Yardım et al., 2017). To determine human papillomavirus E6 gene on a pencil graphite electrode and by monitoring guanine signals, a DNA biosensor was developed (Campos-Ferreira et al., 2016). This approach needs no time-consuming external label and is highly sensitive. However, it requires a very high potential for direct DNA oxidation, and the background signal is too high in electrochemical processes.

6.5.1.2 DNA biosensors based on electrochemical impedance spectroscopy

As mentioned in Section 6.3.2, EIS is a label-free method of studying biological interactions. If the resistance of the electrode-solution interface is changed when the analyte is captured by the immobilized DNA at the electrode surface, the EIS method can be used to detect the resistance change. Various studies have been done on DNA biosensors based on the EIS method. The label-free detection of DNA hybridization by AC impedance was first reported by Lee and Shim (2001) and Li et al. (2002). The investigation of DNA and drug interactions by EIS has also been reported by Li et al. (2005).

So far, a rather good deal of research has been conducted to increase the sensitivity of DNA biosensors and enhance the density of probes and the orientation of DNA probes on the surface of electrodes. For example, Li et al. (2007) reported an amino-terminated G4 PAMAM dendrimer-modified gold electrode for DNA hybridization analysis by the EIS method. Ionic liquids were also used to develop DNA biosensors. In a study by Zhang et al. (2009b), a nanocomposite membrane containing a hydrophobic room-temperature ionic liquid (1-butyl-3-methylimidazolium hexafluorophosphate, BMIMPF6), nano-sized cerium oxide (CeO_2), and SWCNTs was developed for the electrochemical sensing of DNA immobilization and hybridization.

Nanomaterials have also been used in EIS biosensors to improve the characterization of these biosensors. Peng et al. (2006) used the EIS method to investigate the electrochemical detection of DNA hybridization signals amplified by nanoparticles. Li et al. (2008) developed a simple and sensitive EIS biosensor to determine thrombin. They used electrodeposited gold nanoparticles on the GCE surface to amplify the signals. The electrode surface also served as a platform for immobilization of thiolated aptamers. Ensafi et al. (2011) introduced an impedimetric biosensor to detect chronic lymphocytic leukemia based on a gold nanoparticles/gold-modified electrode. In addition, a label-free DNA-based biosensor was reported for simultaneous detection of Pb^{2+}, Ag^{+}, and Hg^{2+} using the EIS technique (Lin et al., 2011). As can be seen in Fig. 6.7, in this biosensor, Pb^{2+} interacts with G-rich DNA strands to form G-quadruplex, C-C mismatch interacts with Ag^{+} to form C-Ag^{+}-C complex, and Hg^{2+} interacts with T-T mismatch to form T-Hg^{2+}-T complex. When immobilized DNA interacts with Pb^{2+}, Ag^{+}, and Hg^{2+} individually, the charge transfer resistance (R_{ct}) decreases. To detect these ions simultaneously in a sample, one ion was measured but the other ions were masked with appropriate materials. EDTA was used to mask Pb^{2+} and Hg^{2+} for detecting Ag^{+}, cysteine was applied

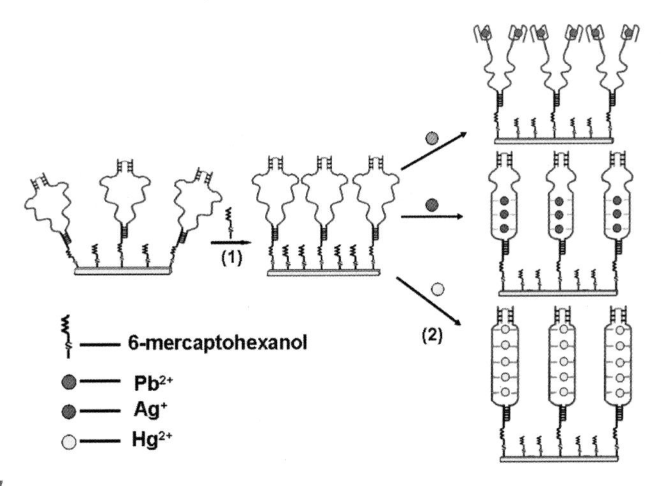

FIGURE 6.7

Schematic diagram of the fabrication of an impedimetric DNA biosensor for the simultaneous detection of Pb^{2+}, Ag^{+}, and Hg^{2+}.

Reprinted from Lin, Z., Li, X., Kraatz, H.-B., 2011. Impedimetric immobilized DNA-based sensor for simultaneous detection of Pb^{2+}, Ag^{+}, and Hg^{2+}. Analytical Chemistry 83(17), 6896–6901 with permission.

to mask Ag^{+} and Hg^{2+} for detecting Pb^{2+}, and a mixture of G-rich and C-rich DNA strands was used to mask Pb^{2+} and Ag^{+} for detecting Hg^{2+}. In another research, Asadzadeh-Firouzabadi and Zare (2017) developed a nanogenosensor based on cysteamine-capped gold nanoparticles to detect the miRNA (miR-25) related to lung cancer.

The above examples indicate that the EIS method needs no particular markers and is amenable to label-free operations. However, impedimetric biosensors are less repeatable than potentiometric and amperometric biosensors.

6.5.1.3 DNA biosensors based on field-effect transistors

Hybridization on the surface of an electrode leads to a change in the properties of the electrode surface. This change forms the fundament of how FET-based DNA biosensors function. Star et al. (2006) studied DNA hybridization using carbon nanotube network FETs. Gao et al. (2013) reported a novel strategy for the detection of hepatitis B virus DNA by means of a silicon nanowire FET combined with rolling circle amplification. A sensor was made by Wang and Jia (2018) for a graphene solution-gated field-effect DNA transistor. Also, using a directional transfer technique based on CVD-grown graphene, an ultrasensitive field-effect DNA transistor biosensor was developed by Zheng et al. (2015). Ion-sensitive FETs, as a type of potentiometric devices, have recently been used extensively in DNA-based biosensors owing to their high stability and small dimensions (Yano et al., 2014).

6.5.2 Electrochemical DNA biosensors based on an indirect detection method

In the indirect or label-based detection approach, DNA labeling can be done with various materials such as redox mediators (e.g. polypyridyl complexes of Ru(II) and Os(II)), redox active compounds (e.g. methylene blue (MB) and ferrocene), semiconductor nanomaterials (e.g. QDs of CdS, ZnS, PbS), and enzymes. Also, redox active species can interact with DNA in one of the modes of electrostatic interaction, intercalation binding, or grooves binding and play the role of a label.

6.5.2.1 Redox mediator–based electrochemical detection

Redox mediators oxidize the target DNA indirectly. Mediators such as bipyridine complexes of transition metals, e.g., Ru (II) and Os (II), make electrons move from DNA (mainly guanine residues for oxidation) to the electrode surface. Ru(II) is first oxidized to Ru(III) at the electrode surface at a potential near the guanine oxidation potential. Then, the guanine residues of the DNA reduce Ru(III) to Ru(II) again, and, thus, an electrocatalytic cycle is formed as shown in reactions (1) and (2) as well as in Fig. 6.8. In this process, when the concentration of Ru(II) is constant, the electrocatalytic signal is proportional to the concentration of guanine residues and, consequently, to the target DNA.

$$Ru(bpy)_3^{2+} \rightarrow Ru(bpy)_3^{3+} + e \qquad 6.1$$

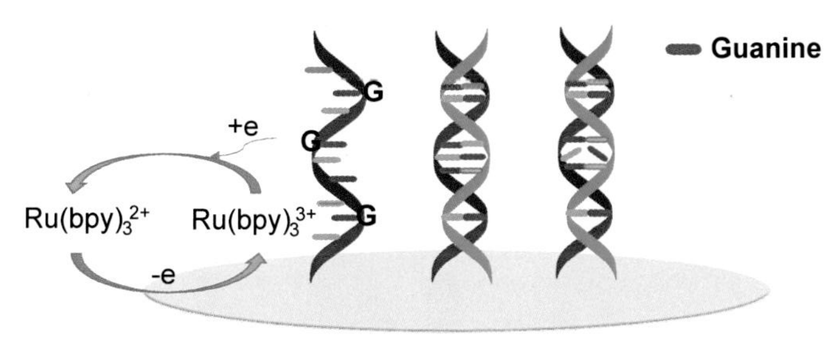

FIGURE 6.8

$Ru(bpy)_3^{2+}$ as a redox mediator for the electrocatalytic detection of DNA.

$$Ru(bpy)_3^{3+} + DNA \rightarrow DNA_{OX} + Ru(bpy)_3^{2+} \qquad 6.2$$

Johnston et al. (1995) applied this method to identify small disorders in the DNA structure and could detect single-stranded DNA (ssDNA), double helix DNA (dsDNA), and single-base mismatch. In another study, Napier et al. (1997) replaced guanine with inosine in the dsDNA probe, and, after hybridizing the DNA probe by the target DNA, they provided a high catalytic current. Moreover, Sistare et al. (1999) used $Ru(bpy)_3^{3+/2+}$ to study the electrocatalytic oxidation of guanine in DNA and oligonucleotides through cyclic voltammetry and chronoamperometry techniques. Detection of chemically induced DNA damage by $Ru(bpy)_3^{2+}$ was also reported by Zhou and Rusling (2001). However, the results of various studies attest the fact that the redox-mediated DNA oxidation method has high sensitivity and does not require complex instrumentations.

6.5.2.2 DNA-mediated charge transport platforms

In this approach, redox reporter compounds that are intrinsically and noncovalently associated with the double helix are used for DNA analysis by electrochemical methods. For the first time, Millan and Mikkelsen (1993) and Millan et al. (1994) used redox indicators for the electrochemical detection of DNA. Since then, various types of redox-active indicators have been used for electrochemical DNA biosensors. For example, researchers have used organic dyes, e.g., MB, $Co(phen)_3^{3+}$, $Os(bpy)_2Cl_2$, and $Ru(NH_3)_6^{3+}$ as metal complexes, and anticancer drugs such as daunomycin and doxorubicin. In these applications, DNA has been a mediator. Obtaining the variation in the electrochemical responses of the probe and the target is the main goal in the use of label-based electrochemical DNA biosensors. Good redox indicators should be able to interact with ssDNA and dsDNA possessing different affinities. Electrochemical responses, thus, change when the hybridization process is done and double helix is formed on the electrode surface. Furthermore, redox-active molecules should have a reversible electrochemical behavior, low overpotential as well as high selectivity and sensitivity. Various voltammetric methods such as linear scan voltammetry, square-wave voltammetry, and constant current chronopotentiometry can be used to detect the interaction of redox markers with

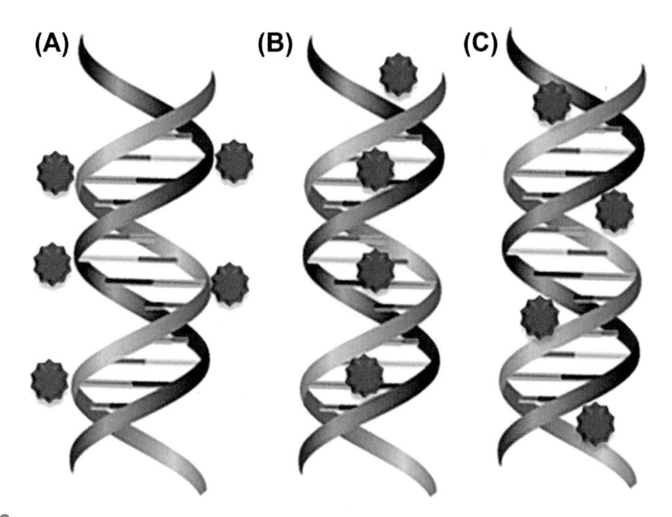

FIGURE 6.9

A schematic display of different modes of interaction between redox indicators and DNA: (A) electrostatic interaction, (B) intercalation interaction, and (C) minor and major groove interaction.

DNA. The interaction of redox indicators with DNA molecules is usually performed through three modes of intercalation, groove, and electrostatic (Fig. 6.9).

6.5.2.2.1 Electrostatic mode

In this method of electrochemical detection, some redox-active molecules, such as cationic metal complexes, can bind to DNA through the electrostatic interaction between the negatively charged phosphate backbone of DNA and the positive charge of redox-active molecules. Steel et al. (1998) used cobalt(III) tris-bipyridyl and ruthenium(III) hexaammine as cationic redox molecules to interact with the negative charge of the DNA phosphate backbone. They investigated the surface density of the DNA immobilized on a gold electrode as well as DNA hybridization by redox markers. Guo et al. (2013) also used $Ru(NH_3)_6^{3+}$ response to determine the DNA strands available at the electrode surface. The response to $Ru(NH_3)_6^{3+}$ increases with an increase in the concentration of the DNA target. The increase in the response is due to the increased presence of the phosphate group after hybridization, which results in the adsorption of more $Ru(NH_3)_6^{3+}$ into duplex DNA via an electrostatic interaction. In another study, Sheng-Zhen et al. (2012) used the electroactive indicator of $Co(phen)_3^{3+}$ to study DNA hybridization on a screen-printed electrode surface. $Co(phen)_3^{3+}$ interacted with ssDNA or dsDNA through electrostatic adsorption. The results indicated that, after the hybridization of the DNA probe with a target, the peak current of the cyclic voltammogram increased, indicating that $Co(phen)_3^{3+}$ has a higher affinity to dsDNA vs. a probe. Application of these cation metal complexes has provided an accurate, robust, and reliable method to determine DNA values on the electrode surface. Nevertheless, electrostatic attraction has some limitations such as nonspecific interaction and weak bonding.

6.5.2.2.2 Intercalation mode

Intercalators are a group of compounds that bind reversibly to DNA by incorporating a planar and aromatic substituent between the base pairs, simultaneously prolonging and unwinding the DNA helix. The intercalation mode needs two adjacent base pairs that separate from each other to create a binding pocket for the ligand. The π-π and hydrophobic interactions between the aromatic ring of the intercalator and the DNA bases serve as the intercalation binding force. Cationic molecules with planar aromatic rings are usually bound to DNA by intercalation. The positive charge can be a part of the ring system or lie on a substituent. Various compounds including metal ion complexes containing iron, cobalt, ruthenium, or osmium, organic dyes such as MB, daunomycin, or anthraquinone derivatives, and anticancer substances such as epirubicin chinomycine interact with the dsDNA structure via intercalation between bases pairs. This method is suitable for detecting the variations in DNA such as damages, mistakes, mismatches, and protein bindings. Lerman (1961) was the first one who proposed the intercalation mode in his pioneering research on mutagens. Piedade et al. (2002) investigated the interaction of adriamycin, a cancerostatic anthracycline antibiotic, with DNA using an electrochemical DNA biosensor. The results indicated that adriamycin intercalates into DNA and reacts with DNA bases. The researchers also studied DNA damage caused by the interactions between DNA and adriamycin. The utilization of ethidium bromide as a hybridization indicator was investigated by Castro et al. (2014). In this research, a DNA probe was immobilized on a modified graphite electrode and then hybridized by the hepatitis B virus DNA sequence. Differential pulse voltammograms (DPVs) of the ethidium bromide accumulated on the modified electrode surface were used to detect DNA hybridization. The results showed that, after the hybridization process occurs, DPVs current increase due to the intercalation of ethidium bromide at the G-C base pairs of dsDNA.

MB, a phenothiazine dye, has been used as a redox hybridization indicator for electrochemical DNA detection in various studies. For example, Kelley et al. (1997) investigated the electrochemistry of intercalated MB at the surface of a DNA-modified gold electrode. The Barton group also developed a biosensor to detect single-base mismatches that are needed for the routine monitoring of genetic mutations and diseases. The group used different redox-active intercalators noncovalently bound to DNA-modified surfaces (Kelley et al., 1999). Erdem et al. (2000) demonstrated MB as a redox hybridization indicator for the first time. Since then, MB has attracted a great deal of attention owing to its unique advantages such as great ability to identify the binding affinities between ssDNA and dsDNA as well as being environmentally friendly. MB can interact with DNA in three ways. As some studies have indicated, MB has a specific affinity for the guanine bases of ssDNA, whereas it has a much lower affinity for dsDNA. This is because dsDNA hinders the binding of MB to guanine and reduces the electrochemical signals

(Jin et al., 2007; Yang et al., 2002). Meanwhile, MB can intercalate between G-C pairing bases. It can also bind to DNA through an electrostatic interaction between MB and the negative charge of the phosphate groups in the inherent structure of DNA (Kara et al., 2002). The intercalation of MB into DNA is nonspecific and commonly occurs outside the DNA helix along its strands. Moreover, MB was used for the electrocatalytic reduction of $Fe(CN)_6^{3-}$ to sensitively detect single-base mismatches (Kelley et al., 1999).

Some redox intercalators such as ethidium bromide, osmium (II/III), and ruthenium (II/III) ions are toxic and expensive, which limits their application in biosensors. Therefore, many researches have tried to introduce new intercalators that are chemically stable, inexpensive, and environmentally friendly and have excellent redox reversibility. For example, Wong and Gooding (2006) used anthraquinonemonosulfonic acid as a redox-active intercalator to monitor real-time DNA hybridization. Also, the interaction of some flavonoid derivatives with ssDNA and dsDNA was studied by Kowalczyk et al. (2010) and Whittemore et al. (1999). Hematoxylin was introduced by Nasirizadeh et al. (2011) as an electroactive label to detect a specific DNA sequence of the human papilloma virus. Chlorogenic acid was used by Asadzadeh-Firouzabadi et al. (2016) as a novel electroactive indicator for the determination of the helicobacter pylori cagE gene. The step-by-step approach to fabricate a DNA hybridization biosensor to determine the helicobacter pylori cagE gene is presented in Fig. 6.10. The alkanethiol DNA probe was self-assembled on a gold electrode and then hybridized by the target DNA. Afterward,

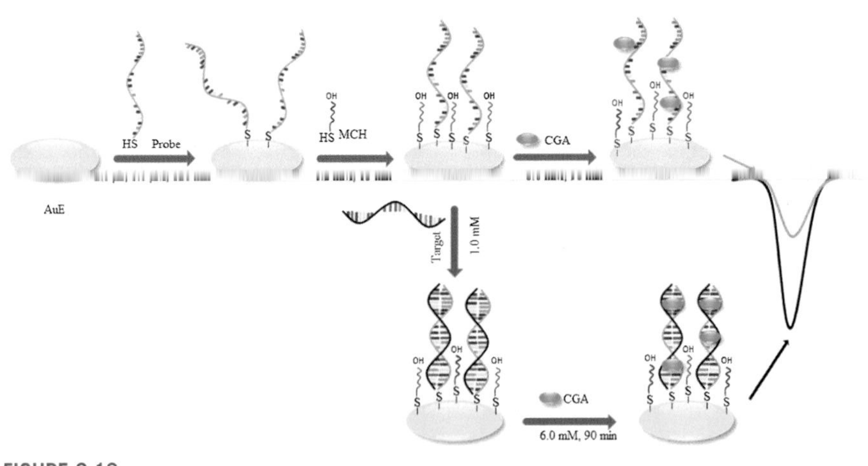

FIGURE 6.10

Schematic representation of the procedures of constructing DNA biosensors to determine *Helicobacter pylori* cagE gene.

Reproduced from Asadzadeh-Firouzabadi, A., Zare, H.R., Nasirizadeh, N., 2016. Electrochemical biosensor for detection of target DNA sequence and single-base mismatch related to helicobacter pylori using chlorogenic acid as hybridization indicator. Journal of the Electrochemical Society 163(3), B43–B48 with permission.

it was immersed into a chlorogenic acid solution, and the DPV technique was applied to measure the chlorogenic acid accumulated at the electrode surface.

Limoges group studied the real-time electrochemical monitoring of DNA amplification during PCR using different redox intercalators (Defever et al., 2011). They investigated various redox intercalators such as MB, Os $[(bpy)_2phen]^{2+}$, $Os[(bpy)_2DPPZ]^{2+}$, $Os[(4,4'\text{-dimethyl-bpy})_2DPPZ]^{2+}$, and Os $[(4,4'\text{-diamino-bpy})_2DPPZ]^{2+}$ (with bpy = 2,2′-bipyridine, phen = phenanthroline, and DPPZ = dipyrido[3,2-a:2′,3′-c]phenazine). The results indicated that, among these probes, $Os[(bpy)_2DPPZ]^{2+}$ is able to interact strongly with dsDNA without significantly inhibiting PCR.

A DNA-mediated charge transfer provides special conditions for mutation detection. It serves as an electrocatalytic assay and a completely different approach to DNA sensing.

6.5.2.2.3 Groove mode

Some indicators can interact with DNA bases by forming hydrogen bonds or van der Waals bonds in a major or minor groove of DNA. When two DNA strands are coiled together, they form two different grooves in the parallel backbone of the dsDNA structure, namely the major groove and the minor groove. The major groove is wider and deeper than the minor one. The bindings of ligand molecules in the two grooves are different. Proteins and large molecules such as oligonucleotides usually bind to the major groove, whereas small molecules can bind to both grooves and preferably to the minor groove. Because adenine-thymine (AT)-rich regions are narrower and slightly deeper than cytosine-guanine (CG)-rich regions, many minor groove binders prefer to bind to AT-rich sequences. For therapeutics, the groove-binding mode is preferred to intercalative binding. This is due to the higher specificity and, generally, lower toxicity of groove binders. Some redox indicators bind to the minor and major grooves of DNA double helix, which can be used in electrochemical DNA biosensors. For example, the ferrocene redox marker (Fc) binds to the major grooves of dsDNA (Ribeiro Teles et al., 2007), or Hoechst 33258 drug binds to dsDNA in the minor groove (Han et al., 2005). In this regard, Baraldi et al. (2004) reviewed DNA minor groove binders as potential antitumors and antimicrobial agents. They stated that DNA minor groove binders form an important class of compounds for cancer treatment. Moreover, minor groove binding agents are not generally mutagenic for therapy. Wang et al. (2011) studied the effect of distamycin (a polyamide-antibiotic), as a minor groove binder, on the conformational changes in three target DNA sequences with alternating AT sites of different sizes. They found that distamycin binds strongly to five or six AT base pairs of dsDNA through the minor groove binding mode.

6.5.2.3 Enzyme label–based electrochemical detection

Enzymes are one of the labels commonly used in electrochemical biosensors. They are favored owing to their ability to amplify electrochemical signals and, consequently, increase the sensitivity of DNA biosensors. They are proteins that serve as highly selective catalysts to increase the rate of reactions. An enzymatic activity can be transduced into electrochemical signals by the catalytic conversion of a substrate into an electroactive product or by redox-mediated electrocatalytic conversion. Various redox-active enzyme labels such as HRP, glucose dehydrogenase (pyrroloquinoline-quinone), alkaline phosphatase (ALP), glucose oxidase (GOx), bilirubin oxidase, and esterase have been used to detect low values of nucleic acids.

Because signal amplification by enzymatic reactions is not adequately high, enzymatic reactions have been combined by an extra signal amplification process such as the use of multiple enzymes per binding events (Munge et al., 2005), recycling or accumulating the reaction product (Dominguez et al., 2004), and coupling of two enzyme labels via substrate or cosubstrate regeneration (Nebling et al., 2004). Redox cycling is a process that repeatedly generates species of signal producers or signal consumers (i.e., molecules or electrons) in the presence of reversible redox species. This process can be simply coupled with enzymatic amplification (Akanda and Ju, 2017; Liu et al., 2008; Walter et al., 2011). HRP (Dong et al., 2011, 2012), GOx (Won et al., 2008), and ALP (Xia et al., 2015) are the enzymatic labels typically used in connection with the electrochemical screening of biocatalytic reaction products in enzyme-amplified systems. Among them, ALP is a nonoxidoreductase enzyme, which has no electrochemical activity when it hydrolyzes the substrate to a redox product. An enzyme label–based electrochemical technique is used to detect DNA hybridization by a direct or sandwich mode. Wang et al. (2002) reported a dual enzyme electrochemical coding technique to detect DNA hybridization. In their work, two enzyme labels (ALP and β-galactosidase) were used to identify two DNA target signals.

Haronikova et al. (2015) used ALP in an enzyme-linked electrochemical process to produce an electroactive species, 1-naphthol, from a nonelectroactive species, 1-naphthyl phosphate, and then detect it by a voltammetry method. Rochelet-Dequaire et al. (2009) used AP as an enzyme label. They amplified its electrochemical response by a diaphorase (DI) secondary enzyme to sensitively determinate an amplified 406-base pair human cytomegalovirus DNA sequence.

An enzyme-based sandwich-type electrochemical biosensor was developed by Zhang et al. (2009a) to detect the breast cancer–associated BRCA-1 gene based on HRP as an enzyme label. The researchers amplified the response and electrodeposited the $Os^{2+/3+}$ redox polymer as an electron transfer mediator. An electrochemical-based DNA-sensing device was introduced to detect Ebola virus DNA through an enzyme amplification process (Ilkhani and Farhad, 2018). To prepare this biosensor, a thiolated probe was immobilized on the surface of a gold screen–printed electrode. Then, hybridization occurred between the capture probe strand DNA and the biotinylated target strand DNA. In the electrochemical detection step, the biotinylated target strand DNA interacted with streptavidin alkaline phosphatase enzyme via a biotin-streptavidin conjugation bond (Fig. 6.11).

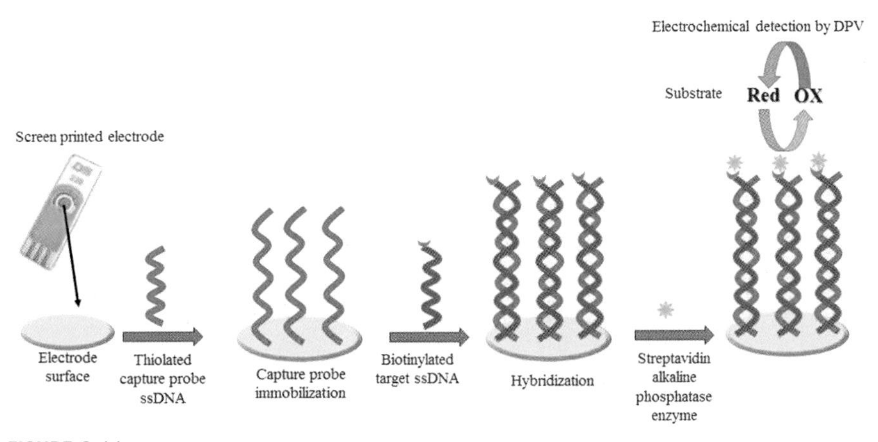

FIGURE 6.11

Schematic representation of different steps for the fabrication of a DNA biosensor for Ebola virus detection.

Reproduced from Ilkhani, H., Farhad, S., 2018. A novel electrochemical DNA biosensor for Ebola virus detection. Analytical Biochemistry 557, 151–155. with permission.

6.5.2.4 Nanoparticles label–based electrochemical detection

In recent decades, nanotechnology has attracted much attention for its capability in designing and development of biosensors. Using nanomaterials in the development of biosensors can improve the properties of those sensors. Nanomaterials can be used for modification of substrates or as labels in the design of biosensors. Moreover, there are new signal transduction approaches that are based on the use of nanomaterials in biosensors. Detection of nanoparticle label–based DNA is done by two approaches. The first is the direct oxidation of nanoparticle labels at a modified electrode surface, and the second is the dissolution of nanoparticles under oxidative treatment and then measurement of the dissolved nanoparticles by the anodic stripping voltammetry method. Because of their excellent electroactivity and easy bioconjugation, metallic and metallic oxide nanoparticles are used extensively in DNA biosensors as labels. Nanoparticle-based assays have significant advantages over conventional diagnostic systems, such as higher sensitivity and selectivity. Various studies have been conducted on the sensitive detection of nanoparticle tags in bioassays. One of the best nanoparticles used in the development of biosensors is gold nanoparticles (AuNPs). This is owing to their unique properties including high surface-to-volume ratio, excellent conductivity, high electrocatalytic activity, and good stability (Authier et al., 2001; Kerman et al., 2004). Rasheed and Sandhyarani (2015) developed an electrochemical genosensor to detect breast cancer 1 gene using functionalized gold nanoparticles as electrochemical labels. In this biosensing system, a probe was immobilized on the surface of a modified gold electrode and then hybridized by the target DNA. Afterward, the free end of the target

DNA was hybridized with AuNPs-labeled reporter DNA, and the hybridization process was measured using chronoamperometry and EIS. Because of the high conductivity of AuNPs, it was possible to report the attomolar concentration of the target DNA reported by this biosensor because of the high conductivity of AuNPs. Moreover, AuNPs can act as carriers for other electroactive labels (Cui et al., 2015; Wang et al., 2008; Zhang et al., 2006). Pinijsuwan et al. (2008) used AuNPs-functionalized latex signal amplification for the subfemtomolar detection of DNA hybridization. They immobilized the target DNA on a screen-printed carbon electrode, and then, hybridization was carried out with *Escherichia coli* DNA probes modified with biotin. In this way, AuNPs-streptavidin complexes were attached to the latex microspheres by a streptavidin-biotin bond. To detect the probe hybridization, Au^{3+} ions were measured by anodic stripping voltammetry after the AuNPs were chemically dissolved.

Silver nanoparticles (AgNPs) can also be used to generate an electrochemical signal. Mehrgardi and Ahangar (2011) used silver nanoparticles as reporter labels in DNA hybridization directly. In that research, the target strand DNA was immobilized on the surface of a gold electrode, and, after hybridization by the probe strand DNA, AgNPs were immobilized on the top of a recognition layer. Their oxidation signals were followed by DPV. The DNA target and the single-base mismatch DNA were detectable by this biosensor.

In another work, Asadzadeh-Firouzabadi and Zare (2018) developed a genosensor to detect miRNA related to lung cancer using AgNPs/SWCNTs nanohybrid as a redox label. They immobilized the DNA probe on the GCE surface and then hybridized the target miRNA. AgNPs/SWCNTs nanohybrid was incubated on the genosensor surface before and after the probe hybridization with the target. The difference in the current of the differential pulse voltammograms of the AgNPs/SWCNTs label before and after the probe was hybridized with the miRNA target was used for the quantitative determination of the miRNA target. According to the researchers, the SWCNTs in an AgNPs/SWCNTs nanohybrid interact with the probe strand, and the AgNPs in a nanohybrid generate electroanalytical signals.

QDs are also widely used as DNA tags because of their remarkable properties. For example, CdS QDs were used as an electroactive label for miR-16 detection (Wang et al., 2013). In that work, the locked nucleic acid-molecular beacon (LNA-MB) probe was immobilized on the surface of an Au electrode. Then, the LNA-MB-modified electrode reacted with miR-16, and MB was opened. Thereafter, the target miRNA was hybridized with the opened loop of the LNA-MB probe, and the RAC template was bound to the released primer. The RCA reaction started upon the introduction of DNA polymerase, and the products of the reaction, i.e., thousands of repeated DNA sequences, were hybridized with the QDs-labeled probe. Finally, anodic stripping voltammetry was performed to determine the Cd^{2+} ions released from the dissolved CdS QDs. The miR-16 detection by this biosensor is schematically presented in Fig. 6.12.

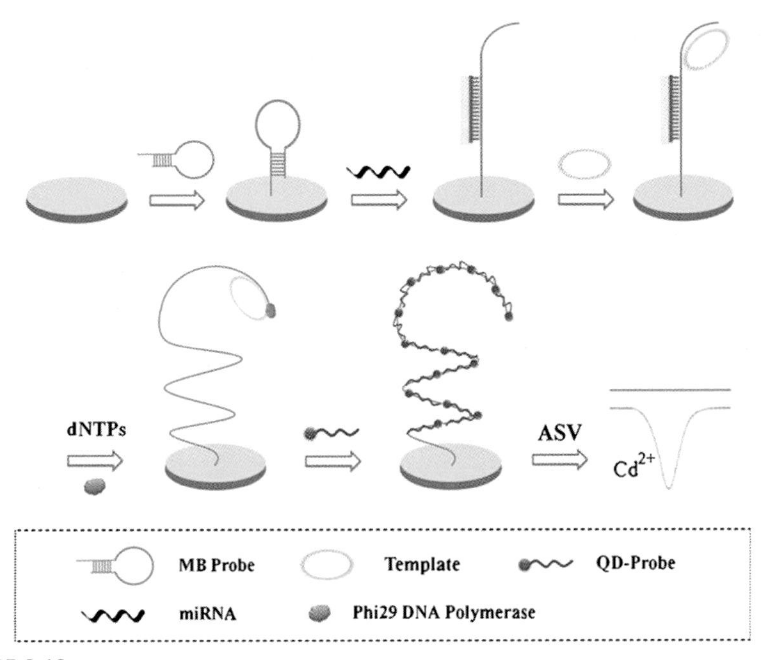

FIGURE 6.12

Schematic diagram of a biosensor designed for miRNA detection.

Reproduced from Wang, D., Hu, L., Zhou, H., Abdel-Halim, E., Zhu, J.-J., 2013. Molecular beacon structure mediated rolling circle amplification for ultrasensitive electrochemical detection of microRNA based on quantum dots tagging. Electrochemistry Communications 33, 80–83 with permission.

6.6 Conclusion

The signals produced on biosensors can be monitored by direct and indirect detection methods. The direct detection mode is based on the detection of targets without using a label, whereas the indirect detection mode applies a label to detect targets. Simplicity, high speed, and low cost are the advantages of the direct detection approach, but the indirect detection approach is complicated and time-consuming owing to the use of labels. It should not be ignored, however, that the indirect detection approach is very sensitive and provides lower detection limits than the direct detection method does.

References

Abbaspour, A., Noori, A., 2008. Electrochemical studies on the oxidation of guanine and adenine at cyclodextrin modified electrodes. Analyst 133 (12), 1664–1672.

Akanda, M.R., Ju, H., 2017. An integrated redox cycling for electrochemical enzymatic signal enhancement. Analytical Chemistry 89 (24), 13480–13486.

Asadzadeh-Firouzabadi, A., Zare, H.R., 2017. Application of cysteamine-capped gold nanoparticles for early detection of lung cancer-specific miRNA (miR-25) in human blood plasma. Analytical Methods 9 (25), 3852–3861.

Asadzadeh-Firouzabadi, A., Zare, H.R., 2018. Preparation and application of AgNPs/SWCNTs nanohybrid as an electroactive label for sensitive detection of miRNA related to lung cancer. Sensors and Actuators B: Chemical 260, 824–831.

Asadzadeh-Firouzabadi, A., Zare, H.R., Nasirizadeh, N., 2016. Electrochemical biosensor for detection of target DNA sequence and single-base mismatch related to helicobacter pylori using chlorogenic acid as hybridization indicator. Journal of the Electrochemical Society 163 (3), B43–B48.

Authier, L., Grossiord, C., Brossier, P., Limoges, B., 2001. Gold nanoparticle-based quantitative electrochemical detection of amplified human cytomegalovirus DNA using disposable microband electrodes. Analytical Chemistry 73 (18), 4450–4456.

Baraldi, P.G., Bovero, A., Fruttarolo, F., Preti, D., Tabrizi, M.A., Pavani, M.G., Romagnoli, R., 2004. DNA minor groove binders as potential antitumor and antimicrobial agents. Medicinal Research Reviews 24 (4), 475–528.

Brabec, V., 1981. Nucleic acid analysis by voltammetry at carbon electrodes. Journal of Electroanalytical Chemistry and Interfacial Electrochemistry 128, 437–449.

Brabec, V., Dryhurst, G., 1978. Electrochemical oxidation of polyadenylic acid at graphite electrodes. Journal of Electroanalytical Chemistry and Interfacial Electrochemistry 91 (2), 219–229.

Butt, H.-J., 1996. A sensitive method to measure changes in the surface stress of solids. Journal of Colloid and Interface Science 180 (1), 251–260.

Campos-Ferreira, D.S., Souza, E.V., Nascimento, G.A., Zanforlin, D.M., Arruda, M.S., Beltrao, M.F., Melo, A.L., Bruneska, D., Lima-Filho, J.L., 2016. Electrochemical DNA biosensor for the detection of human papillomavirus E6 gene inserted in recombinant plasmid. Arabian Journal of Chemistry 9 (3), 443–450.

Castro, A.C.H., Franca, E.G., de Paula, L.F., Soares, M.M., Goulart, L.R., Madurro, J.M., Brito-Madurro, A.G., 2014. Preparation of genosensor for detection of specific DNA sequence of the hepatitis B virus. Applied Surface Science 314, 273–279.

Cui, H.-F., Xu, T.-B., Sun, Y.-L., Zhou, A.-W., Cui, Y.-H., Liu, W., Luong, J.H., 2015. Hairpin DNA as a biobarcode modified on gold nanoparticles for electrochemical DNA detection. Analytical Chemistry 87 (2), 1358–1365.

Defever, T., Druet, M., Evrard, D., Marchal, D., Limoges, B., 2011. Real-time electrochemical PCR with a DNA intercalating redox probe. Analytical Chemistry 83 (5), 1815–1821.

Dominguez, E., Rincon, O., Narvaez, A., 2004. Electrochemical DNA sensors based on enzyme dendritic architectures: an approach for enhanced sensitivity. Analytical Chemistry 76 (11), 3132–3138.

Dong, H., Zhu, Z., Ju, H., Yan, F., 2012. Triplex signal amplification for electrochemical DNA biosensing by coupling probe-gold nanoparticles–graphene modified electrode with enzyme functionalized carbon sphere as tracer. Biosensors and Bioelectronics 33 (1), 228–232.

Dong, X.-Y., Mi, X.-N., Wang, B., Xu, J.-J., Chen, H.-Y., 2011. Signal amplification for DNA detection based on the HRP-functionalized Fe_3O_4 nanoparticles. Talanta 84 (2), 531–537.

Ensafi, A.A., Taei, M., Rahmani, H., Khayamian, T., 2011. Sensitive DNA impedance biosensor for detection of cancer, chronic lymphocytic leukemia, based on gold nanoparticles/gold modified electrode. Electrochimica Acta 56 (24), 8176–8183.

Erdem, A., Kerman, K., Meric, B., Akarca, U.S., Ozsoz, M., 2000. Novel hybridization indicator methylene blue for the electrochemical detection of short DNA sequences related to the hepatitis B virus. Analytica Chimica Acta 422 (2), 139–149.

Ferapontova, E.E., Dominguez, E., 2003. Direct electrochemical oxidation of DNA on polycrystalline gold electrodes. Electroanalysis 15 (7), 629–634.

Gao, A., Zou, N., Dai, P., Lu, N., Li, T., Wang, Y., Zhao, J., Mao, H., 2013. Signal-to-noise ratio enhancement of silicon nanowires biosensor with rolling circle amplification. Nano Letters 13 (9), 4123–4130.

Gao, F., Du, L., Zhang, Y., Zhou, F., Tang, D., 2016. A sensitive sandwich-type electrochemical aptasensor for thrombin detection based on platinum nanoparticles decorated carbon nanocages as signal labels. Biosensors and Bioelectronics 86, 185–193.

Gauglitz, G., Homola, J., 2015. Direct optical detection. Analytical and Bioanalytical Chemistry 407, 3881–3882.

Gong, X., Zhao, R., Yu, X., 2018. A 3-D-silicon nanowire FET biosensor based on a novel hybrid process. Journal of Microelectromechanical Systems 27 (2), 164–170.

Grate, J.W., Warner, M.G., Ozanich Jr., R.M., Miller, K.D., Colburn, H.A., Dockendorff, B., Antolick, K.C., Anheier Jr., N.C., Lind, M.A., Lou, J., 2009. Renewable surface fluorescence sandwich immunoassay biosensor for rapid sensitive botulinum toxin detection in an automated fluidic format. Analyst 134 (5), 987–996.

Guo, Y., Su, S., Wei, X., Zhong, Y., Su, Y., Huang, Q., Fan, C., He, Y., 2013. A silicon-based electrochemical sensor for highly sensitive, specific, label-free and real-time DNA detection. Nanotechnology 24 (44), 444012.

Gupta, V.K., Yola, M.L., Qureshi, M.S., Solak, A.O., Atar, N., Ustundag, Z., 2013. A novel impedimetric biosensor based on graphene oxide/gold nanoplatform for detection of DNA arrays. Sensors and Actuators B: Chemical 188, 1201–1211.

Gutés, A., Lee, B.-Y., Carraro, C., Mickelson, W., Lee, S.-W., Mabouduan, R., 2013. Impedimetric graphene-based biosensors for the detection of polybrominated diphenyl ethers. Nanoscale 5 (13), 6048–6052.

Han, F., Taulier, N., Chalikian, T.V., 2005. Association of the minor groove binding drug Hoechst 33258 with d(CGCGAATTCGCG)2: volumetric, calorimetric, and spectroscopic characterizations. Biochemistry 44 (28), 9785–9794.

Haronikova, L., Spacek, J., Plucnara, M., Horakova, P., Pivonkova, H., Havran, L., Erdem, A., Fojta, M., 2015. Enzyme-linked electrochemical detection of DNA fragments amplified by PCR in the presence of a biotinylated deoxynucleoside triphosphate using disposable pencil graphite electrodes. Monatshefte für Chemie 146 (5), 849–855.

Homola, J., Dostalek, J., Chen, S., Rasooly, A., Jiang, S., Yee, S.S., 2002. Spectral surface plasmon resonance biosensor for detection of staphylococcal enterotoxin B in milk. International Journal of Food Microbiology 75 (1–2), 61–69.

Ilkhani, H., Farhad, S., 2018. A novel electrochemical DNA biosensor for Ebola virus detection. Analytical Biochemistry 557, 151–155.

Jakob, M.H., Dong, B., Gutsch, S., Chatelle, C., Krishnaraja, A., Weber, W., Zacharias, M., 2017. Label-free S_nO_2 nanowire FET biosensor for protein detection. Nanotechnology 28 (24), 245503.

Jin, Y., Yao, X., Liu, Q., Li, J., 2007. Hairpin DNA probe based electrochemical biosensor using methylene blue as hybridization indicator. Biosensors and Bioelectronics 22 (6), 1126–1130.

Johansson, A., Blagoi, G., Boisen, A., 2006. Polymeric cantilever-based biosensors with integrated readout. Applied Physics Letters 89 (17), 173505.

Johnston, D.H., Glasgow, K.C., Thorp, H.H., 1995. Electrochemical measurement of the solvent accessibility of nucleobases using electron transfer between DNA and metal complexes. Journal of the American Chemical Society 117 (35), 8933–8938.

Joo, J., Kwon, D., Yim, C., Jeon, S., 2012. Highly sensitive diagnostic assay for the detection of protein biomarkers using microresonators and multifunctional nanoparticles. ACS Nano 6 (5), 4375–4381.

Kara, P., Kerman, K., Ozkan, D., Meric, B., Erdem, A., Ozkan, Z., Ozsoz, M., 2002. Electrochemical genosensor for the detection of interaction between methylene blue and DNA. Electrochemistry Communications 4 (9), 705–709.

Kausaite-Minkstimiene, A., Ramanaviciene, A., Ramanavicius, A., 2009. Surface plasmon resonance biosensor for direct detection of antibodies against human growth hormone. Analyst 134 (10), 2051–2057.

Kaushik, A., Solanki, P.R., Kaneto, K., Kim, C., Ahmad, S., Malhotra, B.D., 2010. Nanostructured iron oxide platform for impedimetric cholesterol detection. Electroanalysis 22 (10), 1045–1055.

Kelley, S.O., Barton, J.K., Jackson, N.M., Hill, M.G., 1997. Electrochemistry of methylene blue bound to a DNA-modified electrode. Bioconjugate Chemistry 8 (1), 31–37.

Kelley, S.O., Boon, E.M., Barton, J.K., Jackson, N.M., Hill, M.G., 1999. Single-base mismatch detection based on charge transduction through DNA. Nucleic Acids Research 27 (24), 4830–4837.

Kerman, K., Saito, M., Morita, Y., Takamura, Y., Ozsoz, M., Tamiya, E., 2004. Electrochemical coding of single-nucleotide polymorphisms by monobase-modified gold nanoparticles. Analytical Chemistry 76 (7), 1877–1884.

Kiening, M., Niessner, R., Weller, M.G., 2005. Microplate-based screening methods for the efficient development of sandwich immunoassays. Analyst 130 (12), 1580–1588.

Kowalczyk, A., Nowicka, A.M., Jurczakowski, R., Niedzialkowski, P., Ossowski, T., Stojek, Z., 2010. New anthraquinone derivatives as electrochemical redox indicators for the visualization of the DNA hybridization process. Electroanalysis 22 (1), 49–59.

Kuzin, Y., Porfireva, A., Stepanova, V., Evtugyn, V., Stoikov, I., Evtugyn, G., Hianik, T., 2015. Impedimetric detection of DNA damage with the sensor based on silver nanoparticles and neutral red. Electroanalysis 27 (12), 2800–2808.

Lee, T.-Y., Shim, Y.-B., 2001. Direct DNA hybridization detection based on the oligonucleotide-functionalized conductive polymer. Analytical Chemistry 73 (22), 5629–5632.

Lerman, L., 1961. Structural considerations in the interaction of DNA and acridines. Journal of Molecular Biology 3 (1), 18–30.

Li, A., Yang, F., Ma, Y., Yang, X., 2007. Electrochemical impedance detection of DNA hybridization based on dendrimer modified electrode. Biosensors and Bioelectronics 22 (8), 1716–1722.

Li, C.-Z., Liu, Y., Luong, J.H., 2005. Impedance sensing of DNA binding drugs using gold substrates modified with gold nanoparticles. Analytical Chemistry 77 (2), 478–485.

Li, C., Shi, S., Maracas, G., Choong, V.-e., 2002. Reporterless Genosensors Using Electrical Detection Methods. Google Patents.

Li, X., Shen, L., Zhang, D., Qi, H., Gao, Q., Ma, F., Zhang, C., 2008. Electrochemical impedance spectroscopy for study of aptamer–thrombin interfacial interactions. Biosensors and Bioelectronics 23 (11), 1624–1630.

Lin, C., Guo, Y., Zhao, M., Sun, M., Luo, F., Guo, L., Qiu, B., Lin, Z., Chen, G., 2017a. Highly sensitive colorimetric immunosensor for influenza virus H5N1 based on enzyme-encapsulated liposome. Analytica Chimica Acta 963, 112–118.

Lin, T.-C., Li, Y.-S., Chiang, W.-H., Pei, Z., 2017b. A high sensitivity field effect transistor biosensor for methylene blue detection utilize graphene oxide nanoribbon. Biosensors and Bioelectronics 89, 511–517.

Lin, Z., Li, X., Kraatz, H.-B., 2011. Impedimetric immobilized DNA-based sensor for simultaneous detection of Pb^{2+}, Ag^{+}, and Hg^{2+}. Analytical Chemistry 83 (17), 6896–6901.

Liu, G., Wan, Y., Gau, V., Zhang, J., Wang, L., Song, S., Fan, C., 2008. An enzyme-based E-DNA sensor for sequence-specific detection of femtomolar DNA targets. Journal of the American Chemical Society 130 (21), 6820–6825.

Liu, H., Piret, G., Sieber, B., Laureyns, J., Roussel, P., Xu, W., Boukherroub, R., Szunerits, S., 2009. Electrochemical impedance spectroscopy of ZnO nanostructures. Electrochemistry Communications 11 (5), 945–949.

Liu, Q., Liu, Y., Wu, F., Cao, X., Li, Z., Alharbi, M., Abbas, A.N., Amer, M.R., Zhou, C., 2018. Highly sensitive and wearable In_2O_3 nanoribbon transistor biosensors with integrated on-chip gate for glucose monitoring in body fluids. ACS Nano 12 (2), 1170–1178.

Lucarelli, F., Marrazza, G., Palchetti, I., Cesaretti, S., Mascini, M., 2002. Coupling of an indicator-free electrochemical DNA biosensor with polymerase chain reaction for the detection of DNA sequences related to the apolipoprotein E. Analytica Chimica Acta 469 (1), 93–99.

Marques, I., Pinto da Costa, J., Justino, C., Santos, P., Duarte, K., Freitas, A., Cardoso, S., Duarte, A., Rocha-Santos, T., 2017. Carbon nanotube field effect transistor biosensor for the detection of toxins in seawater. International Journal of Environmental Analytical Chemistry 97 (7), 597–605.

Masarik, M., Kizek, R., Kramer, K.J., Billova, S., Brazdova, M., Vacek, J., Bailey, M., Jelen, F., Howard, J.A., 2003. Application of Avidin– Biotin technology and adsorptive transfer stripping square-wave voltammetry for detection of DNA hybridization and avidin in transgenic avidin maize. Analytical Chemistry 75 (11), 2663–2669.

Mehrgardi, M.A., Ahangar, L.E., 2011. Silver nanoparticles as redox reporters for the amplified electrochemical detection of the single base mismatches. Biosensors and Bioelectronics 26 (11), 4308–4313.

Meric, B., Kerman, K., Ozkan, D., Kara, P., Ozsoz, M., 2002. Indicator-free electrochemical DNA biosensor based on adenine and guanine signals. Electroanalysis 14 (18), 1245–1250.

Millan, K.M., Mikkelsen, S.R., 1993. Sequence-selective biosensor for DNA based on electroactive hybridization indicators. Analytical Chemistry 65 (17), 2317–2323.

Millan, K.M., Saraullo, A., Mikkelsen, S.R., 1994. Voltammetric DNA biosensor for cystic fibrosis based on a modified carbon paste electrode. Analytical Chemistry 66 (18), 2943–2948.

Munge, B., Liu, G., Collins, G., Wang, J., 2005. Multiple enzyme layers on carbon nanotubes for electrochemical detection down to 80 DNA copies. Analytical Chemistry 77 (14), 4662–4666.

Napier, M.E., Loomis, C.R., Sistare, M.F., Kim, J., Eckhardt, A.E., Thorp, H.H., 1997. Probing biomolecule recognition with electron transfer: electrochemical sensors for DNA hybridization. Bioconjugate Chemistry 8 (6), 906–913.

Nasirizadeh, N., Zare, H.R., Pournaghi-Azar, M.H., Hejazi, M.S., 2011. Introduction of hematoxylin as an electroactive label for DNA biosensors and its employment in detection of target DNA sequence and single-base mismatch in human papilloma virus corresponding to oligonucleotide. Biosensors and Bioelectronics 26 (5), 2638–2644.

Nebling, E., Grunwald, T., Albers, J., Schäfer, P., Hintsche, R., 2004. Electrical detection of viral DNA using ultramicroelectrode arrays. Analytical Chemistry 76 (3), 689–696.

Oliveira-Brett, A.M., Diculescu, V., Piedade, J., 2002a. Electrochemical oxidation mechanism of guanine and adenine using a glassy carbon microelectrode. Bioelectrochemistry 55 (1–2), 61–62.

Oliveira-Brett, A.M., Silva, L.A.d., Brett, C.M., 2002b. Adsorption of guanine, guanosine, and adenine at electrodes studied by differential pulse voltammetry and electrochemical impedance. Langmuir 18 (6), 2326–2330.

Ozkan-Ariksoysal, D., Tezcanli, B., Kosova, B., Ozsoz, M., 2008. Design of electrochemical biosensor systems for the detection of specific DNA sequences in PCR-amplified nucleic acids related to the catechol-O-methyltransferase Val108/158Met polymorphism based on intrinsic guanine signal. Analytical Chemistry 80 (3), 588–596.

Paleček, E., 1980. Reaction of nucleic acid bases with the mercury electrode: determination of purine derivatives at submicromolar concentrations by means of cathodic stripping voltammetry. Analytical Biochemistry 108 (1), 129–138.

Pang, D.W., Qi, Y.P., Wang, Z.L., Cheng, J.K., Wang, J.W., 1995. Electrochemical oxidation of DNA at a gold microelectrode. Electroanalysis 7 (8), 774–777.

Patil, M., Ramanathan, M., Shanov, V., Kumta, P.N., 2015. Carbon nanotube-based impedimetric biosensors for bone marker detection. Advances in Materials Science for Environmental and Energy Technologies IV: Ceramic Transactions 253, 187–193.

Peng, H., Soeller, C., Cannell, M.B., Bowmaker, G.A., Cooney, R.P., Travas-Sejdic, J., 2006. Electrochemical detection of DNA hybridization amplified by nanoparticles. Biosensors and Bioelectronics 21 (9), 1727–1736.

Piedade, J., Fernandes, I., Oliveira-Brett, A., 2002. Electrochemical sensing of DNA–adriamycin interactions. Bioelectrochemistry 56 (1–2), 81–83.

Pinijsuwan, S., Rijiravanich, P., Somasundrum, M., Surareungchai, W., 2008. Sub-femtomolar electrochemical detection of DNA hybridization based on latex/gold nanoparticle-assisted signal amplification. Analytical Chemistry 80 (17), 6779–6784.

Pournaghi-Azar, M.H., Hejazi, M.S., Alipour, E., 2007. Detection of human interleukine-2 gene using a label-free electrochemical DNA hybridization biosensor on the basis of a non-inosine substituted probe. Electroanalysis 19 (4), 466–472.

Rasheed, P.A., Sandhyarani, N., 2015. Attomolar detection of BRCA1 gene based on gold nanoparticle assisted signal amplification. Biosensors and Bioelectronics 65, 333–340.

Ribeiro Teles, F.R., França Dos Prazeres, D.M., De Lima-Filho, J.L., 2007. Electrochemical detection of a dengue-related oligonucleotide sequence using ferrocenium as a hybridization indicator. Sensors 7 (11), 2510–2518.

Rochelet-Dequaire, M., Djellouli, N., Limoges, B., Brossier, P., 2009. Bienzymatic-based electrochemical DNA biosensors: a way to lower the detection limit of hybridization assays. Analyst 134 (2), 349–353.

Rushworth, J.V., Ahmed, A., Griffiths, H.H., Pollock, N.M., Hooper, N.M., Millner, P.A., 2014. A label-free electrical impedimetric biosensor for the specific detection of Alzheimer's amyloid-beta oligomers. Biosensors and Bioelectronics 56, 83–90.

Savran, C.A., Knudsen, S.M., Ellington, A.D., Manalis, S.R., 2004. Micromechanical detection of proteins using aptamer-based receptor molecules. Analytical Chemistry 76 (11), 3194–3198.

Schulein, J., Graßl, B., Krause, J., Schulze, C., Kugler, C., Muller, P., Bertling, W., Hassmann, J., 2002. Solid composite electrodes for DNA enrichment and detection. Talanta 56 (5), 875–885.

Shekari, Z., Zare, H.R., Falahati, A., 2017. Developing an impedimetric aptasensor for selective label–free detection of CEA as a cancer biomarker based on gold nanoparticles loaded in functionalized mesoporous silica films. Journal of the Electrochemical Society 164 (13), B739–B745.

Sheng-Zhen, C., Qiang, C., Fang-Yi, P., Huang, X.-X., Yu-Ling, J., 2012. Screen-printed electrochemical biosensor for detection of DNA hybridization. Chinese Journal of Analytical Chemistry 40 (8), 1194–1200.

Shiravand, T., Azadbakht, A., 2017. Impedimetric biosensor based on bimetallic AgPt nanoparticle-decorated carbon nanotubes as highly conductive film surface. Journal of Solid State Electrochemistry 21 (6), 1699–1711.

Shojaei, T.R., Salleh, M.A.M., Tabatabaei, M., Ekrami, A., Motallebi, R., Rahmani-Cherati, T., Hajalilou, A., Jorfi, R., 2014. Development of sandwich-form biosensor to detect *Mycobacterium tuberculosis* complex in clinical sputum specimens. Brazilian Journal of Infectious Diseases 18 (6), 600–608.

Shukla, S., Deshpande, S.R., Shukla, S.K., Tiwari, A., 2012. Fabrication of a tunable glucose biosensor based on zinc oxide/chitosan-graft-poly (vinyl alcohol) core-shell nanocomposite. Talanta 99, 283–287.

Singhal, P., Kuhr, W.G., 1997. Direct electrochemical detection of purine-and pyrimidine-based nucleotides with sinusoidal voltammetry. Analytical Chemistry 69 (17), 3552–3557.

Sistare, M.F., Holmberg, R.C., Thorp, H.H., 1999. Electrochemical studies of polynucleotide binding and oxidation by metal complexes: effects of scan rate, concentration, and sequence. Journal of Physical Chemistry B 103 (48), 10718–10728.

Star, A., Tu, E., Niemann, J., Gabriel, J.-C.P., Joiner, C.S., Valcke, C., 2006. Label-free detection of DNA hybridization using carbon nanotube network field-effect transistors. Proceedings of the National Academy of Sciences of the United States of America 103 (4), 921–926.

Steel, A.B., Herne, T.M., Tarlov, M.J., 1998. Electrochemical quantitation of DNA immobilized on gold. Analytical Chemistry 70 (22), 4670–4677.

Steenken, S., Jovanovic, S.V., 1997. How easily oxidizable is DNA? One-electron reduction potentials of adenosine and guanosine radicals in aqueous solution. Journal of the American Chemical Society 119 (3), 617–618.

Stern, D., Pauly, D., Zydek, M., Müller, C., Avondet, M.A., Worbs, S., Lisdat, F., Dorner, M.B., Dorner, B.G., 2016. Simultaneous differentiation and quantification of ricin and agglutinin by an antibody-sandwich surface plasmon resonance sensor. Biosensors and Bioelectronics 78, 111–117.

Subramanian, P., Motorina, A., Yeap, W.S., Haenen, K., Coffinier, Y., Zaitsev, V., Niedziolka-Jonsson, J., Boukherroub, R., Szunerits, S., 2014. An impedimetric immunosensor based on diamond nanowires decorated with nickel nanoparticles. Analyst 139 (7), 1726–1731.

Tang, J., Tang, D., Li, Q., Su, B., Qiu, B., Chen, G., 2011. Sensitive electrochemical immunoassay of carcinoembryonic antigen with signal dual-amplification using glucose oxidase and an artificial catalase. Analytica Chimica Acta 697 (1–2), 16–22.

Tennico, Y.H., Hutanu, D., Koesdjojo, M.T., Bartel, C.M., Remcho, V.T., 2010. On-chip aptamer-based sandwich assay for thrombin detection employing magnetic beads and quantum dots. Analytical Chemistry 82 (13), 5591–5597.

Ting, B.P., Zhang, J., Gao, Z., Ying, J.Y., 2009. A DNA biosensor based on the detection of doxorubicin-conjugated Ag nanoparticle labels using solid-state voltammetry. Biosensors and Bioelectronics 25 (2), 282–287.

Tu, J., Gan, Y., Liang, T., Hu, Q., Wang, Q., Ren, T., Sun, Q., Wan, H., Wang, P., 2018. Graphene FET array biosensor based on ssDNA aptamer for ultrasensitive Hg^{2+} detection in environmental pollutants. Frontiers in Chemistry 6, 1–9.

Walter, A., Wu, J., Flechsig, G.-U., Haake, D.A., Wang, J., 2011. Redox cycling amplified electrochemical detection of DNA hybridization: application to pathogen *E. coli* bacterial RNA. Analytica Chimica Acta 689 (1), 29–33.

Wang, D., Hu, L., Zhou, H., Abdel-Halim, E., Zhu, J.-J., 2013. Molecular beacon structure mediated rolling circle amplification for ultrasensitive electrochemical detection of micro-RNA based on quantum dots tagging. Electrochemistry Communications 33, 80–83.

Wang, J., Cai, X., Jonsson, C., Balakrishnan, M., 1996. Adsorptive stripping potentiometry of DNA at electrochemically pretreated carbon paste electrodes. Electroanalysis 8 (1), 20–24.

Wang, J., Cai, X., Wang, J., Jonsson, C., Palecek, E., 1995. Trace measurements of RNA by potentiometric stripping analysis at carbon paste electrodes. Analytical Chemistry 67 (22), 4065–4070.

Wang, J., Kawde, A.-N., 2001. Pencil-based renewable biosensor for label-free electrochemical detection of DNA hybridization. Analytica Chimica Acta 431 (2), 219–224.

Wang, J., Kawde, A.-N., Musameh, M., Rivas, G., 2002. Dual enzyme electrochemical coding for detecting DNA hybridization. Analyst 127 (10), 1279–1282.

Wang, J., Wang, L., Zhu, Y., Zhang, J., Liao, J., Wang, S., Yang, J., Yang, F., 2016. A high accuracy cantilever array sensor for early liver cancer diagnosis. Biomedical Microdevices 18 (6), 110.

Wang, J., Zhu, X., Tu, Q., Guo, Q., Zarui, C.S., Momand, J., Sun, X.Z., Zhou, F., 2008. Capture of p53 by electrodes modified with consensus DNA duplexes and amplified voltammetric detection using ferrocene-capped gold nanoparticle/streptavidin conjugates. Analytical Chemistry 80 (3), 769–774.

Wang, S., Munde, M., Wang, S., Wilson, W.D., 2011. Minor groove to major groove, an unusual DNA sequence-dependent change in bend directionality by a distamycin dimer. Biochemistry 50 (35), 7674–7683.

Wang, Y., Zhang, Y., Su, Y., Li, F., Ma, H., Li, H., Du, B., Wei, Q., 2014. Ultrasensitive non-mediator electrochemical immunosensors using Au/Ag/Au core/double shell nanoparticles as enzyme-mimetic labels. Talanta 124, 60–66.

Wang, Z., Jia, Y., 2018. Graphene solution-gated field effect transistor DNA sensor fabricated by liquid exfoliation and double glutaraldehyde cross-linking. Carbon 130, 758–767.

Whittemore, N.A., Mullenix, A.N., Inamati, G.B., Manoharan, M., Cook, P.D., Tuinman, A.A., Baker, D.C., Chambers, J.Q., 1999. Synthesis and electrochemistry of anthraquinone-oligodeoxynucleotide conjugates. Bioconjugate Chemistry 10 (2), 261–270.

Won, B.Y., Yoon, H.C., Park, H.G., 2008. Enzyme-catalyzed signal amplification for electrochemical DNA detection with a PNA-modified electrode. Analyst 133 (1), 100–104.

Wong, E.L., Gooding, J.J., 2006. Charge transfer through DNA: a selective electrochemical DNA biosensor. Analytical Chemistry 78 (7), 2138–2144.

Wu, S., Ye, W., Yang, M., Taghipoor, M., Meissner, R., Brugger, J., Renaud, P., 2015. Impedance sensing of DNA immobilization and hybridization by microfabricated alumina nanopore membranes. Sensors and Actuators B: Chemical 216, 105–112.

Xia, N., Zhang, Y., Wei, X., Huang, Y., Liu, L., 2015. An electrochemical microRNAs biosensor with the signal amplification of alkaline phosphatase and electrochemical–chemical–chemical redox cycling. Analytica Chimica Acta 878, 95–101.

Xie, S., Wang, F., Wu, Z., Joshi, L., Liu, Y., 2016. A sensitive electrogenerated chemiluminescence biosensor for galactosyltransferase activity analysis based on a graphitic carbon nitride nanosheet interface and polystyrene microsphere-enhanced responses. RSC Advances 6 (39), 32804–32810.

Xu, S., Mutharasan, R., 2009. Cantilever biosensors in drug discovery. Expert Opinion on Drug Discovery 4 (12), 1237–1251.

Yagati, A.K., Pyun, J.-C., Min, J., Cho, S., 2016. Label-free and direct detection of C-reactive protein using reduced graphene oxide-nanoparticle hybrid impedimetric sensor. Bioelectrochemistry 107, 37–44.

Yang, W., Ozsoz, M., Hibbert, D.B., Gooding, J.J., 2002. Evidence for the direct interaction between methylene blue and guanine bases using DNA-modified carbon paste electrodes. Electroanalysis 14 (18), 1299–1302.

Yano, M., Koike, K., Mukai, K., Onaka, T., Hirofuji, Y., Ogata, K.i., Omatu, S., Maemoto, T., Sasa, S., 2014. Zinc oxide ion-sensitive field-effect transistors and biosensors. Physica Status Solidi A 211 (9), 2098–2104.

Yardım, Y., Vandeput, M., Celebi, M., Senturk, Z., Kauffmann, J.M., 2017. A reduced graphene oxide-based electrochemical DNA biosensor for the detection of interaction between cisplatin and DNA based on guanine and adenine oxidation signals. Electroanalysis 29 (5), 1451–1458.

Yin, H., Wang, M., Li, B., Yang, Z., Zhou, Y., Ai, S., 2015. A sensitive electrochemical biosensor for detection of protein kinase A activity and inhibitors based on Phos-tag and enzymatic signal amplification. Biosensors and Bioelectronics 63, 26–32.

Yin, H., Zhou, Y., Ma, Q., Ai, S., Ju, P., Zhu, L., Lu, L., 2010. Electrochemical oxidation behavior of guanine and adenine on graphene–Nafion composite film modified glassy carbon electrode and the simultaneous determination. Process Biochemistry 45 (10), 1707–1712.

Yue, M., Stachowiak, J.C., Lin, H., Datar, R., Cote, R., Majumdar, A., 2008. Label-free protein recognition two-dimensional array using nanomechanical sensors. Nano Letters 8 (2), 520–524.

Zhang, J., Song, S., Zhang, L., Wang, L., Wu, H., Pan, D., Fan, C., 2006. Sequence-specific detection of femtomolar DNA via a chronocoulometric DNA sensor (CDS): effects of nanoparticle-mediated amplification and nanoscale control of DNA assembly at electrodes. Journal of the American Chemical Society 128 (26), 8575–8580.

Zhang, J., Ting, B.P., Khan, M., Pearce, M.C., Yang, Y., Gao, Z., Ying, J.Y., 2010. Pt nanoparticle label-mediated deposition of Pt catalyst for ultrasensitive electrochemical immunosensors. Biosensors and Bioelectronics 26 (2), 418–423.

Zhang, L., Wan, Y., Zhang, J., Li, D., Wang, L., Song, S., Fan, C., 2009a. The enzyme-amplified amperometric DNA sensor using an electrodeposited polymer redox mediator. Science in China, Series B 52 (6), 746–750.

Zhang, W., Yang, T., Zhuang, X., Guo, Z., Jiao, K., 2009b. An ionic liquid supported CeO_2 nanoshuttles–carbon nanotubes composite as a platform for impedance DNA hybridization sensing. Biosensors and Bioelectronics 24 (8), 2417–2422.

Zheng, C., Huang, L., Zhang, H., Sun, Z., Zhang, Z., Zhang, G.-J., 2015. Fabrication of ultrasensitive field-effect transistor DNA biosensors by a directional transfer technique based on CVD-grown graphene. ACS Applied Materials & Interfaces 7 (31), 16953–16959.

Zhou, L., Ji, F., Zhang, T., Wang, F., Li, Y., Yu, Z., Jin, X., Ruan, B., 2019. An fluorescent aptasensor for sensitive detection of tumor marker based on the FRET of a sandwich structured QDs-AFP-AuNPs. Talanta 197, 444–450.

Zhou, L., Rusling, J.F., 2001. Detection of chemically induced DNA damage in layered films by catalytic square wave voltammetry using Ru $(Bpy)_3^{2+}$. Analytical Chemistry 73 (20), 4780–4786.

Zhu, Q., Gao, F., Yang, Y., Zhang, B., Wang, W., Hu, Z., Wang, Q., 2015. Electrochemical preparation of polyaniline capped Bi2S3 nanocomposite and its application in impedimetric DNA biosensor. Sensors and Actuators B: Chemical 207, 819–826.

CHAPTER

7 Enzyme-based electrochemical biosensors

Aso Navaee[1], Abdollah Salimi[2]

Postdoctoral researcher, Department of Chemistry, Nanotechnology Research Center, University of Kurdistan, Sanandaj, Iran[1]; Professor, Department of Chemistry, Nanotechnology Research Center, University of Kurdistan, Sanandaj, Iran[2]

7.1 Introduction

Basically, a *sensor* is a device that displays a countable signal in response to an exciter. Sensor is distinct from detector, where detectors simply designate the presence of the analytes, whereas sensors quantify the changes in the analyte (Kalantar-zadeh & Fry, 2008). The subclass of physical and chemical sensors is sometimes the same, and commonly all of chemical, physical, or biological principles have contributed in a sensing device. There are different classifications for sensors, which have been classified based on the detection method, technical aspect of sensor, fields of application, transducer (conversion phenomenon), and so on (White, 1987). However, according to the operating principle of the transducer, they may be classified into optical, electrochemical, mass sensitive device, electrical, thermal, magnetic sensors (Hulanicki et al., 1991) as summarized in the following flowchart along with subcategories (Scheme 7.1).

The materials in used in electrical sensors can be appropriate for fabrication of electrochemical sensors, and occasionally, they are classified into a subgroup. However, the transducers used in these sensors distinguish them. There are some other chemical sensors based on other physical phenomena such as β, Γ, X radiation, etc. with less degree of importance, which have not been listed in the above flowchart. Electrochemical sensors are simply adaptable in their ability to deal with a different field of application from environmental pollutant to biological systems. A respectable sensor should clearly identify specific substances, even at low concentration levels with adequate accurate according to its conventional standards. Thousands of electrochemical sensors have been presented for detection of ions, inorganic or organic materials in environmental monitoring, food control, and pharmaceutical research along with medical care in biological organisms. The latter case is termed biosensor, which senses the chemical transformation of biological species into identified specific electrical signals.

Electrochemical Biosensors. https://doi.org/10.1016/B978-0-12-816491-4.00007-3

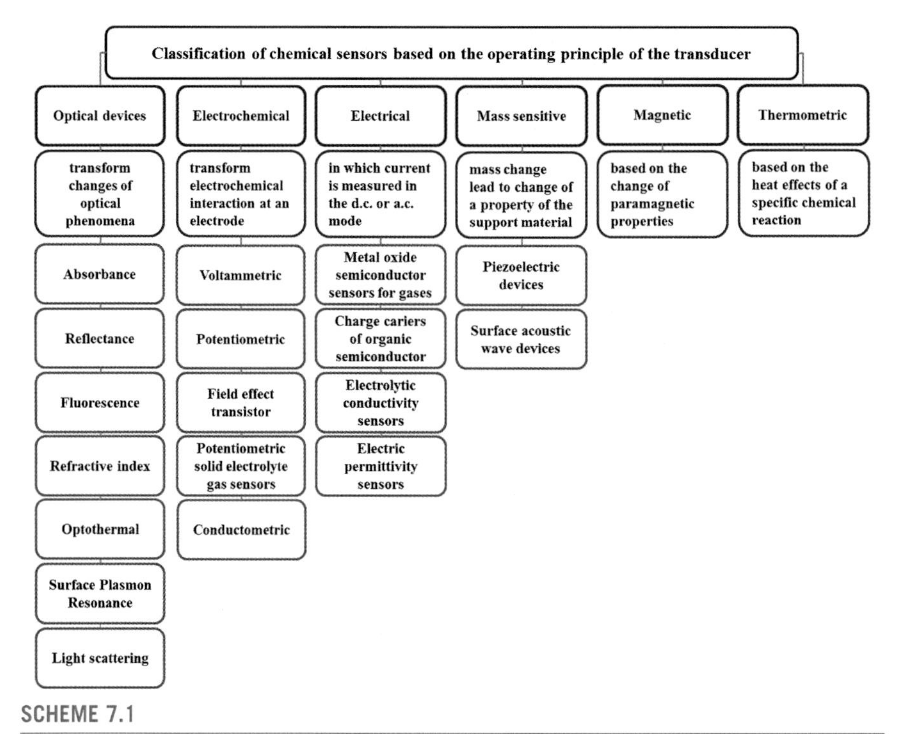

SCHEME 7.1

Sensor classification based on the principle of the transducer and their subdivision.

7.2 What is biosensor?

Basically, based on how biomaterial involved in sensing, biosensor can be defined with two general sights: (1) In an opinion, a biosensor can be defined as a special measurement system designed for monitoring of informal concentration of desired species in human body (by means of biological materials or nanomaterials) that signs disease. (2) Besides the above description, the biomolecule(s) (such as enzyme or aptamer) applied as a receptor in a sensing device in outer of biological system such as toxicity control in food or environmental pollutant by using the biological interactions is termed biosensor. All of biosensors contain a physicochemical transducer to transfer the response of sensing element to the recorder part. Accordingly, there are many different categories of biosensors, based on the fundamental technology (especially, the signal transducer) and what they measure. The arrangement and essential compartment of biosensors from measuring samples to signal processing is summarized in the following flowchart (Scheme 7.2).

Taking of bioreceptor is depended on the samples which are analyzed. For example, antibodies are the best bioreceptor for detecting of proteins. Even, the antibody-conjugated enzyme has been extensively used for analyzing of proteins in a known procedure called ELISA (enzyme-linked immunosorbent assay) that

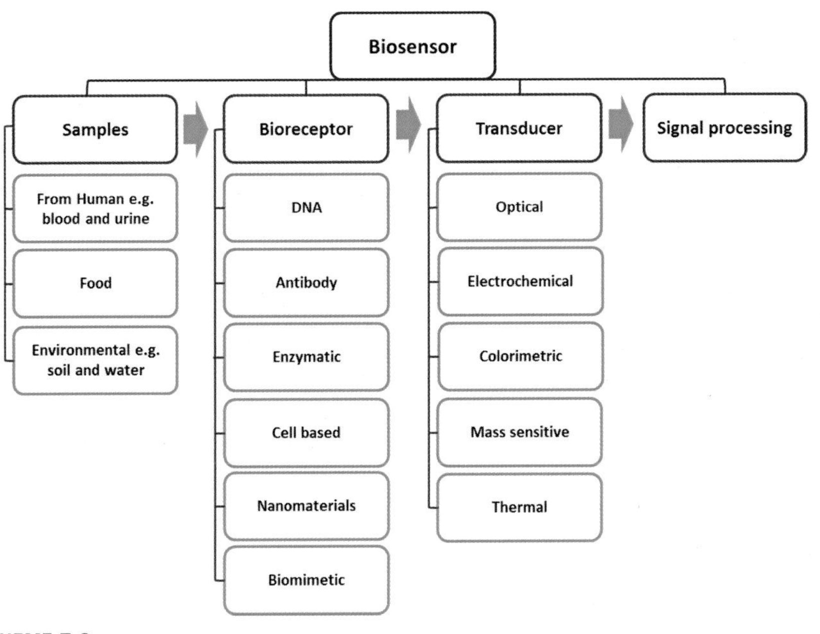

SCHEME 7.2

Basic building of a sensor (biosensor) and types of each element.

also called immunosensor. ELISA is the standard method for quantifying the amount of protein expression in biological systems based on complex formation between an analyte and an enzyme-conjugated antibody. Following that conjugation, the enzymatic reaction is risen up and transduces the detectable signals. Hence, accuracy of other methods of biomedical measurement is compared with the ELISA system. Among the addressed bioreceptors, there are wide varieties of enzymes that have been extensively proposed for fabrication of biosensors based on different transducers, especially in the field of electrochemistry. However, biosensors are still rarely performed in real samples because they cannot practically applicable for a long time.

7.3 Classification of enzymes-based bioreceptors

The significance of enzymes in sensor fabrication is their specificity toward particular molecules, which is the key of selectivity of the biosensor. Enzymes are large molecules with 10–400 kDa. Generally, enzymes are classified into six major categories based on acting on organic substrates, including the following:

- Oxidoreductases catalyze the oxidation/reduction reactions.
- Transferases catalyze transfer of molecular groups from one molecule to another.

- Hydrolases catalyze the hydrolysis of molecules.
- Lyases, cleavage the C—C, C—O, or C—N bonds through adding a functional group to the molecule.
- Ligases catalyze the coupling of two molecules.
- Isomerases catalyze the addition of a functional group to a molecule to form an isomeric form (Scheller and Schubert, 1992; Monošík et al., 2012).

Oxidoreductases such as glucose oxidases and glucose dehydrogenases are the most attractive bioreceptors even in the commercial aspect. In the following of this chapter, oxidoreductases will extensively be discussed. All of enzyme categories include a variety of subgroups, which are summarized in the following flowchart (Scheme 7.3). They are rarely implemented in sensing approaches. However, in the latest review papers related to enzyme inhibition-based biosensors, many reports based on aforementioned enzymes are explained (La et al., 2015; El Harrad et al., 2018). For example, methyltransferase, as a subgroup of transferase, has been developed for DNA analysis (Yin et al., 2013). Glutathione-s-transferase has been proposed for detection of captan, in which captan inhibits the catalytic activity of enzyme (Singha et al., 2009). Hexokinase as another subclass of transferase

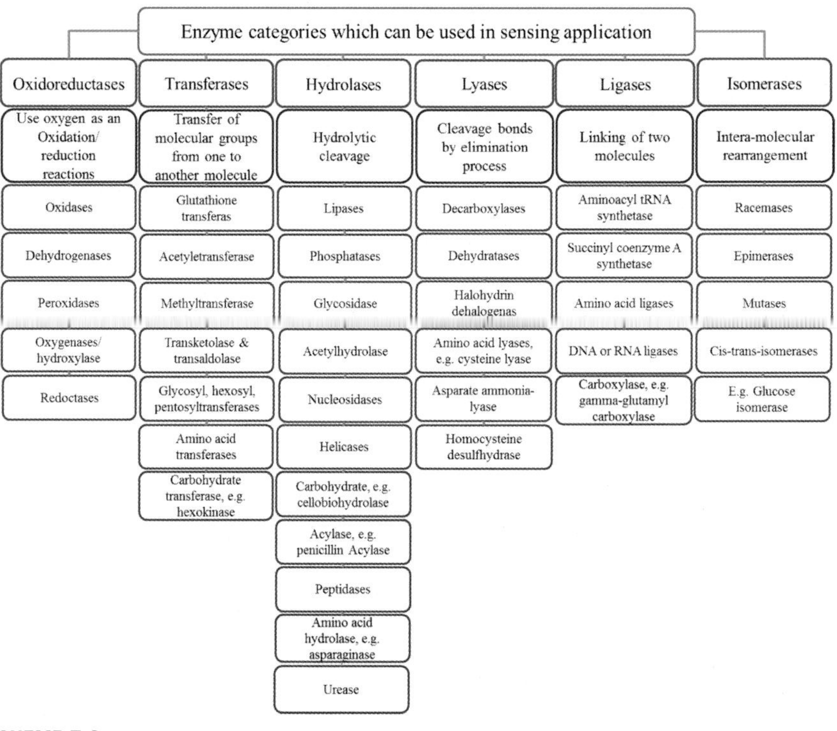

SCHEME 7.3

Six major categories of enzymes based on their functions along with their subclass.

catalyzes the phosphorylation of hexoses (six-carbon sugars such as glucose), and forming hexose phosphate and glucose-6-phosphate, is the good candidate for adenosine triphosphate (ATP) detection. So, it has been successfully developed for that purpose based on amperometric (Kucherenko et al., 2014) and conductometric (Kucherenco et al., 2016) techniques. The principle of phosphorylation of hexose by hexokinase has also been applied for designing a spectrochemical standard method for detection of glucose assisted by glucose-6-phosphate dehydrogenase. We will discuss more about this subject in the following of this chapter.

After oxidoreductase, hydrolases are attracted highest attention in enzymatic electrochemistry. The most approaches are enclosed around detection of organophosphate pesticide using acetylcholinesterase (acetylhydrolase) (Liu and Lin, 2006; Du et al., 2010), organophosphorus hydrolase receptors (Lee et al., 2010), or alkaline phosphatase (Mazzei et al., 2004) based on the inhibition concept. Proteases (peptidases), as other subgroups of hydrolase, are considered in revealing of different proteins (Mahmoud and Kraatz, 2007) such as HIV-related protein or DNA. These types of biosensors have been designed based on enzyme inhibition occurrences. DNA has also been analyzed by ligases-based electrochemical biosensors (Wan et al., 2009). But, based on our knowledge, the electrochemical biosensors based on lyases are rarely reported. Kim (2006) has proposed the ion-selective electrochemical sensor for detection of citrate based on citrate lyase. In another report, spectroscopic approach assisted by lyases, such as application of dehydratase, has been used for lead detection (Ogunseitan et al., 2000). At the meantime, isomerases have not directly been used in sensing application until now. But, few reports related to glucose isomerase in coupling with glucose oxidase have been reported (Scheller and Schubert, 1992). In those studies, the other forms of monosaccharides such as fructose are rearranged to glucose and then determined by glucose oxidase. However, these enzymes may have a high future potential to apply in sensing applications.

The performance of enzyme in biosensing is influenced by pH, temperature, amount and type of enzyme loading. But the important factor in enzymatic electrochemistry is the chemical cofactor that supplements the utility of the enzyme. The most popular enzymes that are widely performed as bioreceptors are the oxidoreductase categories, which they operate assisted by different coenzymes.

Oxidoreductases are a class of enzymes that catalyze the oxidation/reduction reaction through simplifying electron transfer from reductants (electron donors) to oxidants (electron acceptors). The general reaction is as follows: $A^- + B \rightarrow A + B^-$ where A is the reductant and B is the oxidant. They act on the different functional groups such as $-CHO$, $-CHOH$, $CH_2{-}NH_2$, $-CH{=}NH$, or $-CH_2{-}CH_2-$ assisted by inorganic and organic cofactors. In most of oxidation reactions, electrons are finally transported to convert ADP to ATP. Cofactors are the nonprotein chemical compounds or metallic ions that are required for enzyme activity. The complex organic cofactors are called coenzymes. Several of the coenzymes are identified, and vitamins are the major categories of cofactors.

Nicotine amide adenine dinucleotide with or without phosphate moiety (NAD(P)H), flavin adenine dinucleotide-hydrogen ($FAD(H_2)$), flavin mononucleotide-hydrogen ($FMN(H_2)$), pyrroloquinoline quinone (PQQ), and heme are the most known cofactors that they are the electrochemical active sites for the catalytic process of enzymes, and extensively studied in electrochemical fields. Among them, the abundant cofactor in all living cells is NAD(P)H, which is responsible for reaction of more than 400 oxidoreductase enzymes (Gorton and Domínguez, 2007) in both oxidation and reduction reactions. Both oxidized form ($NAD(P)^+$) and reduced form (NAD(P)H) are present in nature, and NADPH is produced when another phosphate group binds to NADH.

In the plants and liver tissues, electron transfer reactions between biomolecules catalyzed by oxidoreductase (especially, by NAD(P)H) play essential roles in metabolisms and catalysis conversion of different materials, such as adenosine diphosphate (ADP) to ATP cycles, ethanol to acetaldehyde, glucose to gluconolactone, pyruvate to lactate, and so on. Moreover, the reducing power of NAD(P)H also is required for other metabolic pathways that catalyzed by other cofactors (comprehensive approach) of enzyme and coenzymes. Moreover, in some cases, the action of oxidoreduction reaction depends on the some metal ions such as iron (e.g., heme and non−heme-dependent oxygenases), copper (e.g., bilirubin oxidase, laccase), zinc (e.g., alcohol dehydrogenases), selenium (e.g., xanthine or nicotinate dehydrogenases), etc. (Blank et al., 2010; Rocchitta et al., 2016). To sum up, such visions in natural phenomena stimulate the application of oxidoreductase in different bioelectrochemical approaches such as biological electron transfer processes in biosensors or biofuel cell fields. Accordingly, electron transfer is identified directly as electrical current with passing of time or potential.

Molybdopterin, iron-sulfur (Fe-S) clusters and glutathione as other cofactors are in the next importance for enzymatic reactions. Glutathione-dependent enzymes and their structure-function relationships have been comprehensively reviewed by Deponte (2013). But few reports have been published related to glutathione-based enzymatic reactions. In one attempt, it has been used for detection of nitrosothiols (Musameh et al., 2006). The most abundant and well-known heme-containing protein is hemoglobin. As we know, hemoglobin is oxygen carrier in mammalian bodies. So it can be used for oxygen detection as it is reported by Cao et al. (2003) through an electrochemical procedure. Although nonelectrochemical approaches are not the scope of this book, but the reference to the recent fluorescence-based research (Nomata and Hisabori, 2018) may disclose the role of heme in oxygen detection. More information will be coming about molybdopterin and Fe-S clusters as important cofactors in catalytic reactions by dehydrogenase and reductase enzymes.

Other cofactors with less importance could be declared as coenzyme Q and lipoamide. Quinone derivatives are very interested in electrochemistry because they display the distinct voltammetric redox peaks. Coenzyme Q reaches here name from the quinone molecule, which is the basic moiety of coenzyme for transport of electron and proton in enzymatic reactions. The redox chemistry of different coenzyme Q are reviewed by Gulaboski et al. (2016). In Fig. 7.1, the chemical structures of oxidized and reduced forms of the aforementioned cofactors along with

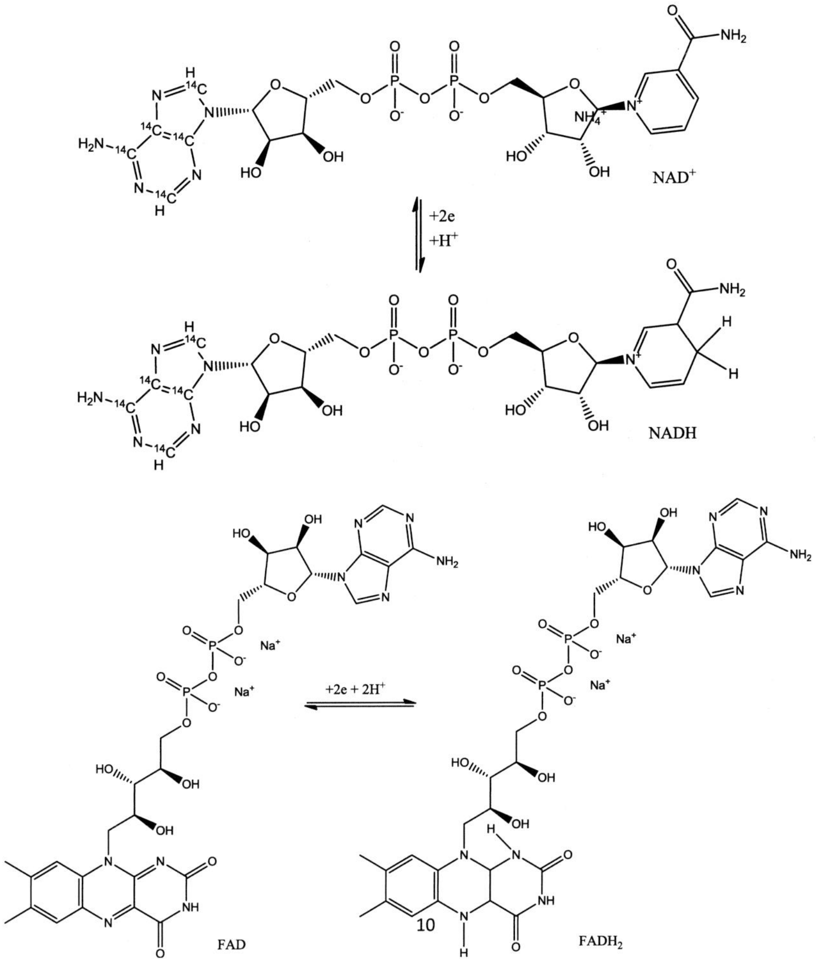

FIGURE 7.1

Chemical structures of important organic cofactors with the reduced and oxidized forms.

number of electron and proton transportation are presented. Some of other cofactors such as coenzyme A (CoA) have the significance importance in biochemistry. But, CoA catalyze the transferring acetyl group in the enzymatic reaction instead of electron, and consequently, it cannot be followed by electrochemistry. Therefore, one reason that oxidoreductases are widely spread out than other enzyme categories in biosensing over the past three decays is the coenzymes that cause the catalytic ability and make detectable sensations. Another reason return to the importance of substrate catalyzed by oxidoreductase, which practically signs a disease in the human body.

Currently, the important issues that influence on the electron transfer between a solid surface and an enzyme have almost been known. Consequently, biological and

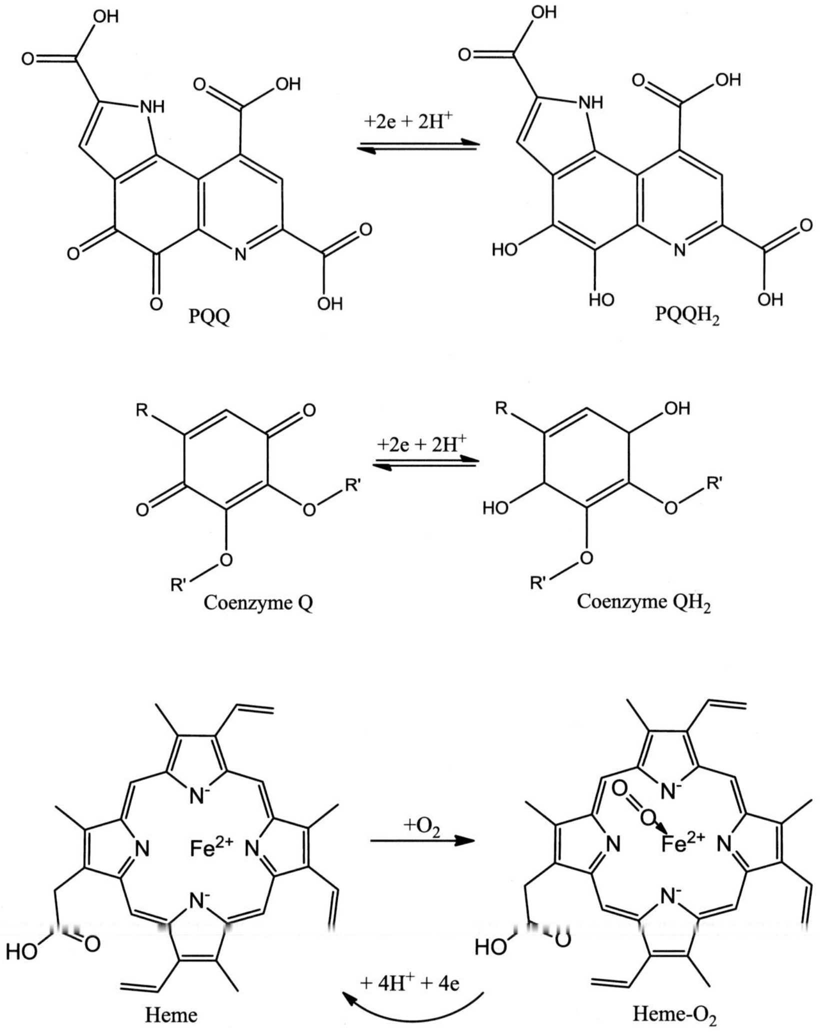

PQQ $\xrightarrow{+2e + 2H^+}$ $PQQH_2$

Coenzyme Q $\xrightarrow{+2e + 2H^+}$ Coenzyme QH_2

Heme $\xrightarrow{+O_2}$ Heme-O_2 ($+ 4H^+ + 4e$)

FIGURE 7.1 cont'd.

electrochemical science has been combined in order to design different modified electrodes through various chemical interactions aimed at efficient electron transfer. On the other hand, significant progresses in instrumentation such as atomic force microscopy (AFM), scanning tunneling microscopy (STM), scanning electron microscope (SEM), infrared and Raman spectroscopy, electron paramagnetic spectroscopy (EPR), and so on for in situ study of modified surfaces have been done. These efforts have led to manage the effective immobilization or orientation of biomolecules on electrode surfaces for efficient electron transfer. As mentioned

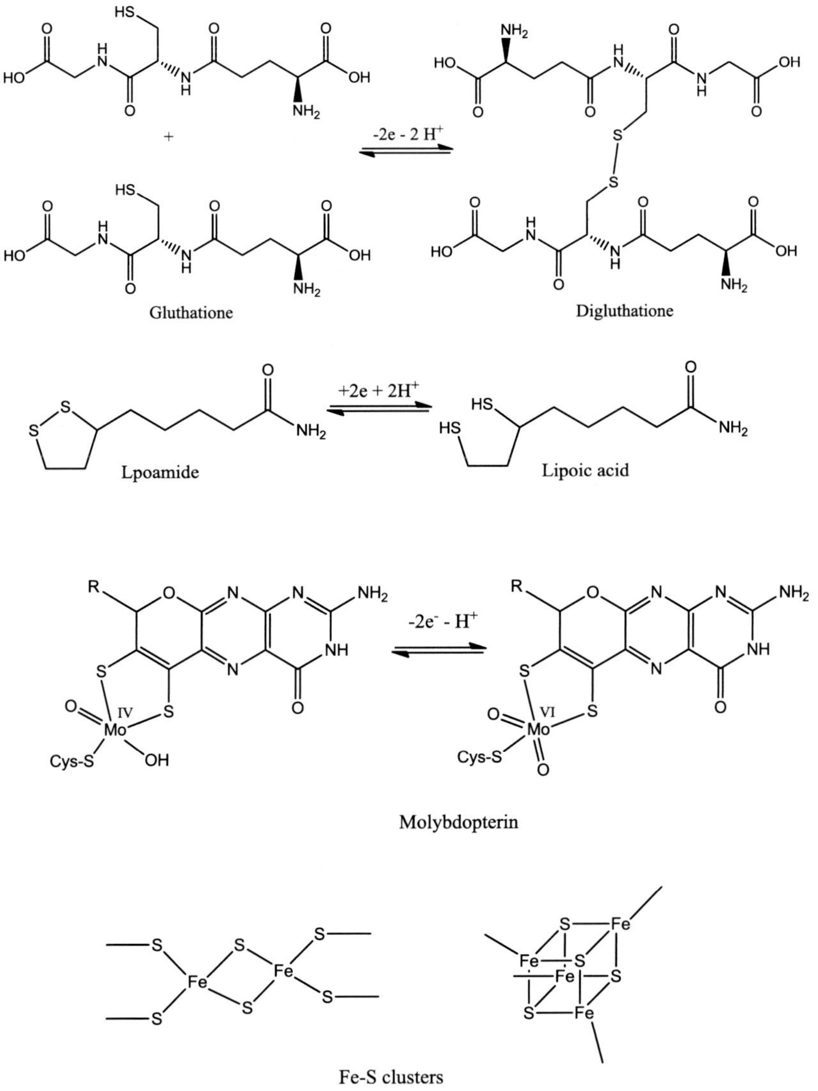

Gluthatione

Digluthatione

Lpoamide

Lipoic acid

Molybdopterin

Fe-S clusters

FIGURE 7.1 cont'd.

above, the mostly applicable enzymes that have extensively been utilized in sensing design are the oxidoreductases. In the following, the major types of oxidoreductases beside recent progress in immobilization and substrate detection are discussed.

There are two outlooks about electron transfer by enzymes: direct and indirect aspects. Also, there are two viewpoints related to direct (mediatorless) electron

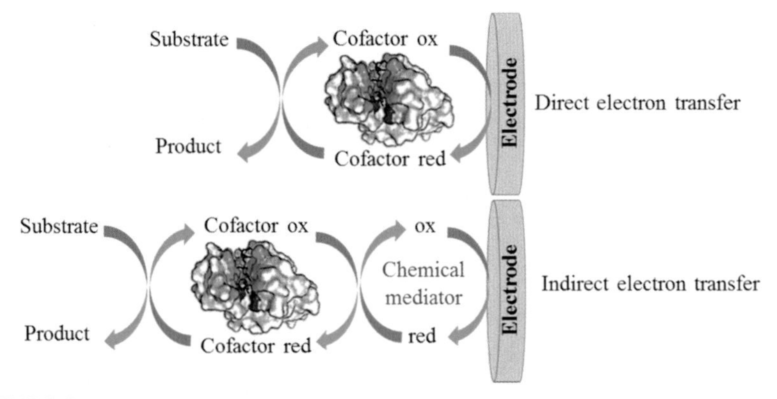

SCHEME 7.4

Direct and indirect electron transfer catalyzed by enzymes.

transfer. In one view, if any electrochemical signal related to the cofactor of enzyme be seen during voltammetric procedure, it cab be said that direct electron transfer (DET) is achieved from enzyme. In another viewpoint, electrochemical signal of cofactor has not be seen, but catalytic reaction by enzyme is occurred without necessary of extrachemical substances (for mediating electron transfer); it is also called DET. Oppositely, when transfer of electron from enzyme to electrode needs an external chemical mediator, it is called indirect or mediator-based electron transfer (Scheme 7.4). Commonly, chemical mediators increase the kinetic of electron transfer, but DET is more attracted for scientist, as it has more simplicity and stability. However, there are still challenges in the description of direct and indirect ET. For example, as mentioned, the criterion of DET has been defined as electrochemical signal of redox cofactor of enzyme, even though the substrate oxidation has been done in presence of an external chemical mediator (Salimi et al., 2007b).

7.4 Oxidoreductase subclasses

Base on the function of enzymes, oxidoreductases are divided into five major categories including oxidases, dehydrogenases, peroxidases, oxygenases/hydroxylases, and reductases. The activity of oxidase, peroxidase and oxygenase depends on the presence of oxygen or hydrogen peroxide, whereas the action of dehydrogenases and reductases does not depend on those oxidants (electron acceptors). For catalytic activity of oxidases, molecular oxygen acts as an acceptor of electrons or hydrogen. Peroxidases are the important enzyme categories in reduction and removing hydrogen peroxide, which is toxic for liver cells. Catalytic oxidation of substrates by oxygenases/hydroxylases is occurred through incorporation of oxygen with the form of hydroxyl groups to the substrates. In contrast, dehydrogenases and reductases are enzymes that oxidize and reduce substrates, respectively, when a cofactor such as $NAD(P)^+$ or NAD(P)H acts as electron transfer mediator. Each of these five categories also involve several subclasses that are summarized in Scheme 7.4. However, all of the aforementioned enzymes are not used in a comparable level for constructing electrochemical biosensors. For example, because of bad diet, the diabetic diagnostic is widespread, and every day an increase in the number of patients is seen. Accordingly, glucose biosensors fabricated by glucose oxidases (GOs) and glucose dehydrogenases (GDHs) are in the first row of the enzyme-based biosensor researches in human (Yoo and Lee, 2010) or food (Cock et al., 2009) samples.

Oxidases catalyze the oxidation of substrate, while molecular oxygen acts as electron acceptor and is reduced to hydrogen peroxide or water.

Glucose oxidase (GOx) is the well-known oxidase enzyme in electrochemistry because of importance of the glucose substrate. Among the different catalytic oxidation of glucose (enzymatic, microbial, and abiotic systems), the enzymatic oxidation has displayed higher selectivity in the mild temperature and pH conditions. After the first development in electrochemical glucose sensing by Clark Jr. and Lyons (1962), several enzymatic glucose biosensors have been reported (Yoo and Lee, 2010; Cock et al., 2009; Clark Jr. and Lyons, 1962; Ferri et al., 2011), where they differ in the types of enzyme and electrode, methods of immobilization, and direct or mediator-based electron transfer. The last three factors will be discussed in this chapter. Besides GOx, glucose dehydrogenases (GDHs), with highest importance (Ferri et al., 2011) and hexokinase, with minor importance, are other popular enzymes in glucose sensing. Hexokinase was discussed above, and GDHs will be discussed in the following of this section. FAD is the only cofactor for catalysis reaction of GOx. The mechanism of glucose oxidation catalyzed by GOx is well known elsewhere and can be drawn as follows (Scheme 7.6):

In one way, electrode (transducer) may be sensitive to the O_2 or H_2O_2 concentration (Scheme 7.6A). These types of sensors are called the first generation sensors. In the other way, in the presence of a chemical mediator, electron can be transferred to the electrode (Scheme 7.6B). This type is classified as second-generation sensors. There is another type of biosensor that is called third-generation sensor, which is based on the DET, without requiring oxygen or mediator.

When electrode senses the O_2 concentration, depletion in signal of O_2 reduction is directly related to the glucose concentration. In contrary, sensitive electrodes toward H_2O_2 are more preferable route for glucose detection, as there are some difficulties in controlling of O_2 flow or concentration. Hence, to attain efficient catalytic activity, enzymes are immobilized on the solid surfaces (electrode). Different modified electrodes based on conductive polymer films or nanomaterials, which effectively attract GOx, and are also sensitive to oxygen or H_2O_2, have been developed in recent years (Schuhmann et al., 1991; Cai and Chen, 2004; Salimi et al., 2009a; Mamyabi et al., 2013; Salimi and Noorbakhsh, 2011; Liang et al., 2015; Baghayer et al., 2017; du Toit and Di Lorenzo, 2014). Alike other enzymes, carbon nanotubes have attracted more attention aimed at GOx immobilization (Cai and Chen, 2004; Salimi et al., 2009a; Mamyabi et al., 2013). Many efforts have been

Oxidoredoctase enzymes catalyze the following reaction: $A_{red} + B_{ox} \rightarrow A_{ox} + B_{red}$

Oxidases	Dehydrogenases	Peroxidases	Oxygenases/ hydroxylase	Redoctases
Oxygen acts as electron acceptor, but not incorporate into the substrate	Unlike oxidases not depend on oxygen present, just cofactor is enough	Use H_2O_2 as an electron acceptor	Incorporate oxygen into the organic substrate: mono & dioxygenases	Catalyze the reduction reaction assisted by cofactors such as NADH
Glucose oxidase	Alcohol dehydrogenase	Catalas	Heme-dependent: cytochrome P450	5α-Reductase
Glutamate oxidase	Glucose dehydrogenase	Horseradish peroxidases	Flavin-dependent	5β-Reductase
Alcohol oxidase	Steroid dehydrogenase	Lactoperoxidase	NAD(P)H-dependent	Aldo-keto, e.g. aldose aldehyde reductase
Lactate oxidase	Lactate dehydrogenase	Glutathione Peroxidases	Copper-dependent	HMG-CoA reductase
Amino acid oxidases	Aldehyde dehydrogenase	Ascorbate peroxidase	Non-heme iron-dependent	Methemoglobin reductase
Cholesterol oxidase	Glutamate dehydrogenase	Soybean peroxidase	Pterin-dependent	Ribonucleotide reductase
Choline Oxidase	Pyruvate dehydrogenase	Cytochrome c peroxidase	ATP-dependent, e.g. Luciferase	Thioredoxin reductase
NADPH oxidase	Aldehyde dehydrogenase	Thyroid peroxidase	E.g.: catechol dioxygenases	Nitroreductase
Oxalate oxidase	Sorbitol dehydrogenase	Peroxiredoxin	E.g.: amino acid oxygenase	Folate reductases
Cytochrome oxidase	Formate dehydrogenase	Myeloperoxidase	Cofactor-independent	Cytochrome P450 reductase
Bilirubin oxidase	Cellobiose dehydrogenases	Lignin peroxidases		Disulfide and sulfite Reductase
Laccase	Amino acid dehydrogenases	Xylem peroxidase		Cytochrome c reductase
Tyrosinase & hemocyanin	Succinate dehydrogenases	Amino acid peroxidases		Imine & amino acid reductases
Galactose oxidase	Aldose dehydrogenases	Superoxide Dismutases		Fumarate reductase
Azurine	Sulfite dehydrogenases	Other heme-peroxidases		Carboxylic acid reductase
Ascorbate oxidase	Fructose dehydrogenases	NADH peroxidase		DMSO (e.g. perchlorate) reductase
	Pyranose dehydrogenase	Haloperoxidases		Other inorganic anion oxide reductase
	Gluconate dehydrogenase	Thiol peroxidase		

SCHEME 7.5

Five major categories of oxidoreductase enzymes and their subclasses based on their function.

developed to achieve DET from GOx, and the redox peaks related to the reaction (7.1) have been clearly observed at different electrodes (Salimi et al., 2007b; Schuhmann et al., 1991; Cai and Chen, 2004; Salimi et al., 2009a; Mamyabi et al., 2013; Salimi and Noorbakhsh, 2011; Liang et al., 2015; Baghayer et al., 2017; du Toit and Di Lorenzo, 2014):

$$GO_X(FADH_2) \rightleftharpoons GO_X(FAD) + 2e^- + 2H^+ \tag{7.1}$$

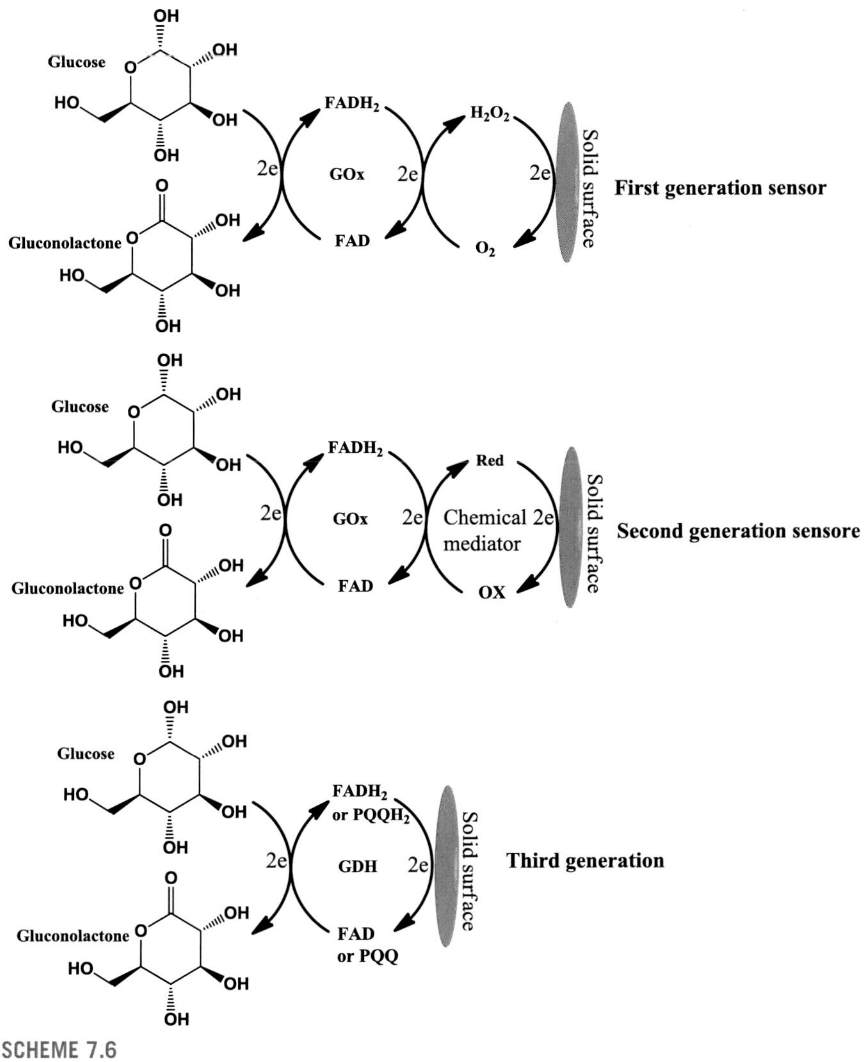

SCHEME 7.6

Mechanisms of glucose oxidation reactions by GOx on three generated electrochemical sensors.

But DET from GOx has no electrocatalytic activity toward glucose oxidation, and as mentioned above, the glucose detection is based on the three concepts: (1) recognition of oxygen reduction reaction (Fig. 7.2A) or (2) H_2O_2 oxidation reaction, which are, respectively, consumed and produced during glucose oxidation, (3) withdrawing electron from $FADH_2$ by a chemical mediator and transfer to the electrode and vice versa (Fig. 7.2B).

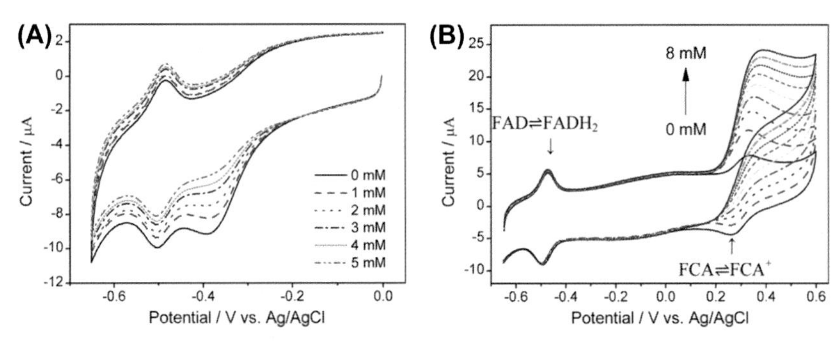

FIGURE 7.2

Cyclic voltammograms recorded by GOx immobilized on reduced graphene oxide/electrode in phosphate buffer saturated by air (A) and N_2-saturated solution containing 0.5 mM ferrocene carboxylic acid (FCA) (B) during addition of different concentrations of glucose.

Reprinted with permission from Liang, B., Guo, X., Fang, L., Hua, Y., Yang, G., Zhu, Q., Wei, J., Ye, X., 2015. Study of direct electron transfer and enzyme activity of glucose oxidase on graphene surface. Electrochemistry Communications 50, 1–5, Elsevier.

7.5 Other flavoprotein-dependent enzymes

Alcohol, lactate, amino acid, cholesterol, NADPH, oxalate and choline oxidases oxidize the same substrate that have taken their name from it. The catalytic electron transfers are arisen using flavoproteins of FAD or FMN. Alcohol oxidases (AOx) catalyze the oxidation of alcohols to ketones in the liver tissues and have been applied in the development of various industrial processes and products such as biosensors and production of various industrially useful carbonyl compounds (Ozimek et al., 2005; Goswami et al., 2013). Lactate oxidases (Lox) catalyze the oxidation of L-lactate to pyruvate using FMN cofactor (Leiros et al., 2006; Taurino et al., 2013). Steroid oxidases such as cholesterol oxidase (ChOx) catalyze the oxidation of steroid substrates, which have a hydroxyl group at the 3-beta position, by means of FAD cofactor (Vrielink and Ghisla, 2009; Salimi et al., 2009b). Choline oxidase catalyzes the two-electron oxidation of choline to betaine aldehyde, as intermediate, and then to glycine betaine with another two electrons transportation (Fan and Gadda, 2005).

FAD-based amino acid oxidases (AAOx) oxidize L-amino acids and D-amino acids to imino acids intermediate and finally to α-keto acids (Pollegioni et al., 2013; Bhagavan, 2002; Kacaniklic et al., 1994). Oxalate oxidase (OOx) is another type of oxidase, which can be applicable for oxalate determination, especially in urine samples (Milardović et al., 2008). Lastly, NADPH oxidases generate reactive oxygen species (ROS) for many purposes such as killing foreign microbes and bacteria (Slauch, 2011). But the excess of ROS may be dangerous for body, and its detection is significant in monitoring some diseases (Rawson et al., 2015). In

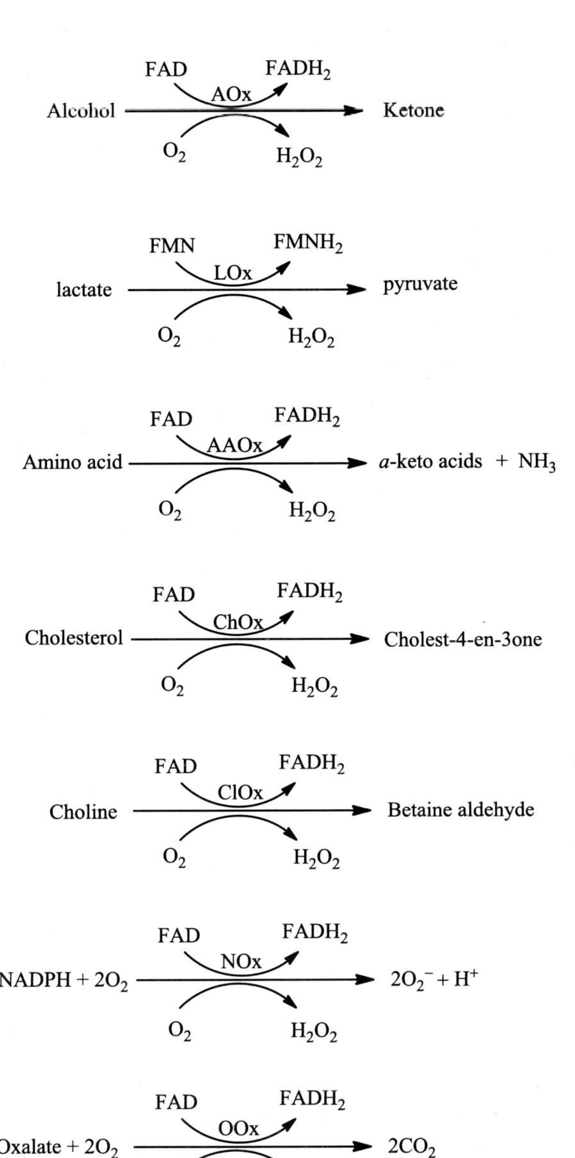

SCHEME 7.7

Catalytic reaction pathways by different flavoprotein-dependent enzymes in liver tissues.

Scheme 7.7 the enzymatic reactions catalyzed by enzyme assisted by FAD or FMN cofactors is summarized.

Multicopper cofactor-dependent oxidase enzymes: cytochrome c oxidase, ascorbate oxidase, bilirubin oxidase, laccase, tyrosinase, hemocyanine, azurin

and galactose oxidase are eight major copper-containing enzymes that oxidize the oxidation of a variety of substances, such as phenols, aromatic amines, L-ascorbate as well as inorganic ions assisted by copper cluster and heme cofactors in liver organisms. During these reactions, electrons are transferred to molecular oxygen, and reduced to water through four-electron transfer (Shleev et al., 2005, 2006; Sakurai and Kataoka, 2007; Singh et al., 2012; Csiffáry et al., 2016). Cytochrome c oxidase has a rather complicated electron transfer mechanism, as it has the heme (an iron complex) cofactor beside copper centers (Lucas et al., 2011). Shleev et al. (2005) have published a comprehensive review about direct electron transfer between copper-containing oxidases, and some of the possible reactions are shown in Scheme 7.8. On one hand, the addressed biological process have been suggested for direct detection of destructive chemical ingredients in human body or controlling

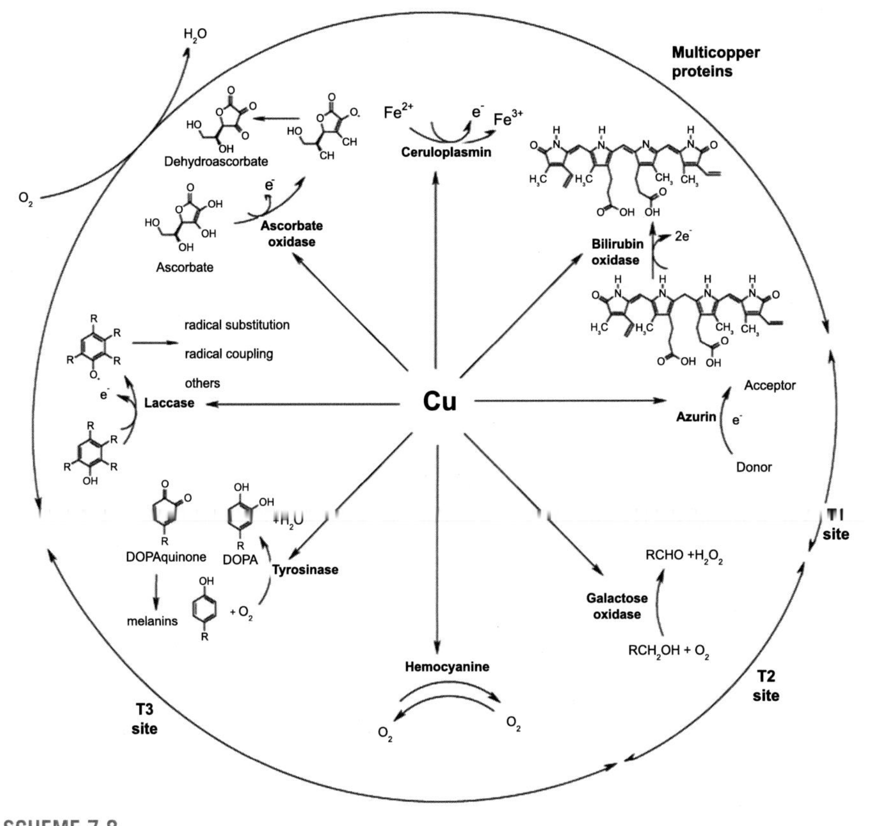

SCHEME 7.8

The essential metabolic functions of copper-containing enzymes.

Reprinted from Elsevier, Shleev, S., Tkac, J., Christenson, A., Ruzgas, T., Yaropolov, A.I., Whittaker, J.W., Gorton, L., 2005. Direct electron transfer between copper-containing proteins and electrodes. Biosensors and Bioelectronics 20, 2517–2554, with permission.

environmental pollutions (Shleev et al., 2005, 2006; Csiffáry et al., 2016; Aghamiri et al., 2018).

For example, a recent publication by Aghamiri et al. (2018) has reviewed a number of trends in immobilization of cytochrome c oxidase aimed at detection of hydrogen peroxide, superoxide (reactive oxygen), and nitric oxide. Laccase or tyrosinase has been suggested for amperometric detection of polyphenol at graphene oxide/carbon nanotube electrode (Vlamidis et al., 2017). Moreover, the oxygen reduction phenomena in the catalytic reaction process have been proposed to apply the immobilized multicopper oxidase enzymes as cathode electrodes in setting up the biofuel cells. Accordingly, the resulted biofuel cells can be practically performed for sensing applications and detection of desired analytes (Fu et al., 2018).

Dehydrogenases catalyze the oxidation of substrates by using various cofactors as electron acceptors. Hence, unlike oxidases, the catalytic reactions of dehydrogenases not depend on the oxygen existence. Also, dehydrogenases sometimes catalyze the reverse reactions such as producing alcohol from acetaldehyde or glutamate from α-keto glutamate (Berg et al., 2002).

Hundreds of dehydrogenases are discovered in the nature. Some important of them are listed in Scheme 7.5. The nonprotein part that acts as electron acceptor may be NAD(P), PQQ, FAD or FMN, and molybdopterin (Ferri et al., 2011; Berg et al., 2002; Hille et al., 2014). Unlike other cofactors, NAD(P) not connected to the enzymes, and it is required to be added when the corresponding enzymes are used in electrochemistry. FAD, PQQ and other cofactors are tightly bound to the active moiety of enzymes, and integrated structures can be used in electrochemistry. Beside aforementioned cofactors, dehydrogenase enzymes commonly contain heme complex, or in some cases, Fe-S cluster and metal ions such as zinc (in alcohol dehydrogenase) for sequential electron transfer between substrate and enzymes (Xu and Minteer, 2013; Yoshida et al., 2015; Rozeboom et al., 2015; Bollella et al., 2018b). Alcohol, glucose, and lactate dehydrogenases have attracted tremendous interest in electrochemistry, as the controlling of their substrates in human body and food samples is of great importance (Bollella et al., 2018b; Schubert et al., 1990; Yamanaka and Mascini, 1992; Jena and Ra, 2006; Monosík et al., 2012; Lin et al., 2018; Navaee and Salimi, 2018). There are few reports about other dehydrogenases (Ikeda et al., 1993; Cruys-Bagger et al., 2012; Ludwig et al., 2013; Cruys-Baggera et al., 2014; Takeda et al., 2016), but addressing to all of them is not feasible in this chapter. For example, cellobiose dehydrogenase is used for detecting cellobiohydrolase (Cruys-Bagger et al., 2012) and biorecognition of carbohydrates (such as cellulose), quinones and catecholamines (Ludwig et al., 2013). Additionally, pyranose dehydrogenase shows the similar activity toward kinetic studding of hydrolases and detection of carbohydrates (Cruys-Baggera et al., 2014; Takeda et al., 2016).

In glucose sensing expertise, GDHs are more favorable, as its catalytic reaction is not depending on oxygen. Also, PQQ and FAD-based GDHs are appropriate for fabrication of third-generation (mediatorless) biosensors. In earlier studies, generation of NADH by GDH through glucose oxidation was the basic concept of glucose

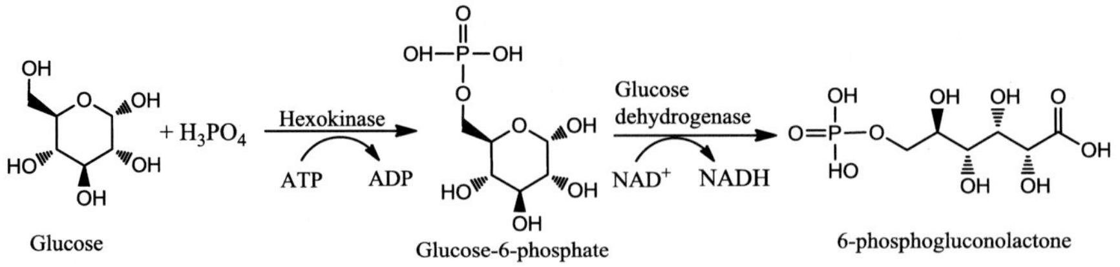

SCHEME 7.9

The catalytic reaction pathway of phosphorylation and oxidation by hexokinase and glucose-6-phosphate dehydrogenase, respectively.

sensing. Accordingly, glucose is firstly phosphorylated by hexokinase enzyme; in which, ATP is reduced to ADP (this reaction in body is the only source of energy for muscles actions). Then, the resulted glucose-6-phosphate is subsequently converted to 6-phosphogluconolactone through enzymatic reaction by glucose-6-phosphate dehydrogenase in presence of NAD^+ (Scheme 7.9). NADH is spectrochemically detected at 340 nm as an indication of glucose concentration (Peterson and Young, 1968; Slein, 1963). Still, this strategy is used as a standard method for glucose assay in various clinical laboratories. Also, electrochemical detection of NADH has been widely studied, as its concentration can be associated to the substrates of enzymes such as alcohol or glucose (Yamanaka and Mascini, 1992; Jafari et al., 2014a,b). They have utilized diverse artificial mediators for transferring of electrons to the electrode.

However, recent studies are focused on direct electron transfer of enzymes, which enable the simple and low-cost biosensing assays. FAD-based dehydrogenase enzymes are versatile types rather than other dehydrogenases, and they currently have attracted greater attention as the DET offers a relatively lower redox potential, more selectivity and sensitivity and eliminates the need for toxic artificial electron mediators (Ferri et al., 2011). Achieving DET from enzymes is associated to the efficient immobilization on the solid surface electrodes. Accordingly, nanomaterials are extensively developed for this purpose. Among different nanomaterials, carbon-based nanomaterials, especially carbon nanotubes (Ludwig et al., 2013; Yang et al., 2015) and gold nanoparticles (Saha et al., 2012) have been further developed because of their excellent conductivity and stability. Also, they have displayed the noble characteristics with more compatibility toward biomolecules. The π-π interaction of aromatic moieties of biomolecules with the similar rings in the rigid structure of carbon nanotubes may be a key factor for effective immobilization. In the case of gold structures, it has displayed a good degree of biocompatibility, since it has been used for repairing the rotting teeth from ancient times. The affinity of gold nanostructures toward thiol moieties of biomolecules leads to proper interaction. Furthermore, the nanostructures of gold derivatives can act as redox metal oxide to mediate the electrons between electrode and redox cofactors (Scheme 7.10). Other dehydrogenases almost follow the same mechanism of GDHs and are widely applied for detection of their substrates.

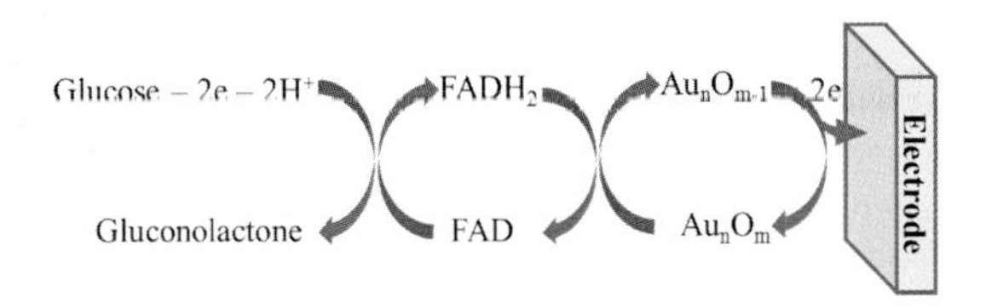

SCHEME 7.10

Proposed electron transfer mechanism from GDH to electrode surface by gold nanoparticles.

Reprinted with permission from Gholami, F., Navaee, A., Salimi, A., Ahmadi, R., Korani, A., Hallaj, R., 2018. Direct enzymatic glucose/O_2 biofuel cell based on poly-thiophene carboxylic acid alongside gold nanostructures substrates derived through bipolar electrochemistry. Scientific Reports 8, 15103, nature publishing groups.

Peroxidases catalyze the wide variety of electron transfer reactions in natural and synthetic substrates assisted by hydrogen peroxide. They are extensively studied by electrochemistry for detection of hydrogen peroxide, halide or metal ions and organic molecules in presence of hydrogen peroxide. They are used in either DET or indirect ET through direct immobilization on the solid support or as a labeling agent with other biomaterial or artificial substances.

Fe (III)-heme is the common cofactor for transfer of electrons, which is presented in Scheme 7.11 (Poulos, 1993; Valentine, 1994). Accordingly, Fe(III)-

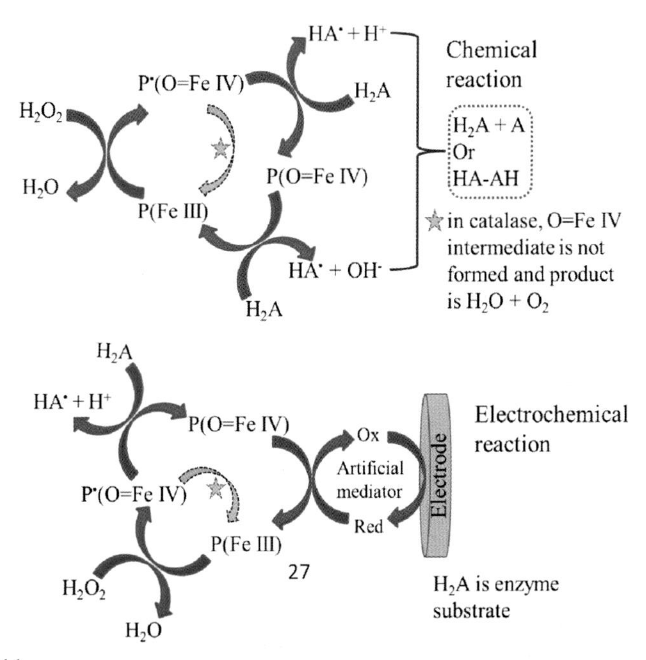

SCHEME 7.11

The mechanism of chemical and electrochemical electron transfer catalyzed by heme-containing peroxidase.

porphyrin is oxidized and an oxygenated cation radical of Fe(IV)-porphyrin is generated. Such radical cation is able to oxidize the substrate in two steps and finally returns to primary Fe(III)-porphyrin. Catalase is the special enzyme for disproportionation of H_2O_2, since O=Fe IV intermediate is not formed, whereas other peroxidases can act on a variety of substances. As shown in the bottom of Scheme 7.11, in an electrochemical reaction, the withdrawing electrons from catalytic oxidation of peroxidase enzymes can be directly or using an artificial chemical mediator are transferred to the electrode, and an electrochemical signal is appeared. Based on such phenomena, several enzymatic biosensors have been developed.

Horseradish peroxidase (HRP), catalase and cytochrome c peroxidase are the well-known peroxidase enzymes, and structure, function, biocatalytic activity, and their perspective have been extensively studied during the past two decades (Poulos, 1993; Valentine, 1994; Dunford, 1999; Leger and Bertrand, 2008; Volkov et al., 2011). In some of peroxidases, beside iron-porphyrin, the catalytic center includes other metal ions. For example, lignin peroxidases such as manganese peroxidase require Mn^{2+} to form Mn^{3+} chelates as diffusing oxidizers for completing the catalytic cycle (Martinez, 2002). Additionally, there are some peroxidase enzymes that have diverse cofactor instead of heme. For example, haloperoxidases usually have the vanadium cofactor in the catalytic center (Wever and Hemrika, 2000). Haloperoxidases are enzymes that catalyze the incorporation of halogen atoms into organic molecules and oxidation reactions (Franssen, 1994). It can be utilized for determination of halides (Kulys and Vidziunaite, 1998) and vanadium. However, the mechanism of electron transfer in all of peroxidase enzymes is almost similar to Scheme 7.10 with slight differences.

Most of reported peroxidase-based biosensors are used to detect H_2O_2 in physiological and food samples (Gaspar et al., 2000; Lindgren et al., 2000; Xu et al., 2003; Salimi et al., 2007a; Bollella et al., 2018a), and lots of them have been proposed for detection of their substrate and related substances (Leger and Bertrand, 2008; Volkov et al., 2011; Rahimi and Joseph, 2019; Pandey et al., 2017; Silwana et al., 2013). Alongside remarkable trends in direct utilization of peroxidase for electrochemical biosensors, additional approaches such as inhibition concept (Silwana et al., 2013; Ghanavati et al., 2014; Sun et al., 2016) and arranging of peroxidases with other enzymes to take the sequence reactions in a unit medium (Gu et al., 2010; Khan and Park, 2015) have been also developed. In the case of inhibition, the inhibitor can compete with substrate for the enzyme-binding site or directly interact with substrate. Similar phenomenon is occurred in various electrochemical reactions such as fouling of receptor by competitive species existing in the reaction matrix or passivation of electrode surface by product(s) of electrochemical reactions. The inhibition model has been suggested for determination of metal ions (Silwana et al., 2013), cyanide (Ghanavati et al., 2014), and sulfide (Sun et al., 2016). In the last case, sulfide ions compete with *o*-phenylenediamine (OPD) substrate (Scheme 7.12) in catalytic reaction of HRP. However, the selectivity of this model still remains as a challenge, as several of species can be oxidized by peroxidases.

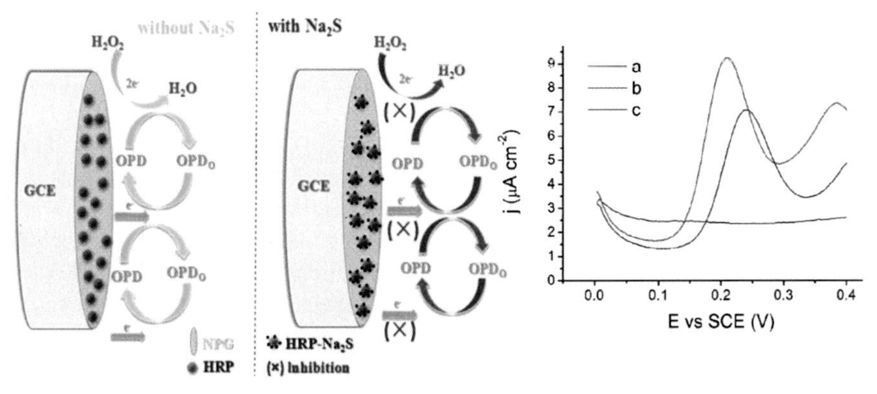

SCHEME 7.12

Electrochemical reaction of HRP/nanoporous gold/glassy carbon electrode (HRP/NPG/GCE) with OPD as a substrate along with differential pulse voltammetry (DPVs) of HRP/NPG/GCE bioelectrode at different conditions: (A) in deaerated PBS (50 mM, pH 7.0); (B) same as (A) with the presence of 100 μM OPD and 0.1 mM H_2O_2; (C) same as (B) with 10 μM sulfide.

Reprinted with permission from Sun, H., Liu, Z., Wu, C., Xu, P., Wang, X., 2016. Amperometric inhibitive biosensor based on horseradish peroxidasenanoporous gold for sulfide determination. Scientific Reports 6, 30905, Nature publishing group.

The sequence enzymes (bienzymes) in an electrochemical system have also been proposed for analysis of lots of substances, specially, glucose. The related studies earlier of 1992 are reviewed by Scheller and Schubert (1992) in which the sequence of NAD^+-dependent steroid dehydrogenase and HRP have been used for assay of 7α-hydroxysteroids. In another effort immobilized luciferase or HRP with oxidases has been suggested for uric acid or cholesterol measurement. More recently, the GOx/HRP binary enzymes have been suggested for glucose detection based on the drawn mechanism in Scheme 7.13 with improved sensitivity than monoenzyme (Gu et al., 2010; Khan and Park, 2015).

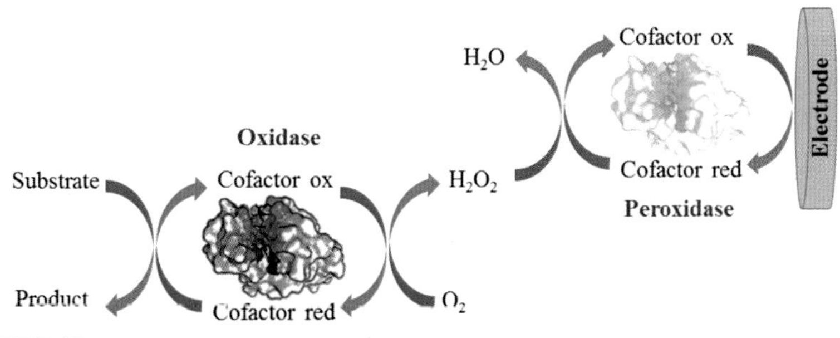

SCHEME 7.13

The possible mechanism of bienzyme composed of oxidase and peroxidase.

With the great attention to the peroxidase enzymes, a lot of natural or synthetic materials have been explored with peroxidase-like activity. Hemoglobin is one example of heme-containing biomolecule that has displayed the peroxidase-like activity and has been immobilized on different nanomaterials aimed at various analysis applications (Wang and Hu, 2001; Saadati et al., 2014) such as nitrite detection in environmental samples (Saadati et al., 2014). Moreover, in recent years, nanostructured enzyme mimics, such as metal oxide nanoparticles, metal organic framework, and so on, have attracted great interest, as promising alternatives for costly and unstable enzymes in biomedical and catalytic applications (Dutta et al., 2012; Wang et al., 2016; Jia et al., 2016; Cao et al., 2017; Nasir et al., 2017). Still, the lower selectivity of enzyme mimic compared with natural protein enzymes remains as a challenge.

Oxygenases oxidize their substrates through incorporating oxygen to them. They are divided into two main groups: monooxygenases and dioxygenases.

Monooxygenases, also called **hydroxylases** or mixed-function oxidases, incorporate a single oxygen atom from molecular oxygen into substrate, and the other oxygen is reduced to water through transferring electrons from substrate.

Dioxygenases or oxygen transferases incorporate both atoms of molecular oxygen into their substrates during an enzymatic reaction.

Oxygenases, a big family of oxidoreductase enzymes, have been widely studied because of selective biooxidation of organic compounds and catalyze a wide variety of organic reactions (Urlacher and Schmid, 2006; Bhagavan and Ha, 2015; Torres Pazmino et al., 2010). Alike oxidases, oxygenases are also divided into different subclasses based on their cofactors, as some of important categories are listed in Scheme 7.5. Oxygenase enzymes may utilize a common reaction intermediate, a ferric hydroperoxide species at the cross-road of mechanism with two parallel mechanistic pathways beyond of that (Badyal et al., 2009). As schematically shown in Scheme 7.14, they commonly contain the heme prosthetic group and work in collaboration with other oxidoreductases containing flavin redox cofactors to transfer electrons from NAD(P)H, as a source of electron, to the heme moiety of oxygenase enzyme (Miller, 2005).

Cytochrome P-450 (CyP450s) monooxygenases, termed for the absorption band at 450 nm, are the most well-known and versatile biocatalysts that are responsible for breaking down several chemicals in the body and catalyze the regio- and stereospecific oxidation of nonactivated hydrocarbons, steroid hormones, certain fats (cholesterol and other fatty acids), and acids used to digest fats (bile acids) in the liver and adrenal cortex (Werck-Reichhart and Feyereisen, 2000). They also lead to detoxify some substances by adding hydroxyl groups that make more water-soluble chemical compounds and more susceptible for further reactions. Alike most enzymatic functions, the energy of oxidation-reduction reactions is finally applied to the ATP, as the more available energy to other reactions. Even though the electrochemistry of CyP450 has been extensively studied (Kazlauskaite et al., 1996; Shukla et al., 2005; Sultana et al., 2005; Sadeghi et al., 2011) and the electrochemical biosensor based on it has been reviewed as well (Yarman et al., 2013), still

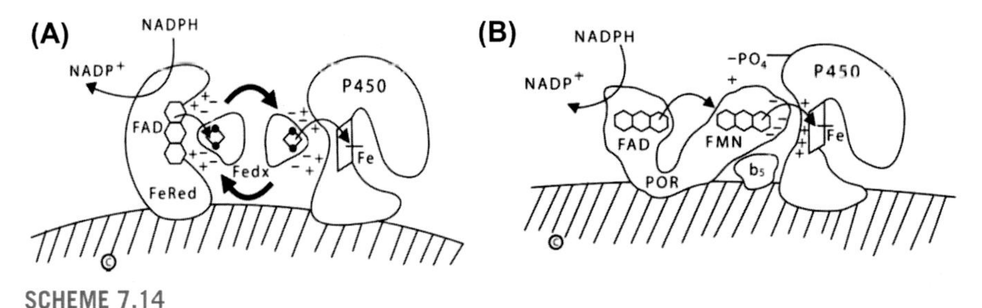

SCHEME 7.14

(A) Diagram of electron transfer by mitochondrial (type I) CyP450 enzymes. Ferredoxin reductase (FeRed), which is bound to the inner mitochondrial membrane, transfers a pair of electrons from NADPH to CyP450. (B) Diagram of electron transfer by microsomal (type II) P450 enzymes. CyP450 oxidoreductase transfer electrons via FAD and FMN moieties to CyP450.

other oxygenases are rarely utilized in electrochemical sensing approaches. Commonly, CyP450 enzymes have displayed a preferable potential in pharmaceutical approaches, as some categories of them are responsible in inactivation of the majority of used drug. For example, they have been used to investigate the polymorphic response in drug metabolism via amperometric techniques (Panicco et al., 2011). Also, a biosensor based on a type of layer-by-layer CyP450 immobilized on PAMA dendrimer and gold nanoparticles has been suggested for detection of caffeine (Müller et al., 2016). Tryptophan-2-monooxygenase and tryptophan-2,3-dioxygenase are the other known oxygenase enzymes, which participate in the catabolism of tryptophan and have attracted attention for biosensing applications. Therefore, they have been applied for electrochemical determination of L-tryptophan (Simonian et al., 1997) and β-triketone herbicides based on the inhibition concept (Rocaboy-Faquet et al., 2016), where a competitive inhibitor has increased the selectivity. However, the inhibition concept or bienzyme methodology has the great potential in developing of biosensors in the future.

Reductases catalyze the reduction of substrate through transferring of electrons from NAD(P)H, as electron donor, to substrate. They catalyze the chemoselective reduction of a wide variety of organic substances with higher oxidized states, such as carbonyl group, imine, carboxylic acid, disulfide, sulfoxide as well as inorganic anion oxides, such as perchlorate, nitrate, nitrite, sulfite, carbonate, arsenate, and so on, to lower oxidized states (Setya et al., 1996; Roldán et al., 2008; Yang et al., 2008; Goldberg et al., 2007; Itoh et al., 2002; Khan et al., 2010; Bardiya and Bae, 2011; Del Giudicea et al., 2013). In liver organisms, these oxidized substances are commonly produced through activity of other aforementioned enzymes to harvest energy (ATP). Notably, the excess of these chemicals sometimes has the destructive effect on the body. Thus, in a healthy tissue, during a systematic and sequential enzymatic reaction, the equilibrium of chemical substances is well adjusted. An example of reductase activity can be referred to the intracellular activity of first two categories listed in flowchart 5, 5α- or 5β-reductase, which catalyze the reduction of 5α- or 5β-steroids to 5α- or 5β-dihydrosteroids (conversion of testosterone into dihydrotestosterone in the prostate gland reduce the size of the prostate volume) (Drury et al., 2009). Some of reductases catalyze the conversion of their special substrate, and probably similar substrates. For example, pentaheme cytochrome c nitrite reductase catalyzes the six-electron reduction of nitrite to ammonia and also sulfite to sulfide (Lukat et al., 2008).

With respect to bioactivity of reductase enzymes, lots of researches have been presented related to electrochemistry of reductases and their biosensing ability. Similar to other oxidoreductase enzymes, the direct electrochemistry of reductases such as fumarate (Sucheta et al., 1993), nitrate (Anderson et al., 2001; Kalimuthu et al., 2016) and CyP450 (Sultana et al., 2007) has been investigated, and they displayed the direct electron transfer at solid electrodes. Among them, more attentions have been paid on CyP450 and nitrate reductases. CyP450 reductases contribute in many oxidoreductase reactions and utilize both FAD and FMN as tightly bound cofactors (Murataliev et al., 2004). As mentioned in Scheme 7.13, CyP450 sometimes acts as electron shuttle between NAD(P)H, as electron donor, and oxygenase enzymes. On the other hand, nitrate substrate is widely used as fertilizer in cultivation, and its high dose in food samples or human body may be toxic. Moreover, nitrate reductase can be sensitive to other chemicals and may be used for fabrication of several biosensors.

The homogenous electron transfer by nitrate reductase can be suggested a like Scheme 7.15A (Kalimuthu et al., 2016), in which the nitrate reductase is known as a multicofactor enzyme. The molybdenum cofactor-binding pterin play an important role in attracting and reducing nitrate to nitrite. Beside Mo, nitrate reductases also contain FAD and heme cofactors, and they transfer electrons from NAD(P)H to the Mo center. Finally, through redox reaction of Mo(IV) to Mo(VI), NO_3- is converted to NO_2-. Perchlorate and nitrate reductase as multicofactor enzymes are belonging to dimethyl sulfoxide (DMSO) reductase family (Bender et al., 2005). But instead of FAD, they contain Fe-S cluster and also depends on the selenide ions.

The heterogeneous electron transfer mediated by hydroquinone (AQH_2) is also presented in Scheme 7.15B, in which electrode acts as electron donor instead of NAD(P)H, and AQH_2 transfers electrons from electrode to Mo(V) intermediate (Kalimuthu et al., 2015). According to this mechanism of electron transfer, a lot of biosensor has been fabricated for detection of nitrate in environmental and physiological samples (Kalimuthu et al., 2015; Glazier, 1998; Quan et al., 2005; Jadán et al., 2017). It has been suggested that xanthine dehydrogenase has the molybdopterin cofactor similar to nitrate and perchlorate reductases (Kalimuthu et al., 2012). But unlike reductases, electrons are transfered on the contrary direction in dehydrogenase enzymes, as substrate is oxidized during reduction of $NAD(P)^+$ to NAD(P)H. Interestingly, catalytic center of two types of known nitrite reductase is completely different from nitrate reductases, as it has the multiheme and multicopper active sites

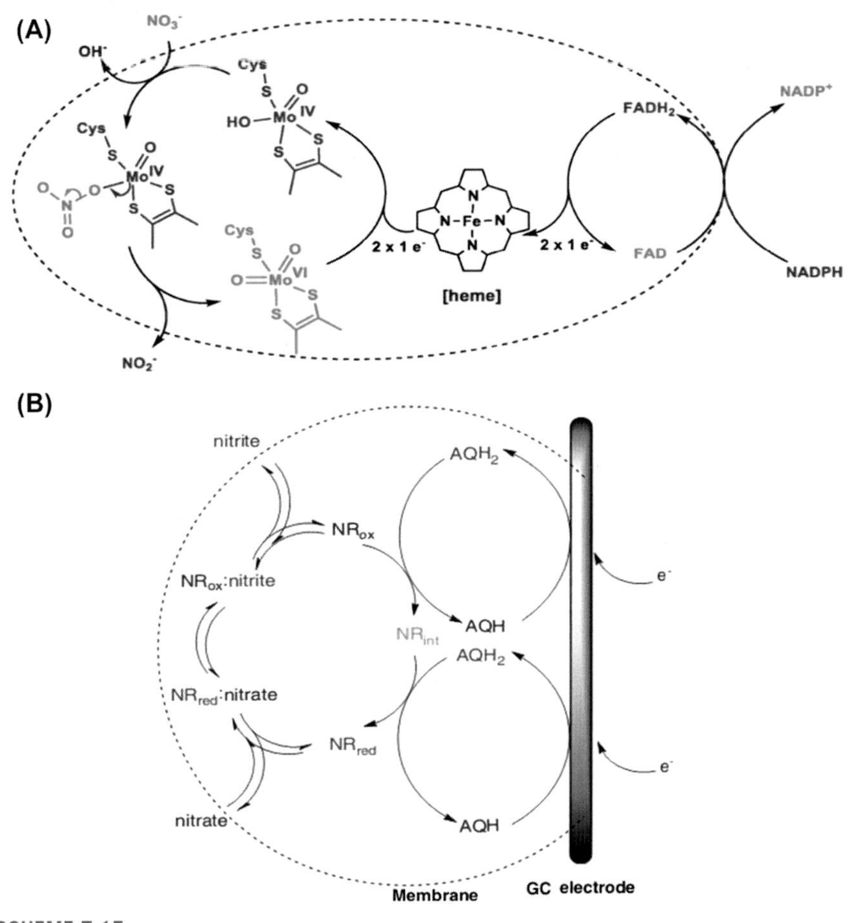

SCHEME 7.15

Electron transfer mechanism by eukaryotic nitrate reductase (A) and immobilized nitrate reductase on a glassy carbon electrode (B).

(A) From Kalimuthu, P., Ringel, P., Kruse, T., Bernhardt, P.V., 2016. Direct electrochemistry of nitrate reductase from the fungus Neurospora crassa. Biochimica et Biophysica Acta 1857, 1506–1513, Elsevier, with permission. (B) From Kalimuthu, P., Fischer-Schrader, K., Schwarz, G., Bernhardt, P.V., 2015. A sensitive and stable amperometric nitrate biosensor employing Arabidopsis thaliana *nitrate reductase. Journal of Biological Inorganic Chemistry 20, 385–393 Springer, with permission.*

(Averill, 1996). The multiheme one has been immobilized beside viologen and successfully applied for nitrite detection (Quan et al., 2006).

There are few reports about other reductases-based biosensors. For example, perchlorate reductase has been applied for monitoring of perchlorate (Okeke et al., 2007), as an environmental pollutions reagent. In another research, 3-hydroxy-3-methyl-glutaryl-coenzyme A reductase (HMG-CoA reductase) has been suggested for detection of mevalonic acid, which is of great interest in health monitoring (Poo-arporn et al., 2017). However, a lot of reductase enzymes such as aldehyde reductases, carboxylic acid reductases, chromate reductase, arsenate reductases, sulfate reductases, and so on are discovered in the nature, and they have a powerful prospective to apply them as receptor in biosensing application, if extraction and purification of enzymes be obtained in a facile and cost-effective way.

7.6 Methods for immobilization of enzymes

Instability of enzymes often limits their usages in sensor devices. To fabricate an enzyme-based electrochemical biosensor, enzymes should be immobilized on a solid surface (electrode) for efficient electron transfer between chemical substrate and electrode surface, which effectively influences on the final success of the biosensors. Various immobilization strategies have been proposed to improve the storage and thermal stability and activity of enzymes. An efficient immobilization technique should create a robust sensor with high sensitivity to the desired analyte, high stability in the ambient and testing media, and be applicable multiple times. Toward this goal, usually an appropriate enzyme stacking on an electrode is obtained when enzyme oriented in a proper direction on electrode surface to offer a facile electron transfer. Additionally, the extramaterial used for stacking enzyme on electrode surface should not hinder the electron transfer between electrode and substrate. There are four main types of enzyme immobilization techniques, which are known as adsorption, covalent bonding, entrapment, and cross-linking methods (Scheme 7.16). A number of review and research papers have been published related to enzyme immobilization techniques along with comparison, advantages, and disadvantages of them (Zhang et al., 2011; Dominguez-Benetton et al., 2013; Yan et al., 2015;

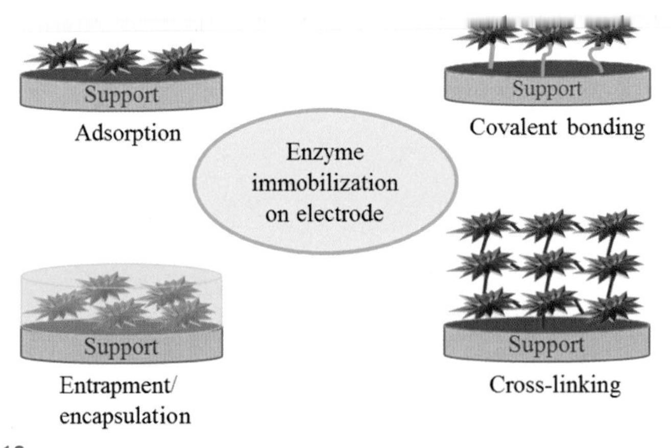

SCHEME 7.16

Common types of enzyme immobilization on electrode surface.

Ronkainen et al., 2010; Milton et al., 2016; O'Conghaile et al., 2013; Nguyen and Kim, 2017; Datta et al., 2013).

7.6.1 Adsorption

The simple way for enzyme immobilization is the adsorption, which is mainly based on the intramolecular forces, such as van der Waals, ionic, π-π or hydrogen bonding forces between enzymes and conductive materials. Carbon-based materials such as graphite, carbon nanotubes, fullerene, graphene and carbon nanofibers along with noble metals, such as Au, Pt or Ag and metal oxide nanoparticles, are used as common conductive support for enzyme immobilization. Generally, enzymes have different functional groups such as thiol-terminated groups with strong affinity to metal surfaces, especially Au structures (du Toit and Di Lorenzo, 2014). However, to improve the efficiency, the electrode surface should be modified to achieve the higher enzyme loading. For example, the layer-by-layer stacking, in which enzymes are sandwiched between two layers of conductive polyions with opposite charges (Scheme 7.17), is the known strategy to improve the immobilization efficiency (Anzai et al., 1998; Zheng et al., 2004; Iost and Crespilho, 2012; Barsana et al., 2014). In view of that, different polyions such as chitosan, polyethyleneimine, polystyrene sulfonate, poly-butanylviologen and similar materials have been widely used for layer-by-layer adsorption of enzymes on electrode surface. On the other hand, some of enzymes need to be oriented on the surface of electrode. Subsequently, the hydrophobicity or hydrophilicity of electrode surface should be altered through primary functionalization by desired chemical substances, as many strategies have been reviewed by Hitaishi et al. (2018). Additionally, these prefunctionalization approaches can be used for better adsorption of enzyme on electrode surface (Wang et al., 2012).

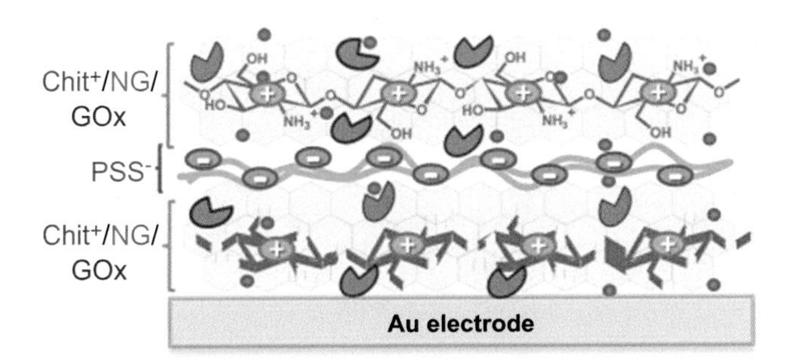

SCHEME 7.17

Layer-by-layer strategy for enzyme immobilization using chitosan (Chit) as polycation and polystyrene sulfonate (PSS) as polyanion.

Reprinted from Barsana, M.M., David, M., Florescu, M., Tugule, L., Brett, C.M.A., 2014. A new self-assembled layer-by-layer glucose biosensor based on chitosan biopolymer entrapped enzyme with nitrogen doped graphene. Bioelectrochemistry 99, 46–52, Elsevier, with permission.

7.6.2 Covalent bonding

This strategy is in the next of research interest because it leads to stabilization of the enzyme. It, beside adsorption phenomena, leads to maximal bonding of enzyme subunits with electrode surface. But, the solid support needs to be modified for desired chemical interactions with enzyme. Several chemical compounds for covalent immobilization of enzymes have been proposed for biochemical performance of enzymes, but a few of them are applicable in bioelectrochemical approaches, because the chemical modifier may decrease the activity of enzyme and hinder the electron transfer. Moreover, it may be a challenge that the covalent bonding exactly took place or not. On the other hand, compared with the adsorption phenomena, chemical bonding is usually irreversible and electrode is not reusable. However, some electroactive modifiers such as ferrocenecarboxylic acid retain the electron transfer, where the amide bond between amine of enzyme and carboxylic acid of ferrocene acts as electron-relaying centers for transferring of electrons (Degani and Heller, 1987). A number of chemical linkers, such as organosilanes or organic molecules containing $-COOH$, $-NH_2$, $-SH$, $-OH$ terminal moieties, such as glutaraldehyde, thiol-containing amino acids, polypyrrole, 1,6-hexanedithiol, and so on, can been used for direct covalent bonding between enzyme and electrode (Salimi and Noorbakhsh, 2011; Dominguez-Benetton et al., 2013; Schuhmann et al., 1990; Lötzbeyer et al., 1994; Wu et al., 1999; Xu et al., 2006; Zucca and Sanjust, 2014).

An example of covalent bonding is shown in Scheme 7.18, which has been proposed some decades ago and still can be implemented with some modification.

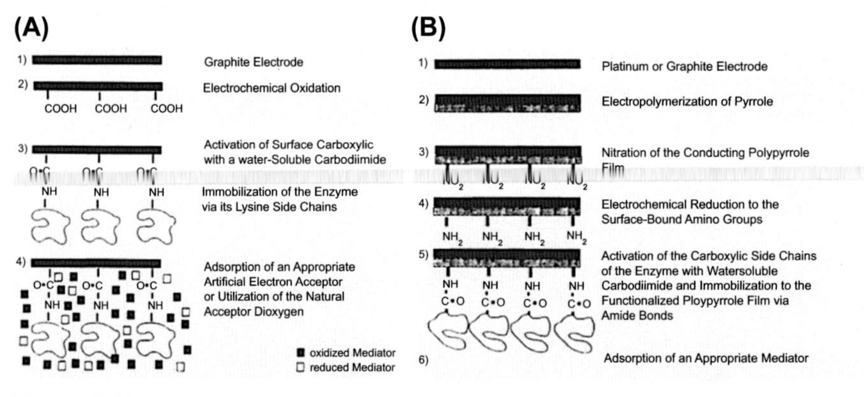

SCHEME 7.18

Preparation of graphite electrode through electrochemical treatment to create the carboxyl functional groups aimed at postfunctionalization with enzyme (A) and electropolymerization of pyrrole followed by electrochemical reduction of $-NO_2$ to $-NH_2$ and then incubated by enzyme containing activated carboxylic groups to form amid bonds (B).

Reprinted with permission from Schuhmann, W., Lammert, R., Uhe, B., Schmidt, H.-L., 1990. Polypyrrole, a new possibility for covalent binding of oxidoreductases to electrode surfaces as a base for stable biosensors. Sensors and Actuators B: Chemical, 537–541 Elsevier.

Accordingly, graphite or other carbon-based materials can be chemically or electrochemically treated to generate the carboxyl functional groups (Scheme 7.18A). Then, the resulted functional groups can be activated by 1-ethyl-3-(3-dimethylaminopropyl) carbodiimide/N-hydroxysuccinimide (EDC)/NHS) or N,N′-dicyclohexylcarbodiimide (DCC) followed by enzyme incubation until the covalent bonding is taking place. It can be performed in a parallel way, such as polymerization of an organic compound containing a functional group on the surface of electrode followed by chemical or electrochemical treatment, activation and incubation by enzyme (Scheme 7.18B).

7.6.3 Entrapment/encapsulation

The third method is entrapment or sometimes called encapsulation, in which a gel-like or membrane-like stuff help to hold enzyme on the electrode surface. Several strategies have been used for proper entrapment and encapsulation into hydrogels, conductive polymers, sol-gels, etc. Essentially, enzyme is physically entrapped and stabilized in every direction within the polymer lattice. The gel matrix should have an efficient conductivity and must be permeable to substrate and product to ensure continuous transformation. Compared with the covalent bonding, this method is simpler and leads to preserve intrinsic properties of enzyme on the electrode surface for a longer time. However, the orientation of enzyme in the gel lattice is not controllable, and consequently, this method may not be applicable for some of the enzymes. Moreover, disjunction of some part of enzyme from electrode and each other could decrease the corresponding activity of immobilized enzyme unit.

An example of sol-gel-based method is the entrapment of glucose oxidase into tetraethoxyorthosilicate and Triton X-100, which has led to a highly stable electrochemical glucose biosensor (Salimi et al., 2004). Polyamines and polyalcohols such as polyvinyl alcohol, acrylamide, alginate, and similar materials that mimic the natural environment of the enzymes have been widely used for developing the hydrogel aimed at enzyme entrapment. But they are not valuable for enzyme electrode because they suffer from low conductivity. However, the lake of their conductivity can be compensated through integration with noble metal nanoparticle. But the utilization of conducting polymer for hydrogel assembly (Scheme 7.19A) may be more appropriate (Li et al., 2015). Another interesting way is the encapsulating of enzyme into microsphere materials or composite of graphene-metal oxide nanostructures. For example, enzyme has been encapsulated into hollow TiO_2-reduced graphene oxide (rGO) microsphere on the electrode surface (Liu et al., 2017). To enhance integration of composite, the entire enzyme/composite is encapsulated by Nafion (Scheme 7.19B). Nafion is interested as proton-conducting polymer and widely used as a membrane material; also it can be used for stabilizing and encapsulating of enzymes (Liu et al., 2017; Meredith et al., 2012) with negligible hindrance effect on substrate diffusion and electron transfer.

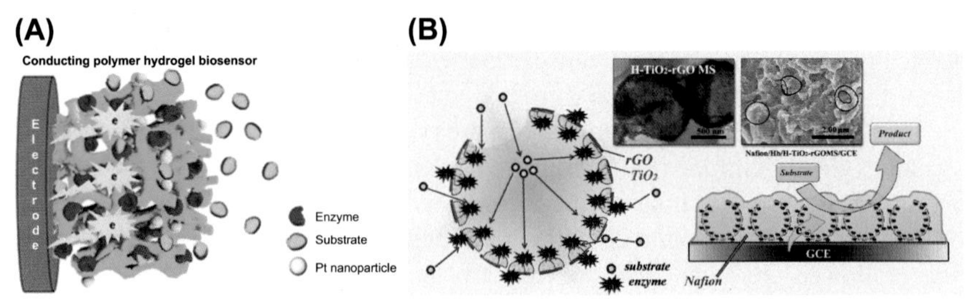

SCHEME 7.19

(A) Schematic of a biosensor based on conducting polymer hydrogel in which the electron transfer is improved through incorporating polymer gel with platinum nanoparticles. (B) Encapsulation of hemoglobin into hollow TiO_2-rGO microsphere incorporated by Nafion (Nafion/Hb/TiO_2-rGO MS/GCE).

(A) Reprinted with permission from Li, L., Shi, Y., Pan, L., Shi, Y., Yu, G., 2015. Rational design and applications of conducting polymer hydrogels as electrochemical biosensors. Journal of Materials Chemistry B 3, 2920–2930, RSC (B) Reprinted with permission from Liu, H., Guo, K., Duan, C., Dong, X., Gao, J., 2017. Hollow TiO_2 modified reduced graphene oxide microspheres encapsulating hemoglobin for a mediator-free biosensor. Biosensors and Bioelectronics 87, 473–479, Elsevier.

7.6.4 Cross-linking

The latest technique that will be discussed here is the cross-linking method or sometimes called cross-linked enzyme crystals (CLECs) in which a bifunctional chemical linker is used to cross-link the adjacent enzyme molecules through covalent bond formation. Compared with others, this method relatively enhances the surface concentration and stabilization of enzyme and decreases the enzyme leakage from solid surfaces. The bifunctional groups can be identical (such as glutaraldehyde, thiourea, glyoxal, and imidoesters) or nonidentical (such as polypyrrole, L-cysteine, DNA) reactive groups. Among them, glutaraldehyde is the most common bifunctional reagent for protein cross-linking.

However, this method has some disadvantages such as complicated experimental processes and difficulties in mass transfer of substrate. The cross-linker agents are also used for integrating enzyme with electrode surface. Alike other methods, the electrode can be primarily modified to enhance immobilization and stability. A number of electrochemical-based cross-linking methods have been proposed. For example, immobilization of bilirubin oxidase assisted by polypyrrole (Chauhan et al., 2016), immobilization of acetylcholinesterase assisted by graphene oxide/silver nanocluster/chitosan for fabrication of inhibition-based biosensor toward organophosphorus pesticides detection (Zhang et al., 2015) and GOx immobilization cross-linked by glutaraldehyde fort glucose detection (Tang et al., 2014; Palod and Singh, 2015). As illustrated in Scheme 7.20, the proposed cross-linking procedures have significantly increased the enzyme loading and consequently enhance the selectivity of biosensor.

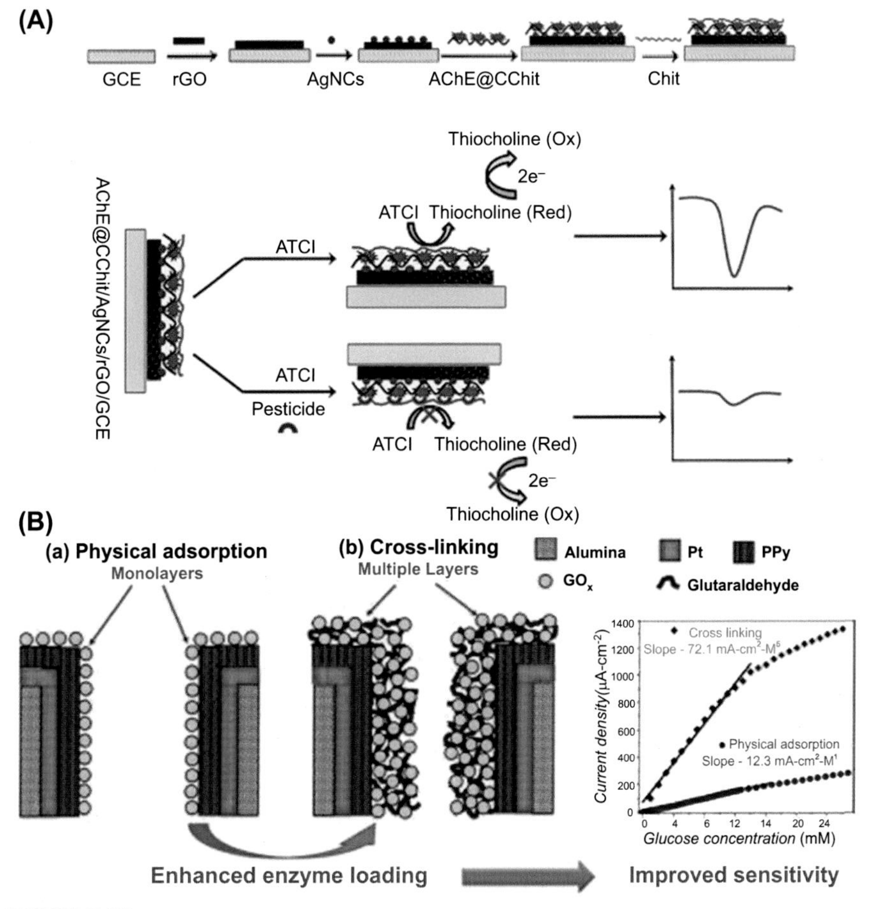

SCHEME 7.20

Two typical electrochemical enzymatic biosensors based on enzyme cross-linking. (A) Inhibition-based biosensor for detection of organophosphorus pesticides using acetylcholinesterase cross-linked by carboxylic chitosan immobilized on rGO/Ag nanoparticles. (B) Immobilization of GOx on polypyrrole nanotubes cross-linked by glutaraldehyde.

(A) Reprinted with permission from Zhang, Y., Liu, H., Yang, Z., Ji, S., Wang, J., Pang, P., Feng, L., Wang, H., Wu, Z., Yang, W., 2015. An acetylcholinesterase inhibition biosensor based on a reduced graphene oxide/silver nanocluster/chitosan nanocomposite for detection of organophosphorus pesticides. Analytical Methods 7, 6213–6219, RSC. (B). Reprinted with permission from Palod, P.A., Singh, V., 2015. Improvement in glucose biosensing response of electrochemically grown polypyrrole nanotubes by incorporating crosslinked glucose oxidase. Materials Science and Engineering: C 55, 420–430, Elsevier.

Some strategies have been suggested based on the coupling of two or more immobilization methods. For example, El Kaoutit et al. (2007) have prepared three amperometric biosensors based on immobilization of tyrosinase on the sonogel-carbon electrode for detection of phenols and polyphenols. They have compared the simple adsorption of the enzyme on the sonogel-carbon electrode with the sandwich configuration electrodes prepared through adsorption followed by entrapment of more layers of enzyme into polyethylene glycol or ion-exchanger Nafion. Interestingly, they saw more activity from Nafion/tyrosinase/sonogel-carbon toward phenols and polyphenols in real samples, demonstrating the more electrochemical compatibility of Nafion than polyethylene glycol.

The immobilized enzymes on conductive surface can be performed in three measurement systems (Rocchitta et al., 2016; Castillo et al., 2004):

1. Off-line procedure, in which biological samples are collected and target analytes are measured using biosensor-based analytical equipment. Therefore, samples may be stored for a long time when the sensing equipment is available.
2. In vivo procedure, in which the enzymatic biosensors are used as implanted device and concentrations of the analytes in extracellular medium are continuously detected. Such biosensors should have high selectivity and sensitivity and cannot be applicable for those analytes that need pre-preparations.
3. Online procedure, in which a sampling device connected to the body or biological samples directly transfers the samples to the biosensor element. This method is sometimes called the "flow-through system."

7.7 Kinetics of immobilized enzyme

An enzymatic reaction can be described as two sequential steps, in which the first step is reversible and an intermediate of enzyme substrate (E-S) complex is formed (Eq. 7.2). Then, in the last step, the product(s) (P) is released from the enzyme by a constant velocity of k_2:

$$\mathrm{E} + \mathrm{S} \underset{k_{-1}}{\overset{k_1}{\rightleftarrows}} ES \overset{k_2}{\rightarrow} E + P \tag{7.2}$$

The constant rate of this reaction has been defined as Eq. (7.3), and it is known as Michaelis-Menten constant (K_M). The velocity of substrate consumption by enzyme in a chemical reaction is shown in Fig. 7.3A, and its relationship to the K_M can be given by Michaelis-Menten Eq. (7.4) (Berg et al., 2002):

$$K_M = \frac{k_2 + k_{-1}}{k_1} \tag{7.3}$$

$$V_0 = \frac{V_{max[S]}}{K_M + [S]} \tag{7.4}$$

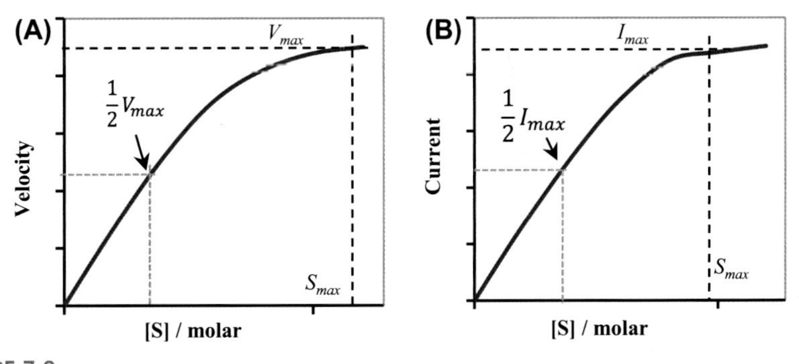

FIGURE 7.3

Velocity of an enzymatic reaction as a function of substrate concentration in a chemical reaction (A) and in an electrochemical reaction (B).

where V_0 is the initial reaction rate, V_{max} is the maximum reaction rate, and $[S]$ is the initial substrate concentration. Hence, K_M represents the amount of substrate required for the enzyme to obtain half of its maximum rate of reaction. Based on the above discussion, Eq. (7.5) can be adapted from Eq. (7.4) and Fig. 7.3B for the electrochemical reactions, which is widely recommended in many literature as the effective kinetic factor of enzymatic reaction at electrode surface (Volkov, 2003; Zhang et al., 2008; Cosnier, 2013; Luz et al., 2014):

$$I_{ss} = \frac{I_{max[S]}}{K_M^{app} + [S]} \tag{7.5}$$

where I_{ss} is the steady-state currents of amperometric measurement for successive addition of substrate, I_{max} is the maximum current under saturated substrate condition, S is the bulk concentration of substrate, and K_M^{app} represents the apparent Michaelis-Menten constant, as the catalyst is the enzyme/electrode, not enzyme itself. K_M^{app} can be simply determined by taking concentration value corresponding to the half of maximum current and has a unit of concentration. Based on the above equations, low K_M value designates the larger attraction of substrate by enzyme and higher rate of reaction. Oppositely, the high value of K_M indicates that substrate does not efficiently bind to the enzyme, and more concentration of substrate is required for reaching the maximum velocity. Hence, the rate of enzymatic reaction may be affected by immobilization technique and solid substrates, which are used for enzyme loading. The solid substrate must have the adequate conductivity, chemical stability, and compatibility with biomolecules and easily be adjusted for desired electron transfer. Sometimes, hydrophobicity or hydrophilicity of solid substrate should be adjusted for desired orientation of enzyme aimed at facilitation of the electron transfer. As previously declared, carbon-based nanomaterials and gold nanostructures are in the first relevant solid substrates, as they have the high conductivity and have shown the enzyme-friendly environment.

Additionally, they can be easily functionalized with different chemical compounds to prepare the desired substrate depending on the preferred immobilization technique.

Conclusion

As a conclusion to this chapter, it should be mentioned that there are a much number of enzymes in the nature, which catalyze the complicated reaction, which mechanism is still unknown. But difficulties in extraction, purification, and storage of enzymes are the main challenges in developing of enzyme-based biosensors technologies. However, in a chemical viewpoint, tremendous efforts have been done to offer the low-cost procedures for preparation of enzymes. In the electrochemical viewpoint, a number of methods have been developed toward the appropriate immobilization and enhancement of stability and sensitivity of enzymatic biosensors. Simplicity, cost, processing time, and reproducibility of the immobilization method have to be taken into consideration. The choice of the most appropriate and judicious immobilization technique tailored designing a more sensitive and stabling biosensor. Nanomaterials with unique physical and chemical properties have an advantage of high surface area for loading enzymes, which allow developing effective biosensors. From a practical point of view, stability and reproducibility are important factors influencing the performances of enzymatic biosensors. In recent years, nanostructures-based enzyme mimics have attracted an increasing attention because of their cost-effectiveness and stability in the ambient temperatures. Moreover, among the various enzymatic biosensors, glucose biosensors is grown more rapidly, whereas progress in other electrochemical biosensors is still in the half way stage. Furthermore, miniaturization and integration of enzymatic biosensor in portable instruments could simplify and extend the applicability of such biosensors. Hence, there is a great prospective in studying and developing enzymes and enzyme mimics toward biosensing in environmental, food industry or biological fields.

References

Aghamiri, Z.S., Mohsennia, M., Rafiee-Pour, H.A., 2018. Immobilization of cytochrome *c* and its application as electrochemical biosensors. Talanta 176, 195–207.

Anderson, L.J., Richardson, D.J., Butt, J.N., 2001. Catalytic protein film voltammetry from a respiratory nitrate reductase provides evidence for complex electrochemical modulation of enzyme activity. Biochemistry 40, 11294–11307.

Anzai, J., Kobayashi, Y., Suzuki, Y., Takeshita, H., Chen, Q., Osa, T., Hoshi, T., Du, X., 1998. Enzyme sensors prepared by layer-by-layer deposition of enzymes on a platinum electrode through avidin-biotin interaction. Sensors and Actuators B 52, 3–9.

Averill, B.A., 1996. Dissimilatory nitrite and nitric oxide reductase. Chemical Reviews 96, 2951–2964.
Badyal, S.K., Eaton, G., Mistry, S., Pipirou, Z., Basran, J., Metcalfe, C.L., Gumiero, A., Handa, S., Moody, P.C.E., Raven, E.L., 2009. Evidence for heme oxygenase activity in a heme peroxidase. Biochemistry 48, 4738–4746.
Baghayer, M., Veisi, H., Ghanei-Motlagh, M., 2017. Amperometric glucose biosensor based on immobilization of glucose oxidase on a magnetic glassy carbon electrode modified with a novel magnetic nanocomposite. Sensors and Actuators B: Chemical 249, 321–330.
Bardiya, N., Bae, J.H., 2011. Dissimilatory perchlorate reduction: a review. Microbiological Research 166, 237–254.
Barsana, M.M., David, M., Florescu, M., Tugule, L., Brett, C.M.A., 2014. A new self-assembled layer-by-layer glucose biosensor based on chitosan biopolymer entrapped enzyme with nitrogen doped graphene. Bioelectrochemistry 99, 46–52.
Bender, K.S., Shang, C., Chakraborty, R., Achenbach, L.A., Belchik, S.M., Coates, J.D., 2005. Identification, characterization, and classification of genes encoding perchlorate reductase. Journal of Bacteriology 187, 5090–5096.
Berg, J.M., Tymoczko, J.L., Stryer, L., 2002. Biochemistry, fifth ed. Freeman and Company.
Bhagavan, N.V., Ha, C.E., 2015. Essentials of Medical Biochemistry. Elsevier. Chapters 13 and 15.
Bhagavan, N.V., 2002. Protein and amino acid metabolism. In: Chapter 17 of Medical Biochemistry, fourth ed. Elsevier, pp. 331–363.
Blank, L.M., Ebert, B.E., Buehler, K., Buhler, B., 2010. Redox biocatalysis and metabolism: molecular mechanisms and metabolic network analysis. Antioxidants and Redox Signaling 13, 349–394.
Bollella, P., Medici, L., Tessema, M., Poloznikov, A.A., Hushpulian, D.M., Tishkov, V.I., Andreu, R., Leech, D., Megersa, N., Marcaccio, M., Gorton, L., Antiochia, R., 2018a. Highly sensitive, stable and selective hydrogen peroxide amperometric biosensors based on peroxidases from different sources wired by Ospolymer: a comparative study. Solid State Ionics 314, 178–186.
Bollella, P., Gorton, L., Antiochia, R., 2018b. Direct electron transfer of dehydrogenases for development of 3rd generation biosensors and enzymatic fuel cells. Sensors 18, 1319.
Cai, C., Chen, J., 2004. Direct electron transfer of glucose oxidase promoted by carbon nanotubes. Analytical Biochemistry 332, 75–83.
Cao, D., He, P., Hu, N., 2003. Electrochemical biosensor utilizing electron transfer in heme protein immobilized on Fe_3O_4 nanoparticles. The Analyst 128, 1268–1274.
Cao, G.J., Jiang, X., Zhang, H., Croley, T.R., Yin, J.J., 2017. Mimicking horseradish peroxidase and oxidase using ruthenium nanomaterials. RSC Advances 7, 52210–52217.
Castillo, J., Gáspár, S., Leth, S., Niculescu, M., Mortari, A., Bontidean, I., Soukharev, V., Dorneanu, S.A., Ryabov, A.D., Csoregi, E., 2004. Biosensors for life quality: design, development and applications. Sensors and Actuators B: Chemical 102, 179–194.
Chauhan, N., Rawal, R., Hooda, V., Jain, U., 2016. Electrochemical biosensor with graphene oxide nanoparticles and polypyrrole interface for the detection of bilirubin. RSC Advances 6, 63624–63633.
Clark Jr., L.C., Lyons, C., 1962. Electrode systems for continuous monitoring in cardiovascular surgery. Annals of the New York Academy of Sciences 102, 29–45.
Cock, L.S., Arenas, A.M.Z., Aponte, A.A., 2009. Use of enzymatic biosensors as quality indices: a synopsis of present and future trends in the food industry. Chilean Journal of Agricultural Research 69, 270–280.

A Comprehensive Approach of Life Science, Part 1 (Chapter 4), ATP and Enzymes. University of Tokyo.

Cosnier, S. (Ed.), 2013. Electrochemical Biosensors. Taylor and Francis group, LLC.

Cruys-Bagger, N., Ren, G., Tatsumi, H., Baumann, M.J., Spodsberg, N., Andersen, H.D., Gorton, L., Borch, K., West, P., 2012. An amperometric enzyme biosensor for real-time measurements of cellobiohydrolase activity on insoluble cellulose. Biotechnology and Bioengineering 109, 3199–3204.

Cruys-Baggera, N., Badino, S.F., Tokin, R., Gontsarik, M., Fathalinejad, S., Miguel, K.J., Trine, D.T., Sørensen, H., Borch, K., Tatsumic, H., Väljamäed, P., Westh, P., 2014. A pyranose dehydrogenase-based biosensor for kinetic analysis of enzymatic hydrolysis of cellulose by cellulases. Enzyme and Microbial Technology 58–59, 68–74.

Csiffáry, G., Fűtő, P., Adányi, N., Kis, A., 2016. Ascorbate oxidase-based amperometric biosensor for l-ascorbic acid determination in beverages. Food Technology and Biotechnology 54, 31–35.

Datta, S., Christena, L.R., Rajaram, Y.R.S., 2013. Enzyme immobilization: an overview on techniques and support materials. 3 Biotech 3, 1–9.

Degani, Y., Heller, A., 1987. Direct electrical communication between chemically modified enzymes and metal electrodes. 1. Electron transfer from glucose oxidase to metal electrodes via electron relays, bound covalently to the enzyme. Journal of Physical Chemistry 91, 1285–1289.

Del Giudicea, I., Limauroa, D., Pedonea, E., Bartoluccia, S., Fiorentino, G., 2013. A novel arsenate reductase from the bacterium Thermus thermophilus HB27: its role in arsenic detoxification. Biochimica et Biophysica Acta 1834, 2071–2079.

Deponte, M., 2013. Glutathione catalysis and the reaction mechanisms of glutathione-dependent enzymes. Biochimica et Biophysica Acta (BBA) - General Subjects 1830, 3217–3266.

Dominguez-Benetton, X., Srikanth, S., Satyawali, Y., Vanbroekhoven, K., Pant, D., 2013. Enzymatic electrosynthesis: an overview on the progress in enzyme electrodes for the production of electricity, fuels and chemicals. Journal of Microbial & Biochemical Technology S6.

Drury, J.E., Di Costanzo, L., Penning, T.M., Christianson, D.W., 2009. Inhibition of human steroid 5beta-reductase (AKR1D1) by finasteride and structure of the enzyme inhibitor complex. Journal of Biological Chemistry 284, 19786–19790.

Du, D., Ye, X., Cai, J., Liu, J., Zhang, A., 2010. Acetylcholinesterase biosensor design based on carbon nanotube-encapsulated polypyrrole and polyaniline copolymer for amperometric detection of organophosphates. Biosensors and Bioelectronics 25, 2503–2508.

du Toit, H., Di Lorenzo, M., 2014. Glucose oxidase directly immobilized onto highly porous gold electrodes for sensing and fuel cell applications. Electrochimica Acta 138, 86–92.

Dunford, B.H., 1999. Heme Peroxidases. John Wiley & Sons, New York.

Dutta, A.K., Maji, S.K., Srivastava, D.N., Mondal, A., Biswas, P., Paul, P., Adhikary, B., 2012. Synthesis of FeS and FeSe nanoparticles from a single source precursor: a study of their photocatalytic activity, peroxidase-like behavior, and electrochemical sensing of H_2O_2. ACS Applied Materials and Interfaces 4, 1919–1927.

El Harrad, L., Bourais, I., Mohammadi, H., Amine, A., 2018. Recent advances in electrochemical biosensors based on enzyme inhibition for clinical and pharmaceutical applications. Sensors 18, 164.

El Kaoutit, M., Naranjo-Rodriguez, I., Temsamani, K.R., de Cisneros, J.L.H.-H., 2007. The Sonogel–Carbon materials as basis for development of enzyme biosensors for phenols and polyphenols monitoring: a detailed comparative study of three immobilization matrixes. Biosensors and Bioelectronics 22, 2958–2966.

Fan, F., Gadda, G., 2005. On the catalytic mechanism of choline oxidase. Journal of the American Chemical Society 127, 2067–2074.

Ferri, S., Kojima, K., Sode, K., 2011. Review of glucose oxidases and glucose dehydrogenases: a bird's eye view of glucose sensing enzymes. Journal of Diabetes Science and Technology 5, 1068–1076.

Franssen, M.C.R., 1994. Halogenation and oxidation reactions with haloperoxidases. Biocatalysis 10, 87–111.

Fu, L., Liu, J., Hu, Z., Zhou, M., 2018. Recent advances in the construction of biofuel cells based self-powered electrochemical biosensors: a review. Electroanalysis 30, 2535–2550.

Gaspar, S., Popescu, I.C., Gazaryan, I.G., Bautista, A.G., Sakharov, I.Y., Mattiasson, B., Csoregi, E., 2000. Biosensors based on novel plant peroxidases: a comparative study. Electrochimica Acta 46, 255–264.

Ghanavati, M., Azad, R.R., Mousavi, S.A., 2014. Amperometric inhibition biosensor for the determination of cyanide. Sensors and Actuators B: Chemical 190, 858–864.

Gholami, F., Navaee, A., Salimi, A., Ahmadi, R., Korani, A., Hallaj, R., 2018. Direct enzymatic glucose/O_2 biofuel cell based on poly-thiophene carboxylic acid alongside gold nanostructures substrates derived through bipolar electrochemistry. Scientific Reports 8, 15103.

Glazier, S.A., 1998. Construction and characterization of nitrate reductase-based amperometric electrode and nitrate assay of fertilizers and drinking water. Analytical Chemistry 70, 1511–1515.

Goldberg, K., Schroer, K., Lütz, S., Liese, A., 2007. Biocatalytic ketone reduction–a powerful tool for the production of chiral alcohols–part I: processes with isolated enzymes. Applied Microbiology and Biotechnology 76, 237–248.

Gorton, L., Domínguez, E., 2007. Electrochemistry of NAD(P)+/NAD(P)H. In: Bard, A.J. (Ed.), Encyclopedia of Electrochemistry. Wiley-VCH Verlag GmbH & Co. KGaA, pp. 67–143.

Goswami, P., Chinnadayyala, S.S.R., Chakraborty, M., Kumar, A.K., Kakoti, A., 2013. An overview on alcohol oxidases and their potential applications. Applied Microbiology and Biotechnology 97, 4259–4275.

Gu, M., Wang, J., Tu, Y., Di, J., 2010. Fabrication of reagentless glucose biosensors: a comparison of mono-enzyme GOD and bienzyme GOD–HRP systems. Sensors and Actuators B: Chemical 148, 486–491.

Gulaboski, R., Markovski, V., Jihe, Z., 2016. Redox chemistry of coenzyme Q—a short overview of the voltammetric features. Journal of Solid State Electrochemistry 20, 3229–3238.

Hille, R., Hall, J., Basu, P., 2014. The mononuclear molybdenum enzymes. Chemical Reviews 114, 3963–4038.

Hitaishi, V.P., Clement, R., Bourassin, N., Baaden, M., de Poulpiquet, A., Sacquin-Mora, S., Ciaccafava, A., Lojou, E., 2018. Controlling redox enzyme orientation at planar electrodes. Catalysts 8, 192.

Hulanicki, A., Geab, S., Ingman, F., 1991. Chemical sensors: definitions and classification. Pure and Applied Chemistry 63, 1247–1250.

Ikeda, T., Miyaoka, S., Miki, K., 1993. Enzyme-catalysed electrochemical oxidation of D-gluconate at electrodes coated with D-gluconate dehydrogenase, a membrane-bound flavohemoprotein. Journal of Electroanalytical Chemistry 352, 267–278.

Iost, R.M., Crespilho, F.N., 2012. Layer-by-layer self-assembly and electrochemistry: applications in biosensing and bioelectronics. Biosensors and Bioelectronics 31, 1–10.

Itoh, N., Matsuda, M., Mabuchi, M., Dairi, T., Wang, J., 2002. Chiral alcohol production by NADH-dependent phenylacetaldehyde reductase coupled with in situ regeneration of NADH. European Journal of Biochemistry 269, 2394–2402.

Jadán, F., Aristoy, M.C., Toldrá, F., 2017. Biosensor based on immobilized nitrate reductase for the quantification of nitrate ions in dry-cured ham. Food Analytical Methods 10, 3481–3486.

Jafari, F., Salimi, A., Navaee, A., 2014a. Electrochemical and photoelectrochemical sensing of NADH and ethanol based on immobilization of electrogenerated chlorpromazine sulfoxide onto graphene-CdS quantum dot/ionic liquid nanocomposite. Electroanalysis 26, 530–540.

Jafari, F., Salimi, A., Navaee, A., 2014b. Electrochemical and photoelectrochemical sensing of dihydronicotinamide adenine dinucleotide and glucose based on noncovalently functionalized reduced graphene oxide-cadmium sulfide quantum dots/poly-nile blue nanocomposite. Electroanalysis 26, 1782–1793.

Jena, B.K., Ra, C.R., 2006. Electrochemical biosensor based on integrated assembly of dehydrogenase enzymes and gold nanoparticles. Analytical Chemistry 78, 6332–6339.

Jia, H., Yang, D., Han, X., Cai, J., Liu, H., He, W., 2016. Peroxidase-like activity of the Co_3O_4 nanoparticles used for biodetection and evaluation of antioxidant behavior. Nanoscale 8, 5938–5945.

Kacaniklic, I., Johansson, K., Marko-Varga, G., Gorton, L., Jonsson-Pettersson, G., Csoregi, E., 1994. Amperometric biosensors for detection of L- and D-amino acids based on coimmobilized peroxidase and L- and D-amino acid oxidases in carbon paste electrodes. Electroanalysis 6, 381–390.

Kalantar-zadeh, K., Fry, B., 2008. Sensor Characteristics and Physical Effects. Springer.

Kalimuthu, P., Leimkuhler, S., Bernhardt, P.V., 2012. Catalytic electrochemistry of xanthine dehydrogenase. Journal of Physical Chemistry B 116, 11600–11607.

Kalimuthu, P., Fischer-Schrader, K., Schwarz, G., Bernhardt, P.V., 2015. A sensitive and stable amperometric nitrate biosensor employing *Arabidopsis thaliana* nitrate reductase. Journal of Biological Inorganic Chemistry 20, 385–393.

Kalimuthu, P., Ringel, P., Kruse, T., Bernhardt, P.V., 2016. Direct electrochemistry of nitrate reductase from the fungus Neurospora crassa. Biochimica et Biophysica Acta 1857, 1506–1513.

Kazlauskaite, J., Westlake, A.C.G., Wong, L.L., Hil, H.A.O., 1996. Direct electrochemistry of cytochrome P450cam. Chemical Communications 0, 2189–2190.

Khan, M., Park, S.Y., 2015. Glucose biosensor based on GOx/HRP bienzyme at liquid–crystal/aqueous interface. Journal of Colloid and Interface Science 457, 281–288.

Khan, M.S., Haas, F.H., Samami, A.A., Gholami, A.M., Bauer, A., Fellenberg, K., Reichelt, M., Hänsch, R., Mendel, R.R., Meyer, A.J., Wirtz, M., Hell, R., 2010. Sulfite reductase defines a newly discovered bottleneck for assimilatory sulfate reduction and is essential for growth and development in *Arabidopsis thaliana*. The Plant Cell Online 22, 1216–1231.

Kim, M., 2006. Determining citrate in fruit juices using a biosensor with citrate lyase and oxaloacetate decarboxylase in a flow injection analysis system. Food Chemistry 99, 851–857.

Kucherenco, I.S., Kucherenco, D.Y., Soldatkin, O.O., Lagarde, F., Dzyadevych, S.V., Soldatkin, A.P., 2016. A novel conductometric biosensor based on hexokinase for determination of adenosine triphosphate. Talanta 150, 469–475.

Kucherenko, I.S., Didukh, D.Y., Soldatkin, O.O., Soldatkin, A.P., 2014. Amperometric biosensor system for simultaneous determination of adenosine-5′-triphosphate and glucose. Analytical Chemistry 86, 5455–5462.

Kulys, J., Vidziunaite, R., 1998. The determination of halides and pseudohalides by the vanadium haloperoxidase based biosensor. Analytical Letters 31, 2607–2623.

La, M., Zhao, X.Y., Peng, Q.L., Chen, C.D., Zhao, G.Q., 2015. Electrochemical biosensors for probing of protease activity and screening of protease inhibitors. International Journal of Electrochemical Science 10, 3329–3339.

Lee, J.H., Yeon, J., Min, P.K., Cha, H.J., Choi, S.S., Yoo, Y.J., 2010. A novel organophosphorus hydrolase-based biosensor using mesoporous carbons and carbon black for the detection of organophosphate nerve agents. Biosensors and Bioelectronics 25, 1566–1570.

Leger, C., Bertrand, P., 2008. Direct electrochemistry of redox enzymes as a tool for mechanistic studies. Chemical Reviews 108, 2379–2438.

Leiros, I., Wang, E., Rasmussen, T., Oksanen, E., Repo, H., Petersen, S.B., Heikinheimo, P., Hough, E., 2006. The 2.1 A structure of Aerococcus viridans L-lactate oxidase (LOX). Acta Crystallographica F62, 1185–1190.

Li, L., Shi, Y., Pan, L., Shi, Y., Yu, G., 2015. Rational design and applications of conducting polymer hydrogels as electrochemical biosensors. Journal of Materials Chemistry B 3, 2920–2930.

Liang, B., Guo, X., Fang, L., Hua, Y., Yang, G., Zhu, Q., Wei, J., Ye, X., 2015. Study of direct electron transfer and enzyme activity of glucose oxidase on graphene surface. Electrochemistry Communications 50, 1–5.

Lin, C., Pratt, B., Honikel, M., Jenish, A., Ramesh, B., Alkhan, A., Belle, J.T.L., 2018. Toward the development of a glucose dehydrogenase-based saliva glucose sensor without the need for sample preparation. Journal of Diabetes Science and Technology 12, 83–89.

Lindgren, A., Ruzgas, T., Gorton, L., Csoregi, E., Ardila, G.B., Sakharov, I.Y., Gazaryan, I.G., 2000. Biosensors based on novel peroxidases with improved properties in direct and mediated electron transfer. Biosensors and Bioelectronics 15, 491–497.

Liu, G., Lin, Y., 2006. Biosensor based on self-assembling acetylcholinesterase on carbon nanotubes for flow injection/amperometric detection of organophosphate pesticides and nerve agents. Analytical Chemistry 78, 835–843.

Liu, H., Guo, K., Duan, C., Dong, X., Gao, J., 2017. Hollow TiO_2 modified reduced graphene oxide microspheres encapsulating hemoglobin for a mediator-free biosensor. Biosensors and Bioelectronics 87, 473–479.

Lötzbeyer, T., Schuhmann, W., Katz, E., Falter, J., Schmidt, H.L., 1994. Direct electron transfer between the covalently immobilized enzyme microperoxidase MP-11 and a cystamine-modified gold electrode. Journal of Electroanalytical Chemistry 377, 291–294.

Lucas, M.F., Rousseau, D.L., Guallar, V., 2011. Electron transfer pathways in cytochrome *c* oxidase. Biochimica et Biophysica Acta (BBA) - Bioenergetics 1807, 1305–1313.

Ludwig, R., Ortiz, R., Schulz, C., Harreither, W., Sygmund, C., Gorton, L., 2013. Cellobiose dehydrogenase modified electrodes: advances by materials science and biochemical engineering. Analytical and Bioanalytical Chemistry 405, 3637–3658.

Lukat, P., Rudolf, M., Stach, P., Messerschmidt, A., Kroneck, P.M.H., Simon, J., Einsle, O., 2008. Binding and reduction of sulfite by cytochrome c nitrite reductase. Biochemistry 47, 2080–2086.

Luz, R.A.S., Pereira, A.R., de Souza, J.C.P., Sales, F.C.P.F., Crespilho, F.N., 2014. Enzyme biofuel cells: thermodynamics, kinetics and challenges in applicability. ChemElectChem 1, 1751–1777.

Mahmoud, K.A., Kraatz, H.B., 2007. A bioorganometallic approach for the electrochemical detection of proteins: a study on the interaction of ferrocene–peptide conjugates with papain in solution and on Au surfaces. Chemistry - A European Journal 13, 5885.

Mamyabi, M.A., Hajari, N., Tuner, A.P.F., Tiwari, A., 2013. A high-performance glucose biosensor using covalently immobilised glucose oxidase on a poly(2,6-diaminopyridine)/carbon nanotube electrode. Talanta 116, 801–808.

Martinez, A.T., 2002. Molecular biology and structure-function of lignin-degrading heme peroxidases. Enzyme and Microbial Technology 30, 425–444.

Mazzei, F., Botre, F., Montilla, S., Pilloton, R., Podesta, E., Botre, C., 2004. Alkaline phosphatase inhibition based electrochemical sensors for the detection of pesticides. Journal of Electroanalytical Chemistry 574, 95–100.

Meredith, S., Xu, S., Meredith, M.T., Minteer, S.D., 2012. Hydrophobic salt-modified nafion for enzyme immobilization and stabilization. Journal of Visualized Experiments 65, e3949.

Milardović, S., Kereković, I., Nodilo, M., 2008. A novel biamperometric biosensor for urinary oxalate determination using flow-injection analysis. Talanta 77, 222–228.

Miller, W.L., 2005. Minireview: regulation of steroidogenesis by electron transfer. Endocrinology 146, 2544–2550.

Milton, R.D., Wang, T., Knoche, K.L., Minteer, S.D., 2016. Tailoring biointerfaces for electrocatalysis. Langmuir 32, 2291–2301.

Monošík, R., Streďanský, M., Šturdík, E., 2012. Biosensors-classification, characterization and new trends. Acta Chimca Slovaca 5, 109–120.

Monosík, R., Stredansky, M., Luspai, K., Magdolen, P., Sturdík, E., 2012. Amperometric glucose biosensor utilizing FAD-dependent glucose dehydrogenase immobilized on nanocomposite electrode. Enzyme and Microbial Technology 50, 227–232.

Müller, M., Agarwal, N., Kim, J., 2016. A cytochrome P450 3A4 biosensor based on generation 4.0 PAMAM dendrimers for the detection of caffeine. Biosensors 6, 44.

Murataliev, M.B., Feyereisen, R., Ann Walker, F., 2004. Electron transfer by diflavin reductases. Biochimica et Biophysica Acta 1698, 1–26.

Musameh, M., Moezzi, N., Schauman, L.M., Meyerhof, M.E., 2006. Glutathione peroxidase-based amperometric biosensor for the detection of S-nitrosothiols. Electroanalysis 18, 2043–2048.

Nasir, M., Nawaz, M.H., Latif, U., Yaqub, M., Hayat, A., Rahim, A., 2017. An overview on enzyme-mimicking nanomaterials for use in electrochemical and optical assays. Microchimica Acta 184, 323.

Navaee, A., Salimi, A., 2018. FAD-based glucose dehydrogenase immobilized on thionine/AuNPs frameworks grafted on amino-CNTs: development of high power glucose biofuel cell and biosensor. Journal of Electroanalytical Chemistry 815, 105–113.

Nguyen, H.H., Kim, M., 2017. An overview of techniques in enzyme immobilization. Applied Science and Convergence Technology 26, 157–163.

Nomata, J., Hisabori, T., 2018. Development of heme protein based oxygen sensing indicators. Scientific Reports 8, 11849.

O'Conghaile, P., Poller, S., MacAodha, D., Schuhmann, W., Leech, D., 2013. Coupling osmium complexes to epoxy-functionalised polymers to provide mediated enzyme electrodes for glucose oxidation. Biosensors and Bioelectronics 43, 30–37.

Ogunseitan, O.A., Yang, S., Ericson, J., 2000. Microbial-aminolevulinate dehydratase as a biosensor of lead bioavailability in contaminated environments. Soil Biology and Biochemistry 32, 1899–1906.

Okeke, B.C., Ma, G., Cheng, Q., Losi, M.E., Frankenberger Jr., W.T., 2007. Development of a perchlorate reductase-based biosensor for real time analysis of perchlorate in water. Journal of Microbiological Methods 68, 69–75.

Ozimek, P., Veenhuis, M., van der Klei, I.J., 2005. Alcohol oxidase: a complex peroxisomal, oligomeric flavoprotein. FEMS Yeast Research 5, 975–983.

Palod, P.A., Singh, V., 2015. Improvement in glucose biosensing response of electrochemically grown polypyrrole nanotubes by incorporating crosslinked glucose oxidase. Materials Science and Engineering: C 55, 420–430.

Pandey, V.P., Awasthi, M., Singh, S., Tiwari, S., Dwivedi, U.N., 2017. A comprehensive review on function and application of plant peroxidases. Biochemistry & Analytical Biochemistry 6, 1000308.

Panicco, P., Dodhia, V.R., Fantuzzi, A., Gilardi, G., 2011. Enzyme-based amperometric platform to determine the polymorphic response in drug metabolism by cytochromes P450. Analytical Chemistry 83, 2179–2186.

Peterson, J.I., Young, D.S., 1968. Evaluation of the hexokinase/glucose-6-phosphate dehydrogenase method of determination of glucose in urine. Analytical Biochemistry 23, 301–316.

Pollegioni, L., Motta, P., Molla, G., 2013. L-amino acid oxidase as biocatalyst: a dream too far? Applied Microbiology and Biotechnology 97, 9323–9341.

Poo-arporn, R.P., Waraho-Zhmayev, D., Pakapongpan, S., Poo-arporn, Y., Khownarumit, P., Surareungchai, W., 2017. Development of mevalonic acid biosensor using amperometric technique based on nanocomposite of nicotinamide adenine dinucleotide and carbon nanotubes. Journal of the Electrochemical Society 164, B349–B355.

Poulos, T.L., 1993. Peroxidases. Current Opinion in Biotechnology 4, 484–489.

Quan, D., Shim, J.H., Kim, J.D., Park, H.S., Cha, G.S., Nam, H., 2005. Electrochemical determination of nitrate with nitrate reductase-immobilized electrodes under ambient air. Analytical Chemistry 77, 4467–4473.

Quan, D., Min, D.G., Cha, G.S., Nam, H., 2006. Electrochemical characterization of biosensor based on nitrite reductase and methyl viologen co-immobilized glassy carbon electrode. Bioelectrochemistry 69, 267–275.

Rahimi, P., Joseph, Y., 2019. Enzyme-based biosensors for choline analysis: a review. Trends in Analytical Chemistry 110, 367–374.

Rawson, F.J., Hicks, J., Dodd, N., Abate, W., Garrett, D.J., Yip, N., Fejer, G., Downard, A.J., Baronian, K.H.R., Jackson, S.K., Mendes, P.M., 2015. Fast, ultrasensitive detection of reactive oxygen species using a carbon nanotube based-electrocatalytic intracellular sensor. ACS Applied Materials and Interfaces 7, 23527–23537.

Rocaboy-Faquet, E., Barthelmebs, L., Calas-Blanchard, C., Noguer, T., 2016. A novel amperometric biosensor for ß-triketone herbicides based on hydroxyphenylpyruvate dioxygenase inhibition: a case study for sulcotrione. Talanta 146, 510–516.

Rocchitta, G., Spanu, A., Babudieri, S., Latte, G., Madeddu, G., Galleri, G., Nuvoli, S., Bagella, P., Demartis, M.I., Fiore, V., Manetti, R., Serra, P.A., 2016. Enzyme biosensors for biomedical applications: strategies for safeguarding analytical performances in biological fluids. Sensors 16, 780.

Roldán, M.D., Pérez-Reinado, E., Castillo, F., Moreno-Vivián, C., 2008. Reduction of polynitroaromatic compounds: the bacterial nitroreductases. FEMS Microbiology Reviews 32, 474–500.

Ronkainen, N.J., Halsall, H.B., Heineman, W.R., 2010. Electrochemical biosensors. Chemical Society Reviews 39, 1747–1763.

Rozeboom, H.J., Yu, S., Mikkelsen, R., Nikolaev, I., Mulder, H.J., Dijkstra, B.W., 2015. Crystal structure of quinone-dependent alcohol dehydrogenase from Pseudogluconobacter saccharoketogenes. A versatile dehydrogenase oxidizing alcohols and carbohydrates. Protein Science 24, 2044–2054.

Saadati, S., Salimi, A., Hallaj, R., Rostami, A., 2014. Direct electron transfer and electrocatalytic properties of immobilized hemoglobin onto glassy carbon electrode modified with ionic-liquid/titanium-nitride nanoparticles: application to nitrite detection. Sensors and Actuators B: Chemical 191, 625–633.

Sadeghi, S.J., Fantuzzi, A., Gilardi, G., 2011. Breakthrough in P450 bioelectrochemistry and future perspectives. Biochemical and Biophysical Acta 1814, 237–248.

Saha, K., Agasti, S.S., Kim, C., Li, X., Rotello, V.M., 2012. Gold nanoparticles in chemical and biological sensing. Chemical Reviews 112, 2739–2779.

Sakurai, T., Kataoka, K., 2007. Structure and function of type I copper in multicopper oxidases. Cellular and Molecular Life Sciences 64, 2642–2656.

Salimi, A., Noorbakhsh, A., 2011. Layer by layer assembly of glucose oxidase and thiourea onto glassy carbon electrode: fabrication of glucose biosensor. Electrochimica Acta 56, 6097–6105.

Salimi, A., Compton, R.G., Hallaj, R., 2004. Glucose biosensor prepared by glucose oxidase encapsulated sol-gel and carbon-nanotube-modified basal plane pyrolytic graphite electrode. Analytical Biochemistry 333, 49–56.

Salimi, A., Sharifi, E., Noorbakhsh, A., Soltanian, S., 2007a. Direct electrochemistry and electrocatalytic activity of catalase immobilized onto electrodeposited nano-scale island of nikel oxide. Biophysical Chemistry 125, 540–548.

Salimi, A., Sharifi, E., Noorbakhsh, A., Soltanian, S., 2007b. Immobilization of glucose oxidase on electrodeposited nickel oxide nanoparticles: direct electron transfer and electrocatalytic activity. Biosensors and Bioelectronics 22, 3146–3153.

Salimi, A., Kavosi, B., Hallaj, R., Babaei, A., 2009a. Fabrication of a highly sensitive glucose biosensor based on immobilization of osmium complex and glucose oxidase onto carbon nanotubes modified electrode. Electroanalysis 21, 909–917.

Salimi, A., Hallaj, R., Soltanian, S., 2009b. Fabrication of a sensitive cholesterol biosensor based on cobalt-oxide nanostructures electrodeposited onto glassy carbon electrode. Electroanalysis 21, 2693–2700.

Scheller, F., Schubert, F., 1992. *Biosensors*, Chapter 11 from *Techniques and Instrumentation in Analytical Chemistry*, Elsevier.

Schubert, F., Scheller, F., Krasteva, N.G., 1990. Lactate-dehydrogenase-based biosensors for glyoxylate and NADH determination: a novel principle of analyte recycling. Electroanalysis 2, 347–351.

Schuhmann, W., Lammert, R., Uhe, B., Schmidt, H.-L., 1990. Polypyrrole, a new possibility for covalent binding of oxidoreductases to electrode surfaces as a base for stable biosensors. Sensors and Actuators B: Chemical 537–541.

Schuhmann, W., Ohara, T.J., Schmidt, H.L., Heller, A., 1991. Electron transfer between glucose oxidase and electrodes via redox mediators bound with flexible chains to the enzyme surface. Journal of the American Chemical Society 113, 1394–1397.

Setya, A., Murillo, M., Leustek, T., 1996. Sulfate reduction in higher plants: molecular evidence for a novel 5′-adenylylsulfate reductase. Proceedings of the National Academy of Sciences of the United States of America 93, 13383–13388.

Shleev, S., Tkac, J., Christenson, A., Ruzgas, T., Yaropolov, A.I., Whittaker, J.W., Gorton, L., 2005. Direct electron transfer between copper-containing proteins and electrodes. Biosensors and Bioelectronics 20, 2517–2554.

Shleev, S., Wetterö, J., Magnusson, K.E., Ruzgas, T., 2006. Electrochemical characterization and application of azurin-modified gold electrodes for detection of superoxide. Biosensors and Bioelectronics 22, 213–219.

Shukla, A., Gillam, E.M., Mitchell, D.J., Bernhardi, P.V., 2005. Direct electrochemistry of enzymes from the cytochrome P450 2C family. Electrochemistry Communications 7, 437–442.

Silwana, B., van der Horst, C., Iwuoha, E., Somerset, V., 2013. Inhibitive Determination of Metal Ions Using a Horseradish Peroxidase Amperometric Biosensor (Chapter 5). Intech Open Science.

Simonian, A.L., Rainina, E.I., Fitzpatrick, P.F., Wild, J.R., 1997. A tryptophan-2-monooxygenase based amperometric biosensor for L-tryptophan determination: use of a competitive inhibitor as a tool for selectivity increase. Biosensors and Bioelectronics 12, 363–371.

Singh, V., Mani, I., Chaudhary, D.K., 2012. Analysis of the multicopper oxidase gene regulatory network of Aeromonas hydrophila. Systems and Synthetic Biology 6, 51–59.

Singha, R.P., Kim, Y.J., Oh, B.K., Choi, J.W., 2009. Glutathione-s-transferase based electrochemical biosensor for the detection of captan. Electrochemistry Communications 11, 181–185.

Slauch, J.M., 2011. How does the oxidative burst of macrophages kill bacteria? Still an open question. Molecular Microbiology 80, 580–583.

Slein, M.W., 1963. D-glucose: determination with hexokinase and glucose-6-phosphate dehydrogenase. In: Method of Enzymatic Analysis. Elsevier, Academic Press, pp. 117–130.

Sucheta, A., Cammack, R., Weiner, J., Armstrong, F.A., 1993. Reversible electrochemistry of fumarate reductase immobilized on an electrode surface. Direct voltammetric observations of redox centers and their participation in rapid catalytic electron transport. Biochemistry 32, 5455–5465.

Sultana, N., Schenkman, J.B., Rusling, J.F., 2005. Protein film electrochemistry of microsomes genetically enriched in human cytochrome P450 monooxygenases. Journal of the American Chemical Society 127, 13460–13461.

Sultana, N., Schenkman, J.B., Rusling, J.F., 2007. Direct electrochemistry of cytochrome P450 reductases in surfactant and polyion films. Electroanalysis 19, 2499–2506.

Sun, H., Liu, Z., Wu, C., Xu, P., Wang, X., 2016. Amperometric inhibitive biosensor based on horseradish peroxidasenanoporous gold for sulfide determination. Scientific Reports 6, 30905.

Takeda, K., Matsumura, H., Ishida, T., Yoshida, M., Igarashi, K., Samejima, M., Ohno, H., Nakamura, N., 2016. pH-dependent electron transfer reaction and direct bioelectrocatalysis of the quinohemoprotein pyranose dehydrogenase. Biochemical and Biophysical Research Communications 477, 369–373.

Tang, W., Li, L., Wu, L., Gong, J., Zeng, X., 2014. Glucose biosensor based on a glassy carbon electrode modified with polythionine and multiwalled carbon nanotubes. PLoS One 9, e95030.

Taurino, I., Reiss, R., Richter, M., Fairhead, M., Thöny-Meyer, L., De Micheli, G., Carrara, S., 2013. Comparative study of three lactate oxidases from Aerococcus viridans for biosensing applications. Electrochimica Acta 93, 72–79.

Torres Pazmino, D.E., Winkler, M., Glieder, A., Fraaije, M.W., 2010. Monooxygenases as biocatalysts: classification, mechanistic aspects and biotechnological applications. Journal of Biotechnology 146, 9–24.

Urlacher, V.B., Schmid, R.D., 2006. Recent advances in oxygenase-catalyzed biotransformations. Current Opinion in Chemical Biology 10, 156–161.

Valentine, J.S., 1994. Dioxygen reactions. In: Chapter 5 from Bioinorganic Chemistry. University Science Books, Mill Valley, CA, ISBN 0-935702-57-1.

Vlamidis, Y., Gualandi, I., Tonell, D., 2017. Amperometric biosensors based on reduced GO and MWCNTs composite for polyphenols detection in fruit juices. Journal of Electroanalytical Chemistry 799, 285–292.

Volkov, A.N., Nicholls, P., Worrall, J.A.R., 2011. The complex of cytochrome c and cytochrome c peroxidase: the end of the road? Biochimica et Biophysica Acta 1807, 1482–1503.

Volkov, A.G. (Ed.), 2003. Interfacial Catalysis. Marcel Dekker Inc.

Vrielink, A., Ghisla, S., 2009. Cholesterol oxidase: biochemistry and structural features. FEBS Journal 276, 6826–6843.

Wan, Y., Zhang, J., Liu, G., Pan, D., Wang, L., Song, S., Fan, C., 2009. Ligase-based multiple DNA analysis by using an electrochemical sensor array. Biosensors and Bioelectronics 24, 1209–1212.

Wang, L., Hu, N., 2001. Direct electrochemistry of hemoglobin in layer-by-layer films with poly(vinyl sulfonate) grown on pyrolytic graphite electrodes. Bioelectrochemistry 53, 205–212.

Wang, Y., Du, J., Li, Y., Shan, D., Zhou, X., Xue, Z., Lu, X., 2012. A amperometric biosensor for hydrogen peroxide by adsorption of horseradish peroxidase onto single-walled carbon nanotubes. Colloids and Surfaces B: Biointerfaces 90, 62–67.

Wang, C., Liu, C., Luo, J., Tian, Y., Zhou, N., 2016. Direct electrochemical detection of kanamycin based on peroxidase-like activity of gold nanoparticles. Analytica Chimica Acta 936, 75–82.

Werck-Reichhart, D., Feyereisen, R., 2000. Cytochromes P450: a success story. Genomic Biology 1, 3003.1–3003.9.

Wever, R., Hemrika, W., 2000. In: Messerschmidt, A., Huber, R., Poulos, T., Wieghardt, K. (Eds.), Vanadium Haloperoxidases from Handbook of Metalloproteins. John Wiley & Sons, Ltd, Chichester.

White, R.M., 1987. A sensor classification Scheme. In: IEEE Transaction, Ultrasonics, Ferroelectrics, and Frequency Control, vol. 34. UFFC, pp. 124–126.

Wu, J., Suls, J., Sansen, W., 1999. Amperometric glucose sensor with enzyme covalently immobilized by sol-gel technology. Analytical Sciences 15, 1029–1032.

Xu, S., Minteer, S.D., 2013. Investigating the impact of multi-heme pyrroloquinoline quinone-aldehyde dehydrogenase orientation on direct bioelectrocatalysis via site specific enzyme immobilization. ACS Catalysis 3, 1756–1763.

Xu, J.Z., Zhu, J.J., Wu, Q., Hu, Z., Chen, H.Y., 2003. An amperometric biosensor based on the coimmobilization of horseradish peroxidase and methylene blue on a carbon nanotubes modified electrode. Electroanalysis 15, 219–224.

Xu, Z., Chen, X., Dong, S., 2006. Electrochemical biosensors based on advanced bioimmobilization matrices. Trends in Analytical Chemistry 25, 899–908.

Yamanaka, H., Mascini, M., 1992. NADH electrochemical sensor coupled with dehydrogenase enzymes. Analytical Letters 25, 983–997.

Yan, E.-K., Cao, H.-L., Zhang, C.-Y., Lu, Q.-Q., Ye, Y.-J., He, J., Huang, L.-J., Yin, D.-C., 2015. Cross-linked protein crystals by glutaraldehyde and their applications. RSC Advances 5, 26163–26174.

Yang, Z.H., Zeng, R., Yang, G., Wang, Y., Li, L.Z., Lv, Z.S., Yao, M., Lai, B., 2008. Asymmetric reduction of prochiral ketones to chiral alcohols catalyzed by plants tissue. Journal of Industrial Microbiology and Biotechnology 35, 1047–1051.

Yang, C., Denno, M.E., Pyakurel, P., Venton, B.J., 2015. Recent trends in carbon nanomaterial-based electrochemical sensors for biomolecules: a review. Analytical Chemica Acta 887, 17–37.

Yarman, A., Wollenberger, U., Scheller, F.W., 2013. Sensors based on cytochrome P450 and CYP mimicking systems. Electrochimica Acta 110, 63–72.

Yin, H., Xu, Z., Wang, M., Zhang, X., Ai, S., 2013. An electrochemical biosensor for assay of DNA methyltransferase activity and screening of inhibitor. Electrochimica Acta 89, 530–536.

Yoo, E.H., Lee, S.Y., 2010. Glucose biosensors: an overview of use in clinical practice. Sensors 10, 4558–4576.

Yoshida, H., Sakai, G., Mori, K., Kojima, K., Kamitori, S., Sode, K., 2015. Structural analysis of fungus-derived FAD glucose dehydrogenase. Scientific Reports 5, 13498.

Zhang, X., Ju, H., Wang, J. (Eds.), 2008. Electrochemical Sensors, Biosensors and Their Biomedical Applications. Elsevier.

Zhang, Y.-H.P., Myung, S., You, C., Zhu, Z., Rollin, J.A., 2011. Toward low-cost biomanufacturing through in vitro synthetic biology: bottom-up design. Journal of Materials Chemistry 21, 18877–18886.

Zhang, Y., Liu, H., Yang, Z., Ji, S., Wang, J., Pang, P., Feng, L., Wang, H., Wu, Z., Yang, W., 2015. An acetylcholinesterase inhibition biosensor based on a reduced graphene oxide/silver nanocluster/chitosan nanocomposite for detection of organophosphorus pesticides. Analytical Methods 7, 6213–6219.

Zheng, H., Okada, H., Nojima, S., Suye, S., Hori, T., 2004. Layer-by-layer assembly of enzymes and polymerized mediator on electrode surface by electrostatic adsorption. Science and Technology of Advanced Materials 5, 371–376.

Zucca, P., Sanjust, E., 2014. Inorganic materials as supports for covalent enzyme immobilization: methods and mechanisms. Molecules 19, 14139–14194.

CHAPTER

Aptamer-based electrochemical biosensors

8

Seyedeh Malahat Shadman, MSc, Marzieh Daneshi, MSc, Fatemeh Shafiei, BSc, Maryam Azimimehr, MSc, Mehrdad Rayati Khorasgani, BSc, Mehdi Sadeghian, MSc, Hasan Motaghi, PhD, Masoud Ayatollahi Mehrgardi, PhD

Department of Chemistry, University of Isfahan, Isfahan, Iran

8.1 Introduction

We are dealing with the quantitative determinations of the vast range of molecules, compounds, pharmaceutics, drugs, contaminants, heavy metals, even microorganisms and pathogens in our daily lives. Electrochemical biosensors have opened a new horizon for these determinations in the fields of chemical analysis, environmental monitoring, forensic sciences, quality controls, biomedical application, and early diagnosis of genetic disease and cancers.

Most of these targets present in the very complex matrixes and need very tedious separation procedures for the elimination of interferences. Pretreatments of these samples before the analysis are very time-consuming. This issue can be resolved by the integration of selective or even specific biorecognition layers to the biosensors. Various biomolecules including, enzymes, antibodies, and aptamers have been used for the modification of electrodes.

Aptamers, nucleic acid analogs of antibodies, have been used as the recognition layers of the biosensors extensively. They offer the advantages of high affinity, specificity, long-term stability at ambient temperature, the simplicity of synthesis, and finally rapid production processes. More importantly, aptamers can be easily modified by adding functional groups, useful for their labeling and for their surface immobilization (Pang et al., 2014). Also, electrochemical transduction offers high sensitivity and specificity and the possibility to miniaturize the required instrumentation providing compact and portable analysis devices (Ronkainen et al., 2010). On the other hand, nanotechnology and the nanoscale modification of electrode surfaces with nanomaterials (NMs) have paved the way in designing integrated nanobioelectrochemical systems, which provide novel and sometimes unique properties owing to the ability of the intrinsic properties of the NMs used and control the architecture of the electrode interface at the nanoscale (Dridi et al., 2017).

These characteristics make aptamers as ideal candidates for the recognition layers of electrochemical biosensors. Several review articles and monographs reviewed the applications of aptamers in bioanalysis (Mascini, 2008; Mascini

Electrochemical Biosensors. https://doi.org/10.1016/B978-0-12-816491-4.00008-5

et al., 2012; Minunni et al., 2004; Tombelli and Mascini, 2009, 2010; Tombelli et al., 2005, 2007). In this chapter, we will classify the aptamer-based electrochemical biosensors based on their targets to small molecules, environmental contaminants, and cancer cells and will review the latest advances and achievements in the field.

8.2 Electrochemical aptasensors against small molecules

In this section, the recent achievements in the field of aptamer-based electrochemical biosensors for the quantification of small molecules, including heavy metals, antibiotics, neurotransmitters, nucleotides, pesticides, and drugs, have been discussed.

8.2.1 Heavy metals ion

Heavy metals are generally referred to the metals with relatively high atomic weights in the range of 63.5–200.6 g mol^{-1} and the densities more than 5 gr cm^{-3}(Srivastava and Majumder, 2008). Although the body requires low amounts of some heavy metal ions such as Zn^{2+}, Mn^{2+}, Fe^{2+}, Cu^{2+} for growth and body health, high levels of them are toxic (Valko et al., 2005). On the other hand, some of them including (Pb^{2+}, Hg^{2+}, As^{3+}, Cd^{2+}) have considered as contaminants and toxic substances even at very low concentrations, which can be regarded as a serious risk to the human health and environment (Bagal-Kestwal et al., 2008; Patrick, 2006). Heavy metal poisoning can be caused by industrial exposure, contaminated air, water, food cycles, medicines, etc. (Verma and Kaur, 2016).

Accumulation of heavy metal ions in the living organisms and soft tissues can affect on the normal functioning of neurological, immune, and cardiovascular systems and so on (Gumpu et al., 2015; Mao et al., 2015). Therefore, in recent years, the determination of trace amounts of heavy metal ions has become very important issue. Using specific interaction between aptamers and target molecules, along with the advantages of electrochemical methods, aptamer-based electrochemical biosensors have become powerful tools for the detection of these ions. Nowadays, different strategies have been used to determine heavy metal ions using aptamer-based electrochemical methods (Table 8.1).

8.2.2 Antibiotics

Antibiotics are powerful medicines that destroy or slow down the growth of bacteria. They are used for therapeutic purposes or as food additives to promote animal growth (Gaugain-Juhel et al., 2009; Arias and Murray, 2009). However, the abuse of antibiotics leads to sustainable side effects to public health and the environment such as bacterial resistance against the antibiotic, the emergence of hypersensitivity reaction, aplastic anemia in several sensitive humans (Lee et al., 2001; Berruga et al., 2011). Therefore, there is an urgent need for the careful monitoring of antibiotic levels to minimize the side effects of antibiotics. Several aptamer-based

Table 8.1 Electrochemical aptasensors against small molecules.

Target	Electrochemical techniques	Strategy	Linearity range	Detection limit	References
Heavy metal ions					
Cd^{2+}	ACV and CV	MB-modified aptamer/gold electrode	250 nM–1 μM	92 nM	Zhad et al. (2017)
Cu^{2+}	SWV	DNA duplex/AuNP/pATP/gold electrode	0.1 nM–10 μM	0.1 pM	Chen et al. (2011)
Hg^{2+}	DPV	PtNAs/CF electrode/cDNA/thionine-labeled Fe_3O_4/rGO nanoprobes/rDNA	0.1–100 nM	30 pM	Luo et al. (2018)
Hg^{2+}	ECL	Fe_3O_4@SiO_2/dendrimers/CdTe@CdS QDs-DNA probe/AuNPs-modified ssDNA (S2)	20 aM–2 μM	2 aM	Babamiri et al. (2018)
As^{3+}	DPV	PDDA/thiolated As-3 aptamer/Au nanoparticle/SPE	0.2–100 nM	0.15 nM	Cui et al. (2016)
Pb^{2+}	DPV	Thionin/graphene/SH-aptamer/Au electrode	0.16 pM–0.16 nM	0.32 fM	Gao et al. (2016b)
Pb^{2+}	PEC	RGO/CdS/aptamer/AuNP-labeled DNA	0.1–50 nM	0.05 nM	Zang et al. (2014)
Antibiotics					
Kanamycin	DPV	Target/aptamer/GR-CO-NH-CH_2-CH_2-NH_2/MWCNTs-BMIMPF6/GCE	0.001–100 μM	0.87 nM	Qin et al. (2015)
Chloramphenicol	SWV	Target/aptamer/GA/poly(1,5-DAN)/EPPG	50–500 fM	11 fM	Goyal and Shim (2015)
Kanamycin	EIS	Target/aptamer/4-CP/SPEs	1.2–75 ng mL^{-1}	0.11 ng mL^{-1}	Sharma et al. (2017)
Oxytetracycline	SWV	Target/DNA-2 (aptamer)/MCH/Fc-labeled DNA-1/Au	10–600 ng mL^{-1}	9.8 ng mL^{-1}	Zheng et al. (2013)
Bleomycin	EIS	Target/AgNCs/Apt@CuFe@FeFe/AE	0.01 fg mL^{-1}–0.1 pg mL^{-1}	0.0082 fg mL^{-1}	Zhou et al. (2018)
Kanamycin	EBFCs	SiO_2@AuNPs-csDNA bioconjugate/aptamer/GOx/AuNPs/CP bioanode & laccase/PDA/AuNPs/CP biocathode	10 pM–100 nM	3 pM	Gai et al. (2017)

Continued

Table 8.1 Electrochemical aptasensors against small molecules.—*cont'd*

Target	Electrochemical techniques	Strategy	Linearity range	Detection limit	References
Neurotransmitters					
Dopamine	DPV	Target/aptamer/PEDOT/rGO/GCE	1 pM–160 nM	78 fM	Wang et al. (2015c)
Dopamine	DPV	Target/MB/aptamer(ssDNA2)/ssDNA1/CNTs COOH/PEI/AuNPs/Au	5–300 nM	2.1 nM	Azadbakht et al. (2016b)
Dopamine	DPV	Target/aptamer/MB/ssDNA1/Au-PtNPs/CNTs-COOH/GC	1–30 nM	0.22 nM	Azadbakht et al. (2016a)
Dopamine	DPV	GO/CNTs/AgNPs/aptamer/GC	3–110 nM	700 ± 19.23 pM	Bahrami et al. (2016)
Dopamine	DPV	Target/aptamer/CNTs/PB/AuNPs/GC	0.5–50 nM	200 pM	Beiranvand et al. (2016)
Dopamine	SWV	Target/aptamer/GR-PANI/GCE	0.007–90 nM	0.00198 nM	Liu et al. (2012)
Dopamine	DPV	Th/AuNPs/DNA1/CNPs/GE	30 nM–6.0 μM	10 nM	Xu et al. (2015)
Dopamine	Amperometry	Aptamer/nano-Au/GCE	5 nM–0.5 μM	1.8 nM	Liu et al. (2016)
Nucleotides					
ATP	CV/DPV	NPGE/F1/ATP/F2/DABA	Up to 3.0 mM	1.0 mM	Kashefi-Kheyrabadi and Mehrgardi (2013)
ATP	SWV	Fc-P/MCH/MB-P/Au electrode	0.1 nM–0.1 mM	1.9 nm	Wu et al. (2013)
ATP	SWV	Fc-P1/MGP/Th-P2 probe, release of redox tag-conjugated aptamers	1.0 pM–500 μM	0.1 pM	Tang et al. (2011)
ATP	ECL	Au NRs/EXO I/aptamer-Fc/capture DNA/ABA/GCE	0.1 pg mL^{-1}–0.5 μg mL^{-1}	3.1 pmol/cell	Xu et al. (2018b)
ATP	DPV	GN/ABA/Au electrode	0.05 nM–1.0 μM	11.7 pM	Wu et al. (2015)
ATP	PEC	S1/MCH/capture DNA/C60-Au NPs/MoS_2/GCE	10 fM–100 nM	3.30 fM	Li et al. (2017b)
ATP	PEC	C DNA/CdS:Mn@Ru(bpy)2(dcbpy)/pDNA/Au/TiO_2/ITO	0.5 pM–5 nM	0.18 pM	Fan et al. (2015)

Pesticides					
Chlorpyrifos	CV, DPV	CuO Fs, SWCNT	0.1–150 ng mL^{-1}	70 pg mL^{-1}	Xu et al. (2018a)
Carbofuran	CV, DPV, EIS	GO-MWCNTs	0.03–0.81 μg L^{-1}	0.015 ng L^{-1}	Li et al. (2017d)
Carbaryl	CV, DPV, EI	AchE/e-pGON	0.3–6.1 μg L^{-1}	0.15 ng L^{-1}	Li et al. (2017c)
Chlorpyrifos	CV	AChE/ZrO_2/RGO	10^{-13}–10^{-9} and 10^{-9}–10^{-4} M	0.1 pM	Mogha et al. (2016)
Carbaryl		GR/PANI	38–194 ng mL^{-1}	20 ng mL^{-1}	Li et al. (2016b)
Methidathion, chlorpyrifos-ethyl	CV	NA/Ag@rGO-NH_2/AChE/GCE	0.012–0.105 and 0.021–0.122 mg mL^{-1}	9.5 and 14 ng mL^{-1}	Guler et al. (2017b)
Methyl parathion and malathion	CV/DPV	CS/AuNPs	0.19–760 and 1.5–1513.5 nM	0.19 and 1.51 nM	Bao et al. (2015a)
Cocaine	DPV	TiNTA/AuNP	10–600 nM	5.2 nM	Li and Tang (2017)
Codeine	CV/EIS	PAMAM/GA/chitosan/AuNPs/SPEs	1 pM–100 nM	0.3 pM	Niu et al. (2016)
Cocaine	AFSD	AuNP	1.0–10 nM	0.138 nM	Guler et al. (2017a)
Cocaine	CV/EIS	AuNPs/HDT/gold electrode	1.0–15.0 mM	0.5 mM	Li et al. (2008)

electrochemical sensors have been developed to determine various antibiotics, such as chloramphenicol, streptomycin, kanamycin, penicillin, ampicillin, and tetracycline.

An electrochemical biosensor for sensitive detection of ampicillin (AMP) was designed based on combination polymerase-assisted target recycling amplification with strand displacement amplification using polymerase and nicking endonuclease (Wang et al., 2015a). With the signal amplification strategy, this aptasensor can detect AMP with excellent sensitivity, and the LOD can be as low as attomole level of AMP.

A visible-light photoelectrochemical (PEC)-based aptasensor for tetracycline (TET) was developed in which graphitic carbon nitride (g-C_3N_4) coupled with CdS quantum dots (g-C_3N_4-CdS) was utilized as photoactive species in a PEC sensor (Liu et al., 2015). In this approach, TET-binding aptamer was immobilized on to g-C_3N_4-CdS-modified FTO. During the PEC sensing, the aptamer would specifically capture TET present in the sample, creating a photocurrent signal through the reaction between the captured TET and photogenerated holes. The sensor shows a linear response to TET in the concentration ranges from 10 to 250 nM with a detection limit of 5.3 nM.

In another research, a multiplex electrochemical aptasensor for the simultaneous detection of chloramphenicol (CAP) and oxytetracycline (OTC) was developed based on high-capacity magnetic hollow porous nanotracers coupling exonuclease-assisted target recycling (Yan et al., 2016). As Fig. 8.1 illustrates, the amplification process relies on the exonuclease-assisted target recycling amplification and metal ions encoded (Cd^{2+} and Pb^{2+}) magnetic hollow porous nanoparticles (MHPs) to create voltammetry signals. A linear relationship was obtained between the peak currents for lead and cadmium and concentrations of CAP and OTC in the linear range over 0.0005 to 50 ng mL^{-1} with detection limits of 0.15 and 0.10 pg mL^{-1} for CAP and OTC, respectively. The other recent advances in the quantifications of antibiotics using aptamer-based electrochemical biosensors have been listed in Table 8.1.

8.2.3 Dopamine

Neurotransmitters are endogenous chemicals that act as messengers and transmit signals from one neuron to another or between non-neuronal body cells across chemical synapses, exchanging information throughout the brain and body. Major neurotransmitters are divided into two categories: small-molecule transmitters and neuropeptide. Small-molecule transmitters can be further differentiated into monoamines such as dopamine and amino acids such as glutamate (Freberg, 2009).

Dopamine (DA) is a small neurotransmitter molecule discovered in the 1950s by Carlsson et al. (1957) and plays a key role in the central nervous system that controls movement, endocrine function, reward behavior, and memory processes (Beaulieu and Gainetdinov, 2011). The unstable regulation in neuronal release and uptake of DA causes various diseases and disorders, such as Alzheimer's disease, Parkinson's

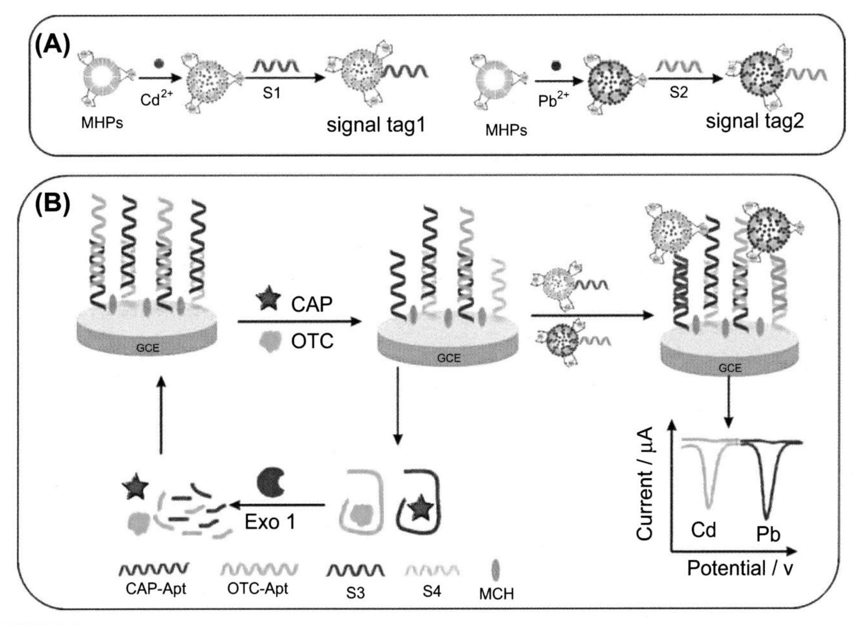

FIGURE 8.1

Schematic illustration of simultaneous electrochemical detection of CAP and OTC based on MHPs cascade multiple amplification strategy (Yan et al., 2016).

Adopted with permission from Elsevier.

disease, Huntington's disease, epilepsy, pheochromocytoma, and neuroblastoma (Liu et al., 2013; Robinson et al., 2008). Therefore, the quantitative determinations of DA appear to be important for diagnosis, monitoring disease state, and therapeutic interventions. Since the discovery of dopamine aptamer, several methods have been developed based on the electrochemical aptasensor for the measurement of this transmitter (Table 8.1).

8.2.4 Adenosine-5'-triphosphate

Adenosine-5'-triphosphate (ATP) is a multifunctional nucleotide, which plays a vital role in many biological processes, including synthesis and degradation of important cellular compounds, and transporting chemical energy in every living cell during the process of metabolism. It serves as an important indicator of cell viability and injury, proliferation, inhibition induced by various biological agents or small molecule drugs and microbial contamination in the food industry (He et al., 2012). The concentration changes of ATP in cells are in connection with many diseases, for instance, malignant tumors, angiocardiopathy, and Parkinson's diseases. The normal concentrations of ATP in serum are between 0.1 and 3 mM. ATP has been acknowledged as a powerful and reliable indicator for the diagnosis of disease, monitoring of disease progression and evaluation of prognosis (Li et al., 2017b). Therefore, the determination of ATP is essential in biochemical studies, clinical diagnosis, and food safety. Numerous aptasensors have been fabricated for ATP quantification using fluorescence, colorimetry, liquid chromatography, electrochemiluminescence (ECL), UV-vis reflectance spectrum, and electrochemistry as detection techniques (Hu et al., 2016). Despite their wide range of applications, these methods usually have some limitations, such as expensive laboratory instruments, require skilled technicians, and require time-consuming testing process. Nevertheless, electrochemical aptasensors offer the advantages of low cost and fast response, miniaturization capability, and minimal power requirement, which are proved to be an excellent tool for ATP detection (Kashefi-Kheyrabadi and Mehrgardi, 2012). These advantages lead to the attraction of many interests of researchers to design and fabricate electrochemical aptasensors for the determination of ATP (Table 8.1). For instance, a sensitive and non-enzymatic with signal amplification strategy for the detection of ATP has been developed using silver nanoparticle–decorated graphene oxide (AgNPs-GO) as a redox probe. For this purpose, the first amino-labeled fragment of aptamer (F1) was immobilized on the surface of a graphite screen-printed electrode. Subsequently, the F1-modified surface was interacted by AgNPs-GO via π-π stacking. In the presence of ATP, the second fragment of the aptamer, F2, forms an associated complex with the immobilized F1 and causes AgNPs-GO leave the surface. Consequently, a remarkable decrease in the oxidation signal of the AgNPs was followed by an analytical signal, which delivers a linear response over the range of 10.0 ($\pm$0.6) to 850 ($\pm$5) nM with a detection limit of 5.0 ($\pm$0.2) nM (Mashhadizadeh et al., 2017) (Fig. 8.2).

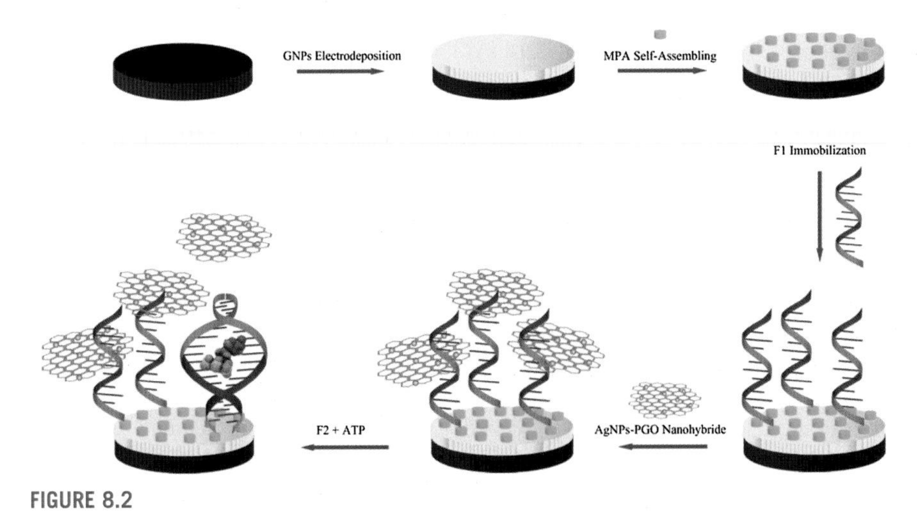

FIGURE 8.2

Schematic illustration of the aptasensor based on GO as signal indicator for the detection of ATP by flowing oxidation signal of the AgNPs (Mashhadizadeh et al., 2017).

Adopted with permission from Springer.

8.2.5 Toxins

Food quality control and environmental monitoring have become one of the main concerns. Accordingly, the development of the sensitive and easy-to-use method in precise detection of toxins (e.g., pesticides, toxins, drugs, and heavy metal) has been taken an efficient step in maintaining healthy society and environment. Obviously, advent accidentally or deliberately chemical and biological hazardous contaminants in food commodities represent a major threat for human health with serious consequences on the healthcare system and economic productivity (Ye et al., 2018; Slikker et al., 2018; Tang et al., 2016). Although there are many conventional methods such as HPLC, which are selective and sensitive, the efforts are continued to identify easy, cheap, and fast techniques with at least detection limit. The exploit of the aptamer, as recognition element in the designing of new generations of biosensors, has attracted many interests because of the high stability, large dynamic concentration range, prolonged shelf life, and low cross-reactivity (Rapini and Marrazza, 2016; Yao et al., 2010). Furthermore, aptamer sequences are selected from the library in SELEX according to their affinity toward a target molecule with high dissociation constants (K_d), which play a sufficient role in achieving the best selectivity and specificity and improving dynamic ranges (Fan et al., 2013).

Food toxins are a large variety of natural substances generated by fungi, algae, plants, or bacteria metabolism, which affects adversely on humans or environment even at very low doses. Generally, toxins are classified into three main groups: mycotoxins, algal toxins, and plant/bacteria toxins (Dridi et al., 2017).

An electrochemical aptasensor has been designed based on target-induced immobilization of gold nanoparticles (AuNPs) on the surface of electrode and methylene blue as redox probe for sensitive and selective detection of aflatoxin M_1 (AFM_1) and the hairpin-shaped structure of AFM_1 aptamer (Apt). This aptasensor allowed determination of AFM_1 with a detection limit of 0.9 ng L^{-1}.

Another electrochemical biosensor (Fig. 8.3) is designed for ultrasensitive detection of hydroxylated polychlorinated biphenyl (OH-PCB); concentration range is between 2.9×10^{-11} M and 2.9×10^{-7} M with immobilizing Au-S bond on the surface of the modified glassy carbon electrode owing to incorporation of gold nanoparticles/polydimethyl diallyl ammonium chloride-graphene composite (Yang et al., 2018).

8.2.6 Pesticides

Pesticides are a group of chemical substances used in agriculture to inhibit the growth of infecting species (e.g., weeds, insects, fungi); hence these chemical substances release in the environment and their toxicity remains for a long time (Hassani et al., 2017; Verdian, 2018). Accordingly, aptasensors play a pivotal role in detecting and measuring these hazardous substances. Table 8.1 shows a summary of some electrochemical aptasensors for detection of pesticides substances.

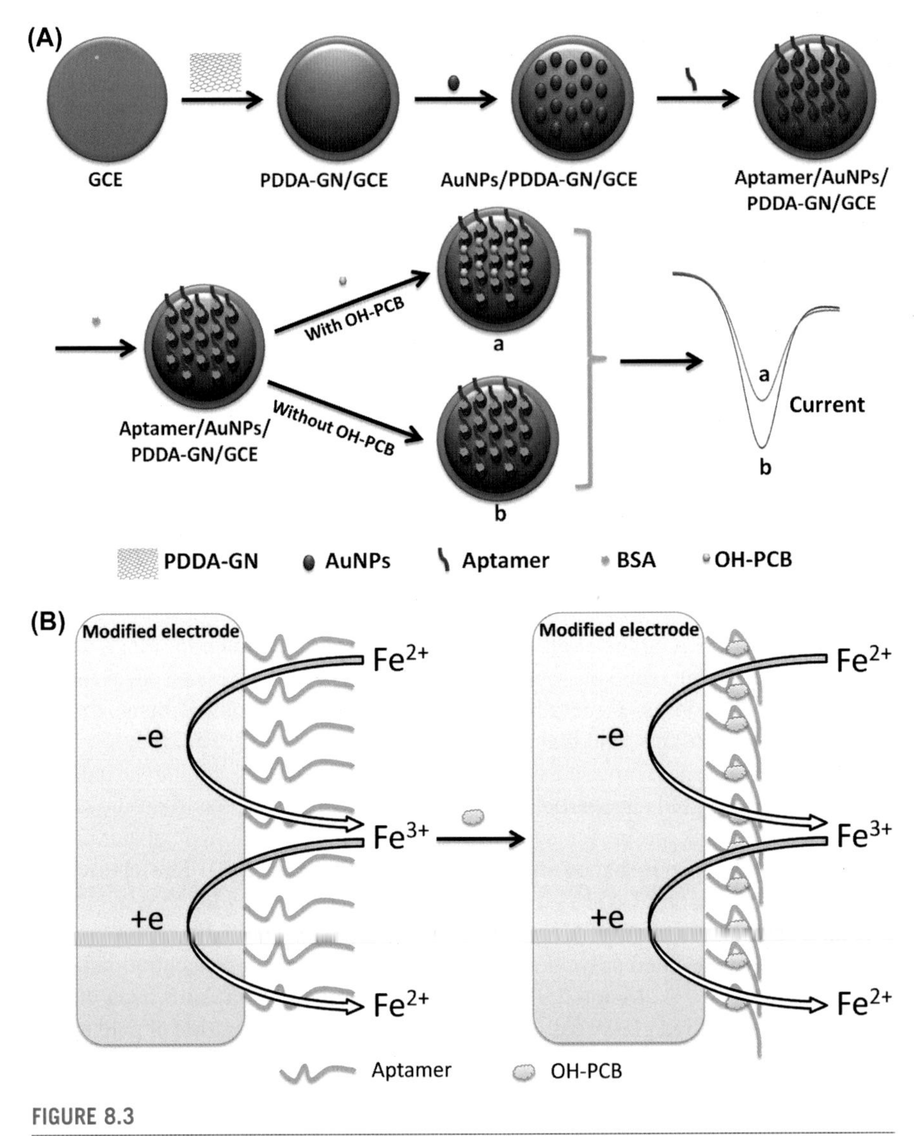

FIGURE 8.3

Schematic illustration of (A) the electrochemical aptasensor structure (B) and the redox reactions at electrode surface (Yang et al., 2018).

Adopted with permission from Elsevier.

8.2.7 Drugs

Opioid use in chronic pain treatment is complex. These materials can be both beneficial and harmful. Given the increasing use of drugs, it is difficult to identify people who use these drugs. That is why we are looking for a quick and easy way to diagnose these materials. In recent years, drug use has been high, which is uncomfortable (Vowles et al., 2015). Cocaine is one of the most important types of drugs that have determined using electrochemical aptasensors in recent years. The World Health Organization (WHO) estimated that 0.7% of the disease in 2004 was ascribed to cocaine use (Taylor et al., 2017). The high use of cocaine causes depression and dependence. Drug abuse has a negative impact on society (Kerr et al., 2005). Cocaine can be easily detected by specific binding interactions between cocaine and aptamer strands. This specific attachment can be reported by different electrochemical methods.

Construction of an electrochemical aptasensor begins with immobilization of an aptamer on a conducting substrate. Aptamers have been immobilized on these electrodes, preferably gold or carbon-based electrodes, either by direct chemisorption or through the functionalization and surface modification of the electrode surface by chemical cross-linking. Typically, the electrode surface was first modified with polythiophene bearing polyalanine homopeptide side chains (PT-Pala). Subsequent cocaine aptamer was conjugated onto the obtained PT-Pala film, which has greater sensitivity compared with directly conjugated cocaine aptamer onto the gold electrode. The aptasensor platform presents good results, and no interference, as well as sample matrix effect, was seen in the presence of benzoylecgonine (BE) and enables to detect trace amounts of cocaine with linear range of 2.5−10 nM and BE with linear range of 0.5−50 μM (Bozokalfa et al., 2016).

Single-walled carbon nanotubes (SWCNTs)−based biosensor was also developed for cocaine detection. The detection method is based on the selectivity of aptamer toward its target and the stronger interaction of SWCNTs with single-stranded DNA (ssDNA) than double-stranded DNA (dsDNA). In the absence of cocaine complimentary strand of aptamer (CS), modified electrode is intact and the SWCNTs could not bind to Apt-CS strongly, resulting a weak electrochemical signal. In the presence of target, aptamer binds to its target, leaves the CS, and the SWCNTs could bind to CS, leading to enhancement of electrochemical signal. A detection limit as low as 105 pM for cocaine has been achieved. Furthermore, this aptasensor was applied for detection of cocaine in serum with a limit of detection as low as 136 pM (Taghdisi et al., 2015).

Another highly sensitive electrochemical aptasensor for detecting cocaine using both electrochemical impedance spectroscopy and differential pulse voltammetry within the broad concentration range of 0.001−1.0 ng mL^{-1} and the low limit of detection of 1.29 and 2.22 pM was reported by Su et al. (2017a). Fig. 8.4 illustrates two-dimensional zirconium-based metal-organic framework nanosheets embedded with Au nanoclusters, which show good electrochemical activity and physicochemical stability and possessed high specific surface area and were used to immobilized cocaine aptamer strands as biosensitive platform. The other recent advances in the field of determination of drugs have been listed in Table 8.1.

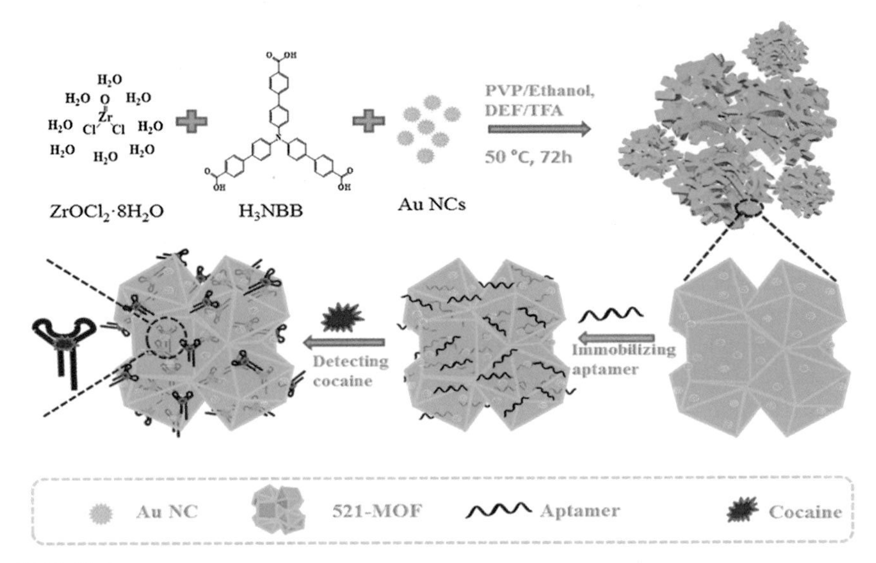

FIGURE 8.4

Steps of fabrication of an electrochemical aptasensor for detection of cocaine based on AuNCs@Zr-MOF-based nanosheets (Su et al., 2017a).

Adopted with permission from American Chemical Society.

8.3 Electrochemical aptasensors against proteins

Proteins play a vital role in constructing and repairing muscles and other tissues of the body such as blood, bone, cartilage, and skin. Hence, quantification of specific proteins surely has a direct bearing in early clinical diagnosis, treatment, and even prevention diseases (Bertolotti, 2018; Wu et al., 2018). To meet such demands for highly sensitive detection of proteins, efforts have been made to the design rapid, suitable, and reliable sensing systems to detect many proteins associated to different diseases (Li et al., 2013).

A simple and precise label-free method for electrochemical detection of thrombin (TB), as one kind of serine protease, was developed by rolling-circle amplification (RCA) triggering polyadenine production for adsorption on gold nanoparticles with depositing of aptamer I on gold electrode as the capture-probe via poly adenine-Au and immobilizing thrombin and the aptamer II as a primer hybridized with the RCA template to initiate the RCA process. Herein, the linear dynamic range of 0.1 pM–10 nM was obtained (Fan et al., 2018).

A host-guest-recognition-based electrochemical aptasensor for thrombin detection was also reported (Fan et al., 2012). Thrombin-binding aptamer (TBA) was dually labeled with a thiol at its 3′ end and a 4-((4-(dimethylamino) phenyl)azo) benzoic acid (dabcyl) at its 5′ end, respectively, which was previously immobilized on one Au electrode surface by Au S bond and used as the thrombin probe during the protein sensing procedure. One special electrochemical marker was prepared by modifying CdS nanoparticle with β-cyclodextrins (CdS-CDs), which was used as electrochemical signal provider and would conjunct with the thrombin probe–modified electrode through the host-guest recognition of CDs to dabcyl. The detection limit of 4.6 pM was obtained.

In another research (Yan et al., 2018), a simple and sensitive photoelectrochemical aptasensor for the detection of prion protein was developed based on β-CD and Rhodamin B (RhB). Because of the blocking effect of prion protein to the 16 transfer channel of the electron donor (AA) provided by the inner cavity of β-CD and dyes-sensitizing effect of RhB to TiO_2, the prion can be detected sensitively with a wide linear response range from 200 fM to 2000 fM and a low detection limit of 50.9 fM (3σ).

Zheng et al. (2007) developed an ultrasensitive electrochemical aptasensor through network-like thiocyanuric acid/gold nanoparticles for detecting thrombin. A sandwich format of magnetic nanoparticle-immobilized aptamer I, thrombin, and gold nanoparticle-labeled aptamer II was formed. The detection limit was down to 7.82 aM.

On the other side, proteins can act as biomarkers for the early diagnosis of various diseases. The National Cancer Institute defines the biomarker as "a biological molecule found in blood, other body fluids, or tissues that is a sign of a normal or abnormal process or of a condition or disease." Biomarkers are divided into tumor markers, cardiac markers, inflammatory markers, hepatic markers, and others. Tumor markers are one of the most valuable tools of early cancer detection, classification, staging, progression monitoring, and assessment of resistance to chemotherapy (Labib et al., 2016). Various biomarkers have been detected using electrochemical aptasensors. For example, an amplified electrochemical aptasensor for the detection of PDGF-BB has been developed through the hybridization of DNA-functionalized silver nanoparticles (AgNPs). The aggregation of AgNPs as the analytical signal was followed using differential pulse stripping voltammetry (DPSV) technique. Two kinds of DNA-modified AgNPs were simultaneously added on the surface of SPE array chip electrode and sandwiched PDGF-BB. Aggregated tags showed a strong electroactivity for the signal amplification through stripping detection of silver after preoxidation. There is a linear relationship between the peak currents and the logarithm of concentrations of PDGF-BB in the range 5 pg mL^{-1}–1000 ng mL^{-1} with a limit of detection of 1.6 pg mL^{-1}. Furthermore, PDGF-BB and thrombin proteins have been detected in a single-step strategy. The proposed aptasensor can be used for the determination of different proteins simultaneously (Song et al., 2014). Fig. 8.5 schematically

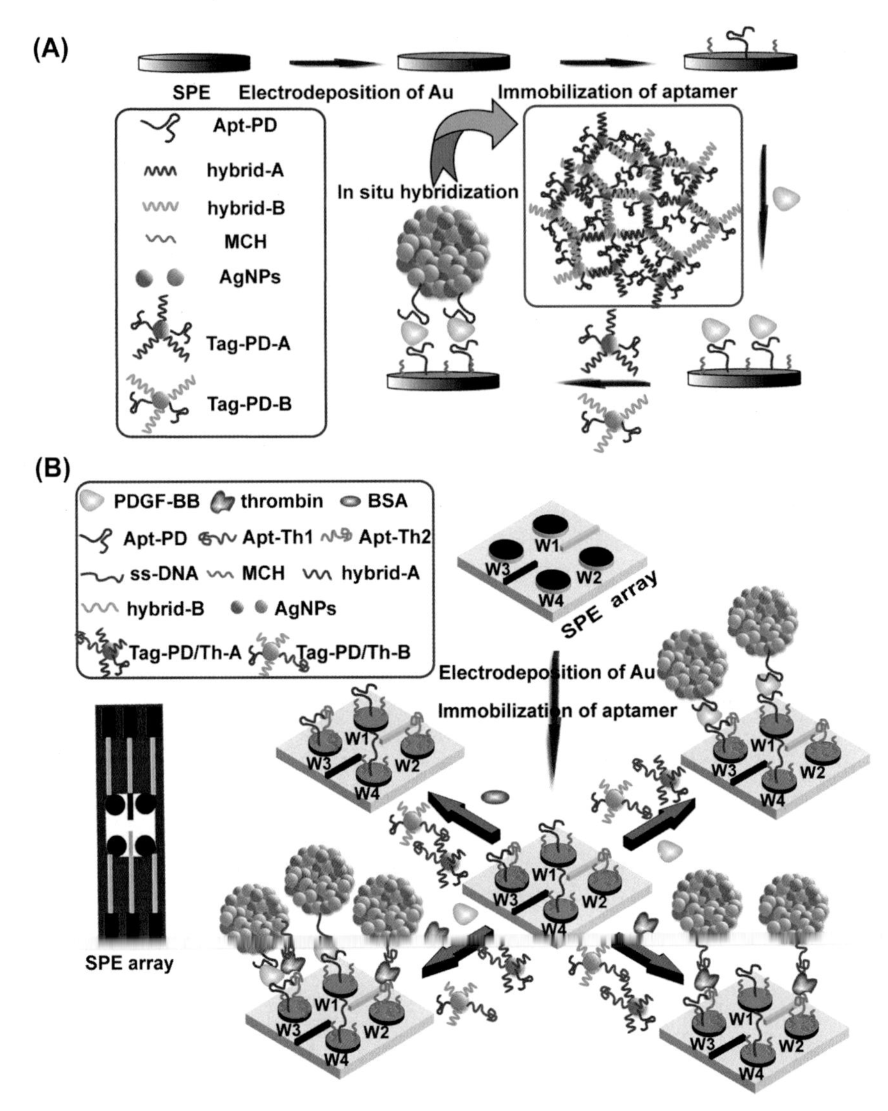

FIGURE 8.5

Schematic illustration of the electrochemical assay (A) and multiplexed assay (B) of protein with signal amplification through AgNPs aggregate induced by in situ hybridization on SPE array (Song et al., 2014).

Reprinted with permission from American Chemical Society.

illustrates the applied strategy for the fabrication of this electrochemical aptasensor.

Table 8.2 summarizes the recent developments in the field of aptamer-based electrochemical biosensors for the detection of various proteins and biomarkers.

Table 8.2 Recent advances of the aptamer-based electrochemical biosensors for the detection of various proteins and biomarkers.

Target	Electrochemical technique	Strategy	Linearity range	Detection limit	Reference
Thrombin	CV	Au-PANI-Gra hybrid	1.0 pM–30 nM	5.6×10^{-13} M	Bai et al. (2013)
Thrombin	DPV, CV, EIS	Au electrode	5.0 pM–50 nM	0.9 pM	Bao et al. (2015b)
Thrombin	EIS	AuNPs	1.0 pM–10 nM	1.9×10^{-13} M	Cao et al. (2015)
Thrombin	EIS	Graphene oxide/double-stranded DNA nanocomposite	0.1–100 nM	0.06 nM	Li et al. (2016a)
Thrombin	SWV, CV	Aptamer-target-aptamer sandwich structure	1–600 nM	170 pM	Yu et al. (2014)
Thrombin	DPV, CV,EIS	TiO_2/MWCNT/CHIT/SB	0.00005–10 nM	1.0 fM	Heydari-Bafrooe et al. (2016)
Thrombin	DPV	PdNPs/PDDA-G-MoS_2 nanocomposites and enzymatic signal amplification	0.0001–40 nM	0.062 pM	Jing et al. (2015)
Thrombin	DPV	MnPP-catalyzed aerobic oxidation of L-cysteine	0.1–25 nM	0.02 nM	Zheng et al. (2015)
Thrombin	EIS	AuNPs/dsDNA-GO nanocomposite	0.1–100 nM	0.06 nM	Li et al. (2016a)
Thrombin	DPV	Network-like thiocyanuric acid/AuNP	0.074–1.494 fM	7.82 aM	Zheng et al. (2007)
Thrombin	DPV	Host-guest recognition	–	4.6 pM	Fan et al. (2012)
Thrombin	ACV	"Signal-on/off" aptasensor based on the biobarcoded amplification	0.003–30 nM	1.1 pM	Wang et al. (2017b)

Continued

Table 8.2 Recent advances of the aptamer-based electrochemical biosensors for the detection of various proteins and biomarkers.—*cont'd*

Target	Electrochemical technique	Strategy	Linearity range	Detection limit	Reference
Thrombin	DPV	Sandwich-type aptasensor has been based on (Pt NPs decorated carbon nanocages)	0.05 pM–20 nM	10 fM	Gao et al. (2016a)
Thrombin	DPV	Biobarcode amplification assay	40–750 fM	6.2 fM	Zhang et al. (2009)
Thrombin	DPV	Signal amplification strategy based on three-dimensional ordered macroporous	0.20 pM–20 nM	0.02 pM	Tian et al. (2016b)
Thrombin	EIS	Aptasensing platform based on carbon nanofibers –enriched screen-printed electrodes	25–150 μg mL^{-1}	1.8 μg mL^{-1}	Erdem et al. (2015)
Thrombin	ACV	Inverted aptamer beacon	0.5–70 nM	0.21 nM	Zhuang et al. (2014)
Thrombin	DPV	Thiolated thrombin aptamer with benzoquinone-hexanedithiol/Au	50 fM–1.5 nM	20 fM	Su et al. (2017b)
Thrombin	Amperometry	Amplification strategy of Au@GS and DNA-CoPd NPs conjugates	0.01–2.00 ng mL^{-1}	5.0 pg mL^{-1}	Wang et al. (2016)
Thrombin	DPV	Cyclodextrin-functionalized graphene-AuNP	1.6 fM–800 fM	1 fM	Xue et al. (2015a)
Thrombin	SWV, CV	Aptamer-target-aptamer sandwich structure	1–600 nM	170 pM	Yu et al. (2014)
C-reactive protein	CV, EIS, SWV	Gold surface	1–100 pM	–	Jarczewska et al. (2018)

Activated protein C	DPV, CV, EIS	(PS/SPE)	5–12.5 μg mL^{-1}	0.74 μg mL^{-1} (buffer medium), 2.03 μg mL^{-1} (serum)	Erdem and Congur (2014)
VEGF	CV	DNA-templated Ag/Pt bimetallic nanoclusters	6.0–20 pM	4.6 pM	Fu et al. (2016)
VEGF	PEC	g-C_3N_4/Au NPs/HT/S1+S2 aptamer/S2+S3 aptamer/MB/target	100 fM–10 nM	30 fM	Da et al. (2018)
VEGF	CV-EIS	(Ag/AgCl/GE)	1 pg mL^{-1}–1 ng mL^{-1}	0.82 μg mL^{-1}	Cheng et al. (2012)
VEGF	nFIS	GE/aptamer/VEGF/antibody-magnetic bead	5–1000 pg mL^{-1}	–	Qureshi et al. (2015)
PDGF-BB	ACV	Self-assembling ferrocene-labeled aptamer onto Au electrode	20 pg mL^{-1}–200 ng mL^{-1}	10 pg mL^{-1}	Zhang et al. (2015)
PDGF-BB	DPV	Catalase-functional DNA-PtNPs dendrimer	50 fM–35 nM	20 fM	Zhang et al. (2014)
PDGF-BB	EIS	$Co_3(PO_4)_2$-based nanocomposites	0.5–100 ng mL^{-1}	61.5 pg mL^{-1}	He et al. (2016)
PDGF-BB	ECL	MoS_2-AuNPs/Apt1/MCH/PDGF-BB/QDs-Apt2/GCE	0.01–100 pM	1.1 fM	Liu et al. (2014b)
MUC1	DPV, EIS	Au nanoparticle/thiolated aptamer/MUC1 protein/graphite SPE	2.5–15 ng mL^{-1}	3.6 ng mL^{-1}	Florea et al. (2013)
MUC1	CV	MB-anti-MUC1 aptamers/gold electrode	Up to 1.5 μM	50 nM	Ma et al. (2013)
MUC1	DPV	ITO	1.0 pg mL^{-1}–50 ng mL^{-1}	0.40 pg mL^{-1}	Lin et al. (2018)

Continued

Table 8.2 Recent advances of the aptamer-based electrochemical biosensors for the detection of various proteins and biomarkers.—*cont'd*

Target	Electrochemical technique	Strategy	Linearity range	Detection limit	Reference
MUC1	Photoelectrochemical (PEC)	TiO_2 NT/aptamer/c-DNA@CD/MUC1	0.002–0.2 μM	0.52 nM	Tian et al. (2016a)
MUC1	EIS	SPE/carbon nanotube/aptamer/MUC1	0.1–2 U mL^{-1}	0.02 U mL^{-1}	Nawaz et al. (2016)
PSA	ESI Amperometric	11-Amino alkanethiol/Au nanoparticle/SH-aptamer/MCH or FcSH/PSA	1–10 pg mL^{-1}	10 pg mL^{-1}	Jolly et al. (2017)
PSA	DPV	GE/gold nanospheres/aptamer-MB/PSA	0.125–200 ng mL^{-1}	50 pg mL^{-1}	Rahi et al. (2016)
CEA	SWV	GCE/GO-antibody/CEA/HAP-aptamer	0.1 pg mL^{-1} –10 ng mL^{-1}	0.05 pg mL^{-1}	Si et al. (2017)
CEA	DPV	GE/aptamer I/CEA/aptamer II-AuNPs	1–200 ng mL^{-1}	0.5 ng mL^{-1}	Shu et al. (2013)
CEA	DPV	GCE/Au deposit/aptamer I/CEA/Pt@AuNWs-aptamer II-Tb	0.001–80 ng mL^{-1}	0.31 pg mL^{-1}	Xue et al. (2015b)
Myoglobin (Mb)	CV, DPV	Au/RGD/GR-COOH/GCE	0.0001–0.2 g L^{-1}	26.3 ng mL^{-1}	Li et al. (2017a)
Myoglobin	CV-DPV	(Aptamer/AuNP//RGD/GR-COOH/GCE)	0.0001–0.2 g L^{-1}	26.3 ng mL^{-1}	Li et al. (2017a)
Lysozyme	SWV	Aptamer modified with Fc	1–30 nM	0.45 nM	Chen and Guo (2013)
Prion	Photoelectrochemical	Dyes-sensitizing effect of RhB to TiO_2	200–2000 fM	50.9 fM	Yan et al. (2018)
Antidigoxigenin antibody Thrombin	DPV	Allosteric kissing complex-based	1–40 ng mL^{-1} 10 pM–0.2 nM	1 ng mL^{-1} 10 pM	Zhao et al. (2018)

Lysozyme	SWV	GNPs-modified SPEs	1.0–50.0 pg mL^{-1}	0.3 pg mL^{-1}	Xie et al. (2014)
Lysozyme	EIS	Comparison of two different aptamers (COX and TRAN)	0.1–0.8 μM 25 nM–0.8 μM	100 nM 25 nM	Ocaña et al. (2015)
Lysozyme	CV, DPV, EIS	Signal-off architecture	7–30 nM	0.45 nM	Chen and Guo (2013)
Lysozyme	SWV	Aptamer/tetrahexahedral (THH) Au NCs/GCE	0.1 pM–10 nM	0.1 pM	Chen et al. (2013)
Cytochrome C	EIS	Polypyrrole/aptamer/cytochrome c/SPE	10 pM–1 nM	5 pM	Shafaat et al. (2018)
TNFα	CV and ESI	GSPE/Ag@Pt-GRs/apt	0–60 pg mL^{-1}	2.07 pg mL^{-1}	Mazloum-Ardakani et al. (2015)
Nuclear factor kappa B (NF-κB)	DPV	Based on target binding-triggered ratiometric Signal readout and polymerase-assisted protein recycling amplification	0.1 pg mL^{-1}–15 ng mL^{-1}	0.03 pg mL^{-1}	Peng et al. (2018)
Human cardiac troponin I	DPV	GE/gold nanodumbbells/aptamer/troponin I	0.05–500 ng mL^{-1}	8.0 pg mL^{-1}	Negahdary et al. (2017)

8.4 Electrochemical aptasensors against cancer cells and microorganisms

The new estimation of the International Agency for Research on Cancer, GLOBOCAN 2018, shows that global cancer incidence has risen to 18.1 million cases, and unfortunately, 9.6 million cases have died. Lung, breast, colorectal, prostate, and stomach cancer are the most prevalent types. Statistics exhibit one-in-eight men and one-in-eleven women will die from cancer. Also, the early diagnosis of cancers plays a vital role in the effective treatment of cancers (Wu et al., 2007).

For the first time, the aptamer against a whole cell line, U251, was recognized by Gold's team (Daniels et al., 2003). Tan and his research group also developed this new strategy, Cell SELEX, to identify cancer-specific aptamers as effective molecular probes for cancer study (Herr et al., 2006; Shangguan et al., 2006). Subsequently, the first electrochemical aptasensor designed to diagnose leukemia cells based on an aptamer-modified gold electrode (Pan et al., 2009).

Our research team has also focused on the ultrasensitive detection of various cancer cells using electrochemical aptasensors. For instance, because of the high importance of the early stage of liver cancer diagnosis for its effective treatment, a simple sandwich-battlemented impedimetric aptasensor was designed for the detection of a hepatocellular carcinoma cell line, HepG2. Mercaptopropionic acid (MPA) self-assembled monolayer on a gold electrode provides the primary aptamer linkage to the electrode. Liver cancer cells specifically sandwiched between the primary and secondary aptamers, which presented a wide linear range over 1×10^2 to 1×10^6 cell/mL, with an LOD of 2 cell/mL (Kashefi-Kheyrabadi et al., 2014).

In another effort, a closed bipolar electrode system integrated with 3D printed microchannels was fabricated for following ECL signals to diagnose human breast cancer cells (MCF-7). By applying a sandwich strategy, thiolated AS1411 aptamer was immobilized on the anodic pole of bipolar and secondary aptamer was modified with AuNPs. Because of the specific interaction of aptamer with overexpressed membrane proteins, nucleolins, target cells were captured. The treatment of captured cells with gold nanoparticle–modified secondary aptamers leads to the accumulation of GNPs. Afterward, the enhanced ECL of luminol on the gold nanoparticle surfaces was followed as an analytical signal. Two acceptable linear ranges 10–100 and 100–700 cells/mL with an LOD of about 10 cells/mL was obtained (Motaghi et al., 2018). Fig. 8.6 shows the schematic illustration of applied strategy in the designing of this aptasensor.

Electrochemical aptasensors were also developed for the detection of microorganisms. Microorganisms are incredibly diverse, including, bacteria, fungi, planktons, protists, archaea, and some amebae (Debelian et al., 1994). Viruses and

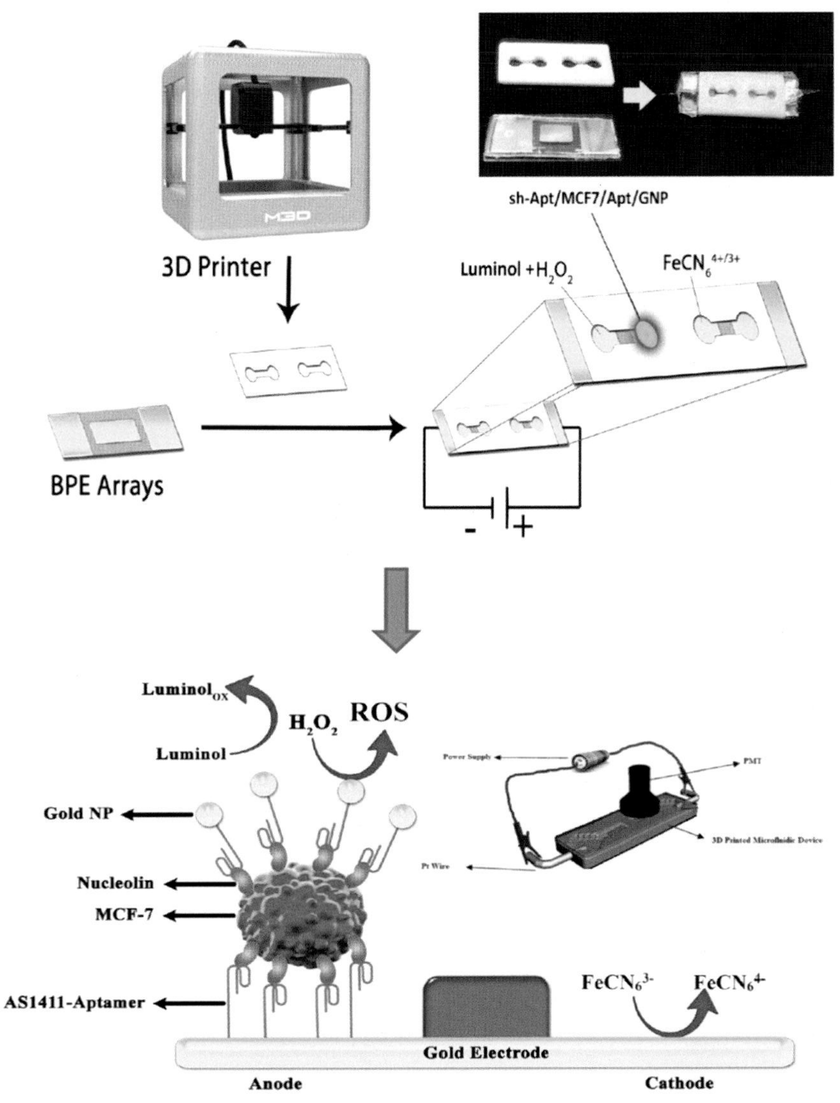

FIGURE 8.6

Schematic illustration of a BPE aptasensor for sensitive detection of human breast cancer cells (Motaghi et al., 2018).

Reprinted with permission from Elsevier.

prions are not considered to be living organisms but are often considered to be microorganisms because of their extremely small size. The developed electrochemical aptasensors for the detection of whole cells and microorganisms have been listed in Table 8.3.

Table 8.3 The electrochemical aptasensors for the detection of whole cells and microorganisms.

Target	Electrochemical techniques	Strategy	Linearity range	Detection limit	Reference
AGS	Amperometery	SPE/MWCNTs-AuNPs/aptamer I/cell/aptamer II–Au@Ag	1.0×10^{1} to 5.0×10^{5}	6	Tabrizi et al. (2017b)
Ramos	ECL	GCE/SWCNTs/aptamer /cell/aptamer II-G5 dendrimer-ECL probe	1.0×10^{2} to 1.0×10^{4}	55	Yang et al. (2016)
CCRF	ECL	GCE/rGO-PAMAP dendrimer/aptamer/cell/Au-nanoprobe	1.0×10^{2} to 1.0×10^{5}	38	Chen et al., (2014)
MCF-7	DPV	Porous graphene oxide-Au composites/aptamer I/cell/aptamer II-porous PtFe alloy	100 to 5.0×10^{7}	38	Yan et al. (2013)
Hela	EIS	GCE/perylene tetracarboxylic acid –functionalized graphene/aptamer/cell	1.0×10^{3} to 1.0×10^{6}	794	Feng et al. (2011)
HepG2	DPV	GCE/AuNPs deposit/aptamer I/cell/aptamer II-nanoprobe	1×10^{2} to 1×10^{7}	15	Sun et al. (2016)
HepG2	DPV	GE/aptamer I/cell/aptamer II-nanoprobe	1×10^{2} to 1×10^{7}	30	Sun et al. (2015)
DLD-1	EIS	GCE/carbon nanospheres/aptamer/cell	1.25×10^{2} to 1.25×10^{6}	40	Cao et al. (2014)
MCF7	CV	GE/aptamer I/cell/aptamer II-HRP	100 to 1×10^{7}	100	Zhu et al. (2013)
Leukemia cancer CCRF-CEM	DPV	SPE/magnetic NPs@Au shell-aptamer/cell	10 to 1×10^{6}	10	Khoshfetrat and Mehrgardi, (2017)

Leukemic Jurkat	EIS	GE/aptamer/cell	50 to 5 × 10^5	105	Bábelová et al. (2018)
MCF7	Chronocoulometry	GE/via DNA walker	0–500	47	Cai et al., (2016)
Colorectal cancer (CT26)	EIS, CV	SPE/SBA-15/AuNPs/aptamer I/cell/aptamer II	10 to 1.0 × 10^5 and 1.0 × 10^5 to 6.0 × 10^6	2	Hashkavayi et al. (2017)
MCF-7	SWV, PL	GCE/bismuth film/cell/QD&MB-aptamer	250 to 10^4	201	Hua et al. (2013)
Prostate cancer	EIS	GE/spacer/streptavidin modified with peptide and aptamer/cell	10^2–10^4	-	Min et al. (2010)
MEAR (CTC)	DPV	GCE/spacer/dual-aptamer/circulating tumor cell	1–14 in 10^9 WBC	1 in 10^9 WBC	Qu et al. (2014)
Human leukemic lymphoblast cancer	Amperometric	SPE/PEDOT-AuNPs-aptamer I/cell/modified aptamer II	1.0 × 10^1 to 5.0 × 10^5	8	Tabrizi et al. (2017a)
VCaP cells SK-BR-3 cells MDA-MB-231 Cells(CTC)	LSV	GE array/anti-EpCAM aptamer/cell/metal NPs-specific aptamer or antibody	2–200	2	Wan et al. (2014)
Hela	DPV	GE/GR-PAH-Fc/PSS(LBL)/aptamer/cell	10–10^6	10	Wang et al. (2015b)
MCF-7	DPV	GE/aptamer I/cell/aptamer II-AuNPs-GO	10–10^5	8	Wang et al. (2017a)
Human non–small-cell lung cancer (NSCLC) A549	DPV	GE/via binding-induced dual catalytic hairpin assembly	50–10^5	30	Zhang et al. (2018)
A549 human NSCLC	Amperometry	GCE/AuNPs/aptamer I/cell/aptamer II-AuNPs-hydrazine	15 to 1 × 10^6	8	Mir et al. (2015)

Continued

Table 8.3 The electrochemical aptasensors for the detection of whole cells and microorganisms.—*cont'd*

Target	Electrochemical techniques	Strategy	Linearity range	Detection limit	Reference
SK-BR-3 breast cancer	Photoelectrochemical (PEC)	ITO/ZnO-GR/AuNPs/aptamer/cell	1×10^2 to 1×10^6	58	Liu et al. (2014a)
Escherichia coli O157: H7	CV/EIS/DPV	MOF/PANI/GA	2.1×10^1 to 2.1×10^7 CFU mL^{-1}	2 CFU mL^{-1}	Shahrokhian and Ranjbar, (2018)
Pathogenic bacteria	CV/EIS	AuNPs branches	10–106 CFU mL^{-1}	7 CFU mL^{-1}	Majdinasab et al. (2018)
Salmonella	CV/DPV	rGO-CHI	10^1–10^6 CFU mL^{-1}	10^1 CFU mL^{-1}	Dinshaw et al. (2017)
E. coli O157: H7	EIS	Covalently immobilized on the 3D-IDEA surface grafted with mercaptosilane	10^1–10^5 CFU mL^{-1}	2.9×10^2 CFU mL^{-1}	Brosel-Oliu et al. (2018)
E. coli O111	DPV	Gold electrode surface via Authiol	2×10^2 to 2×10^6 CFU mL^{-1}	112 CFU mL^{-1}	Luo et al. (2012)
Staphylococcus aureus	CV/EIS	rGO/AuNP	10 to 10^6 CFU mL^{-1}	1 CFU mL^{-1}	Jia et al. (2014)
Salmonella	CV/EIS	ssDNA/MWCNT/ITO electrode	5.5×10^1 and 6.7×10^1 CFU mL^{-1}	5.5×10^1 CFU mL^{-1}	Hasan et al. (2018)
Enterococcus faecalis gene	DPV	Electrodeposited gold nanostructures/toluidine blue	1.0×10^{-17} to 1.0×10^{-10} M	4.7×10–20 M	Nazari-Vanani et al. (2018)
Salmonella	CV/EIS	GCE/GO/GNPs	2.4 to 2.4×10^3 CFU mL^{-1}	3 CFU mL^{-1}	Ma et al., (2014)

8.5 Conclusion

The sensitive and selective detection of biomaterials is still a big challenge in analytical chemistry. Electrochemical biosensors offer a promising strategy for this purpose by applying a biorecognition layer against specific targets. Whereas antibodies have been applied vastly as the recognition layers in the immunosensors, aptamers offer some important advantages rather than antibodies, including higher selectivities and affinities, longer shelf life at ambient temperature, and the simplicity of synthesis, attracted more and more interests of researchers, recently. The electrochemical aptasensors have been developed for a wide range of analytes from small molecules to the whole cells.

Abbreviations

3D printer	Three-dimensional printer
ABA	p-Aminobenzoic acid
AChE	Acetylcholinesterase
ACV	Alternating current voltammetry
AE	Au electrode
AFM1	Aflatoxin M1
AFSD	Aptamer folding-based sensory device
AMP	Ampicillin
Apt	Aptamer
ATP	Adenosine triphosphate
AuNPs	Gold nanoparticles
BMIMPF6	1-Butyl-3-methylimidazolium hexafluorophosphate
BPE	Bipolar electrode
CAP	Chloramphenicol
CEA	Carcinoembryonic antigen
CF	Carbon fibers
CFU	Colony-forming unit
CHI	Chitosan
CNTs	Carbon nanotubes
CNTs-COOH	Acid-oxidized carbon nanotubes
CP	Carbon paper
CS	Complimentary strand of the aptamer
CV	Cyclic voltammetry
DA	Dopamine
DABA	3,4-Aminobenzoic acid
DPSV	Differential pulse stripping voltammetry
DPV	Differential pulse voltammetry
dsDNA	Double-stranded DNA
DTSSP	3,3′-Dithiobis (sulfosuccinimidylpropionate)
EBFCs	Enzyme biofuel cells
ECL	Electrochemiluminescence

EIS	Impedance spectroscopy
e-pGON	Porous graphene oxide network
EPPG	Edge plane pyrolytic graphite
ETA	Acetone ethanolamine
EXO	Exonuclease I
Fc	Ferrocene
Fc-PAH	Ferrocene-appended poly(allylamine hydrochloride)
GA	Glutaraldehyde
GCE	Glassy carbon electrode
GE	Gold electrode
GNPs	Gold nanoparticles
GS	Graphene sheet
GSPE	Graphene screen-printed electrodes
HDT	1,6-Hexanedithiol
HPLC	High-performance liquid chromatography
HRP	Horseradish peroxidase
HT	Hexanethiol
ITO	Indium tin oxide
LBL	Layer by layer
LOD	Limit of detection
MB	Methylene blue
Mb	Myoglobin
MCF7	Michigan Cancer Foundation-7 (breast cancer cells)
MCH	6-Mercapto-1-hexanol
MGP	Magnetic graphene platform
MHPs	Magnetic hollow porous nanoparticles
MOF	Metal organic framework
MPA	Mercaptopropionic acid
MWCNT	Multiwall carbon nanotube
NF-κB	Nuclear factor kappa B
NMs	Nanomaterials
NPGE	Nanoporous gold electrode
OH-PCB	Hydroxylated polychlorinated biphenyl
OTC	Oxytetracycline
PAMAM	Polyamidoamine
PANI	Polyaniline
PATP	4-Aminothiophenol
PB	Prussian blue
PDA	Polydopamine
PDDA	Polydiallyldimethylammonium
PDGF	Platelet-derived growth factor
PdNPs	Palladium nanoparticles
PEC	Photoelectrochemistry
PEDOT	Poly(3,4-ethylenedioxythiophene)
PEI	Polyethyleneimine
PSA	Prostate-specific antigen
PSS	Poly(sodium-p-styrenesulfonate)
PtNAs	Platinum nanotube arrays

PtNPs	Platinum nanoparticles
PT-Pala	Polythiophene bearing polyalanine homopeptide side chains
QD	Quantum dot
RCA	Rolling-circle amplification
RGD	Arginine–glycine–aspartic
rGO	Reduced graphene oxide
SA	Streptavidin
SB	Schiff base
SELEX	Systematic evolution of ligands by exponential enrichment
SPE	Screen-printed electrode
SPGE	Screen-printed gold electrode
ss-DNA	Single-stranded DNA
SWCNTs	Single-wall carbon nanotubes
SWV	Square wave voltammetry
TB	Thrombin
TBA	Thrombin-binding aptamer
TET	Tetracycline
Th	Thionine
THH	Tetrahexahedral
TiNTA	TiO_2 nanotube array
VEGF	Vascular endothelial growth factor
WBC	White blood cell
WHO	World Health Organization
β-CD	β-cyclodextrins

Acknowledgments

The authors would like gratefully to acknowledge the research council of the University of Isfahan.

References

Arias, C.A., Murray, B.E.J.N.E.J.O.M., 2009. Antibiotic-resistant bugs in the 21st century—a clinical super-challenge. New England Journal of Medicine 360, 439–443.

Azadbakht, A., Roushani, M., Abbasi, A.R., Derikvand, Z.J., 2016a. Design and characterization of electrochemical dopamine–aptamer as convenient and integrated sensing platform. Analytical Biochemistry 507, 47–57.

Azadbakht, A., Roushani, M., Abbasi, A.R., Menati, S., Derikvand, Z., 2016b. A label-free aptasensor based on polyethyleneimine wrapped carbon nanotubes in situ formed gold nanoparticles as signal probe for highly sensitive detection of dopamine. Materials Science and Engineering: C 68, 585–593.

Babamiri, B., Salimi, A., Hallaj, R.J.B., 2018. Switchable electrochemiluminescence aptasensor coupled with resonance energy transfer for selective attomolar detection of Hg^{2+} via CdTe@ CdS/dendrimer probe and Au nanoparticle quencher. Biosensors and Bioelectronics 102, 328–335.

Bábelová, L., Sohová, M.E., Poturnayová, A., Buríková, M., Bizík, J., Hianik, T., 2018. Label-free electrochemical aptasensor for jurkat cells detection as a potential diagnostic tool for leukemia. Electroanalysis 30, 1487–1495.

Bagal-Kestwal, D., Karve, M.S., Kakade, B., Pillai, V.K., 2008. Invertase inhibition based electrochemical sensor for the detection of heavy metal ions in aqueous system: application of ultra-microelectrode to enhance sucrose biosensor's sensitivity. Biosensors and Bioelectronics 24, 657–664.

Bahrami, S., Abbasi, A.R., Roushani, M., Derikvand, Z., Azadbakht, A.J., 2016. An electrochemical dopamine aptasensor incorporating silver nanoparticle, functionalized carbon nanotubes and graphene oxide for signal amplification. Talanta 159, 307–316.

Bai, L., Yan, B., Chai, Y., Yuan, R., Yuan, Y., Xie, S., Jiang, L., He, Y., 2013. An electrochemical aptasensor for thrombin detection based on direct electrochemistry of glucose oxidase using a functionalized graphene hybrid for amplification. Analyst 138, 6595–6599.

Bao, J., Hou, C., Chen, M., Li, J., Huo, D., Yang, M., Luo, X., Lei, Y., 2015a. Plant esterase–chitosan/gold nanoparticles–graphene nanosheet composite-based biosensor for the ultrasensitive detection of organophosphate pesticides. Journal of Agricultural and Food Chemistry 63, 10319–10326.

Bao, T., Wen, W., Zhang, X., Wang, S., 2015b. An exonuclease-assisted amplification electrochemical aptasensor of thrombin coupling "signal on/off" strategy. Analytica Chimica Acta 860, 70–76.

Beaulieu, J.-M., Gainetdinov, R.R., 2011. The physiology, signaling, and pharmacology of dopamine receptors. Pharmacological Reviews 63, 182–217.

Beiranvand, S., Abbasi, A.R., Roushani, M., Derikvand, Z., Azadbakht, A., 2016. A simple and label-free aptasensor based on amino group-functionalized gold nanocomposites-Prussian blue/carbon nanotubes as labels for signal amplification. Journal of Electroanalytical Chemistry 776, 170–179.

Berruga, M., Beltrán, M., Novés, B., Molina, A., Molina, M., 2011. Effect of penicillins on the acidification of yogurt made from Ewe's milk during the storage. In: Science and Technology against Microbial Pathogens: Research, Development and Evaluation. World Scientific.

Bertolotti, A., 2018. Importance of the subcellular location of protein deposits in neurodegenerative diseases. Current Opinion in Neurobiology 51, 127–133.

Bozokalfa, G., Akbulut, H., Demir, B., Guler, E., Gumus, Z.P., ODACI Demirkol, D., Aldemir, E., Yamada, S., Endo, T., Coskunol, H., 2016. Polypeptide functional surface for the aptamer immobilization: electrochemical cocaine biosensing. Analytical Chemistry 88, 4161–4167.

Brosel-Oliu, S., Ferreira, R., Uria, N., Abramova, N., Gargallo, R., MUÑOZ-Pascual, F.-X., Bratov, A., 2018. Novel impedimetric aptasensor for label-free detection of *Escherichia coli* O157: H7. Sensors and Actuators B: Chemical 255, 2988–2995.

Cai, S., Chen, M., Liu, M., He, W., Liu, Z., Wu, D., Xia, Y., Yang, H., Chen, J., 2016. A signal amplification electrochemical aptasensor for the detection of breast cancer cell via free-running DNA walker. Biosensors and Bioelectronics 85, 184–189.

Cao, H., Ye, D., Zhao, Q., Luo, J., Zhang, S., Kong, J., 2014. A novel aptasensor based on MUC-1 conjugated CNSs for ultrasensitive detection of tumor cells. Analyst 139, 4917–4923.

Cao, J.-T., Zhang, J.-J., Gong, Y., Ruan, X.-J., Liu, Y.-M., Chen, Y.-H., Ren, S.-W., 2015. A competitive photoelectrochemical aptasensor for thrombin detection based on the use of TiO2 electrode and glucose oxidase label. Journal of Electroanalytical Chemistry 759, 46–50.

Carlsson, A., Lindqvist, M., Magnusson, T., 1957. 3, 4-Dihydroxyphenylalanine and 5-hydroxytryptophan as reserpine antagonists. Nature 180, 1200.

Chen, X., He, Y., Zhang, Y., Liu, M., Liu, Y., Li, J., 2014. Ultrasensitive detection of cancer cells and glycan expression profiling based on a multivalent recognition and alkaline phosphatase-responsive electrogenerated chemiluminescence biosensor. Nanoscale 6, 11196–11203.

Chen, Z., Guo, J., 2013. A reagentless signal-off architecture for electrochemical aptasensor for the detection of lysozyme. Electrochimica Acta 111, 916–920.

Chen, Z., Guo, J., Li, J., Guo, L., 2013. Tetrahexahedral Au nanocrystals/aptamer based ultra-sensitive electrochemical biosensor. RSC Advances 3, 14385–14389.

Chen, Z., Li, L., Mu, X., Zhao, H., Guo, L., 2011. Electrochemical aptasensor for detection of copper based on a reagentless signal-on architecture and amplification by gold nanoparticles. Talanta 85, 730–735.

Cheng, W., Ding, S., Li, Q., Yu, T., Yin, Y., Ju, H., Ren, G., 2012. A simple electrochemical aptasensor for ultrasensitive protein detection using cyclic target-induced primer extension. Biosensors and Bioelectronics 36, 12–17.

Cui, L., Wu, J., Ju, H., 2016. Label-free signal-on aptasensor for sensitive electrochemical detection of arsenite. Biosensors and Bioelectronics 79, 861–865.

Da, H., Liu, H., Zheng, Y., Yuan, R., Chai, Y., 2018. A highly sensitive VEGF165 photoelectrochemical biosensor fabricated by assembly of aptamer bridged DNA networks. Biosensors and Bioelectronics 101, 213–218.

Daniels, D.A., Chen, H., Hicke, B.J., Swiderek, K.M., Gold, L., 2003. A tenascin-C aptamer identified by tumor cell SELEX: systematic evolution of ligands by exponential enrichment. Proceedings of the National Academy of Sciences 100, 15416–15421.

Debelian, G.J., Olsen, I., Tronstad, L., 1994. Systemic diseases caused by oral microorganisms. Dental Traumatology 10, 57–65.

Dinshaw, I.J., Muniandy, S., Teh, S.J., Ibrahim, F., Leo, B.F., Thong, K.L., 2017. Development of an aptasensor using reduced graphene oxide chitosan complex to detect Salmonella. Journal of Electroanalytical Chemistry 806, 88–96.

Dridi, F., Marrakchi, M., Gargouri, M., Saulnier, J., Jaffrezic-Renault, N., Lagarde, F., 2017. Nanomaterial-based Electrochemical Biosensors for Food Safety and Quality Assessment. Nanobiosensors. Elsevier.

Erdem, A., Congur, G., 2014. Dendrimer enriched single-use aptasensor for impedimetric detection of activated protein C. Colloids and Surfaces B: Biointerfaces 117, 338–345.

Erdem, A., Congur, G., Mayer, G., 2015. Aptasensor platform based on carbon nanofibers enriched screen printed electrodes for impedimetric detection of thrombin. Journal of Electroanalytical Chemistry 758, 12–19.

Fan, G.-C., Zhao, M., Zhu, H., Shi, J.-J., Zhang, J.-R., Zhu, J.-J., 2015. Signal-on Photoelectrochemical Aptasensor for adenosine triphosphate detection based on sensitization effect of CdS: Mn@ Ru (bpy) 2 (dcbpy) nanocomposites. Journal of Physical Chemistry C 120, 15657–15665.

Fan, H., Li, H., Wang, Q., He, P., Fang, Y., 2012. A host–guest-recognition-based electrochemical aptasensor for thrombin detection. Biosensors and Bioelectronics 35, 33–36.

Fan, L., Zhao, G., Shi, H., Liu, M., Li, Z., 2013. A highly selective electrochemical impedance spectroscopy-based aptasensor for sensitive detection of acetamiprid. Biosensors and Bioelectronics 43, 12–18.

Fan, T., Du, Y., Yao, Y., Wu, J., Meng, S., Luo, J., Zhang, X., Yang, D., Wang, C., Qian, Y., 2018. Rolling circle amplification triggered poly adenine-gold nanoparticles production for label-free electrochemical detection of thrombin. Sensors and Actuators B: Chemical 266, 9–18.

Feng, L., Chen, Y., Ren, J., Qu, X., 2011. A graphene functionalized electrochemical aptasensor for selective label-free detection of cancer cells. Biomaterials 32, 2930–2937.

Florea, A., Taleat, Z., Cristea, C., MAZLOUM-Ardakani, M., Săndulescu, R., 2013. Label free MUC1 aptasensors based on electrodeposition of gold nanoparticles on screen printed electrodes. Electrochemistry Communications 33, 127–130.

Freberg, L., 2009. Discovering Biological Psychology. Cengage Learning.

Fu, X.-M., Liu, Z.-J., Cai, S.-X., Zhao, Y.-P., Wu, D.-Z., Li, C.-Y., Chen, J.-H., 2016. Electrochemical aptasensor for the detection of vascular endothelial growth factor (VEGF) based on DNA-templated Ag/Pt bimetallic nanoclusters. Chinese Chemical Letters 27, 920–926.

Gai, P., Gu, C., Hou, T., Li, F., 2017. Ultrasensitive self-powered aptasensor based on enzyme biofuel cell and DNA bioconjugate: a facile and powerful tool for antibiotic residue detection. Analytical Chemistry 89, 2163–2169.

Gao, F., Du, L., Zhang, Y., Zhou, F., Tang, D., 2016a. A sensitive sandwich-type electrochemical aptasensor for thrombin detection based on platinum nanoparticles decorated carbon nanocages as signal labels. Biosensors and Bioelectronics 86, 185–193.

Gao, F., Gao, C., He, S., Wang, Q., Wu, A., 2016b. Label-free electrochemical lead (II) aptasensor using thionine as the signaling molecule and graphene as signal-enhancing platform. Biosensors and Bioelectronics 81, 15–22.

Gaugain-Juhel, M., Delépine, B., Gautier, S., Fourmond, M.-P., Gaudin, V., HURTAUD-Pessel, D., Verdon, E., Sanders, P., 2009. Validation of a liquid chromatography-tandem mass spectrometry screening method to monitor 58 antibiotics in milk: a qualitative approach. Food Additives & Contaminants 26, 1459–1471.

Goyal, R.N., Shim, Y.-B., 2015. Glutaraldehyde sandwiched amino functionalized polymer based aptasensor for the determination and quantification of chloramphenicol. RSC Advances 5, 69356–69364.

Guler, E., Bozokalfa, G., Demir, B., Gumus, Z.P., Guler, B., Aldemir, E., Timur, S., Coskunol, H., 2017a. An aptamer folding based sensory platform decorated with nanoparticles for simple cocaine testing. Drug Testing and Analysis 9, 578–587.

Guler, M., Turkoglu, V., Basi, Z., 2017b. Determination of malation, methidathion, and chlorpyrifos ethyl pesticides using acetylcholinesterase biosensor based on Nafion/Ag@ rGO-NH2 nanocomposites. Electrochimica Acta 240, 129–135.

Gumpu, M.B., Sethuraman, S., Krishnan, U.M., Rayappan, J.B.B., 2015. A review on detection of heavy metal ions in water–An electrochemical approach. Sensors and Actuators B: Chemical 213, 515–533.

Hasan, M.R., Pulingam, T., Appaturi, J.N., Zifruddin, A.N., Teh, S.J., Lim, T.W., Ibrahim, F., Leo, B.F., Thong, K.L., 2018. Carbon nanotube-based aptasensor for sensitive electrochemical detection of whole-cell Salmonella. Analytical Biochemistry 554, 34–43.

Hashkavayi, A.B., Raoof, J.B., Ojani, R., Kavoosian, S., 2017. Ultrasensitive electrochemical aptasensor based on sandwich architecture for selective label-free detection of colorectal cancer (CT26) cells. Biosensors and Bioelectronics 92, 630–637.

Hassani, S., Momtaz, S., Vakhshiteh, F., Maghsoudi, A.S., Ganjali, M.R., Norouzi, P., Abdollahi, M., 2017. Biosensors and their applications in detection of organophosphorus pesticides in the environment. Archives of Toxicology 91, 109–130.

He, L., Zhang, S., Ji, H., Wang, M., Peng, D., Yan, F., Fang, S., Zhang, H., Jia, C., Zhang, Z., 2016. Protein-templated cobaltous phosphate nanocomposites for the highly sensitive and selective detection of platelet-derived growth factor-BB. Biosensors and Bioelectronics 79, 553–560.

He, X., Zhao, Y., He, D., Wang, K., Xu, F., Tang, J., 2012. ATP-responsive controlled release system using aptamer-functionalized mesoporous silica nanoparticles. Langmuir 28, 12909–12915.

Herr, J.K., Smith, J.E., Medley, C.D., Shangguan, D., Tan, W., 2006. Aptamer-conjugated nanoparticles for selective collection and detection of cancer cells. Analytical Chemistry 78, 2918–2924.

Heydari-Bafrooei, E., Amini, M., Ardakani, M.H., 2016. An electrochemical aptasensor based on TiO_2/MWCNT and a novel synthesized Schiff base nanocomposite for the ultrasensitive detection of thrombin. Biosensors and Bioelectronics 85, 828–836.

Hu, T., Wen, W., Zhang, X., Wang, S., 2016. Nicking endonuclease-assisted recycling of target–aptamer complex for sensitive electrochemical detection of adenosine triphosphate. Analyst 141, 1506–1511.

Hua, X., Zhou, Z., Yuan, L., Liu, S., 2013. Selective collection and detection of MCF-7 breast cancer cells using aptamer-functionalized magnetic beads and quantum dots based nano-bio-probes. Analytica Chimica Acta 788, 135–140.

Jarczewska, M., Rebiś, J., Górski, Ł., Malinowska, E., 2018. Development of DNA aptamer–based sensor for electrochemical detection of C-reactive protein. Talanta 189, 45–54.

Jia, F., Duan, N., Wu, S., Ma, X., Xia, Y., Wang, Z., Wei, X., 2014. Impedimetric aptasensor for *Staphylococcus aureus* based on nanocomposite prepared from reduced graphene oxide and gold nanoparticles. Microchimica Acta 181, 967–974.

Jing, P., Yi, H., Xue, S., Chai, Y., Yuan, R., Xu, W., 2015. A sensitive electrochemical aptasensor based on palladium nanoparticles decorated graphene–molybdenum disulfide flower-like nanocomposites and enzymatic signal amplification. Analytica Chimica Acta 853, 234–241.

Jolly, P., Zhurauski, P., Hammond, J.L., Miodek, A., Liébana, S., Bertok, T., Tkáč, J., Estrela, P., 2017. Self-assembled gold nanoparticles for impedimetric and amperometric detection of a prostate cancer biomarker. Sensors and Actuators B: Chemical 251, 637–643.

Kashefi-Kheyrabadi, L., Mehrgardi, M.A., 2012. Aptamer-conjugated silver nanoparticles for electrochemical detection of adenosine triphosphate. Biosensors and Bioelectronics 37, 94–98.

Kashefi-Kheyrabadi, L., Mehrgardi, M.A., 2013. Aptamer-based electrochemical biosensor for detection of adenosine triphosphate using a nanoporous gold platform. Bioelectrochemistry 94, 47–52.

Kashefi-Kheyrabadi, L., Mehrgardi, M.A., Wiechec, E., Turner, A.P., Tiwari, A., 2014. Ultrasensitive detection of human liver hepatocellular carcinoma cells using a label-free aptasensor. Analytical Chemistry 86, 4956–4960.

Kerr, T., Small, W., Wood, E., 2005. The public health and social impacts of drug market enforcement: a review of the evidence. International Journal of Drug Policy 16, 210–220.

Khoshfetrat, S.M., Mehrgardi, M.A., 2017. Amplified detection of leukemia cancer cells using an aptamer-conjugated gold-coated magnetic nanoparticles on a nitrogen-doped graphene modified electrode. Bioelectrochemistry 114, 24–32.

Labib, M., Sargent, E.H., Kelley, S.O., 2016. Electrochemical methods for the analysis of clinically relevant biomolecules. Chemical Reviews 116, 9001–9090.

Lee, M., Lee, H., Ryu, P.J.A.-A.J.O.A.S., 2001. Public health risks: chemical and antibiotic residues-review. Asian-Australasian Journal of Animal Sciences 14, 402–413.

Li, C., Li, J., Yang, X., Gao, L., Jing, L., Ma, X., 2017a. A label-free electrochemical aptasensor for sensitive myoglobin detection in meat. Sensors and Actuators B: Chemical 242, 1239–1245.

Li, G.-Z., Tang, D., 2017. Bioresponsive controlled glucose release from TiO_2 nanotube arrays: a simple and portable biosensing system for cocaine with a glucometer readout. Journal of Materials Chemistry B 5, 5573–5579.

Li, H., Wang, C., Wu, Z., Lu, L., Qiu, L., Zhou, H., Shen, G., Yu, R., 2013. An electronic channel switching-based aptasensor for ultrasensitive protein detection. Analytica Chimica Acta 758, 130–137.

Li, M.-J., Zheng, Y.-N., Liang, W.-B., Yuan, R., Chai, Y.-Q., 2017b. Using p-type PbS quantum dots to quench photocurrent of fullerene–Au NP@ MoS_2 composite structure for ultrasensitive photoelectrochemical detection of ATP. ACS Applied Materials and Interfaces 9, 42111–42120.

Li, X., Qi, H., Shen, L., Gao, Q., Zhang, C., 2008. Electrochemical aptasensor for the determination of cocaine incorporating gold nanoparticles modification. Electroanalysis: An International Journal Devoted to Fundamental and Practical Aspects of Electroanalysis 20, 1475–1482.

Li, Y., Shi, L., Han, G., Xiao, Y., Zhou, W., 2017c. Electrochemical biosensing of carbaryl based on acetylcholinesterase immobilized onto electrochemically inducing porous graphene oxide network. Sensors and Actuators B: Chemical 238, 945–953.

Li, Y., Wang, Q., Zhang, Y., Deng, D., He, H., Luo, L., Wang, Z., 2016a. A label-free electrochemical aptasensor based on graphene oxide/double-stranded DNA nanocomposite. Colloids and Surfaces B: Biointerfaces 145, 160–166.

Li, Y., Zhang, Y., Han, G., Xiao, Y., Li, M., Zhou, W., 2016b. An acetylcholinesterase biosensor based on graphene/polyaniline composite film for detection of pesticides. Chinese Journal of Chemistry 34, 82–88.

Li, Y., Zhao, R., Shi, L., Han, G., Xiao, Y., 2017d. Acetylcholinesterase biosensor based on electrochemically inducing 3D graphene oxide network/multi-walled carbon nanotube composites for detection of pesticides. RSC Advances 7, 53570–53577.

Liu, C., Zhang, H., Huang, Y., Chen, Z., Luo, F., Wang, J., Guo, L., Qiu, B., Lin, Z., Yang, H., 2018. Homogeneous electrochemical aptasensor for mucin 1 detection based on exonuclease I-assisted target recycling amplification strategy. Biosensors and Bioelectronics 117, 474–479.

Liu, F., Zhang, Y., Yu, J., Wang, S., Ge, S., Song, X., 2014a. Application of ZnO/graphene and S6 aptamers for sensitive photoelectrochemical detection of SK-BR-3 breast cancer cells based on a disposable indium tin oxide device. Biosensors and Bioelectronics 51, 413–420.

Liu, L., Du, J., Li, S., Yuan, B., Han, H., Jing, M., Xia, N., 2013. Amplified voltammetric detection of dopamine using ferrocene-capped gold nanoparticle/streptavidin conjugates. Biosensors and Bioelectronics 41, 730–735.

Liu, L., Xia, N., Meng, J.-J., Zhou, B.-B., Li, S.-J., 2016. An electrochemical aptasensor for sensitive and selective detection of dopamine based on signal amplification of electrochemical-chemical redox cycling. Journal of Electroanalytical Chemistry 775, 58–63.

Liu, S., Xing, X., Yu, J., Lian, W., Li, J., Cui, M., Huang, J., 2012. A novel label-free electrochemical aptasensor based on graphene–polyaniline composite film for dopamine determination. Biosensors and Bioelectronics 36, 186–191.

Liu, Y.-M., Zhou, M., Liu, Y.-Y., Shi, G.-F., Zhang, J.-J., Cao, J.-T., Huang, K.-J., Chen, Y.-H., 2014b. Fabrication of electrochemiluminescence aptasensor based on in situ growth of gold nanoparticles on layered molybdenum disulfide for sensitive detection of platelet-derived growth factor-BB. RSC Advances 4, 22888–22893.

Liu, Y., Yan, K., Zhang, J., 2015. Graphitic carbon nitride sensitized with CdS quantum dots for visible-light-driven photoelectrochemical aptasensing of tetracycline. ACS Applied Materials and Interfaces 8, 28255–28264.

Luo, C., Lei, Y., Yan, L., Yu, T., Li, Q., Zhang, D., Ding, S., Ju, H., 2012. A rapid and sensitive aptamer-based electrochemical biosensor for direct detection of Escherichia coli O111. Electroanalysis 24, 1186–1191.

Luo, J., Jiang, D., Liu, T., Peng, J., Chu, Z., Jin, W., 2018. High-performance electrochemical mercury aptasensor based on synergistic amplification of Pt nanotube arrays and Fe3O4/rGO nanoprobes. Biosensors and Bioelectronics 104, 1–7.

Ma, F., Ho, C., Cheng, A.K., Yu, H.-Z., 2013. Immobilization of redox-labeled hairpin DNA aptamers on gold: electrochemical quantitation of epithelial tumor marker mucin 1. Electrochimica Acta 110, 139–145.

Ma, X., Jiang, Y., Jia, F., Yu, Y., Chen, J., Wang, Z., 2014. An aptamer-based electrochemical biosensor for the detection of Salmonella. Journal of Microbiological Methods 98, 94–98.

Majdinasab, M., Hayat, A., Marty, J.L., 2018. Aptamer-based assays and aptasensors for detection of pathogenic bacteria in food samples. TRAC Trends in Analytical Chemistry 107, 60–77.

Mao, S., Chang, J., Zhou, G., Chen, J., 2015. Nanomaterial-enabled rapid detection of water contaminants. Small 11, 5336–5359.

Mascini, M., 2008. Aptamers and their applications. Analytical and Bioanalytical Chemistry 390, 987–988.

Mascini, M., Palchetti, I., Tombelli, S., 2012. Nucleic acid and peptide aptamers: fundamentals and bioanalytical aspects. Angewandte Chemie International Edition in English 51, 1316–1332.

Mashhadizadeh, M.H., Naseri, N., Mehrgardi, M.A., 2017. A simple non-enzymatic strategy for adenosine triphosphate electrochemical aptasensor using silver nanoparticle-decorated graphene oxide. Journal of the Iranian Chemical Society 14, 2007–2016.

Mazloum-Ardakani, M., Hosseinzadeh, L., Taleat, Z., 2015. Synthesis and electrocatalytic effect of Ag@ Pt core–shell nanoparticles supported on reduced graphene oxide for sensitive and simple label-free electrochemical aptasensor. Biosensors and Bioelectronics 74, 30–36.

Min, K., Song, K.-M., Cho, M., Chun, Y.-S., Shim, Y.-B., Ku, J.K., Ban, C., 2010. Simultaneous electrochemical detection of both PSMA (+) and PSMA (−) prostate cancer cells using an RNA/peptide dual-aptamer probe. Chemical Communications 46, 5566–5568.

Minunni, M., Tombelli, S., Gullotto, A., Luzi, E., Mascini, M., 2004. Development of biosensors with aptamers as bio-recognition element: the case of HIV-1 Tat protein. Biosensors and Bioelectronics 20, 1149–1156.

Mir, T.A., Yoon, J.-H., Gurudatt, N., Won, M.-S., Shim, Y.-B., 2015. Ultrasensitive cytosensing based on an aptamer modified nanobiosensor with a bioconjugate: detection of human non-small-cell lung cancer cells. Biosensors and Bioelectronics 74, 594–600.

Mogha, N.K., Sahu, V., Sharma, M., Sharma, R.K., Masram, D.T., 2016. Biocompatible ZrO2-reduced graphene oxide immobilized AChE biosensor for chlorpyrifos detection. Materials and Design 111, 312–320.

Motaghi, H., Ziyaee, S., Mehrgardi, M.A., Kajani, A.A., Bordbar, A.-K., 2018. Electrochemiluminescence detection of human breast cancer cells using aptamer modified bipolar electrode mounted into 3D printed microchannel. Biosensors and Bioelectronics 118, 217–223.

Nawaz, M.A.H., Rauf, S., Catanante, G., Nawaz, M.H., Nunes, G., Marty, J.L., Hayat, A., 2016. One step assembly of thin films of carbon nanotubes on screen printed interface for electrochemical aptasensing of breast cancer biomarker. Sensors 16, 1651.

Nazari-Vanani, R., Sattarahmady, N., Yadegari, H., Heli, H., 2018. A novel and ultrasensitive electrochemical DNA biosensor based on an ice crystals-like gold nanostructure for the detection of *Enterococcus faecalis* gene sequence. Colloids and Surfaces B: Biointerfaces 166, 245–253.

Negahdary, M., BEHJATI-Ardakani, M., Sattarahmady, N., Yadegari, H., Heli, H., 2017. Electrochemical aptasensing of human cardiac troponin I based on an array of gold nanodumbbells-Applied to early detection of myocardial infarction. Sensors and Actuators B: Chemical 252, 62–71.

Niu, X., Huang, L., Zhao, J., Yin, M., Luo, D., Yang, Y., 2016. An ultrasensitive aptamer biosensor for the detection of codeine based on a Au nanoparticle/polyamidoamine dendrimer-modified screen-printed carbon electrode. Analytical Methods 8, 1091–1095.

Ocaña, C., Hayat, A., Mishra, R.K., Vasilescu, A., DEL Valle, M., Marty, J.-L., 2015. Label free aptasensor for Lysozyme detection: a comparison of the analytical performance of two aptamers. Bioelectrochemistry 105, 72–77.

Pan, C., Guo, M., Nie, Z., Xiao, X., Yao, S., 2009. Aptamer-based electrochemical sensor for label-free recognition and detection of cancer cells. Electroanalysis: An International Journal Devoted to Fundamental and Practical Aspects of Electroanalysis 21, 1321–1326.

Pang, S., Labuza, T.P., He, L., 2014. Development of a single aptamer-based surface enhanced Raman scattering method for rapid detection of multiple pesticides. Analyst 139, 1895–1901.

Patrick, L., 2006. Lead toxicity part II: the role of free radical damage and the use of antioxidants in the pathology and treatment of lead toxicity. Alternative Medicine Review 11, 114–127.

Peng, K., Xie, P., Yang, Z.-H., Yuan, R., Zhang, K., 2018. Highly sensitive electrochemical nuclear factor kappa B aptasensor based on target-induced dual-signal ratiometric and polymerase-assisted protein recycling amplification strategy. Biosensors and Bioelectronics 102, 282–287.

Qin, X., Guo, W., Yu, H., Zhao, J., Pei, M., 2015. A novel electrochemical aptasensor based on MWCNTs–BMIMPF 6 and amino functionalized graphene nanocomposite films for determination of kanamycin. Analytical Methods 7, 5419–5427.

Qu, L., Xu, J., Tan, X., Liu, Z., Xu, L., Peng, R., 2014. Dual-aptamer modification generates a unique interface for highly sensitive and specific electrochemical detection of tumor cells. ACS Applied Materials and Interfaces 6, 7309–7315.

Qureshi, A., Gurbuz, Y., Niazi, J.H., 2015. Capacitive aptamer–antibody based sandwich assay for the detection of VEGF cancer biomarker in serum. Sensors and Actuators B: Chemical 209, 645–651.

Rahi, A., Sattarahmady, N., Heli, H., 2016. Label-free electrochemical aptasensing of the human prostate-specific antigen using gold nanospears. Talanta 156, 218–224.

Rapini, R., Marrazza, G., 2016. Biosensor Potential in Pesticide Monitoring. Comprehensive Analytical Chemistry. Elsevier.

Robinson, D.L., Hermans, A., Seipel, A.T., Wightman, R.M.J.C.R., 2008. Monitoring rapid chemical communication in the brain, 108, 2554–2584.

Ronkainen, N.J., Halsall, H.B., Heineman, W.R., 2010. Electrochemical biosensors. Chemical Society Reviews 39, 1747–1763.

Shafaat, A., Faridbod, F., Ganjali, M.R., 2018. Label-free detection of cytochrome C by a conducting polymer-based impedimetric screen-printed aptasensor. New Journal of Chemistry 42, 6034–6039.

Shahrokhian, S., Ranjbar, S., 2018. Aptamer immobilization on amino functionalized metal-organic frameworks: an ultrasensitive platform for electrochemical diagnostic of *Escherichia coli* O157: H7. Analyst 143, 3191–3201.

Shangguan, D., Li, Y., Tang, Z., Cao, Z.C., Chen, H.W., Mallikaratchy, P., Sefah, K., Yang, C.J., Tan, W., 2006. Aptamers evolved from live cells as effective molecular probes for cancer study. Proceedings of the National Academy of Sciences 103, 11838–11843.

Sharma, A., Istamboulie, G., Hayat, A., Catanante, G., Bhand, S., Marty, J.L., 2017. Disposable and portable aptamer functionalized impedimetric sensor for detection of kanamycin residue in milk sample. Sensors and Actuators B: Chemical 245, 507–515.

Shu, H., Wen, W., Xiong, H., Zhang, X., Wang, S., 2013. Novel electrochemical aptamer biosensor based on gold nanoparticles signal amplification for the detection of carcinoembryonic antigen. Electrochemistry Communications 37, 15–19.

Si, Z., Xie, B., Chen, Z., Tang, C., Li, T., Yang, M., 2017. Electrochemical aptasensor for the cancer biomarker CEA based on aptamer induced current due to formation of molybdophosphate. Microchimica Acta 184, 3215–3221.

Slikker Jr., W., De Souza Lima, T.A., Archella, D., De Silva Junior, J.B., Barton-Maclaren, T., Bo, L., Buvinich, D., Chaudhry, Q., Chuan, P., Deluyker, H., 2018. Emerging technologies for food and drug safety. Regulatory Toxicology and Pharmacology 98, 115–128.

Song, W., Li, H., Liang, H., Qiang, W., Xu, D., 2014. Disposable electrochemical aptasensor array by using in situ DNA hybridization inducing silver nanoparticles aggregate for signal amplification. Analytical Chemistry 86, 2775–2783.

Srivastava, N., Majumder, C., 2008. Novel biofiltration methods for the treatment of heavy metals from industrial wastewater. Journal of Hazardous Materials 151, 1–8.

Su, F., Zhang, S., Ji, H., Zhao, H., Tian, J.-Y., Liu, C.-S., Zhang, Z., Fang, S., Zhu, X., Du, M., 2017a. Two-dimensional zirconium-based metal–organic framework nanosheet composites embedded with Au nanoclusters: a highly sensitive electrochemical aptasensor toward detecting cocaine. ACS Sensors 2, 998–1005.

Su, Z., Xu, H., Xu, X., Zhang, Y., Ma, Y., Li, C., Xie, Q., 2017b. Effective covalent immobilization of quinone and aptamer onto a gold electrode via thiol addition for sensitive and selective protein biosensing. Talanta 164, 244–248.

Sun, D., Lu, J., Chen, Z., Yu, Y., Mo, M., 2015. A repeatable assembling and disassembling electrochemical aptamer cytosensor for ultrasensitive and highly selective detection of human liver cancer cells. Analytica Chimica Acta 885, 166–173.

Sun, D., Lu, J., Zhong, Y., Yu, Y., Wang, Y., Zhang, B., Chen, Z., 2016. Sensitive electrochemical aptamer cytosensor for highly specific detection of cancer cells based on the hybrid nanoelectrocatalysts and enzyme for signal amplification. Biosensors and Bioelectronics 75, 301–307.

Tabrizi, M.A., Shamsipur, M., Saber, R., Sarkar, S., 2017a. Flow injection amperometric sandwich-type aptasensor for the determination of human leukemic lymphoblast cancer cells using mwcnts-pdnano/ptca/aptamer as labeled aptamer for the signal amplification. Analytica Chimica Acta 985, 61–68.

Tabrizi, M.A., Shamsipur, M., Saber, R., Sarkar, S., Sherkatkhameneh, N., 2017b. Flow injection amperometric sandwich-type electrochemical aptasensor for the determination of adenocarcinoma gastric cancer cell using aptamer-Au@ Ag nanoparticles as labeled aptamer. Electrochimica Acta 246, 1147–1154.

Taghdisi, S.M., Danesh, N.M., Emrani, A.S., Ramezani, M., Abnous, K., 2015. A novel electrochemical aptasensor based on single-walled carbon nanotubes, gold electrode and complimentary strand of aptamer for ultrasensitive detection of cocaine. Biosensors and Bioelectronics 73, 245–250.

Tang, D., Tang, J., Li, Q., Su, B., Chen, G., 2011. Ultrasensitive aptamer-based multiplexed electrochemical detection by coupling distinguishable signal tags with catalytic recycling of DNase I. Analytical Chemistry 83, 7255–7259.

Tang, T., Deng, J., Zhang, M., Shi, G., Zhou, T., 2016. Quantum dot-DNA aptamer conjugates coupled with capillary electrophoresis: a universal strategy for ratiometric detection of organophosphorus pesticides. Talanta 146, 55–61.

Taylor, I.M., Du, Z., Bigelow, E.T., Eles, J.R., Horner, A.R., Catt, K.A., Weber, S.G., Jamieson, B.G., Cui, X.T., 2017. Aptamer-functionalized neural recording electrodes for the direct measurement of cocaine in vivo. Journal of Materials Chemistry B 5, 2445–2458.

Tian, J., Huang, T., Lu, J., 2016a. A photoelectrochemical aptasensor for mucin 1 based on DNA/aptamer linking of quantum dots and TiO_2 nanotube arrays. Analytical Methods 8, 2375–2382.

Tian, R., Chen, X., Li, Q., Yao, C., 2016b. An electrochemical aptasensor electrocatalyst for detection of thrombin. Analytical Biochemistry 500, 73–79.

Tombelli, S., Mascini, M., 2009. Aptamers as molecular tools for bioanalytical methods. Current Opinion in Molecular Therapeutics 11, 179–188.

Tombelli, S., Mascini, M., 2010. Aptamers biosensors for pharmaceutical compounds. Combinatorial Chemistry & High Throughput Screening 13, 641–649.

Tombelli, S., Minunni, M., Mascini, M., 2005. Analytical applications of aptamers. Biosensors and Bioelectronics 20, 2424–2434.

Tombelli, S., Minunni, M., Mascini, M., 2007. Aptamers based assays for diagnostics, environmental and food analysis. Biomolecule Engeneering 24, 191–200.

Valko, M., Morris, H., Cronin, M.J.C.M.C., 2005. Metals, toxicity and oxidative stress. Current Medicinal Chemistry 12, 1161–1208.

Verdian, A., 2018. Apta-nanosensors for detection and quantitative determination of acetamiprid–A pesticide residue in food and environment. Talanta 176, 456–464.

Verma, N., Kaur, G., 2016. Trends on biosensing systems for heavy metal detection. In: Comprehensive Analytical Chemistry. Elsevier.

Vowles, K.E., Mcentee, M.L., Julnes, P.S., Frohe, T., Ney, J.P., Van Der Goes, D.N., 2015. Rates of opioid misuse, abuse, and addiction in chronic pain: a systematic review and data synthesis. Pain 156, 569–576.

Wan, Y., Zhou, Y.G., Poudineh, M., Safaei, T.S., Mohamadi, R.M., Sargent, E.H., Kelley, S.O., 2014. Highly specific electrochemical analysis of cancer cells using multinanoparticle labeling. Angewandte Chemie 126, 13361–13365.

Wang, H., Wang, Y., Liu, S., Yu, J., Xu, W., Guo, Y., Huang, J., 2015a. Target–aptamer binding triggered quadratic recycling amplification for highly specific and ultrasensitive detection of antibiotics at the attomole level. Chemical Communications 51, 8377–8380.

Wang, K., He, M.-Q., Zhai, F.-H., He, R.-H., Yu, Y.-L., 2017a. A novel electrochemical biosensor based on polyadenine modified aptamer for label-free and ultrasensitive detection of human breast cancer cells. Talanta 166, 87–92.

Wang, L., Ma, R., Jiang, L., Jia, L., Jia, W., Wang, H., 2017b. A novel "signal-on/off" sensing platform for selective detection of thrombin based on target-induced ratiometric electrochemical biosensing and bio-bar-coded nanoprobe amplification strategy. Biosensors and Bioelectronics 92, 390–395.

Wang, T., Liu, J., Gu, X., Li, D., Wang, J., Wang, E., 2015b. Label-free electrochemical aptasensor constructed by layer-by-layer technology for sensitive and selective detection of cancer cells. Analytica Chimica Acta 882, 32–37.

Wang, W., Wang, W., Davis, J.J., Luo, X., 2015c. Ultrasensitive and selective voltammetric aptasensor for dopamine based on a conducting polymer nanocomposite doped with graphene oxide. Microchimica Acta 182, 1123–1129.

Wang, Y., Zhang, Y., Yan, T., Fan, D., Du, B., Ma, H., Wei, Q., 2016. Ultrasensitive electrochemical aptasensor for the detection of thrombin based on dual signal amplification strategy of Au@ GS and DNA-CoPd NPs conjugates. Biosensors and Bioelectronics 80, 640–646.

Wu, J., Fu, Z., Yan, F., Ju, H., 2007. Biomedical and clinical applications of immunoassays and immunosensors for tumor markers. TRAC Trends in Analytical Chemistry 26, 679–688.

Wu, L., Xiong, E., Yao, Y., Zhang, X., Zhang, X., Chen, J., 2015. A new electrochemical aptasensor based on electrocatalytic property of graphene toward ascorbic acid oxidation. Talanta 134, 699–704.

Wu, L., Zhang, X., Liu, W., Xiong, E., Chen, J., 2013. Sensitive electrochemical aptasensor by coupling "Signal-on"and "Signal-off"strategies. Analytical Chemistry 85, 8397–8402.

Wu, Y., Li, G., Zou, L., Lei, S., Yu, Q., Ye, B., 2018. Highly active DNAzyme-peptide hybrid structure coupled porous palladium for high-performance electrochemical aptasensing platform. Sensors and Actuators B: Chemical 259, 372–379.

Xie, D., Li, C., Shangguan, L., Qi, H., Xue, D., Gao, Q., Zhang, C., 2014. Click chemistry-assisted self-assembly of DNA aptamer on gold nanoparticles-modified screen-printed carbon electrodes for label-free electrochemical aptasensor. Sensors and Actuators B: Chemical 192, 558–564.

Xu, G., Huo, D., Hou, C., Zhao, Y., Bao, J., Yang, M., Fa, H., 2018a. A regenerative and selective electrochemical aptasensor based on copper oxide nanoflowers-single walled carbon nanotubes nanocomposite for chlorpyrifos detection. Talanta 178, 1046–1052.

Xu, H.-Y., Jin, L.-S., Xu, N., Chen, J.-R., Wu, M.-S., 2018b. Dual-quenching strategy for determination of ATP based on aptamer and exonuclease I-assisted electrochemiluminescence resonance energy transfer. Analytical Methods 10, 2347–2352.

Xu, Y., Hun, X., Liu, F., Wen, X., Luo, X., 2015. Aptamer biosensor for dopamine based on a gold electrode modified with carbon nanoparticles and thionine labeled gold nanoparticles as probe. Microchimica Acta 182, 1797–1802.

Xue, Q., Liu, Z., Guo, Y., Guo, S., 2015a. Cyclodextrin functionalized graphene–gold nanoparticle hybrids with strong supramolecular capability for electrochemical thrombin aptasensor. Biosensors and Bioelectronics 68, 429–436.

Xue, S., Yi, H., Jing, P., Xu, W., 2015b. Dendritic Pt@ Au nanowires as nanocarriers and signal enhancers for sensitive electrochemical detection of carcinoembryonic antigen. RSC Advances 5, 77454–77459.

Yan, M., Sun, G., Liu, F., Lu, J., Yu, J., Song, X., 2013. An aptasensor for sensitive detection of human breast cancer cells by using porous GO/Au composites and porous PtFe alloy as effective sensing platform and signal amplification labels. Analytica Chimica Acta 798, 33–39.

Yan, X., Li, J., Yang, R., Li, Y., Zhang, X., Chen, J., 2018. A new photoelectrochemical aptasensor for prion assay based on cyclodextrin and Rhodamine B. Sensors and Actuators B: Chemical 255, 2187–2193.

Yan, Z., Gan, N., Li, T., Cao, Y., Chen, Y.J.B., 2016. A sensitive electrochemical aptasensor for multiplex antibiotics detection based on high-capacity magnetic hollow porous nanotracers coupling exonuclease-assisted cascade target recycling. Biosensors and Bioelectronics 78, 51–57.

Yang, H., Yang, Q., Li, Z., Du, Y., Zhang, C., 2016. Sensitive electrogenerated chemiluminescence aptasensor for the detection of ramos cells incorporating polyamidoamine dendrimers and oligonucleotide. Sensors and Actuators B: Chemical 236, 712–718.

Yang, K., Li, Z., Lv, Y., Yu, C., Wang, P., Su, X., Wu, L., He, Y., 2018. Graphene and AuNPs based electrochemical aptasensor for ultrasensitive detection of hydroxylated polychlorinated biphenyl. Analytica Chimica Acta 1041, 94–101.

Yao, C., Zhu, T., Qi, Y., Zhao, Y., Xia, H., Fu, W., 2010. Development of a quartz crystal microbalance biosensor with aptamers as bio-recognition element. Sensors 10, 5859–5871.

Ye, Y., Guo, H., Sun, X., 2018. Recent progress on cell-based biosensors for analysis of food safety and quality control. Biosensors and Bioelectronics 126, 389–404.

Yu, P., Zhou, J., Wu, L., Xiong, E., Zhang, X., Chen, J., 2014. A ratiometric electrochemical aptasensor for sensitive detection of protein based on aptamer–target–aptamer sandwich structure. Journal of Electroanalytical Chemistry 732, 61–65.

Zang, Y., Lei, J., Hao, Q., Ju, H., 2014. "Signal-on" photoelectrochemical sensing strategy based on target-dependent aptamer conformational conversion for selective detection of lead (II) ion. ACS Applied Materials and Interfaces 6, 15991–15997.

Zhad, H.R.L.Z., Torres, Y.M.R., Lai, R.Y., 2017. A reagentless and reusable electrochemical aptamer-based sensor for rapid detection of Cd (II). Journal of Electroanalytical Chemistry 803, 89–94.

Zhang, J., Yuan, Y., Chai, Y., Yuan, R., 2014. Amplified amperometric aptasensor for selective detection of protein using catalase functional DNA–PtNPs dendrimer as a synergetic signal amplification label. Biosensors and Bioelectronics 60, 224–230.

Zhang, S., Hu, X., Yang, X., Sun, Q., Xu, X., Liu, X., Shen, G., Lu, J., Shen, G., Yu, R., 2015. Background eliminated signal-on electrochemical aptasensing platform for highly sensitive detection of protein. Biosensors and Bioelectronics 66, 363–369.

Zhang, X., Qi, B., Li, Y., Zhang, S., 2009. Amplified electrochemical aptasensor for thrombin based on bio-barcode method. Biosensors and Bioelectronics 25, 259–262.

Zhang, Y., Luo, S., Situ, B., Chai, Z., Li, B., Liu, J., Zheng, L., 2018. A novel electrochemical cytosensor for selective and highly sensitive detection of cancer cells using binding-induced dual catalytic hairpin assembly. Biosensors and Bioelectronics 102, 568–573.

Zhao, M., Zhang, S., Chen, Z., Zhao, C., Wang, L., Liu, S., 2018. Allosteric kissing complex-based electrochemical biosensor for sensitive, regenerative and versatile detection of proteins. Biosensors and Bioelectronics 105, 42–48.

Zheng, D., Zhu, X., Zhu, X., Bo, B., Yin, Y., Li, G., 2013. An electrochemical biosensor for the direct detection of oxytetracycline in mouse blood serum and urine. Analyst 138, 1886–1890.

Zheng, J., Feng, W., Lin, L., Zhang, F., Cheng, G., He, P., Fang, Y., 2007. A new amplification strategy for ultrasensitive electrochemical aptasensor with network-like thiocyanuric acid/gold nanoparticles. Biosensors and Bioelectronics 23, 341–347.

Zheng, Y., Yuan, Y., Chai, Y., Yuan, R., 2015. A label-free electrochemical aptasensor based on the catalysis of manganese porphyrins for detection of thrombin. Biosensors and Bioelectronics 66, 585–589.

Zhou, N., Yang, L.-Y., Hu, B., Song, Y., He, L., Chen, W., Zhang, Z., Liu, Z., Lu, S., 2018. Core-shell heterostructured CuFe@ FeFe Prussian blue analogue coupling with silver nanoclusters via a one-step bio-inspired approach: efficiently non-label aptasensor for detecting bleomycin in various aqueous environments. Analytical Chemistry 90, 13624–13631.

Zhu, X., Yang, J., Liu, M., Wu, Y., Shen, Z., Li, G., 2013. Sensitive detection of human breast cancer cells based on aptamer–cell–aptamer sandwich architecture. Analytica Chimica Acta 764, 59–63.

Zhuang, J., He, Y., Chen, G., Tang, D., 2014. Binding-induced internal-displacement of inverted aptamer beacon: toward a novel electrochemical detection platform. Electrochemistry Communications 47, 25–28.

CHAPTER

Nucleic acid–based electrochemical biosensors

9

Ayemeh Bagheri Hashkavayi[1,2], Jahan Bakhsh Raoof[1]

Electroanalytical Chemistry Research Laboratory, Department of Analytical Chemistry, Faculty of Chemistry, University of Mazandaran, Babolsar, Iran[1]; Department of Chemistry, Faculty of Sciences, Persian Gulf University, Bushehr, Iran[2]

9.1 Introduction

Nucleic acids are unbranched polymers that consist of four single units called nucleotides. Nucleotides come from three components: phosphate, sugar, and nitrogen-containing nucleobases (Jayaratne et al., 2006). Differences in nucleotides are due to the difference in their bases sequence. Nucleic acids are divided into two general categories based on the type of sugar in their structure. Thus, there are two types of nucleic acids in living organisms:

Ribonucleic acid's (RNA) monomers are ribonucleotides, which have ribose sugar, and deoxyribonucleic acid's (DNA) monomers are deoxynucleotides, which have deoxyribose sugar. Another difference between the two nucleic acids is that there are four bases adenine (A), guanine (G), cytosine (C), and thymine (T) in DNA, whereas in RNA, uracil (U) base has been replaced by thymine (Shamsi and Kraatz, 2013).

A chemical sensor is a device that provides continuous information on the chemical properties of its surroundings. Electrochemical sensors are an important branch of the chemical sensors. Today, the increasing need to detection and measurement of chemical elements and compounds in different parts of the environment, whether living or nonliving, including chemicals in the atmosphere, the environment, the identification of chemical species in fluids and living environments, especially the human body can be seen. This need can lead us to design, build, and use more of the sensors and biosensors. A biosensor is a device that a biologically active layer such as an enzyme, cell, living tissue, antibody or nucleic acid, as a probe stabilized on a substrate, which is produced an analytical measurable signal based on the type of interaction with the target species (Hejazi et al., 2010; Hamidi-Asl et al., 2013). Currently, glucose biosensors are one of the most successful sensors in the market.

The basis of the function of a DNA biosensor or genosensor is to fix a single-stranded oligonucleotide (ssDNA) as a probe on a transducer surface to identify its complementary DNA sequences or target molecule (Hejazi et al., 2008; Fotouhi et al., 2013). When the probe is immobilized on the surface of the transducer, it acts

Electrochemical Biosensors. https://doi.org/10.1016/B978-0-12-816491-4.00009-7

like a biological diagnostic molecule and a transducer is a component that transforms the biologic event into a measurable signal. Electrochemical DNA sensors combine the analytical power of electrochemical methods with the characteristics of the nucleic acid detection process and provide great promise in the detection of DNA hybridization.

Diagnosis of DNA hybridization is done in two ways:

The first method is direct or indicator-free method (using the intrinsic signals of DNA bases), which is done without the use of the electroactive marker. In this method, the guanine or the adenine oxidation signal as the most electroactive bases in DNA is recorded (Chiorcea-Paquim and Oliveira-Brett, 2014; Bagheryan et al., 2016b).

The second method is an indirect method (Hashkavayi and Raoof, 2018; Hashkavayi et al., 2018); the detection of DNA hybridization is done using electroactive compounds such as some of the pharmaceutical compounds, cationic metal complexes, anticancer drugs, organic dyes, and enzymes. The electrochemical markers should differentiate between ssDNA and double-stranded DNA (dsDNA) and has a clear electrochemical response (Kowalczyk et al., 2010; Raoof et al., 2011; Hamidi-Asl et al., 2013).

Electrochemical DNA biosensors can be used in detecting genetic defects, DNA damage, and small molecules (e.g., drugs and carcinogens), measuring various ions, etc., which will be explained in this section (Fig.9.1) (Khairy et al., 2017).

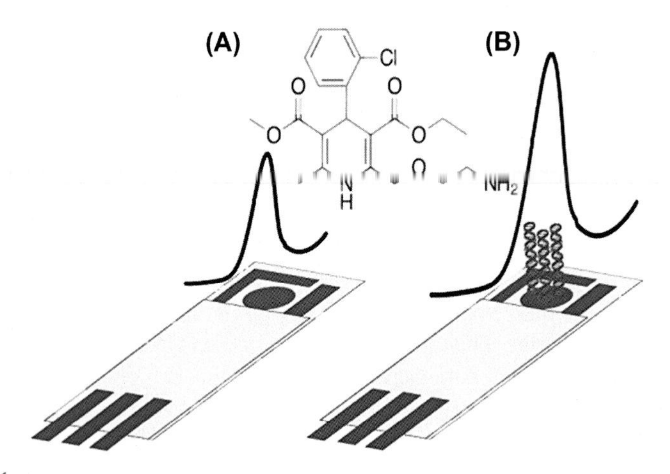

FIGURE 9.1

Simple and sensitive determinations of amlodipine using (A) bare and (B) DNA-modified graphite screen-printed electrodes.

Reprinted with permission from Khairy, M., Khorshed, A. A., Rashwan, F. A., Salah, G. A., Abdel-Wadood, H. M., Banks, C. E., 2017. Sensitive determination of amlodipine besylate using bare/unmodified and DNA-modified screen-printed electrodes in tablets and biological fluids. Sensors and Actuators B: Chemical 239, 768–775.

9.2 Hybridization and effective factors on this process

The hybridization phenomenon is very sensitive and requires precise control. The main factors that affect the rate of hybridization include hybridization time, temperature, and kinetics.

Adjusting the linear dynamic range of the target DNA detection can be done by controlling the time of the hybridization. The short hybridization period causes a wider linear dynamic range (Wang, 2002).

Another factor that greatly affects the speed of DNA hybridization is temperature. The highest rate of hybridization has been reported at 20°C lower than the boiling point of two strand DNA (Millan et al., 1994; Caruana and Heller, 1999). In addition, the kinetic of the hybridization is inversely related to the length of the sequence of the DNA strand. If the probe DNA and the target DNA are long, the sensitivity is reduced due to steric hindrance (Lucarelli et al., 2004).

9.3 Probes and their immobilization on the electrode surface

A very important step in designing of an electrochemical DNA biosensor is the immobilization of the DNA probe on the electrode surface, which has a significant effect on the response and efficiency of a DNA electrochemical biosensor (Paquim et al., 2004; Siddiquee et al., 2010). Immobilization methods depend on the type of electrode surface and the application of the biosensor. Some of the immobilization methods can be mentioned covalent attachment on the functionalized surface, adsorption, avidin/streptavidin-biotin interaction and accommodate in a sol-gel, carbon paste or polymeric structure (Caruana and Heller, 1999; Erdem et al., 1999; Kashefi-Kheyrabadi and Mehrgardi, 2013; Hashkavayi et al., 2017). Some of these methods are briefly reviewed. Fig. 9.2 shows different methods for immobilization of DNA on the electrode surfaces (Rashid et al., 2017).

9.3.1 Immobilization of the DNA probe through adsorption

This method is the easiest way to fix a probe on the surface of an electrode, which in the following a brief explanation is given on how adsorption methods are carried out by physical adsorption, adsorption at the controlled potential, and covalent attachment (Pividori et al., 2000; Nimse et al., 2014).

9.3.1.1 Physical adsorption

A simple method for immobilization of DNA at the surface of a glassy carbon electrode is that the electrode is immersed in the probe solution. The electrode is then allowed to dry. Another way to adsorption on carbon electrodes is dropping the

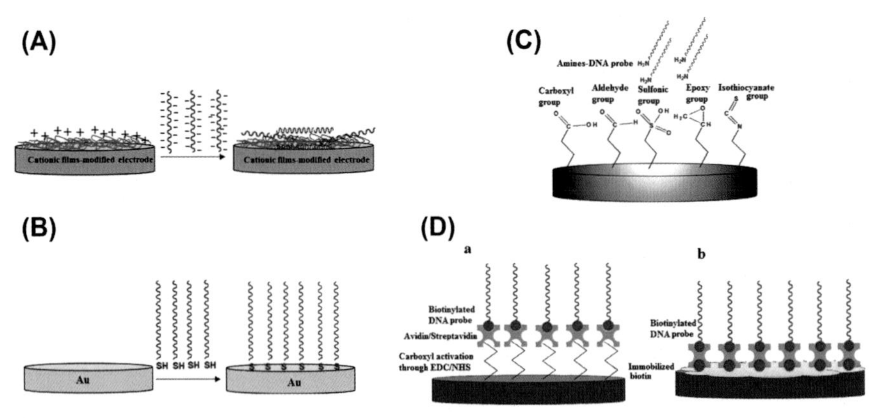

FIGURE 9.2

A schematic of the DNA probe immobilization via electrostatic adsorption (A), chemisorption (B), covalently attachment (C), immobilization of biotinylated DNA probe; (a) avidin/streptavidin-functionalized electrode through carboxyl group activation; (b) biotin/avidin (streptavidin)/biotin sandwiches technique (D).

Reprinted with permission from Rashid, J. I. A., Yusof, N. A., 2017. The strategies of DNA immobilization and hybridization detection mechanism in the construction of electrochemical DNA sensor: a review. Sensing and Bio-Sensing Research 16, 19–31.

probe solution. That is, a small drop of a probe is placed on the surface of the electrode and dried (Pividori et al., 2000).

9.3.1.2 Adsorption at a controlled potential

This method is the easiest way to fix the probe onto pretreated carbon electrodes. In this method, at first, the clean and smooth surface of the carbon electrode is electrochemically pretreated by applying an appropriate potential, during the certain time, which increases the roughness and hydrophilicity of the carbon surface. Then, the electrochemical adsorption of DNA probe is performed with the proper potential (Fig. 9.2A). Therefore, the stability of the immobilized probe was increased through electrostatic interaction between the positively charged carbon substrate and the negatively charged phosphate by applying this potential (Wang and Kawde, 2001; Drummond et al., 2003; Bagheryan et al., 2014).

9.3.1.3 DNA immobilization by covalent attachment

DNA hybridization kinetic directly depends on the availability of the probe on the electrode surface. Immobilization through this method is important owing to the stability of the bond between the DNA molecule and the electrode surface (Fig. 9.2B and C). This means that over the time, DNA will not be separated from the electrode surface to the testing solution. This will increase the hybridization process at the electrode surface (Lucarelli et al., 2004; Bagheryan et al., 2016a).

9.3.1.4 DNA immobilization by the formation of the avidin-biotin complex

Formation of avidin-biotin complex or streptavidin biotin is widely used in the field of DNA biosensors. Avidin and streptavidin are large quaternary proteins that have the same molecular weight and structure and are very stable. Biotin is a vitamin that strongly attaches to the avidin or streptavidin linkage centers. The binding of streptavidin to biotin provides the possibility of implementing a method for the preparation of DNA-coated electrodes, in which a modified surface with avidin is coupled to a biotinized oligonucleotide (Fig. 9.2D).

9.4 Types of DNA interactions with molecules and ions

In general, the interaction of small molecules and different ions (as a ligand) with a DNA molecule occurs in a completely three different ways, including intercalation of compounds between DNA base pairs, attaching to the DNA molecule grooves, and external junction via electrostatic bond, which is described briefly below (Erdem and Ozsoz, 2002; Sirajuddin et al., 2013b).

9.4.1 Intercalation of compounds between DNA base pairs

Most of the compounds that intercalated between base pairs have an aromatic ring (such as anticancer drugs) and flat surfaces (Fig. 9.3A). For this reason, their entry

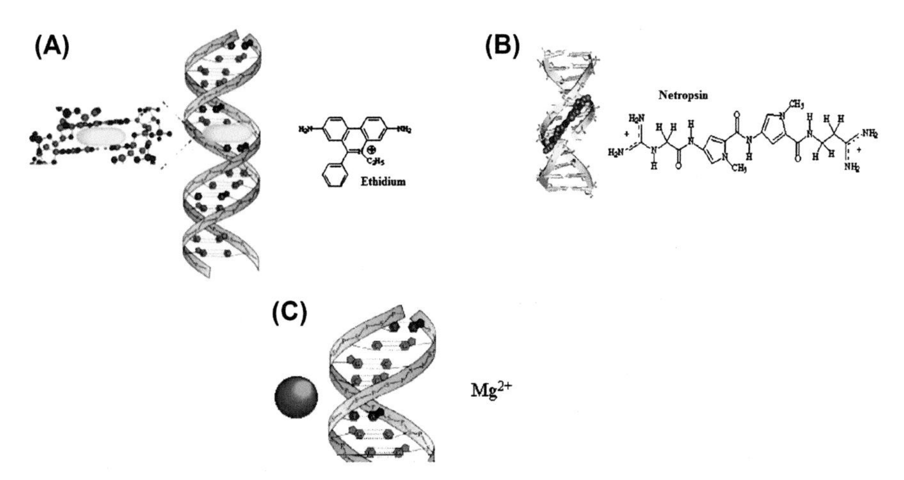

FIGURE 9.3

Types of DNA interactions with molecules and ions: intercalation (A), groove binding (B), and electrostatic (C).

Reprinted with permission from Sirajuddin, M., Ali, S., Badshah, A., 2013. Drug–DNA interactions and their study by UV–Visible, fluorescence spectroscopies and cyclic voltametry. Journal of Photochemistry and Photobiology B: Biology 124, 1–19.

into the DNA structure is easier than other compounds (Minasyan et al., 2006; Sirajuddin et al., 2013a). Intercalating between base pairs requires a change in the rotation angle of the sugar-phosphate, to enter the flat aromatic compounds (Bejune et al., 2003; Li et al., 2008).

9.4.2 Attaching to the DNA molecule grooves

Both large and small grooves in DNA can act as a place to interact with molecules (Fig. 9.3B). Small grooves in the DNA strand are rich in A and T, and the width of this groove is low. The large grooves in the DNA strands are rich in C and G and have a large number of hydrophilic phosphates, which leads to a negative density accumulation in this area. Interaction with molecules or ions in a small groove does not require much change in the structure of DNA; in this interaction, noncovalent molecular bonds are established between the ligands and the groups in that place (Erdem and Ozsoz, 2002; Palecek and Bartosik, 2012).

9.4.3 External junction via electrostatic bond

DNA molecules are charged compounds, in which the phosphate groups present in them have a great influence on their interactions (Fig. 9.3C). As these charged groups in the DNA helix are in a very small space, the repulsive forces between these groups should be reduced for the stability of DNA molecules.

To achieve this, the DNA molecule has a large amount of affinity to positive ions in solution. In fact, cationic compounds reduce the repulsion of negative charges and create a nonspecific external connection (Minasyan et al., 2006).

9.5 Mutations and damages in DNA

Living cells need to properly function of thousands of genes; each of these genes can be damaged by a mutation in a protein-encoding state or mutation in the gene expression control sequences or mRNA processing. Two major sources of mutation are the inaccurate replication of DNA and chemical damage to the genetic material. DNA molecules not only suffer from spontaneous damage, such as removal of bases, but are also attacked by chemicals and rays. These mutations lead to DNA replication errors and DNA damage. Thus, cells should search the entire genome for DNA errors and repair the damage (Baker et al., 2003).

9.5.1 Types of mutations

Mutation involves any imaginable change in the DNA sequence. Mutations that cause a change in single nucleotide are called point mutations. Other mutations lead to a large rearrangement in the chromosomes structure and severe changes in

DNA. DNA in cells is permanently exposed to various physical or chemical agents that may cause chemical changes in DNA molecules (Friedberg, 2000, 2003; Schärer, 2003).

9.5.1.1 DNA alkylation

In the alkylating of DNA, methyl or ethyl groups are transferred to reactive positions on bases and DNA phosphates. One of the most vulnerable positions to alkylation is the oxygen at C-6 of guanine. As a result of this alkylation, when damaged DNA is replicated, the C:G base pair changes to A:T (Watson, 1970).

9.5.1.2 Amine removal from DNA

The removal of the amine from the cytosine bases is the most important type of DNA damage caused by hydrolysis. In physiological conditions, cytosine spontaneously loses its amine group and produces uracil. Adenine and guanine can also lose their amines, which are converted to hypoxanthine and xanthine, respectively. As a result of this mutation, adenine instead of thymine creates a hydrogen bond with cytosine and guanine binds to cytosine with two hydrogen bonds (Watson, 1970).

9.5.1.3 DNA oxidation

DNA is also affected by reactive oxygen species (hydrogen peroxide, hydroxyl radicals). These oxidizers are free radical producers. Guanine oxidation results in the formation of oxo-guanine, which is highly mutated, and leads to the formation of base pairs with both adenine and cytosine. This is the most common type of mutation in human cancers (Watson, 1970).

9.5.1.4 DNA damage, which is caused by ultraviolet radiation

With the thinning of the ozone layer in recent years, the number of people who are suffering from skin cancer is on the rise. This skin damage is caused by ultraviolet radiation on DNA (Häder and Sinha, 2005; Ikehata et al., 2008; Brem et al., 2017). The sun's ultraviolet radiation can create two types of the most abundant mutagenic and toxic damage DNA, such as the cyclobutane pyrimidine dimers and also bivalent isomers, which prevent the movement of DNA polymerase on the DNA template (Perdiz et al., 2000; Narayanan et al., 2010).

9.6 Some applications of electrochemical DNA biosensors

The electrochemical DNA biosensors have been used for obtaining sequence-specific information, detection of clinical diagnosis of genetic or infectious diseases, the properties of some pharmaceutical compounds, environmental monitoring, etc. Therefore, some applications of these electrochemical biosensors were described in the following sections.

9.6.1 Preparation of an electrochemical biosensor to detect DNA damage

Detection and prevention of DNA damage are important in the prevention and treatment of many diseases. Hence, the development of simple and sensitive tools to detect DNA damage has attracted the attention of many researchers. Recently, many studies have been carried out on electrochemical DNA biosensors as effective tools for detecting of DNA damage.

Mousavisani et al. constructed a novel label-free electrochemical biosensor for ultrasensitive detection of DNA damage based on the use of Au nanoparticles–modified screen-printed gold electrode. They immobilized the thiolated DNA on the surface of the modified electrode, and then they used Fenton's reaction to damage of DNA (Mousavisani et al., 2018b). As iron is the most abundant metal in the biological systems, by the reaction of iron ions with hydrogen peroxide (H_2O_2), a highly reactive hydroxyl radical is produced, which phenomenon is called Fenton reaction (Hayes, 1997; Jia et al., 2008). Transition metal ions that usually participate in the Fenton reaction are copper, iron, and chromium. Fenton reactions in live tissues increase the risk of mutation (Burkitt, 2003; Mello et al., 2006; Liang Wu et al., 2012). Hence, the study of DNA damage caused by this system is important. To cause damage in DNA by Fenton reaction, the prepared biosensor was placed in damaging solution (TBS (0.1 M, pH = 7) containing ($CuSO_4 \cdot 5H_2O$ (2.5×10^{-8} M), H_2O_2 (2.5×10^{-4} M), and ascorbic acid (1.0×10^{-6} M)) for 1 h and then the Nyquist plot was examined before and after exposure to the damaging solution. Also, control experiments were carried out by placing biosensors in other solutions. Fig. 9.4 shows the histogram of obtained values ($\Delta R/R$) of Nyquist plots.

The obtained results showed that the value of $\Delta R/R$ of the fabricated sensor did not much change in the TBS (a), TBS contain $CuSO_4 \cdot 5H_2O$ (b), and H_2O_2(c). However, significant changes were observed in the value of $\Delta R/R$ in the TBS containing $CuSO_4 \cdot 5H_2O$ + AA (d), $CuSO_4 \cdot 5H_2O + H_2O_2$ (e), and $CuSO_4 \cdot 5H_2O$ + AA + H_2O_2 (f). The reason for these changes is that in the presence of oxygen, ascorbate ions act as a source of hydrogen peroxide production. Ascorbate oxidation is catalyzed in the presence of metallic ions and leads to the production of hydrogen peroxide. Finally, the reduced form of copper ions can reduce hydrogen peroxide and generate hydroxyl radicals; those obtained radicals lead to DNA damage (Buettner, 1988; Buettner and Jurkiewicz, 1996; Song and Buettner, 2010). In the presence of ascorbate ion, reduced forms of copper ions are produced again and lead to the production of more hydroxyl radicals. However, in the absence of ascorbic acid, this does not happen (Udenfriend et al., 1954; Du et al., 2012).

Also, to investigate the protective effect of deferoxamine (DFO) in preventing DNA damage, Nyquist plots of the prepared biosensor were recorded before and after exposure to the damaging solution, in the absence and presence of DFO. The obtained results indicated the antioxidant effect of DFO in prohibiting of DNA damage.

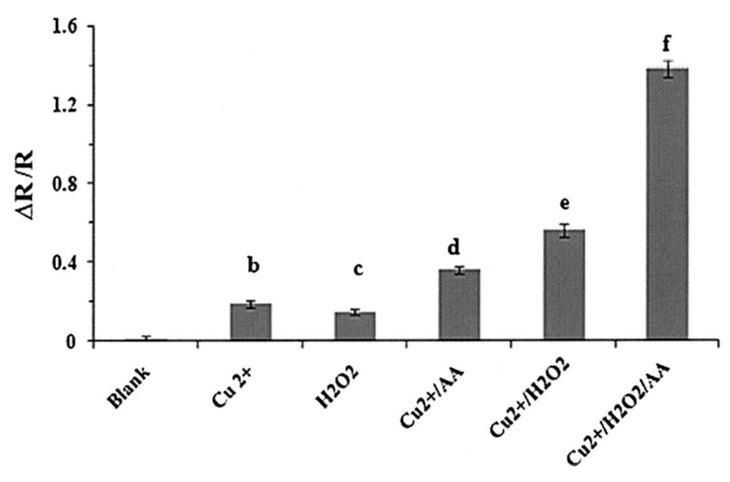

FIGURE 9.4

The ΔR/R values for DNA/AuNPs/SPGE after incubation with TBS (pH 7.0) containing blank (A), 2.5×10^{-8} M $CuSO_4$ (B), 1×10^{-4} M H_2O_2 (C), 2.5×10^{-8} M $CuSO_4$ $+1 \times 10^{-6}$ M AA (D), 2.5×10^{-8} M $CuSO_4$ $+ 1 \times 10^{-4}$ M H_2O_2 (E), and 2.5×10^{-8} M $CuSO_4$ $+ 1 \times 10^{-4}$ M H_2O_2 $+ 1 \times 10^{-6}$ M AA (F).

Reprinted with permission from Mousavisani, S. Z., Raoof, J. B., Ojani, R., Bagheryan, Z., 2018b. An impedimetric biosensor for DNA damage detection and study of the protective effect of deferoxamine against DNA damage. Bioelectrochemistry 122, 142–148.

Kara et al. constructed an electrochemical biosensor for the detection of DNA damage arising from radioactive materials. They constructed DNA biosensor by electrodepositing fish sperm DNA (by applying a potential of 0.50 V to dsDNA solution) on the carbon paste electrode (CPE). They used direct and indirect irradiation methods to create damage in DNA (Kara et al., 2007). In direct methods, they put technetium (Tc-99m) in Eppendorf tubes contained fish sperm dsDNA sample and incubated for 6, 12, and 24 h, but in indirect irradiation, they put the Eppendorf tubes contained dsDNA sample in test tubes which Tc-99m had been put inside it previously. They examined the changes in the guanine oxidation signal before and after the irradiation. Results showed that before irradiation, the intensity of the guanine signal is more than that of the samples, which are treated by indirect and direct irradiation methods, respectively. The decrease in the guanine signal indicates damage in DNA. They also investigated the damage of DNA with other radioactive substances such as iodine (I-131). Results showed that, after direct and indirect radiations, the intensity of the guanine signals decreased, respectively. The obtained results showed that electrochemical methods are very effective in the investigation of DNA damages.

In another study, Mousavisani et al. (2018a). constructed a new DNA biosensor based on immobilization of breast cancer 1 gene on diazonium-modified screen-printed electrode (SPE) (Mousavisani et al., 2018a). To cause damage to DNA,

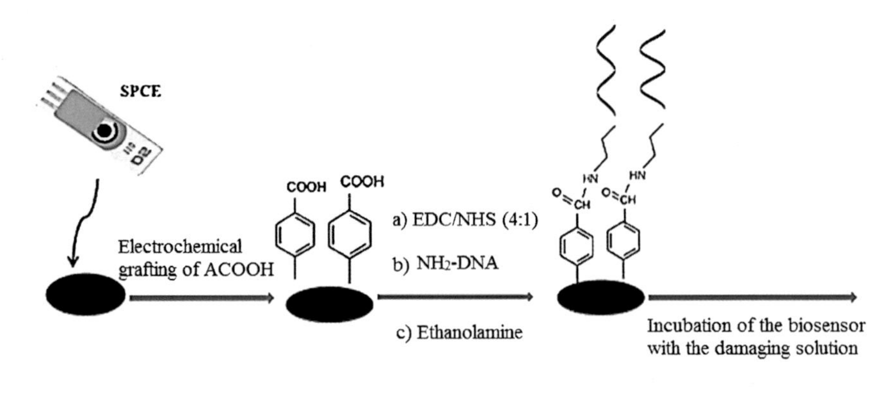

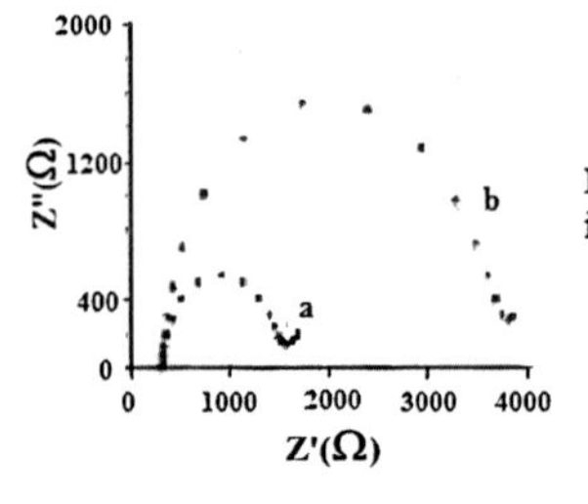

Nyquist plots of DNA/ACOOH/SPCE (a) before and (b) after incubation with the damaging soloution.

FIGURE 9.5

Schematic representation of DNA sensor fabrication and DNA damage detection.

Reprinted with permission from Mousavisani, S. Z., Raoof, J.-B., Turner, A. P., Ojani, R., Mak, W. C., 2018a. Label-free DNA sensor based on diazonium immobilisation for detection of DNA damage in breast cancer 1 gene. Sensors and Actuators B: Chemical 264, 59–66.

they used TBS (0.1 M, pH = 7.0) containing $CuSO_4 \cdot 5H_2O$ (1.0×10^{-3} M), H_2O_2 (1.0×10^{-3} M), and ascorbic acid (1.0×10^{-6} M)) to exposure with DNA for 1 h. Then they investigated the damage by use of Nyquist plots before and after exposure to the damaging solution. As can be seen in Fig. 9.5, following the DNA damage, the electron transfer resistance increased, which indicates a more difficult electron transfer process. This is because the oxidative damage of the DNA can cause a variety of changes, such as dsDNA breaks and its sugars and bases damage.

9.6.2 Use of electrochemical nucleic acid–based biosensors for diagnosis of genetic defects and clinical applications

Recently, scientists have tried to create cheap and easy methods to recognize the point mutations that lead to inherited disease or disorders. The mutation can occur in any base of the DNA strands and lead to changes in the DNA sequence. Therefore, it is necessary to prepare sensors that can easily detect these genetic defects.

Hamidi-Asl et al. (2015) prepared a genosensor based on the microfluidic system for detection of the point mutation in the P53 gene. The DNA probe was functionalized with biotin and immobilized on paramagnetic beads attached to streptavidin.

The basis of fixing the probe on the electrode surface was the covalent bond of biotin-streptavidin. They used three oligonucleotide sequences in this work. One of the oligonucleotides is used as a capture probe, the other as the signaling probe, and the third as the target probe. In this work, an enzyme-linked immunosorbent assay (ELISA) system was used for effective differentiation between sequences of complementary and noncomplementary target DNAs and single-base-mismatched DNA of p53 gene. The ELISA is based on the use of an enzyme, which, during the test, produces a suitable sign for tracing the sample. In this work, the alkaline phosphatase enzyme was used.

The used oligonucleotide strands are biotinylated capture and signaling probe, complementary and noncomplementary DNA, and single-base-mismatched DNA.

From the first, the purchased magnetic beads contain streptavidin, and the capture probe is attached to the biotin from the 5′ end. The hybridization solution for the formation of sandwich structures involves combining signaling probe DNA and target DNA. After that, the hybridization solution was added to the magnetic beads modified by the capture probe DNA. At the end, the magnetic beads on which sandwich hybridization was performed are provided. Whereas signaling probe DNA is connected from one side to the target DNA, there is free biotin on the other side (Fig. 9.6).

For electrochemical detection, graphite SPEs were placed on a strong magnet and a solution of magnetic beads was placed on the electrode surface. Then α-naphthyl phosphate solution was poured onto the entire electrodes. α-Naphthyl phosphate acts as a substrate, which is converted into α-naphthol phosphate by reaction with the enzyme. The produced product of the enzymatic reaction is electroactive, which is detected with a differential pulse voltammetric technique. The current intensity of α-naphthol phosphate is a measure of the efficiency of the hybridization process. The higher the concentration of the target DNA, the greater the amount of signaling probe that will attach to the target DNA. As a result, more of the substrate is converted to the product. The detection limit in this method is calculated equal to 5.90 pM. Also, the function of this sensor was investigated in identifying the complementary DNA from noncomplementary target DNA and the single-base-mismatched DNA.

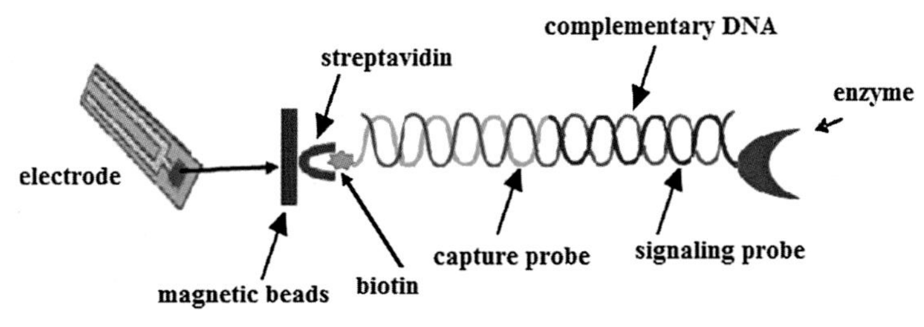

FIGURE 9.6

Preparation of genosensor based on sandwich hybridization method.

All investigations were also carried out in a dynamic mode system, using microfluidic platform device and chronoamperometric method. The results showed that working on a microfluidic platform device revealed more sensitivity and selectivity for the distinction between complementary, noncomplementary, and point mutation target DNAs.

Hartati et al. (2013) have also reported a DNA biosensor for detection of the mitochondrial DNA A3243G tRNA Leu mutation. To investigate the ability of the sensor to detect and distinguish between complementary, noncomplementary, and single-base-mismatched DNAs, they used the differential pulse voltammetric signal of Meldola's Blue (MDB) indicator. This indicator intercalates between two DNA strands. To prepare the genosensor, they immobilized capture probe on a disposable pencil graphite electrode (PGE). The used oligonucleotide strands are as follows: wild-type target (WTT, mismatch containing sequences), wild-type probe (WTP), mutant-type target (MTT, mismatch containing sequences), and mutant-type probe (MTP).

The results showed that the prepared sensor can easily distinguish between types of mentioned sequences. The intensity of the MDB signal notably increased after hybridization with MTT probe compared with MTP. Also, after hybridization with WTT (mismatch containing sequences), a significant decrease in MDB signal was observed. This indicates that there is no interaction between the probe and mismatch DNA. Results showed that the prepared electrochemical DNA biosensor can easily detect the hybridization effect and also can distinguish between match and mismatch DNA.

In another study by Xu et al. (2017), they made a microfluidic channel system based on graphene single-crystal domain on a SiO_2/Si substrate for detection of DNA hybridization. For hybridization, first, they modified the graphene surface with 1-pyrenebutanoic acid succinimidyl ester (PBASE); after that, immobilization of amine-terminated DNA was carried out. The hybridization detection of probe DNA with the target, complementary, noncomplementary, and single-base-pair-mismatched DNAs was evaluated by investigating the binding kinetics (determination of association/dissociation rate constants and equilibrium association constants). The fabricated system showed a detection limit of 10 pM for DNA.

Das et al. (2010) prepared an electrochemical biosensor for *Mycobacterium tuberculosis* detection, which causes tuberculosis. To prepare biosensor, at first, they placed a film of zirconia through electrodeposition on the gold electrode surface. Then, the DNA probe was stabilized on electrode surface through the tendency between the oxygen atoms of phosphate groups and zirconium. They used DPV method for following the hybridization events of DNA probe with genomic DNA of *M. tuberculosis* at the surface of ZrO_2/Au. Results showed that the prepared biosensor can be used for sensitive recognition of *M.* with detection limit of 0.065×10^{-9} g L^{-1}.

Castro et al. (2014) constructed an electrochemical genosensor for detection of the hepatitis B virus DNA sequence. They used poly(4-aminophenol)-modified graphite electrodes for preparation of the genosensor. After immobilizing the

DNA probe on the electrode surface, they used two methods for investigating the detection events: in the first method, by following the DPV signals resulting from the oxidation of DNA bases, and in the second method, by using the signals of ethidium bromide as an electroactive hybridization indicator, before and after interaction with different concentrations of target DNA. Prepared genosensor showed a detection limit of 2.61×10^{-9} M for detection of the hepatitis B virus DNA sequence.

9.6.3 Preparation of electrochemical DNA biosensors for recognition of G-quadruplex structure and the study of its interaction with some stabilizing ligands

Telomeres are a specific nucleoprotein structure at the end of all eukaryotic chromosomes, which play an important role in regulating cell age and stabilizing the genome and the process of getting cells cancerous (Han and Hurley, 2000; Blackburn, 2001). Telomerase is a specialized enzyme targeted at telomere regions, which are commonly found in cancerous cells in large quantities (Greenberg et al., 1999; Zheng et al., 2008). Hence, controlling their performance provides a good aim for designing anticancer drugs.

If telomerase binding to a telomeric sequence is inhibited by the interaction of small molecules with telomeric sequence, then telomerase activity decreases directly. In humans, telomeric DNA consists of a double-stranded structure, which at the end forms the four-strand structure by repeating the TTAGGG sequence (Phan, 2010). The guanine-rich DNA strands form a G-quadruplex molecular structure (Phan and Mergny, 2002; Chiorcea-Paquim and Oliveira-Brett, 2014). Formation of the G-quadruplex structure was shown in Fig. 9.7.

An important factor in the formation of the DNA quadruplex structure is the presence of special cations that are specifically bound to guanine O_6 carbonyl groups between guanine flat plates. Among the cations, alkaline earth cations such as potassium and sodium are the best ones (Qi and Shafer, 2005; Wu and Brosh, 2010).

Indeed, creating a quadruplex structure at the end of the telomeres region prevents the telomerase enzyme activity and, as a result, prevents telomeres prolongation and the formation of cancerous cells (Xu et al., 2009).

Today because of the importance of the role of these structures in the body and the created diseases, researchers have focused on how they can be stable in different conditions and their interaction with a variety of drugs or compounds. Also, they use the results in designing and manufacturing anticancer drugs.

Bagheryan et al. (2016b) prepared a DNA biosensor based on the change in the configuration of DNA G-quadruplex structure, to identify the anticancer compounds. For constructing the DNA biosensor, they modified the SPE with SBA@NPPNSH, and then, DNA sequence was immobilized on the surface of the modified electrode. They examined the voltammetric behavior of the biosensor in different conditions, including different salt concentrations, pH values of the solution, and in the presence of stabilizing ligands.

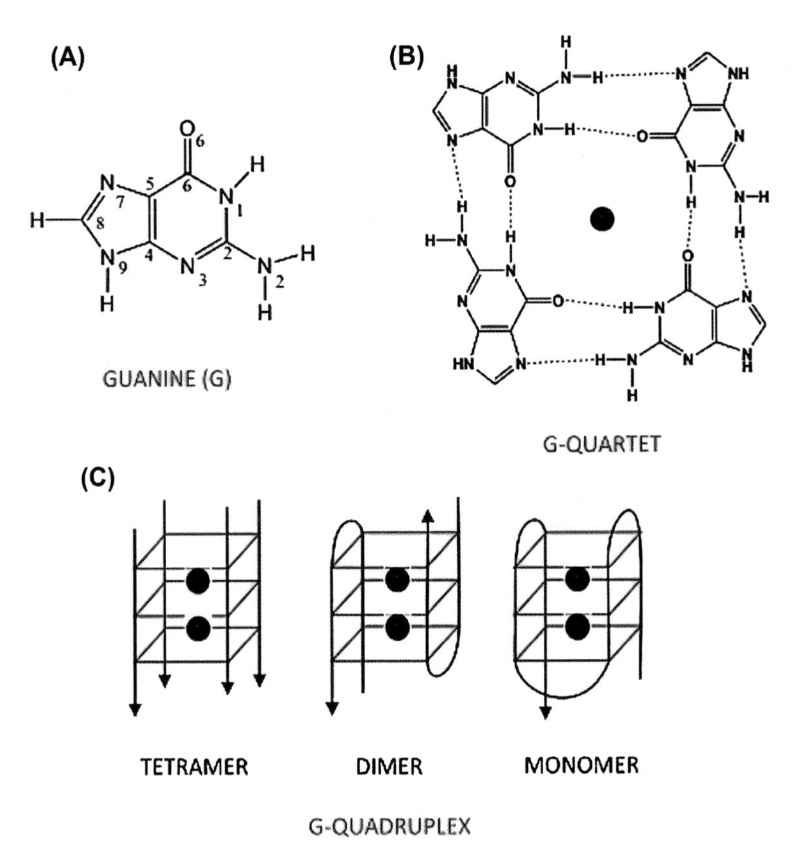

FIGURE 9.7

(A) Chemical structure of guanine (G) base. (B) G-quartet (Gq) showing the hydrogen bonding between four planar G bases. (C) Schematic representation of tetramer, dimer, and monomer G-quadruplexes composed by three staked G-quartets; the cations that stabilize the G-quadruplexes are shown as black balls.

Reprinted with permission from Chiorcea-Paquim, A.-M., Oliveira-Brett, A. M., 2014. Redox behaviour of G-quadruplexes. Electrochimica Acta 126, 162–170.

They investigated the behavior of the G-quadruplex structure of DNA in the absence and presence of different concentrations of KCl at the modified electrode surface. Before adding KCl to the solution, the guanine oxidation peak is observed at a potential of 0.92 V, which is related to the oxidation of free guanines present in the DNA structure. After adding a concentration of 0.01 M of KCl, a new oxidation peak can be seen at a potential of 1.064 V, indicating the formation of the G-quadruplex structure of DNA. With increasing the salt concentration to 1.0 M, the guanine oxidation peak current at 0.92 V decreased and the anodic peak of the G-quadruplex structure of DNA has moved toward more positive potentials

(1.134 V) and its peak current also increased, which indicates the stability of the G-quadruplex structure of DNA in these conditions.

In the next step, the effect of the pH values of the solution on the formation of the G-quadruplex structure was evaluated. The results showed that with increasing pH of the solution to 7.0, guanine oxidation peaks moved to more positive potentials, and the height of the guanine oxidation peak related to the G-quadruplex structure DNA increased, which indicates that this structure is getting more stable. With increasing pH to pH = 9.0, the peak height related to the oxidation of guanine in the G-quadruplex structure of DNA is greatly reduced and the oxidation peak current of free guanine is increased, which indicates the instability of G-quadruplex structure of DNA in this pH value. The highest oxidative peak current of the guanine in the G-quadruplex structure of DNA was observed at pH = 7.0; therefore, pH 7.0 was considered as the appropriate pH for the formation of the G-quadruplex structure of DNA.

Eventually, they examined the effect of two ligands: TMPyP4 (5, 10, 15, 20-tetra-(N-methyl-4-pyridyl) porphyrin) and letrozole on the formation of the G-quadruplex structure of DNA at the surface of SBA@NPPNSH/SPE. The results showed that after the addition of increasing concentrations of TMPyP4 ligand, the guanine oxidation peak potential moved to more positive potential, which represents the formation and stability of the G-quadruplex structure of DNA. Also, by examining the differential pulsed voltammetry signals, letrozole showed a slight ability to stabilize the G-quadruplex structure of DNA.

Yang et al. (2009) have proposed a method for investigating the electrochemical behavior of the complex of G-quadruplex-hemin and examine their potential for reduction of hydrogen peroxide. They followed the signal of G-quadruplex-hemin complex via cyclic voltammetry (CV) technique, which showed a pair of quasi-reversible redox peaks. CV results showed that with increasing the concentration of H_2O_2 in the range of 1.0×10^{-6} to 1.0×10^{-4} M, the intensity of reduction peak current increased. Results showed that the catalytic effect of hydrogen peroxide could be increased up to six times in the presence of human telomerase sequences. Also, they examined the efficacy of some G-quadruplex binders such as PIPER (N, N-bis [2-(1-piperidino)-ethyl]-3, 4, 9, 10-perylene tetracarboxylic diimide) and TMPyP4. The obtained results from spectrometry studies showed that the binding of these binders to the complex of G-quadruplex-hemin adjusts the catalytic ability of the complex.

Gao et al. (2017) have also reported an electrochemical DNA hybridization biosensor, which is constructed based on the formation of G-quadruplex-hemin complex (Fig. 9.8). They used hemin as an electrochemical reporter because it is a porphyrin and has the ability to stabilize the G-quadruplex structure and to form the DNA-hemin complexes. They divided the G-quadruplex DNA sequence into two parts for preparation of the sensor. The first part, which was a guanine-rich thiolated sequence (capture probe), was immobilized on the gold electrode surface. Then a mixture of the target DNA sequence, second-guanine-rich sequence, and hemin was dropped on the electrode surface, which led to the intercalation of hemin into

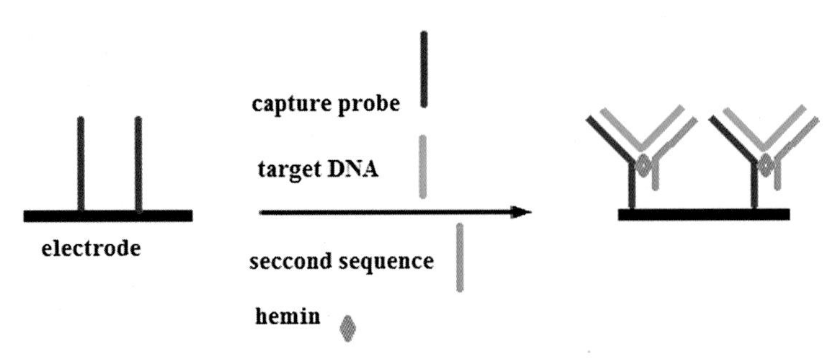

FIGURE 9.8

Schematic of the label-free and signal-on E-DNA sensor, which is based on proximity hybridization-triggered hemin/G-quadruplex formation.

Reprinted with permission from Gao, F., Fan, T., Wu, J., Liu, S., Du, Y., Yao, Y., Zhou, F., Zhang, Y., Liao, X., Geng, D., 2017. Proximity hybridization triggered hemin/G-quadruplex formation for construction a label-free and signal-on electrochemical DNA sensor. Biosensors and Bioelectronics 96, 62–67.

the G-quadruplex structure and formation of sandwich structures. The intensity of the hemin reduction signal depends on the concentration of the target DNA. Results showed that the fabricated electrochemical sensor can detect the target DNA in the range of 10^{-13} to 10^{-9} M with a detection limit of 54 fM.

9.6.4 Preparation of electrochemical biosensors based on nucleic acid, to measure very small amounts of some heavy metal ions

Today, heavy metal ion contamination is one of the most important environmental challenges in human societies, especially in advanced countries. These ions are nondegradable and can accumulate in the body and cause many diseases (Fu and Wang, 2011; Gumpu et al., 2015; Zhan et al., 2016). Therefore, the development of simple and cost-effective methods for the sensitive measurement of toxic heavy metal cations is a critical issue.

Unlike the most of developed analytical methods to the measurement of these ions, the electrochemical measurement methods are considered by many researchers because of their short analysis time, low cost, high sensitivity, and high compatibility for in situ measurements (Cui et al., 2015; Bansod et al., 2017). In electrochemical measurements of heavy metal ions, some inorganic, organic, and biological compounds are used to provide a sensor for measuring metal ions. By using biomaterials in the preparation of sensors, electrochemical biosensors are produced. The biological materials used in the construction of biosensors for measurement of heavy metal ions include enzymes, amino acids, peptides, proteins, cells, and nucleic acids (Cui et al., 2015). The nucleic acid biosensors, which are used for measuring heavy metal ions, are based on the interaction between the relevant metal ion and nucleic acid. The most important of them can be noted to the preparation of electrochemical sensors based on the coordination of cytosine-silver (I) ion—cytosine and thymine-mercury (II) ion—thymine. Also, it can be mentioned to the construction of DNA-zyme and G-quadruplex DNA-based sensors.

Ebrahimi et al. (2015) constructed a novel electrochemical biosensor for ultra-sensitive detection of silver ion (Fig. 9.9). Silver ions can interact with DNA cytosine bases and form the stable structure of C-Ag^{+}-C, unexpectedly causing hybridization of DNA with incomplete C-C bases (Daune et al., 1966). The ethyl green indicator was used as an electroactive DNA hybridization indicator because of a different tendency to ssDNA and dsDNA. So it was able to distinguish between these two structures of DNA (Raoof et al., 2011, 2013). They prepared an electrochemical biosensor based on the interaction of cytosine with silver ion on the

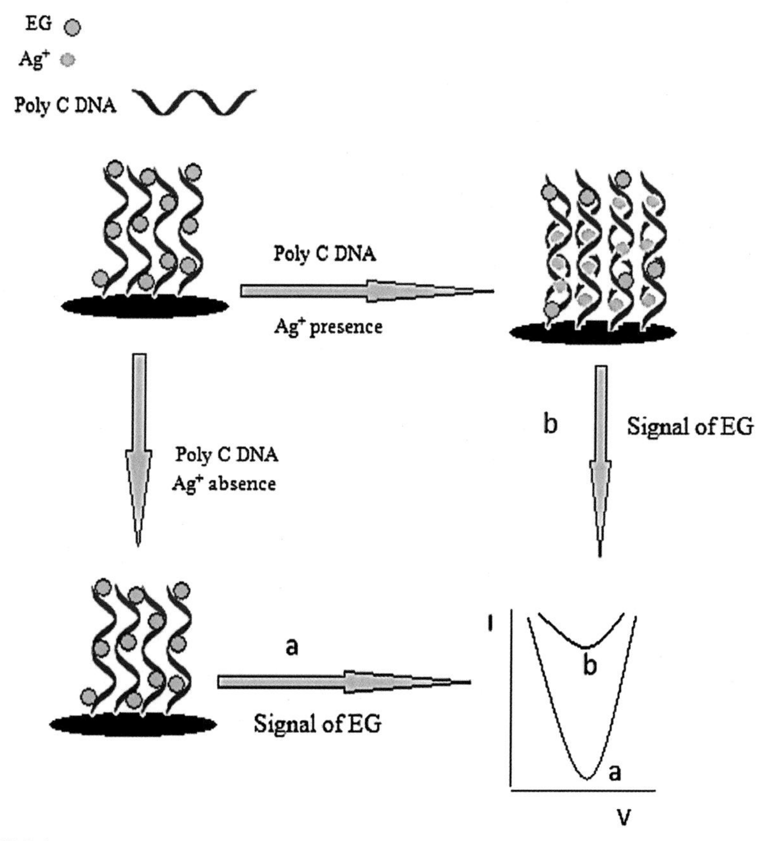

FIGURE 9.9

Fabrication of the electrochemical biosensor for Ag^{+} determination based on Ag^{+}-induced DNA hybridization.

Reprinted with permission from Ebrahimi, M., Raoof, J. B., Ojani, R., 2015 Novel electrochemical DNA hybridization biosensors for selective determination of silver ions. Talanta 144, 619–626.

gold nanoparticles–modified CPE. In which the hybridization of DNA probe and the noncomplementary single-stranded target DNA were achieved (both with the sequence of 5′-CCC CCC CCC C-3 ′), and the ethyl green differential pulse voltammetry signal was the basis of the electrochemical measurement of this ion.

When no silver ions were added to the hybridization solution, the probe DNA and the target DNA cannot be hybridized through noncomplementary bases, but, as soon as the addition of a small amount of silver ion to the test solution, a decrease in intensity of the ethyl green peak current was observed because of the hybridization of probe and target DNA for the formation of the C-Ag^{+}-C complex. This proves that silver ions have effectively been succeeded for hybridization of target DNA and probe DNA with the noncomplementary sequences. The efficiency of fabricated electrochemical biosensor based on C-Ag^{+}-C was investigated for different concentrations of silver ions by differential pulse voltammetry method. Results showed that the fabricated sensor can detect the silver ions in the range of 1.0×10^{-9} to 9.0×10^{-11} M with a detection limit of 2.6×10^{-11} M. The prepared voltammetric biosensor has shown a good selectivity for measuring silver ions against some of the other metal ions and also has an acceptable performance for measurement of Ag^{+} in the real samples.

Lai et al. (2011) prepared an electrochemical DNA biosensor for selective detection of Hg^{2+}. To prepare the sensor, they self-assembled thiol-terminated thymine-containing oligonucleotide strands on the gold electrode surface (Fig. 9.10). Mercury ions can interact with the thymine bases to form a stable T-Hg^{2+}-T

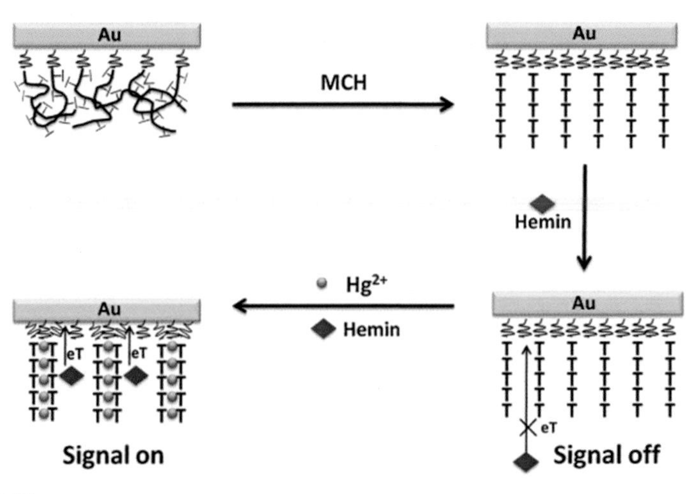

FIGURE 9.10

Schematic illustration for electrochemical detection of Hg^{2+} using hemin as a redox indicator. The sequence of the DNA probe: 5-HS-C6-TTTTT.

Reprinted with permission from Lai, Y., Ma, Y., Sun, L., Jia, J., Weng, J., Hu, N., Yang, W., Zhang, Q. (2011) A highly selective electrochemical biosensor for Hg^{2+} using hemin as a redox indicator. Electrochimica Acta 56(9), 3153–3158.

structure and unexpectedly lead to the stability of two-stranded DNA containing incomplete T T bases (Ono and Togashi, 2004). To evaluate the performance of the sensor, they followed the DPV signal of hemin as an electroactive indicator. The results showed that in the absence of Hg^{2+}, the steric hindrance of single-stranded DNA prevents the adsorption of hemin on the electrode surface. But, in the presence of Hg^{2+} and its interaction with thymine-containing oligonucleotide strands and formation of double-stranded DNA, more space is created for hemin on the surface of the electrode. As a result, with the increase of Hg^{2+} concentration, the intensity of the hemin electrochemical signal also increased. Selectivity of the prepared sensor was evaluated against a large number of elements; results showed that the sensor has a high ability to selective detection of Hg^{2+}.

In another study, Qu et al. (2015) have reported a selective electrochemical DNA biosensor for the detection of cadmium. They immobilized the thiolated ssDNA, which was rich in adenine and guanine on the surface of the gold electrode. To investigate the interaction of cadmium with DNA, they used the cyclic voltammetry signal of methylene blue as an electrochemical indicator. Results showed that with increasing the concentration of cadmium, the intensity of methylene blue signal also increased. They also checked the selectivity of the fabricated sensor against other ions such as copper, zinc, sodium, magnesium, and lead. Finally, the prepared sensor showed a high ability to detect cadmium ion and did not respond to other ions.

9.7 Conclusion

The electrochemical biosensors combine the ability of electrochemical techniques with the specificity of biological detection processes. The DNA-based biosensors are important for the detection of nucleic acid sequences in the fields of gene analysis, diagnosis of pathogens, and criminal sciences. To investigate the interaction of DNA with other molecules, the electroactive indicator or the intrinsic signal of the DNA bases can be used. Nucleic acid–based electrochemical biosensors have many applications that some of them are described in this chapter, including identification of genetic defects through hybridization, examination of DNA damage, investigation of G-quadruplex structures and their stabilization, and selective measurement of heavy metals ions through special interactions with some DNA bases.

References

Bagheryan, Z., Raoof, J.-B., Golabi, M., Turner, A.P., Beni, V., 2016a. Diazonium-based impedimetric aptasensor for the rapid label-free detection of *Salmonella typhimurium* in food sample. Biosensors and Bioelectronics 80, 566–573.

Bagheryan, Z., Raoof, J.-B., Ojani, R., Rezaei, P., 2014. Development of a new biosensor based on functionalized SBA-15 modified screen-printed graphite electrode as a nano-reactor for G-quadruplex recognition. Talanta 119, 24–33.

Bagheryan, Z., Raoof, J., Ojani, R., 2016b. Voltammetric characterization of human telomeric G-quadruplex: a label free method for anticancer drug detection. Bioelectrochemistry 107, 25–29.

Baker, T.A., Watson, J.D., Bell, S.P., Gann, A., Losick, M., Levine, R., 2003. Molecular Biology of the Gene. Benjamin-Cummings Publishing Company.

Bansod, B., Kumar, T., Thakur, R., Rana, S., Singh, I., 2017. A review on various electrochemical techniques for heavy metal ions detection with different sensing platforms. Biosensors and Bioelectronics 94, 443–455.

Bejune, S.A., Shelton, A.H., McMillin, D.R., 2003. New dicationic porphyrin ligands suited for intercalation into B-form DNA. Inorganic Chemistry 42 (25), 8465–8475.

Blackburn, E.H., 2001. Switching and signaling at the telomere. Cell 106 (6), 661–673.

Brem, R., Guven, M., Karran, P., 2017. Oxidatively-generated damage to DNA and proteins mediated by photosensitized UVA. Free Radical Biology and Medicine 107, 101–109.

Buettner, G.R., 1988. In the absence of catalytic metals ascorbate does not autoxidize at pH 7: ascorbate as a test for catalytic metals. Journal of Biochemical and Biophysical Methods 16 (1), 27–40.

Buettner, G.R., Jurkiewicz, B.A., 1996. Catalytic metals, ascorbate and free radicals: combinations to avoid. Radiation Research 145 (5), 532–541.

Burkitt, M.J., 2003. Chemical, biological and medical controversies surrounding the Fenton reaction. Progress in Reaction Kinetics and Mechanism 28 (1), 75–103.

Caruana, D.J., Heller, A., 1999. Enzyme-amplified amperometric detection of hybridization and of a single base pair mutation in an 18-base oligonucleotide on a 7-μm-diameter microelectrode. Journal of the American Chemical Society 121 (4), 769–774.

Castro, A.C.H., França, E.G., de Paula, L.F., Soares, M.M., Goulart, L.R., Madurro, J.M., Brito-Madurro, A.G.J.A.S.S., 2014. Preparation of Genosensor for Detection of Specific DNA Sequence of the Hepatitis B Virus, vol. 314, pp. 273–279.

Chiorcea-Paquim, A.-M., Oliveira-Brett, A.M., 2014. Redox behaviour of G-quadruplexes. Electrochimica Acta 126, 162–170.

Cui, L., Wu, J., Ju, H., 2015. Electrochemical sensing of heavy metal ions with inorganic, organic and bio-materials. Biosensors and Bioelectronics 63, 276–286.

Das, M., Sumana, G., Nagarajan, R., Malhotra, B.J.A.P.L., 2010. Zirconia Based Nucleic Acid Sensor for Mycobacterium Tuberculosis Detection, vol. 96 (13), p. 133703.

Daune, M., Dekker, C.A., Schachman, H.K., 1966. Complexes of silver ion with natural and synthetic polynucleotides. Biopolymers 4 (1), 51–76.

Drummond, T.G., Hill, M.G., Barton, J.K., 2003. Electrochemical DNA sensors. Nature Biotechnology 21 (10), 1192.

Du, J., Cullen, J.J., Buettner, G.R., 2012. Ascorbic acid: chemistry, biology and the treatment of cancer. Biochimica et Biophysica Acta 1826 (2), 443–457.

Ebrahimi, M., Raoof, J.B., Ojani, R., 2015. Novel electrochemical DNA hybridization biosensors for selective determination of silver ions. Talanta 144, 619–626.

Erdem, A., Kerman, K., Meric, B., Akarca, U.S., Ozsoz, M., 1999. DNA electrochemical biosensor for the detection of short DNA sequences related to the hepatitis B virus. Electroanalysis 11 (8), 586–587.

Erdem, A., Ozsoz, M., 2002. Electrochemical DNA biosensors based on DNA-drug interactions. Electroanalysis 14 (14), 965–974.

Fotouhi, L., Hashkavayi, A.B., Heravi, M.M., 2013. Interaction of sulfadiazine with DNA on a MWCNT modified glassy carbon electrode: determination of DNA. International Journal of Biological Macromolecules 53, 101–106.

Friedberg, E.C., 2000. Biological responses to DNA damage: a perspective in the new millennium. Cold Spring Harbor Symposia on Quantitative Biology 65, 593–602.
Friedberg, E.C., 2003. DNA damage and repair. Nature 421 (6921), 436–440.
Fu, F., Wang, Q., 2011. Removal of heavy metal ions from wastewaters: a review. Journal of Environmental Management 92 (3), 407–418.
Gao, F., Fan, T., Wu, J., Liu, S., Du, Y., Yao, Y., Zhou, F., Zhang, Y., Liao, X., Geng, D., 2017. Proximity hybridization triggered hemin/G-quadruplex formation for construction a label-free and signal-on electrochemical DNA sensor. Biosensors and Bioelectronics 96, 62–67.
Greenberg, R.A., Chin, L., Femino, A., Lee, K.-H., Gottlieb, G.J., Singer, R.H., Greider, C.W., DePinho, R.A., 1999. Short dysfunctional telomeres impair tumorigenesis in the INK4aΔ2/3 cancer-prone mouse. Cell 97 (4), 515–525.
Gumpu, M.B., Sethuraman, S., Krishnan, U.M., Rayappan, J.B.B., 2015. A review on detection of heavy metal ions in water – an electrochemical approach. Sensors and Actuators B: Chemical 213, 515–533.
Häder, D.-P., Sinha, R.P., 2005. Solar ultraviolet radiation-induced DNA damage in aquatic organisms: potential environmental impact. Mutation Research/Fundamental and Molecular Mechanisms of Mutagenesis 571 (1), 221–233.
Hamidi-Asl, E., Raoof, J.B., Hejazi, M.S., Sharifi, S., Golabi, S.M., Palchetti, I., Mascini, M., 2015. A genosensor for point mutation detection of P53 gene PCR product using magnetic particles. Electroanalysis 27 (6), 1378–1386.
Hamidi-Asl, E., Raoof, J.B., Ojani, R., Hejazi, M.S., 2013. Indigo carmine as new label in PNA biosensor for detection of short sequence of p53 tumor suppressor gene. Electroanalysis 25 (9), 2075–2083.
Han, H., Hurley, L.H., 2000. G-quadruplex DNA: a potential target for anti-cancer drug design. Trends in Pharmacological Sciences 21 (4), 136–142.
Hartati, Y.W., Topkaya, S.N., Maksum, I.P., Ozsoz, M., 2013. Sensitive detection of mitochondrial DNA A3243G tRNALeu mutation via an electrochemical biosensor using meldola's blue as a hybridization indicator. Advances in Analytical Chemistry 3 (A), 20–27.
Hashkavayi, A.B., Raoof, J.B., 2019. Ultrasensitive and reusable electrochemical aptasensor for detection of tryptophan using of $[Fe(bpy)_3](p\text{-}CH_3C_6H_4SO_2)_2$ as an electroactive indicator. Journal of Pharmaceutical and Biomedical Analysis 163, 180–187.
Hashkavayi, A.B., Raoof, J.B., Ojani, R., 2018. Preparation of epirubicin aptasensor using curcumin as hybridization indicator: competitive binding assay between complementary strand of aptamer and epirubicin. Electroanalysis 30 (2), 378–385.
Hashkavayi, A.B., Raoof, J.B., Ojani, R., Kavoosian, S., 2017. Ultrasensitive electrochemical aptasensor based on sandwich architecture for selective label-free detection of colorectal cancer (CT26) cells. Biosensors and Bioelectronics 92, 630–637.
Hayes, R.B., 1997. The carcinogenicity of metals in humans. Cancer Causes & Control 8 (3), 371–385.
Hejazi, M.S., Pournaghi-Azar, M.H., Alipour, E., Karimi, F., 2008. Construction, electrochemically biosensing and discrimination of recombinant plasmid (pEThIL-2) on the basis of interleukine-2 DNA insert. Biosensors and Bioelectronics 23 (11), 1588–1594.
Hejazi, M.S., Raoof, J.-B., Ojani, R., Golabi, S.M., Asl, E.H., 2010. Brilliant cresyl blue as electroactive indicator in electrochemical DNA oligonucleotide sensors. Bioelectrochemistry 78 (2), 141–146.

Ikehata, H., Kawai, K., Komura, J.-i., Sakatsume, K., Wang, L., Imai, M., Higashi, S., Nikaido, O., Yamamoto, K., Hieda, K., 2008. UVA1 genotoxicity is mediated not by oxidative damage but by cyclobutane pyrimidine dimers in normal mouse skin. Journal of Investigative Dermatology 128 (9), 2289–2296.

Jayaratne, T.E., Ybarra, O., Sheldon, J.P., Brown, T.N., Feldbaum, M., Pfeffer, C.A., Petty, E.M., 2006. White Americans' genetic lay theories of race differences and sexual orientation: their relationship with prejudice toward blacks, and gay men and lesbians. Group Processes and Intergroup Relations 9 (1), 77–94.

Jia, S., Liang, M., Guo, L.-H., 2008. Photoelectrochemical detection of oxidative DNA damage induced by Fenton reaction with low concentration and DNA-associated Fe2+. The Journal of Physical Chemistry B 112 (14), 4461–4464.

Kara, P., Dağdeviren, K., Özsöz, M., 2007. An electrochemical DNA biosensor for the detection of DNA damage caused by radioactive iodine and technetium. Turkish Journal of Chemistry 31 (3), 243–249.

Kashefi-Kheyrabadi, L., Mehrgardi, M.A., 2013. Aptamer-based electrochemical biosensor for detection of adenosine triphosphate using a nanoporous gold platform. Bioelectrochemistry 94, 47–52.

Khairy, M., Khorshed, A.A., Rashwan, F.A., Salah, G.A., Abdel-Wadood, H.M., Banks, C.E., 2017. Sensitive determination of amlodipine besylate using bare/unmodified and DNA-modified screen-printed electrodes in tablets and biological fluids. Sensors and Actuators B: Chemical 239, 768–775.

Kowalczyk, A., Nowicka, A.M., Jurczakowski, R., Niedzialkowski, P., Ossowski, T., Stojek, Z., 2010. New anthraquinone derivatives as electrochemical redox indicators for the visualization of the DNA hybridization process. Electroanalysis 22 (1), 49–59.

Lai, Y., Ma, Y., Sun, L., Jia, J., Weng, J., Hu, N., Yang, W., Zhang, Q., 2011. A highly selective electrochemical biosensor for Hg^{2+} using hemin as a redox indicator. Electrochimica Acta 56 (9), 3153–3158.

Li, F., Chen, W., Tang, C., Zhang, S., 2008. Recent development of interaction of transition metal complexes with DNA based on biosensor and its applications. Talanta 77 (1), 1–8.

Lucarelli, F., Marrazza, G., Turner, A.P., Mascini, M., 2004. Carbon and gold electrodes as electrochemical transducers for DNA hybridisation sensors. Biosensors and Bioelectronics 19 (6), 515–520.

Mello, L.D., Hernandez, S., Marrazza, G., Mascini, M., Kubota, L.T., 2006. Investigations of the antioxidant properties of plant extracts using a DNA-electrochemical biosensor. Biosensors and Bioelectronics 21 (7), 1374–1382.

Millan, K.M., Saraullo, A., Mikkelsen, S.R., 1994. Voltammetric DNA biosensor for cystic fibrosis based on a modified carbon paste electrode. Analytical Chemistry 66 (18), 2943–2948.

Minasyan, S., Tavadyan, L., Antonyan, A., Davtyan, H., Parsadanyan, M., Vardevanyan, P., 2006. Differential pulse voltammetric studies of ethidium bromide binding to DNA. Bioelectrochemistry 68 (1), 48–55.

Mousavisani, S.Z., Raoof, J.-B., Turner, A.P., Ojani, R., Mak, W.C., 2018a. Label-free DNA sensor based on diazonium immobilisation for detection of DNA damage in breast cancer 1 gene. Sensors and Actuators B: Chemical 264, 59–66.

Mousavisani, S.Z., Raoof, J.B., Ojani, R., Bagheryan, Z., 2018b. An impedimetric biosensor for DNA damage detection and study of the protective effect of deferoxamine against DNA damage. Bioelectrochemistry 122, 142–148.

Narayanan, D.L., Saladi, R.N., Fox, J.L., 2010. Ultraviolet radiation and skin cancer. International Journal of Dermatology 49 (9), 978–986.
Nimse, S.B., Song, K., Sonawane, M.D., Sayyed, D.R., Kim, T., 2014. Immobilization techniques for microarray: challenges and applications. Sensors 14 (12), 22208–22229.
Ono, A., Togashi, H., 2004. Highly selective oligonucleotide-based sensor for mercury(II) in aqueous solutions. Angewandte Chemie International Edition 43 (33), 4300–4302.
Palecek, E., Bartosik, M., 2012. Electrochemistry of nucleic acids. Chemical Reviews 112 (6), 3427–3481.
Paquim, A.-M.C., Diculescu, V., Oretskaya, T., Brett, A.O., 2004. AFM and electroanalytical studies of synthetic oligonucleotide hybridization. Biosensors and Bioelectronics 20 (5), 933–944.
Perdiz, D., Gróf, P., Mezzina, M., Nikaido, O., Moustacchi, E., Sage, E., 2000. Distribution and repair of bipyrimidine photoproducts in solar UV-irradiated mammalian cells possible role of dewar photoproducts in solar mutagenesis. Journal of Biological Chemistry 275 (35), 26732–26742.
Phan, A.T., 2010. Human telomeric G-quadruplex: structures of DNA and RNA sequences. The FEBS Journal 277 (5), 1107–1117.
Phan, A.T., Mergny, J.L., 2002. Human telomeric DNA: G-quadruplex, i-motif and Watson–Crick double helix. Nucleic Acids Research 30 (21), 4618–4625.
Pividori, M., Merkoci, A., Alegret, S., 2000. Electrochemical genosensor design: immobilisation of oligonucleotides onto transducer surfaces and detection methods. Biosensors and Bioelectronics 15 (5–6), 291–303.
Qi, J., Shafer, R.H., 2005. Covalent ligation studies on the human telomere quadruplex. Nucleic Acids Research 33 (10), 3185–3192.
Qu, J., Wu, L., Liu, H., Li, J., Lv, H., Fu, X., Song, Y., 2015. A Novel Electrochemical Biosensor Based on DNA for Rapid and Selective Detection of Cadmium.
Raoof, J.B., Ojani, R., Ebrahimi, M., Hamidi-Asl, E., 2011. Developing a nano-biosensor for DNA hybridization using a new electroactive label. Chinese Journal of Chemistry 29 (11), 2541–2551.
Raoof, J.B., Ojani, R., Ebrahimi, M., Hamidi-Asl, E., 2013. Application of Ethyl Green as Electroactive Indicator in Genosensors and Investigation of Its Interaction With DNA.
Rashid, J.I.A., Yusof, N.A., 2017. The strategies of DNA immobilization and hybridization detection mechanism in the construction of electrochemical DNA sensor: a review. Sensing and Bio-Sensing Research 16, 19–31.
Schärer, O.D., 2003. Chemistry and biology of DNA repair. Angewandte Chemie International Edition 42 (26), 2946–2974.
Shamsi, M.H., Kraatz, H.-B., 2013. Interactions of metal ions with DNA and some applications. Journal of Inorganic and Organometallic Polymers and Materials 23 (1), 4–23.
Siddiquee, S., Yusof, N.A., Salleh, A.B., Tan, S.G., Bakar, F.A., Heng, L.Y., 2010. DNA hybridization based on *Trichoderma harzianum* gene probe immobilization on self-assembled monolayers on a modified gold electrode. Sensors and Actuators B: Chemical 147 (1), 198–205.
Sirajuddin, M., Ali, S., Badshah, A., 2013. Drug–DNA interactions and their study by UV–Visible, fluorescence spectroscopies and cyclic voltametry. Journal of Photochemistry and Photobiology B: Biology 124, 1–19.

Song, Y., Buettner, G.R., 2010. Thermodynamic and kinetic considerations for the reaction of semiquinone radicals to form superoxide and hydrogen peroxide. Free Radical Biology and Medicine 49 (6), 919–962.

Udenfriend, S., Clark, C.T., Axelrod, J., Brodie, B.B., 1954. Ascorbic acid in aromatic hydroxylation. Journal of Biological Chemistry 208, 731–738.

Wang, J., 2002. Electrochemical nucleic acid biosensors. Analytica Chimica Acta 469 (1), 63–71.

Wang, J., Kawde, A.-N., 2001. Pencil-based renewable biosensor for label-free electrochemical detection of DNA hybridization. Analytica Chimica Acta 431 (2), 219–224.

Watson, J.D., 1970. Molecular Biology of the Gene, second ed.

Wu, L., Yang, Y., Zhang, H., Zhu, G., Zhang, X., Chen, J., 2012. Sensitive electrochemical detection of hydroxyl radical with biobarcode amplification. Analytica Chimica Acta 756, 1–6.

Wu, Y., Brosh, R.M., 2010. G-quadruplex nucleic acids and human disease. The FEBS Journal 277 (17), 3470–3488.

Xu, S., Zhan, J., Man, B., Jiang, S., Yue, W., Gao, S., Guo, C., Liu, H., Li, Z., Wang, J., 2017. Real-time reliable determination of binding kinetics of DNA hybridization using a multichannel graphene biosensor. Nature Communications 8, 14902.

Xu, Y., Ishizuka, T., Kurabayashi, K., Komiyama, M., 2009. Consecutive formation of G-quadruplexes in human telomeric-overhang DNA: a protective capping structure for telomere ends. Angewandte Chemie International Edition in English 48 (42), 7833–7836.

Yang, Q., Nie, Y., Zhu, X., Liu, X., Li, G., 2009. Study on the electrocatalytic activity of human telomere G-quadruplex–hemin complex and its interaction with small molecular ligands. Electrochimica Acta 55 (1), 276–280.

Zhan, S., Wu, Y., Wang, L., Zhan, X., Zhou, P., 2016. A mini-review on functional nucleic acids-based heavy metal ion detection. Biosensors and Bioelectronics 86, 353–368.

Zheng, G., Daniel, W.L., Mirkin, C.A., 2008. A new approach to amplified telomerase detection with polyvalent oligonucleotide nanoparticle conjugates. Journal of the American Chemical Society 130 (30), 9644–9645.

CHAPTER

10 Peptide-based electrochemical biosensors

Mihaela Puiu[1], Camelia Bala[1,2]

R&D Center LaborQ, University of Bucharest, Bucharest, Romania[1]; Department of Analytical Chemistry, University of Bucharest, Bucharest, Romania[2]

10.1 Introduction

Peptide segments have arisen as versatile building blocks for biohybrid nanomaterials because of their innate capacity to pack in highly ordered 1D, 2D, and 3D supramolecular assemblies or to the strong tethering of nanosized aggregates through electrostatic or covalent bonding. Nanocomposites containing various peptide elements are suitable for electrochemical bio/immunosensing because they are able to merge the high specificity of biomolecular interactions with the excellent conducting features of metallic or carbon nanostructures (Puiu and Bala, 2018a). Twenty natural amino acids (AAs) are the basic units in peptide synthesis. All of them are chiral, save glycine (G), and fall into L-configuration. In essence, they have the same structure and differ only in the R-group at the central carbon (C_α) position of the molecule. Their specific configurations are strongly dependent on the neighborhood of the R groups in the peptidic backbone (Ulijn and Smith, 2008). The type of the R group compels the specific chemistry of each AA residue via covalent or noncovalent interactions: disulfide bridges, hydrogen bonding, π-π stacking, hydrophobic, van der Waals, and electrostatic interactions (Puiu and Bala, 2018b). As an ensemble, these weak interactions yield stable secondary structures, such as α- or 3_{10} helices, β-hairpins, and β-sheets (Lowik et al., 2010; Loo et al., 2012). The helical conformations ensure inherent electron transfer (ET) and conductive properties, leading to the idea that peptide aggregates might be used for building materials with electronic capabilities (Pepe-Mooney and Fairman, 2009). Furthermore, short peptides (up to 10 AA residues) can be easily created using standard synthesis procedures. Thus, the complicated and time-consuming steps required for synthesizing large proteins may be avoided (Gazit, 2007; Lakshmanan et al., 2012). The charged carboxyl and amine groups are responsible for the hydrophilicity of thin peptide films, enabling the decrease of proteins adsorption onto surfaces because of the common hydrophobic interactions (Karimzadeh et al., 2018). In addition, near the isoelectric point, peptide sequences can be used to prevent the nonspecific adsorption of charged proteins onto various electrochemical transducers. Synthetic peptides containing designed motifs (AA sequences correlated to a

Electrochemical Biosensors. https://doi.org/10.1016/B978-0-12-816491-4.00010-3

specific function of a natural protein) are designated as protein-mimetic peptides (PMPs) (Groß et al., 2016; Zhang et al., 2017). PMPs display self-assembling properties, are biocompatible, and can act as effective bio-recognition elements (Wei et al., 2017). Most of bio- and immunosensors are built upon the high selectivity and affinity of enzyme/substrate reactions (Xu et al., 2016), enzyme/inhibitor (Arduini et al., 2010), antibody/antigen (Palaniappan et al., 2013), and receptor/ligand interactions (Hansen et al., 2007). Despite their various applications, from diagnosis to food and drug monitoring, these types of sensors had to face several restrictions: deactivation/denaturation of the recognition biomolecule during the assay (Guida et al., 2018); steric hindrance of analytes of large molecular size such as antibodies, DNA strands, or enzymes (Cheng et al., 2015); and finally, use of expensive biological reagents (Pavan and Berti, 2012). These drawbacks can be overcome by the use of synthetic molecules such as aptamers (Ke et al., 2014), liposomes (Liu and Boyd, 2013), and peptides (Puiu and Bala, 2018b) as recognition elements.

10.2 Creating peptidic interfaces: coating strategies for electrodes

In most peptide-based sensing assays, the peptide motifs are integrated in different hybrid materials designed to cover the transducer's surface. They act as active probes (Shinohara et al., 2015; González-Fernández et al., 2018a) or conductive bridges (Puiu et al., 2014; Jampasa et al., 2014). In the first case, the peptide sequences are used as recognition elements for their own receptors. Antimicrobial peptides (AMPs) (Hoyos-Nogués et al., 2016) and cell-penetrating peptides (Kalafatovic and Giralt, 2017; Guidotti et al., 2017) can interact with cell surface domains acting as peptide ligands for several receptors, such as antigenic peptide sequences (Strzemińska et al., 2016), antibodies (Zaitouna et al., 2015), and bacterial cells (Hoyos-Nogués et al., 2016; Li et al., 2014). The main strategies for electrode interfacing in peptide-based electrochemical sensors are as follows:

- The "bottom-up" method, where small-sized structures self-assemble to yield highly ordered large aggregates (Adler-Abramovich and Gazit, 2014). The self-assembly of peptides is possible when the designed AA sequences mimic the pattern of the conformational units of naturally existing proteins: β-sheets, hairpins and turns, α-/3_{10} helices, and coiled coils (Puiu and Bala, 2018a).
- The "top-down" method, where the reduction of large systems down into smaller pieces produces multifunctional nanoscale frames (Yogeswaran and Chen, 2008). This can be performed either by etching down a bulk material or by maneuvering components into specific locations (Yu et al., 2013).

Whereas the bottom-up method builds up nanomaterials from small units such as atoms or molecules, the top-down approach produces nanostructures by dismantling larger materials by means of lithographic tools (i.e., physical top-down) or through chemical-based procedures (i.e., chemical top-down) (Yu et al., 2013). Amphiphilic and aromatic peptides are the most used building blocks for "smart materials" developed through self-assembling. For example, aromatic dipeptide nanotubes having 50–300 nm diameters and up to microns lengths were developed from diphenyl alanine (Gazit, 2007; Smith et al., 2011). Moreover, amyloid peptides with a core sequence of dipeptide L, L-diphenylalanine can self-assemble to form stable rodlike-shaped structures with diameters in the nanometer range (Bianchi et al., 2014).

10.3 Peptide-modified surfaces and interrogation modes in electrochemical bioassays

Typical electrochemical peptide-based assays are carried out in a three-electrode cell, which contains a working electrode, usually coated with a sensing layer, a counterelectrode, and a reference electrode. For electrodes modified with peptide-decorated materials, a redox reporter is attached to one end of a peptide sequence, whereas the other end is usually anchored either on a thiolic self-assembled monolayer (SAM) or directly chemosorbed onto the electrode's surface. Therefore, the redox reporter is separated from the electrode surface by a distance defined by the thickness of the adsorbed layer (Juhaniewicz et al., 2015). The immobilization of peptide sequence onto carbon-based nanomaterials or gold surfaces, nanoparticles, or nanorods follows several directions:

- Direct chemisorption of lipoic acid–functionalized peptides onto Au surface (Puiu et al., 2014; Gatto et al., 2012);
- Covalent binding of thiol groups of cysteine residues to carboxyl groups of alkylthiolated SAMs using 1-ethyl-3-(3-diaminoprpyl)carbodiimide EDC/N-hydroxysuccinimide/2-(2′ pyridinyl-dithio)ethanamine (PDEA) chemistry (Wong et al., 2009);
- Covalent immobilization through maleimide and acyl halide reaction using EDC/NHS esters and 3-3′N-[ε-maleimidocaproic acid] hydrazide (EMCH) (Lequoy et al., 2016). Carbon nanomaterials with free carboxyl groups can also be activated via EDC/NHS esters to bind free amino groups from peptides (Ding et al., 2013).
- "Click chemistry," where the peptide probe is usually modified at the $N_{terminus}$ with an alkyne moiety, whereas the $C_{terminus}$ contains the redox reporter. The alkyne-modified peptide probes can easily react with azide-modified alkyl SAMs (Gerasimov and Lai, 2011) or azide-modified carbon nanomaterials, using Cu(I) as catalyst (Chen, 2016). Histidine-tagged MB peptides were also reported to reversibly bind metal ions Ni(II) and Zn(II)/imidazole-modified alkylthiolated SAMs (Zaitouna and Lai, 2011).

A spacer is often introduced between the thiol group and the peptide probe to promote enzyme or antibody accessibility and to confer flexibility to the probe (Karimzadeh et al., 2018). Besides, the optimal design of a redox peptide layer requires coadsorption of diluent molecules, with adjusted lengths to remove the interactions between the charged redox centers (Puiu and Bala, 2018b). The most encountered redox reporters in peptide affinity or enzymatic assays are Methylene Blue (MB) and ferrocene (Fc) (Ding et al., 2013; Moore et al., 2013) attached either to peptide probes (Zaitouna et al., 2015) or to peptide bridges (Puiu et al., 2014). Upon applying the potential, ET occurs between the redox reporter and the electrode's surface. The entire process is mediated by the bridge separating the surface from the redox reporter (Juhaniewicz and Sek, 2012). The specific binding of the target analyte to the redox-tagged probe/bridge triggers a change in the manner the ET undergoes through the coating layer (Puiu and Bala, 2018b). Consequently, there are "signaling-on" sensors, when ET rate through the surface increases following the binding of the target, or "signaling-off" sensors, when ET rate is decreased as a result of the target/surface interaction (Puiu and Bala, 2018a).

10.4 Electron transfer across peptide bridges

The capability of a peptide to form compact SAMs, which is decisive for an efficient ET through a peptidic layer, is influenced by the length of the peptide primary structure, by the type of secondary structure, and by the occurrence of aromatic groups on the side chains (Gobbo et al., 2016). Short peptides may populate several conformations, because they are very flexible and rapidly interconverted between the different conformers (Long et al., 2005; Gatto et al., 2007). As a consequence, they form loosely packed SAMs with large degrees of inhomogeneity and up to 15% vacant gold sites (Long et al., 2005). On the other side, long helical peptides fold in well-ordered and densely packed films (Lauz et al., 2012).

The α-helix is the most common secondary structure in peptides and proteins, with the dihedral angles ω, Φ, and Ψ having the values of 180 degrees, −57 degrees, and −47 degrees, respectively. In the case of the 3_{10} helices, the dihedral angles are 180 degrees, −49 degrees, and −27 degrees (Orlowski et al., 2007; Mandal and Kraatz, 2006). The main distinction between a 3_{10} helix and an α-helix resides in the manner in which the hydrogen bonds between the carbonyl and amide groups from the same or different peptide chain are established. For example, in α-helices, the carbonyl group of residue i interacts with the amide group in residue $i+4$ (Vieira-Pires and Morais-Cabral, 2010; Zhang et al., 2013). Thus, the α-helix displays the following features: 3.6 residues per turn, two consecutive residues making an angle of 100 degrees around the helical axis, a helical rise per residue of 1.5 Å, and a helical pitch of 5.4 Å (Smith et al., 2005). By contrast, in 3_{10} helix, the carbonyl group in residue i and the nitrogen of the amide group in residue $i+3$ are hydrogen bonded (Zhang and Hermans, 1994). Canonical 3_{10} helices have three residues per turn, with an angle of 120 degrees between consecutive residues,

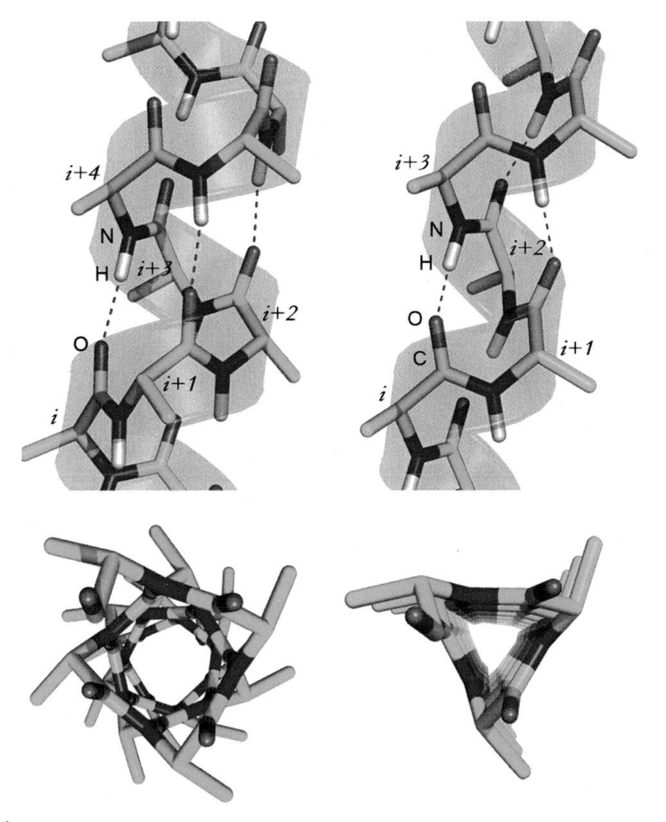

FIGURE 10.1

Side views and views along the axes of a helix in a canonical α-helical conformation (left) and canonical 3_{10} conformation (right).

Reproduced from Vieira-Pires, R.S., Morais-Cabral, J.H., 2010. 3_{10} helices in channels and other membrane proteins. The Journal of General Physiology 136, 585–592; with permission of Rockefeller University Press.

a helical rise per residue of 1.93–2.0 Å, and a helical pitch of 5.8–6 Å. Thus, 3_{10} helix is longer and thinner than the α-helix with the same number of residues (Vieira-Pires and Morais-Cabral, 2010; Bezer et al., 2014) (Fig. 10.1).

Gatto et al. (2012) have demonstrated that six residues-long peptides containing α-aminoisobutyric acid (Aib) residues can be "forced" to fold in helical conformations. Other works reported that oligopeptides containing up to eight residues, modified at $N_{terminus}$ with a lipoyl moiety, are able to form densely packed monolayers (Gatto et al., 2007, 2012; Puiu et al., 2014), mainly owing to their helical secondary structure. The molecular dipole moment strongly affects the rate of the peptide-mediated ET (Kai et al., 2008; Gatto et al., 2007), especially in the α-helical peptides, which exhibit a large macrodipole moment (approximately 3.5 D per residue) (Orlowski et al., 2007), oriented parallel to the molecular axis. 3_{10} helices exhibit high dipole moments, leading to an effective positive charge at the $N_{terminus}$ and a negative charge at $C_{terminus}$. It was demonstrated that the charge transfer through the peptide spacer is affected by the separation between the acceptor (A) and donor (D) sites, the peculiarities of the peptide backbone, along with its secondary structure and the resulting flexibility (Long et al., 2005; Sisido et al., 2001).

The electronic coupling between donor and acceptor is expected to be weak because the distance separating the D/A sites is relatively large (Juhaniewicz and Sek, 2012). Therefore, ET occurs in a nonadiabatic regime, and its rate constant (k_{ET}) can be described with Fermi's golden rule (Siddarth and Marcus, 1990):

$$k_{ET} = \left(\frac{2\pi}{h}\right)|H_{DA}|^2(FC) \tag{10.1}$$

where H_{DA} is the electronic coupling matrix element between D and A, and FC denotes the Franck-Condon factor, accounting the contribution of the nuclear reorganization to the ET rate constant (Siddarth and Marcus, 1990):

$$FC = \frac{1}{(4\pi\lambda k_B T)^{1/2}} \exp\left(-\frac{\left(\Delta G^0 + \lambda\right)^2}{4\pi\lambda k_B T}\right) \tag{10.2}$$

where λ represents the reorganization energy (the energy required to reorient all atoms from an equilibrium state to the product state), ΔG^0 is the standard Gibbs free energy of ET reaction, k_B is the Boltzmann constant, and T is the absolute temperature. Two major mechanisms were proposed to explain the distance dependence of the rate of ET in peptide scaffolds: a superexchange assisted through bridge ET and an electron-hopping mechanism (Orlowski et al., 2007; Polo et al., 2005).

The hoping mechanism states that the electron resides on the bridge for a short time during its passing from D to A and undergoes further sequential tunneling steps between its neighboring units, which act as "hopping" sites (Orlowski et al., 2007); in the superexchange mechanism, the conjugated bridge only serves as a medium to deliver the electron from the donor to the acceptor (Long et al., 2005; Mandal and Kraatz, 2006). The ET rate constant decays exponentially with distance in a superexchange mechanism, whereas a linear decrease with distance is expected for a hopping mechanism (Long et al., 2005). Because $|H_{DA}|$ depends on the overlap of the electronic wave functions of the donor and the acceptor and thus upon the distance r on which electron tunneling takes place (Feldberg, 2010), it can be expressed as:

$$|H_{DA}|^2 = |H_{DA}|^2_{(r=r_0)} \exp(-\beta(r - r_0)) \tag{10.3}$$

where β is the damping factor with typical values below 1.2 $Å^{-1}$ for a through-space ET (Siddarth and Marcus, 1990).

If $r - r_0 = l_{DA}$ is the distance separating the donor from the acceptor, then the ET rate constant displays an exponential decrease with distance, according to the following equation (Marcus and Sutin, 1985):

$$k_{ET} \propto \exp(-\beta l_{DA}) \tag{10.4}$$

The tunneling pathway model predicts decay factors of 1.0 and 1.3 $Å^{-1}$ for β-strand and α-helix conformations, respectively (Gray and Winkler, 2005). Because of the sharp exponential decay of the ET rate constant with the increase of bridge length, the superexchange mechanism is restricted to short distances. The driving force here is the difference between the redox potentials of the donor and acceptor (Fermi level in the case of metal). For low driving forces, the upper limit for tunneling is around 2 nm (Gray and Winkler, 2005). Still, in certain proteins, the ET occurs over long distances, reaching up to 3.5 nm. The multistep mechanism involves specific units/functional groups, which can be reversibly oxidized or reduced, and act as "relay stations," where the electron/hole resides for a short time before it continues to the next hopping step (Ener et al., 2017; Juhaniewicz and Sek, 2012). In these cases, the distance dependence of the ET rate constant is less significant compared with superexchange mechanism and can be depicted as (Nitzan, 2001):

$$k_{ET} \propto \frac{1}{N} \exp\left(-\frac{E_a}{k_B T}\right) \tag{10.5}$$

where N is the number of hopping sites and E_a is the activation energy of ET reaction. The hopping competes with superexchange mechanism in an ET system. There are numerous studies dedicated to ET through peptide bridges, but SAMs of helical peptides (α and 3_{10} configurations) are of particular interest for electrochemical biosensors (Chaudhry et al., 2010; Gatto et al., 2012; Juhaniewicz and Sek, 2012). The 3_{10} helix is presumed to be more conductive than the α-helix (because of the increased number, $i \rightarrow i+3$ vs. $i \rightarrow i+4$ is stronger intramolecular H-bonded). Thus, the 3_{10} conformation may be a potential candidate for the ET active conformer, and the formation rate of this conformation (which is slower in close-packed SAMs compared with that in solutions) controls the overall ET rate in helical peptide SAMs (Mandal and Kraatz, 2012).

10.5 Electrochemical techniques in peptide-based biosensing assays

Electrochemical methods have been extensively used not only for studying the ET through peptide bridges (Eckermann et al., 2010; Chaudhry et al., 2010; Martic et al., 2011; Polo et al., 2005) but also for fabricating peptide-based hybrid materials with conductive properties for analytical purposes. A three-electrode classic cell or screen-printed electrodes are required with a peptide-coated working electrode. Thus, the redox reporter is separated from the electrode surface by a certain distance, given by the thickness of the adsorbed film. Upon applying the potential, the ET occurs between the redox center and the electrode's surface, and the global process is mediated by the molecules forming the monolayer (Juhaniewicz and Sek, 2012). Thus, ET may occur through a two-component SAM having both redox-tagged molecules and passive diluent molecules (Wain et al., 2008; Puiu et al., 2014; Mandal and Kraatz, 2006). Current results emphasize the importance of the monolayers' characteristics for the mediated ET, particularly when considering peptide sequences with a covalently bound redox center, such as Fc (Orlowski et al., 2007) or MB (Juhaniewicz et al., 2015). To ensure an efficient ET in a redox-peptide system, coadsorption of diluent molecules, with adjustable lengths, is required. In this manner, the interactions between the charged redox reporters can be minimized or eliminated. Both ET mechanisms are possible, but the prevailing mechanism is switchable, depending on the D/A distance and the driving force of the molecular system. For short distances or for large driving forces, the tunneling mechanism prevails (Kai et al., 2008). One-component electroactive monolayers may exhibit significantly faster standard rate constants, compared with two-component monolayers containing coadsorbed molecules (Finklea and Hanshew, 1992). In a one-component system, the redox centers are closer to each other, and electron hopping between them may occur. Also, strong electrostatic interactions between charged sites may influence the measured ET rate constant (Kitagawa et al., 2006). Moreover, the molecular movement of the peptide backbone may be obstructed by tight molecular packing in compact layers. Restricted molecular dynamics may slow down or prevent the ET (Long et al., 2005). Therefore, it is difficult to discriminate a purely electrochemical output from an electrochemical signal altered by the molecular motion of the surface-bound molecules. For example, the time scale of the electron movement from a ferrocene Fc to the gold surface mediated by a peptide layer is often smaller than the time scale of molecular motions, especially when an external electric field forcing charged molecules to align themselves within the field gradient, is applied (Wain et al., 2008; Heck et al., 2012; Long et al., 2005; Mandal and Kraatz, 2012). The most common electrochemical methods for measuring k_{ET} through redox-modified SAMs and for analyte detection are cyclic voltammetry (CV), alternating current (AC), square wave (SW), differential pulse (DP) voltammetry, and electrochemical impedance spectroscopy (EIS).

10.5.1 Cyclic and linear sweep voltammetry

In potential sweep methods, the potential is varied and the current resulting from a redox event is measured. CV does not require sophisticated or expensive equipment, and moreover, CV is the electrochemical technique less affected by the kinetic heterogeneity (distribution of ET rates owing to changes in the molecular environment of the redox reporter caused by SAM defects (Eckermann et al., 2010)). Thus, because of kinetic heterogeneity, k_{ET} obtained from electrochemical measurements may contain contributions either from well-organized domains or from those that are disordered (Juhaniewicz et al., 2015). The surface coverage and k_{ET} can be determined from CV measurements and used further to build up electrochemical affinity–based formats. The ET rate from the redox reporter to the electrode surface may be hampered following the specific binding of an analyte, and its variation can be correlated with the analyte's concentration. The surface coverage Γ can be calculated from the slope of the dependence of I_p on the scan rate ν (Wain et al., 2008; Gatto et al., 2007) and compared with a theoretical maximum based on the molecular surface area of the adsorbed species as in Eq. (10.6) (Eckermann et al., 2010):

$$I_p = \frac{n^2 F^2}{4RT} \nu A_{SUR} \Gamma \tag{10.6}$$

where n represents the number of electrons transferred per mole of reaction, A_{SUR} is the surface area, and F is the Faraday constant. The redox potential, E_0, was calculated as the average of the anodic and cathodic peak potentials, E_{pa} and E_{pc}; the peak separation ΔE_p, was defined as $E_{pa} - E_{pc}$. The peak separation increases at high values of the scan rate, whereas at slow scan rates, the peak separation is practically zero, because the redox center is bound to the electrode and diffusion is not involved (Bard and Faulkner, 2000; Zoski, 2007). The overpotential η was defined as the difference $E_p - E_0$ and was used further to estimate k_{ET}. For each scan rate ν, the ET rate constant at a certain value of $\eta(k_s(\eta))$ can be calculated as:

$$k_s = \frac{I_p}{Q} \tag{10.7}$$

where Q is the overall passing charge.

In the experimental setup, the initial potential is selected such that the oxidation state of the redox centers is maintained at the start of a scan. The potential is scanned past the formal potential E_0 to a point at which the current returns to baseline. The scan direction is then reversed and returned to the initial starting potential. In the next experiments, the scan rate is varied to give a range of η values that can be used in the Tafel plot η versus log $k_s(\eta)$ to estimate k_{ET} (Okamoto et al., 2009). CV was often used to verify the insulating properties and the integrity of SAMs (Orlowski et al., 2007; Kitagawa et al., 2006; Gatto et al., 2012) to calculate the influence of the peptide macrodipole over k_{ET} and the surface coverage (Chaudhry et al., 2010; Juhaniewicz and Sek, 2012; Gatto et al., 2007; Mandal and Kraatz, 2006; Heck et al., 2012; Okamoto et al., 2009).

The surface coverage and the rate constant (k_{ET}) can be estimated through CV measurements and used further in affinity-based biosensors, because the ET rate from the redox reporter to the electrode surface may be hindered following the specific binding of a target analyte. It was recently shown that Fc-aminoisobutyric acid (Aib) peptide covalently loaded on double-walled carbon nanotubes (DWCNTs) immobilized onto gold through a thiolic SAM (Fig. 10.2) showed significantly increased peak current compared with their single-walled counterparts (SWCNTs) (Moore et al., 2013). Here, the ET rate capabilities were probed with CV.

This fact was assigned to a higher loading of the Fc-modified peptide to the outer wall of the nanotube, through the presence of a larger number of defects sites within the sp^2 carbon lattice for the DWCNTs (Moore et al., 2013). The apparent ET rate constants of the Aib_5-Fc decorated DWCNTs and SWCNTs were estimated using Laviron's method for ΔE_p <200 mV (Eckermann et al., 2010). The transfer

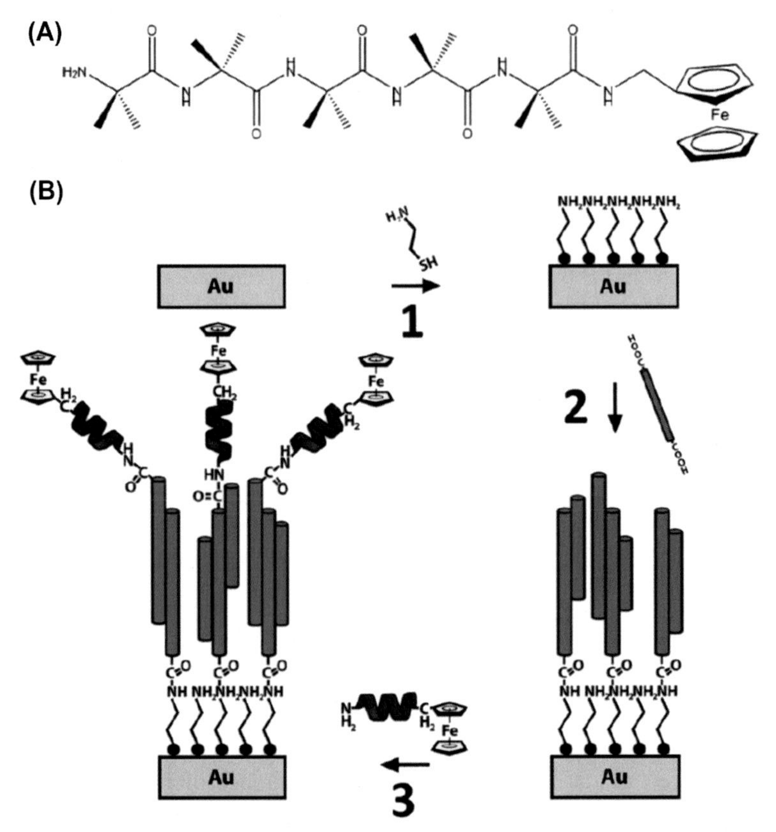

FIGURE 10.2

Design of an Fc-tagged helix peptide—loaded on DWCNTs immobilized on gold with excellent conductive properties: (A) the redox peptide Aib5-Fc and (B) fabrication of conductive layer.

Reproduced from Moore, K.E., Flavel, B.S., Yu, J., Abell, A.D., Shapter, J.G., 2013. Increased redox-active peptide loading on carbon nanotube electrodes. Electrochimica Acta 89, 206–211; with permission of Elsevier.

coefficient, α, was considered 0.3–0.7 and the average ET rate constants, k_{app}, were calculated according to the equation:

$$k_{app} = \frac{mvnF}{RT} \tag{10.8}$$

where m was s a dimensionless parameter related to the peak-to-peak separation (Moore et al., 2013). From the slope of the plot of $v = f(m - 1)$, the term mv was estimated, and hence k_{app} determined. The $k_{app}s$ for the DWCNT and SWCNT electrodes were 31 ± 6 and $23 \pm 2\ s^{-1}$, respectively. These values are comparable with ET rates reported for other redox-bound vertically aligned CNT arrays on cysteamine-modified gold (Flavel et al., 2009). These results indicated that DWCNTs offer a better alternative to SWCNTs in building electrochemical peptide-based sensors.

There are not numerous works reporting peptide-based biosensors using CV, but it is worth mentioning the study of Lee et al. (2017) on a concentric electrode device for the detection of matrix metalloproteinase-9 (MMP-9). The GPLGMWSRC sequence conjugated with MB as a specific substrate to MMP-9 was immobilized on Au electrodes using gold/thiol chemistry (Fig. 10.3). The peptide cleavage by MMP-9 brought the release of the redox reporter MB from the electrodes and caused the decrease of the electrical tunneling current. The analytical performances of this assay were a linear range of 1 pM to 1 nM and a limit of

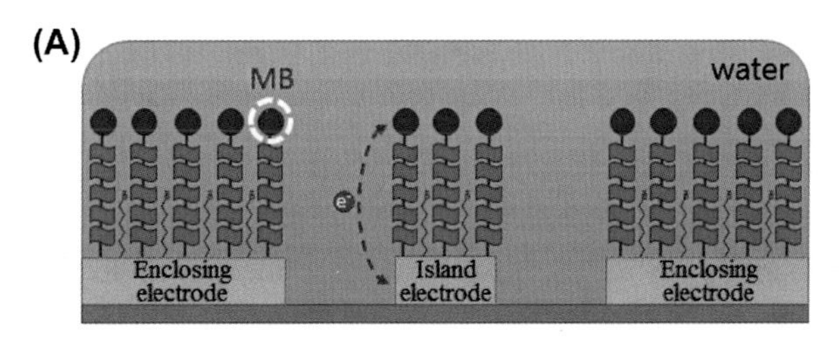

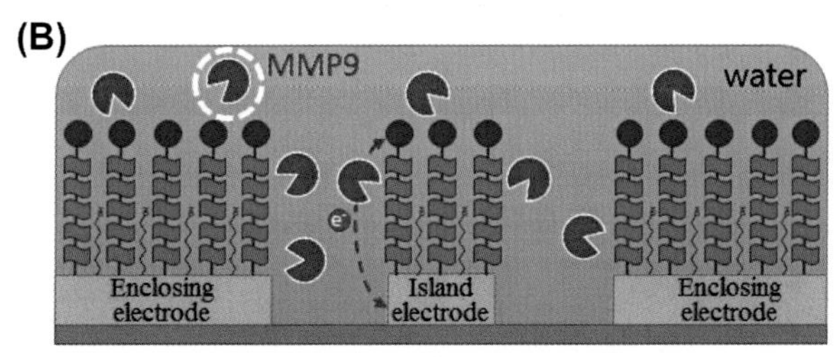

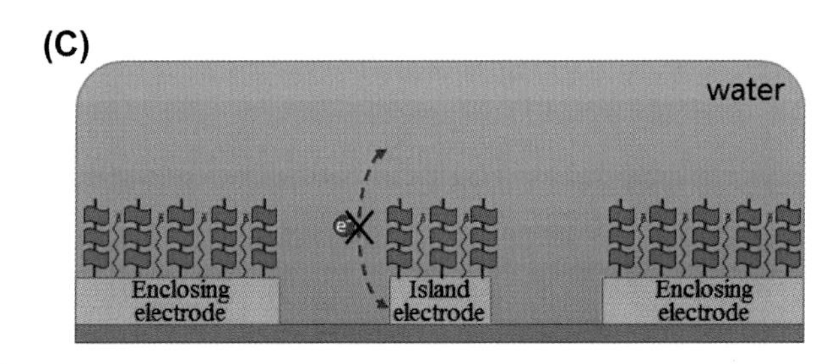

FIGURE 10.3

"Signaling-off" sensing principle of enzymatic peptide cleavage. (A) Electron tunneling in the presence of MB peptide on Au electrode, (B) peptide cleavage reaction of the MB-tagged peptide with MMP-9, (C) decrease in the current of electron tunneling after the peptide cleavage reaction.

Reproduced from Lee, J., Yun, J.Y., Lee, W.C., Choi, S., Lim, J., Jeong, H., Shin, D.-S., Park, Y.J., 2017. A reference electrode-free electrochemical biosensor for detecting MMP-9 using a concentric electrode device. Sensors and Actuators B: Chemical 240, 735–741; with permission of Elsevier.

detection (LOD) of 7 pM in phosphate buffer saline without interferences in the presence of 20 nM bovine serum albumin (BSA).

Linear sweep voltammetry (LSV) is a simpler subset CV, consisting of a single unidirectional voltage sweep. LSV is used instead of CV when there is no relevant information on the return scan of the CV, such as when the ET is followed by a very fast irreversible reaction (Speiser, 2007). Meng et al. (2018) developed a peptide cleavage-based electrochemical method for the detection of prostate-specific antigen (PSA) where the peptide probe was immobilized on the gold electrode surface via cysteine at the $N_{terminus}$ (Fig. 10.4). Graphene oxide (GO) was then attached on the electrode surface through $\pi-\pi$ stacking interactions between the aromatic rings from peptide residues and the graphene oxide rings. GO displayed electrocatalytic activity for the continuous deposition of silver. Silver ions were firstly absorbed through the uniform functional groups on the basal plane and edges of GO by electrostatic interactions. Then, silver nanoparticles (AgNPs) were synthesized "on the fly" with GO as both reducing agent and dispersant agent; finally, a significant silver stripping current of AgNPs was obtained in the absence of the target PSA.

The PSA presence in the sample was firstly converted to the cleavage of peptide probe on the electrode surface. As the cracked peptide was no longer able to assist the immobilization of GO and AgNPs, the electrochemical response changed significantly. In this case, LOD was 0.33 pg mL^{-1} with a linear range from 5 to 2×10^4 pg mL^{-1}. For all that, SWV is by far the voltammetric technique for electrochemical peptide—based biosensors, because it preserves the advantages of CV and LSV in terms of simplicity and cost-effectiveness, but in addition provides enhanced sensitivity (Reeves et al., 1993).

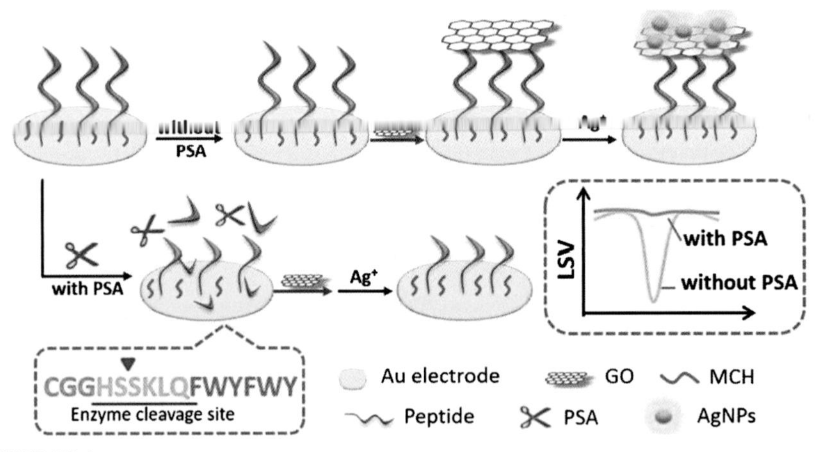

FIGURE 10.4

Schematic representation of a "signaling-off" biosensor for PSA detection, using a peptide/GO/AgNPs system, with 6-mercaptohexanol (MCH) as a spacer molecule. The peptide cleavage by PSA induces a sharp decrease of the anodic stripping current generated by the oxidation of AgNPs (Meng et al., 2019) with permission of Elsevier.

10.5.2 Square wave and differential pulse voltammetry

SWV is a method based on a potential modulation consisting of a staircase potential ramp modified with square-shaped potential pulses (Reeves et al., 1993). To minimize the contribution of the charging current, the current sampling is performed at the end of each potential pulse. The net current (for each potential cycle) is obtained as the difference between the forward and backward currents of a single potential pulse and represents the output of an SWV voltammogram (Mirceski et al., 2007). Another related voltammetric technique is DPV, which uses pulses of constant amplitude superimposed on the potential linear sweep (Osteryoung and Osteryoung, 1985). As in the case of SWV, the plot of the net current versus applied potential represents a symmetrical peak-shaped curve (He et al., 2007). The decrease or the increase of the net peak current Ip during SWV or DPW experiments, as a response to a specific binding effect occurring at the electrode functionalized surface, expresses the basic principle of most peptide-based bio- and immunoassays.

In the work of Ding et al. (2013), self-assembled nanowires of Fc-tagged peptides (Fc-PNW) comprising the phenylalanine-phenylalanine (Phe-Phe) sequence, coated with poly(diallyl dimethyl)ammonium chloride) (PDDA), were functionalized with gold nanoparticles (AuNps) and probe antibodies (Ab_2) for a sandwich immunoassay targeting human IgG antibody. The capture antibody (Ab_1) was attached onto a graphene/gold nanoparticle (AuNPs-GN) composite film. The binding of the target IgG to the capture antibody followed by the binding of Fc-PNW-Ab_2 causes a sharp increase of the peak current in the SWV assay. The Fc peptide electrochemical immunosensor displayed a low detection limit (5 fg mL^{-1}) for human IgG and a wide linear range encompassing four orders of magnitude (from 10 fg mL^{-1} to 100 pg mL^{-1}).

Another "signaling-on" approach provided the proof of concept of a reagentless, nonenzymatic sensor for PSA using the cleavage of immobilized peptide probe onto the surface via click chemistry (Strzemińska et al., 2016). A mixed monolayer, containing peptide-streptavidin (SA) probe and naphthoquinone derivatives, playing the role of electrochemical transducers for the molecular recognition PSA/peptide was immobilized onto glassy carbon electrode. PSA cleavage of the SA-peptide conjugate causes a significant increase of the SWV output owing to the diminution of the steric hindrance at the electrode surface and to the increase of diffusion rates of the counterions (Fig. 10.5). Here, the reported dynamic range was within 1 pM and 1 nM.

A similar approach was recently developed for the detection of human neutrophil elastase (HNE), produced by polymorphonuclear neutrophils (González-Fernández et al., 2018b). The developed signaling-off sensor exploited HNE's proteolytic activity. The MB-tagged peptide-based sensor contained an HNE-specific cleavage sequence incorporated in a ternary peptide/MCH/2,2′-(ethylenedioxy)diethanethiol (PDT) SAMs on gold. To generate an HNE electrochemical response, the MB-labeled peptide probe was modified with a 2-unit ethylene glycol moiety (PEG-2) and a cysteine at $C_{terminus}$, to enable a spontaneous immobilization of the probe

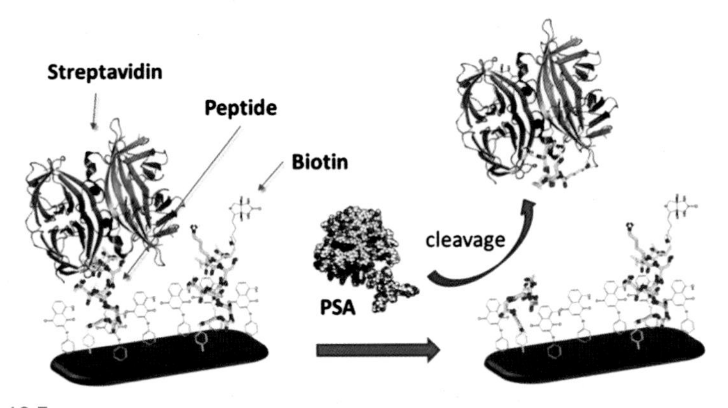

FIGURE 10.5

Working principle of the "signaling-on" peptide-based PSA sensor.

Reproduced from Strzemińska, I., Sainte Rose Fanchine, S., Anquetin, G., Reisberg, S., Noël, V., Pham, M.C., Piro, B., 2016. Grafting of a peptide probe for prostate-specific antigen detection using diazonium electroreduction and click chemistry. Biosensors and Bioelectronics 81:131–137; with permission of Elsevier.

through gold/thiol chemistry (Fig. 10.6A). The chosen negative control probe was the equivalent noncleavable D-amino acid sequence.

The specific enzyme-catalyzed cleavage caused the release of the labeled peptide fragment from the surface into solution (Fig. 10.6B), resulting in a sharp decrease of the SWV signal.

The developed platform enabled HNE detection between 10 and 150 nM in blood samples, by analyzing the signal change as a function of time following a Michaelis-Menten-based kinetic model (Fig. 10.6).

Xu et al. (2016) reported a DPV "signaling-on" assay for matrix metalloproteinase 2 (MMP-2) detection, based on the target-induced cleavage of a biotin-conjugated peptide, by using bimetallic Pt and Pd nanoparticles encapsulated in mesoporous-hollow ceria nanospheres (Pt/Pd/mhCeO_2NS) as nanocarrier electrocatalysts (Fig. 10.7).

The Pt/Pd/mhCeO_2NS were served as nanocarriers to anchor SA and electroactive thionine (Thi) to obtain SA/Thi/Pt/Pd/mhCeO_2NS nanoprobes. The labeled peptide (biotin-P1) was firstly oriented onto the electrode surface electrodeposited with Au. The introduction of target MMP-2 resulted in the specific recognition and cleavage of P1 at a certain site. Through the high affinity of SA to biotin in remained peptides not cleaved by MMP-2, SA/Thi/Pt/Pd/mhCeO_2NS were bound to the resultant electrode surface. The DPV output was a result of the presence of Thi. The redox reaction between Thi and H_2O_2, catalyzed by Pt/Pd/mhCeO_2NS was used to amplify the DPV signal, thus yielding an LOD of 0.078 pg mL^{-1} in spiked serum samples.

Zhao et al. (2018) proposed an electrochemical assay for the detection of intracellular kinase (PKA) by incorporating peptide nanoprobe-assisted signal labeling and signal amplification. PKA-specific peptide P1 containing a serine residue, which could be phosphorylated through PKA catalysis in the presence of

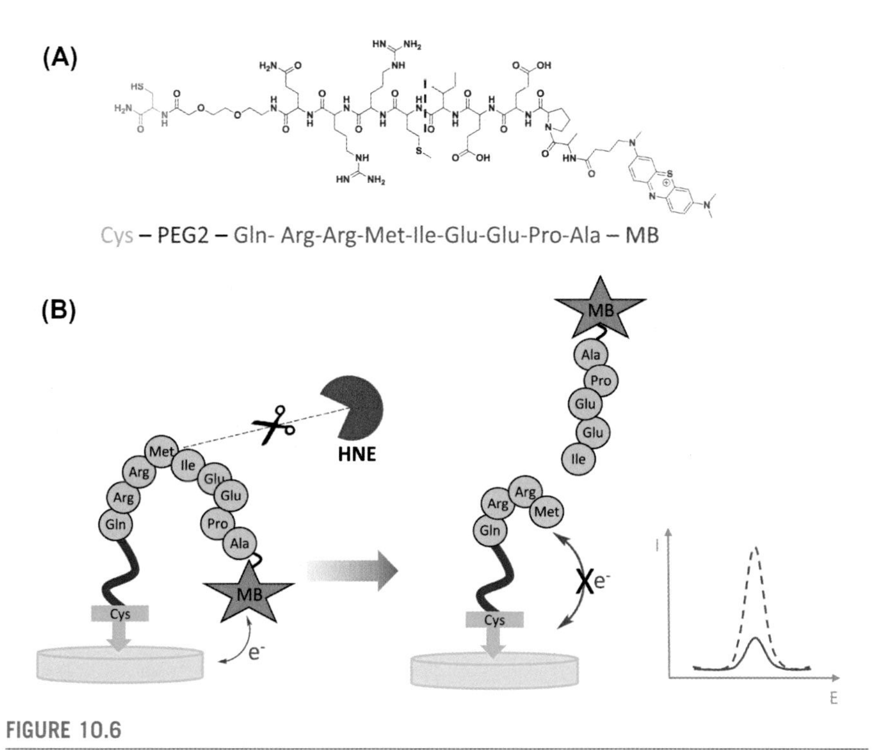

FIGURE 10.6

(A) Design of the MB-labeled peptide probe for HNE. (B) Detection principle: the enzymatic cleavage of the MB-tagged peptide substrate caused a sharp decrease of the SWV signal.

Reproduced with permission from González-Fernández, E., Staderini, M., Yussof, A., Scholefield, E., Murray, A.F., Mount, A.R., Bradley, M., 2018b. Electrochemical sensing of human neutrophil elastase and polymorphonuclear neutrophil activity. Biosensors and Bioelectronics 119, 209–214.

adenosine-5′-triphosphate (ATP), was self-assembled on the surface of a gold electrode. Another artificial peptide P2 contains a short template for preparation of copper nanoparticles-based nanoprobe (P2-CuNPs) and provides arginine residues for specific recognition of phosphorylation site. After PKA-catalyzed phosphorylation, phosphorylated P1 specially binds with P2-CuNPs through ultrastable phosphate-guanidine interaction (Fig. 10.8).

The surface-attached CuNPs released numerous copper ions in an acidic oxidizing environment, which were quantitatively determined by stripping DPV. Here LOD was 0.0019 U/mL, which is comparable with enzyme-linked immunosorbent assay (ELISA) result.

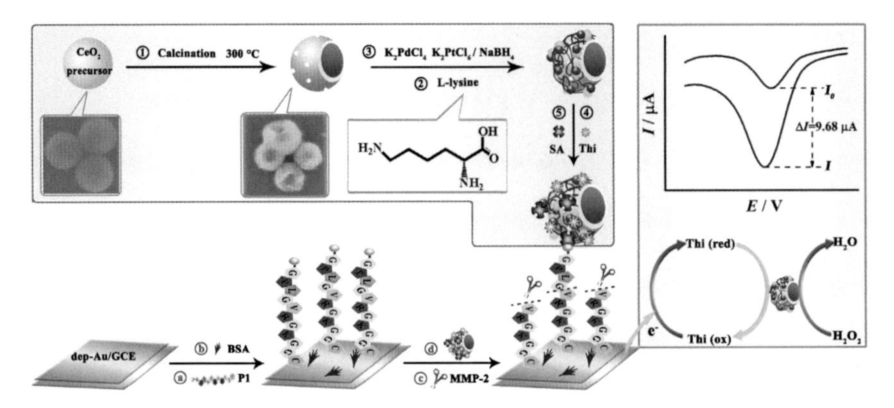

FIGURE 10.7

Fabrication of the peptide-based biosensor for MMP-2 detection; $\Delta I = I - I_0$, where I_0 and I are DPV peak current before and after the addition of H_2O_2.

Reproduced from Xu, W., Jing, P., Yi, H., Xue, S., Yuan, R., 2016. Bimetallic Pt/Pd encapsulated mesoporous-hollow CeO_2 nanospheres for signal amplification toward electrochemical peptide-based biosensing for matrix metalloproteinase 2. Sensors and Actuators B: Chemical 230, 345–352; with permission of Elsevier.

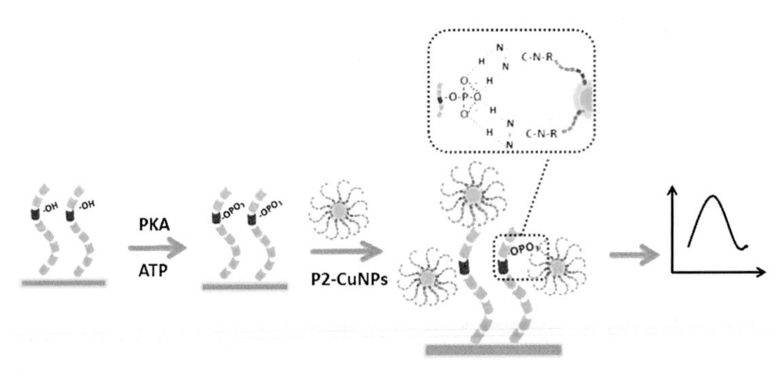

FIGURE 10.8

Schematic illustration of an electrochemical assay of PKA using peptide-templated multifunctional nanoprobe.

Reproduced from Zhao, J., Yang, L., Dai, Y., Tang, Y., Gong, X., Du, D., Cao, Y., 2018. Peptide-templated multifunctional nanoprobe for feasible electrochemical assay of intracellular kinase. Biosensors and Bio-electronics 119, 42–47; with permission of Elsevier.

10.5.3 Alternating current voltammetry and electrochemical impedance spectroscopy

AC voltammetry, similarly to CV, is a potential sweep method (Eckermann et al., 2010). The starting and ending potentials are specified that index the E_0 of the redox species. Additionally, a sinusoidally oscillating AC wave is superimposed on the potential waveform. The frequency of the AC can be varied; the magnitude of the AC oscillations is small compared with the overall voltage variation (Scholz, 2009). The resulting AC is recorded, and the electrochemical response appears as a single peak.

EIS measures the frequency response of a system by measuring the impedance, Z. To do that, a small AC signal is applied over a range of frequencies at a specified potential. Varying the frequency changes the relative contribution of each of the elements in equivalent circuit (Randles circuit model) to the overall impedance. The AC output is a peak-shaped curve like SW- and DPV signals. Compared with these techniques, AC voltammetry and especially EIS are much more time-consuming. Zaitouna and Lai (2014) reported a "signaling-off" peptide-based biosensor for anti-Ara h2 antibodies. The MB-tagged peptide probe was immobilized onto a Ni (II)-nitriloacetic acid (NTA) SAM. In this case, LOD was 1 pM, significantly lower than the LOD obtained for peptide probe immobilized onto thiolated SAMs (200 pM). Xia et al. (2017) reported an EIS signaling-on assay with peptide probes as the receptors of targets and the inducers of gold nanoparticles (AuNPs) assembly on the electrode surface. The human chorionic gonadotropin (hCG) was first detected through binding to its specific peptide ligand. This peptide triggered the aggregation of AuNPs in solution in the absence of hCG. The attachment of hCG onto the surface through the probe-target interaction prevents the peptide to trigger the aggregation of AuNPs, thus leading to an increased charge transfer resistance (Fig. 10.9A). Within this approach, hCG was determined with an LOD of 0.6 mIU mL^{-1}. The Randles circuit is presented in (Fig. 10.9B). Measuring impedances over a wide range of frequencies allows the value of each individual element of the Randles circuit to be determined (Katz and Willner, 2003). Here R_S is the solution resistance between the working and the reference electrode, Q is the monolayer constant phase element (with $Z_{CPE} = 1/Q(i\omega)^n$, ω being the angular frequency), and R_{ET} is the ET resistance (Xia et al., 2017).

Ma et al. (2019) reported two labeled and nonlabeled electrochemical peptide-based formats for the detection of the hemopexin domain of matrix metalloproteinase-14 (PEX-14). Herein, the thiolated PEX-14-binding peptide inhibitor (ISC) or the ISC-tagged with Fc-carboxylic acid (CIS-Fc) was self-assembled onto gold electrode (Fig. 10.10) and checked for interacting with PEX-14 by EIS (label-free assay) or DPV (when ISC was tagged with a redox reporter, Fc). The examination of the analytical performances of the two sensing formats for PEX-14 has shown a lower LOD for the Fc-labeled approach (0.3 pg mL^{-1}) compared with the nonlabeled one (0.03 ng mL^{-1}).

Another label-free assay reports an electrochemical peptide—based sensor for the detection of Dengue virus type 2 NS1 (DENV2 NS1) protein, a specific and sensitive protein biomarker for dengue fever diagnosis (Lim et al., 2018). The NS1-binding peptides were identified by polyvalent phage display (biopanning an M13 random peptide library for DENV2 NS1 protein), and their binding affinities were checked through CV and EIS measurements. Among the tested peptide, the EHDRMHAYYLTR (R3#10) sequence, rich in basic residues, provided the highest decrease of CV output and highest increase in impedance (Fig. 10.11).

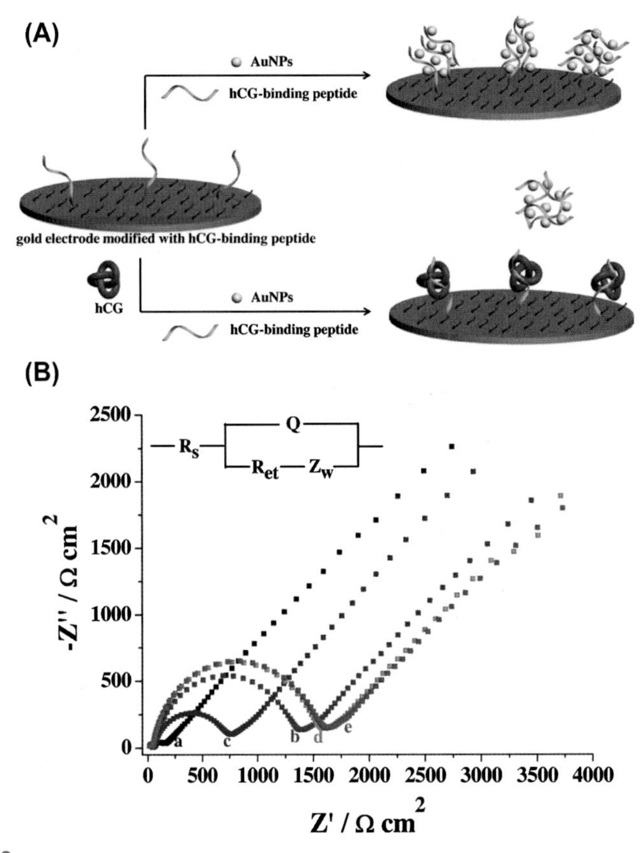

FIGURE 10.9

(A) Schematic representation of a "signaling-on" peptide-based sensor for hCG detection with a peptide probe as the receptor of hCG and the inducer of AuNPs assembly. (B) EIS of the bare electrode (curve a) and peptide/MCH-modified electrode before (curve b) and after (curve c) incubation with AuNPs/peptide. Curve d corresponds to that after incubation with hCG and AuNPs/peptide, and curve e corresponds to that after incubation with hCG only.

Reproduced from Xia, N., Wang, X., Yu, J., Wu, Y., Cheng, S., Xing, Y., Liu, L., 2017. Design of electrochemical biosensors with peptide probes as the receptors of targets and the inducers of gold nanoparticles assembly on electrode surface. Sensors and Actuators B: Chemical 239, 834–840; with permission of Elsevier.

The thermodynamic dissociation constant K_d obtained through EIS measurements was found 3.9 ± 0.09 nM, with an LOD of 0.025 μg mL^{-1}.

10.5.4 Potentiometry

Although this technique based on the measurement of the electromotive force generated in a two-electrode cell as a response of a binding event (Yin and Qin, 2013) is not often encountered in electrochemical peptide–based assay, it is worth

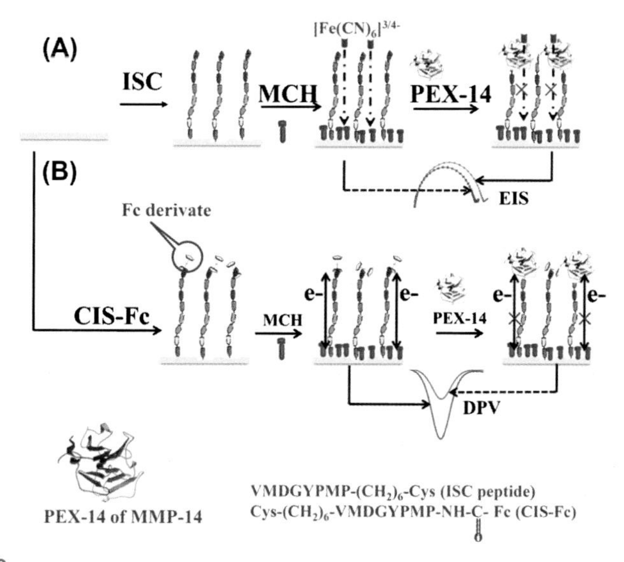

FIGURE 10.10

Two sensing approaches using the same coating procedure of the gold electrode: (A) "signaling-on" detection principle using EIS and (B) "signaling-off" detection principle using DPV and an Fc-tagged peptide probe.

Reproduced from Ma, F., Zhu, Y., Chen, Y., Liu, J., Zeng, X., 2019. Labeled and non-label electrochemical peptide inhibitor-based biosensing platform for determination of hemopexin domain of matrix metalloproteinase-14. Talanta 194, 548–553; with permission of Elsevier.

mentioning the work of Lv et al. (2018) reporting a potentiometric sandwich assay for bacteria involving peptide pairs derived from an AMP ligand.

Ion-selective electrodes (ISEs) with polymeric membranes containing selective carriers (ionophores) are the most used potentiometric sensors. Polymeric liquid membrane containing poly(vinyl chloride) (PVC) and *o*-nitrophenyl octylether (*o*-NPOE) and the cation exchanger sodium tetraphenylborate were used in this assay. An original long peptide, belonging to Leucocin A and possessing activity against *Listeria monocytogenes* (LM) strains, was split into two fragments for a short peptide pair–based sandwich assay. The capture fragment was conjugated to magnetic beads to eliminate the interference from the background solution. Horseradish peroxidase (HRP) was attached to the detection peptide and catalyzed the oxidation of 3,3′,5,5′-tetramethylbenzidine with H_2O_2, thus inducing the potential change on the polymeric membrane ISE (Fig. 10.12).

The magnetic beads and HRP were bound to the split peptide pairs via biotin-SA interaction. In the presence of LM, the two short peptides formed sandwich complex with LM. The conjugates were transferred to the electrochemical cell for potentiometric measurements after the magnetic separation. TMB was oxidized by H_2O_2 in the presence of HRP. The oxidation intermediates (free radicals, charge transfer complexes, and diimine species) were positively charged in a neutral environment.

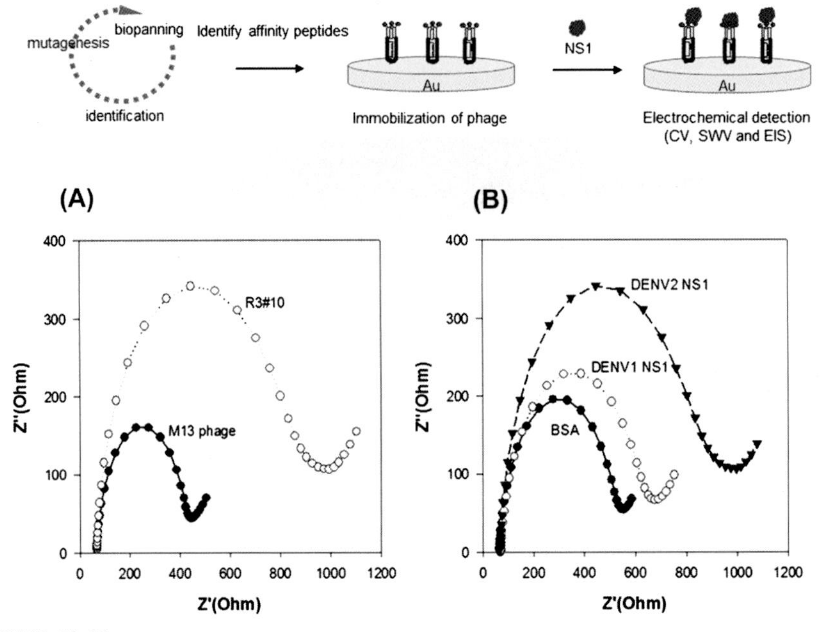

FIGURE 10.11

Schematic representation of an affinity signaling-on sensor for DENV2 NS1 protein (above); Nyquist plot (below) with Z' versus Z'', where Z' and Z'' are real and imaginary parts: (A) increase of R_{et} in the EIS assay with a DENV2 NS1-coated gold electrode previously incubated with EHDRMHAYYLTR (R3#10) phage; (B) selectivity of R3#10-coated electrode toward DENV2 NS1 protein as Ret was significantly lager after sensor's incubation with the target compared with incubation with BSA or Dengue virus type 1 (DENV1NS1) protein.

Reproduced from Lim, J.M., Kim, J.H., Ryu, M.Y., Cho, C.H., Park, T.J., Park, J.P. 2018. An electrochemical peptide sensor for detection of dengue fever biomarker NS1. Analytica Chimica Acta 1026, 109–116; with permission of Elsevier.

The heterogeneous ion exchange of these cationic intermediates in the solution with the Na^+ ions in the membrane induced a large potential response on the cation exchanger–doped polymeric liquid membrane ISE. The detection limit was 10 colony-forming units (CFU)/mL.

10.6 Earnings and drawbacks of electrochemical peptide–based assays

As previously mentioned, in most peptide-based assays, the peptide sequences are tagged either to Fc or MB (Martic et al., 2011) and possess extra functional groups able to covalently or reversibly bind surfaces or coating layers (Wang et al., 2017).

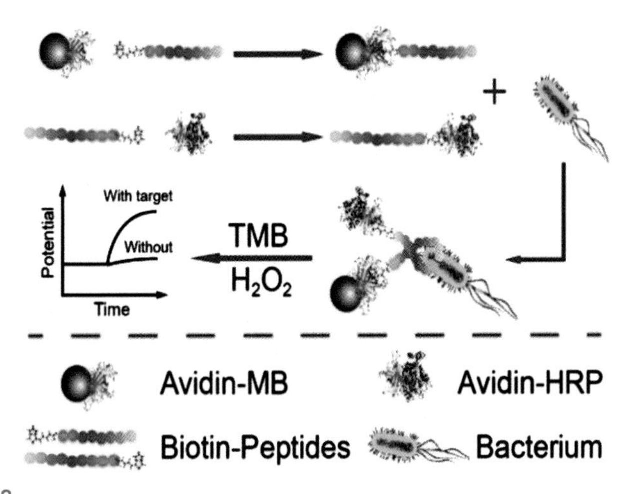

FIGURE 10.12

Schematic illustration of a potentiometric sandwich assay based on short AMP pairs for the detection of *Listeria monocytogenes.* Here MB designates the magnetic beads.

Reproduced from Lv, E., Ding, J., Qin, W., 2018. Potentiometric detection of Listeria monocytogenes via a short antimicrobial peptide pair-based sandwich assay. Analytical Chemistry 90, 13600–13606; with permission of ACS Publications.

Both peptide cleavage and affinity formats use Fc and MB for enzyme and antibodies detection (Gerasimov and Lai, 2010; Puiu et al., 2014). Electrochemical affinity assays are probably the most encountered, compared with the enzymatic cleavage formats, because they involve a smaller number of steps, allow direct detection of analyte in a reagentless manner, and provide reusable and long-term stable coated surfaces. However, the main disadvantages of these approaches are fluctuating voltammetric baseline and capacitance current drifting (because the secondary structure of the peptide chain cannot be controlled as in the case of short DNA oligomers (Puiu et al., 2014)) Moreover, the conjugation of a redox reporter often modifies the specific binding affinity. On the other hand, the enzymatic cleavage formats overcome these drawbacks because the proteolytic activity of the target enzymes is rarely influenced by the attachment of the redox reporter to the free end of the immobilized peptide probe.

Functionalization of peptide nanomaterials with highly ordered and conductive nanostructures, such as graphene, is the advanced tool for signal amplification in typical electrochemical assays (Wang et al., 2017). Particularly, graphene was successfully used both as support for peptide immobilization and as catalyst for AgNPs formation. Stripping voltammetric techniques involving anodic oxidation of AgNPs and CuNPs deposited onto electrode as a result of a binding event are promising tools, which may replace soon the classical redox-labeled peptide assays. A brief overview of the latest findings in electrochemical peptide based assays is summarized in Table 10.1.

Table 10.1 Electrochemical peptide–based sensors with peptide elements for relevant biomolecules.

Target biomolecule	Assay principle	Peptide sequence/ function	Redox reporter	Electrochemical method/format	Analytical performance	Electrode	References
Matrix metalloproteinase-9 (MMP-9)	Enzymatic cleavage	GPLGMWSRC/ probe	MB	CV/signalling-on	Linear range: 1 pM–1 nM LOD = 7 pM	Au	Lee et al. (2017)
Prostate-specific antigen	Affinity cleavage	CGGHSSKL QFWYFWY/probe	AgNPs	LSV/signaling-off	Linear range: $5–2 \times 10^{4}$ pg mL^{-1} LOD = 0.33 pg mL^{-1}	Au modified with mixed thiol/ peptide/GO	Meng et al. (2018)
Matrix metalloproteinase-2 (MMP-2)	Enzymatic cleavage	GPLGVRGK GGC/probe	Thi	DPV/signaling-on	Linear range: 0.1 pg mL^{-1}–10 ng mL^{-1} LOD = 0.078 pg mL^{-1}	GSE with deposited AuNP	Xu et al. (2016)
Matrix metalloproteinase-14 (PEX-14)	Affinity	VMDGYPMP/probe	Fc	DPV/signaling-off	Dynamic range: 1–10 pg mL^{-1} LOD = 0.3 pg mL^{-1}	Au modified with mixed thiol/peptide SAM	Ma et al. (2019)
Prostate-specific antigen	Affinity cleavage	HSSKLQL/probe	Fc	DPV/signaling-on	Linear range: 0.001–30 ng mL^{-1} LOD = 0.78 pg mL^{-1}	GCE modified with CNT-polyamidoamine dendrimers	Xie et al. (2015)
Epidermal growth factor receptor	Affinity	YHWYGYTP QNVI/probe	Fc	DPV/signaling-off	Linear range: $1 \times 10^{-10}–1 \times 10^{-6}$ $g \cdot L^{-1}$ LOD = 3.7×10^{-11} $g \cdot L^{-1}$.	Au modified with mixed thiol/peptide SAM	Li et al. (2013)
Protein kinase (PKA)	Enzymatic phosphorylation	CLRRASLG/probe CCYGGVLRR/ support	CuNps	DPV/signaling-on	Linear range: 1.1–50 $U \cdot mL^{-1}$ 0.0019 $U \cdot mL^{-1}$	Au	Zhao et al. (2018)
Prostate-specific antigen	Affinity cleavage	HSSKLQL/probe	2-Hydroxy-naphthoquinone	SWV/signaling-on	Dynamic range: 1 pM–1 nM	GCE modified with 4-azidoaniline	Strzemińska et al. (2016)
Antigliadin peptide antibody	Affinity	QLQPFPQPELPY PQPQLPYPQPQLP YPPQPQPF /probe YAAAHAEAR/ support	MB	SWV/signaling-off	Dynamic range: 0.22–6.7 nM	Au modified with mixed peptide SAM	Puiu et al. (2014)

Trypsin	Enzymatic cleavage	FRR/probe	MB	SWV/signaling-off	Dynamic range: 0.1–100 nM LOD = 88 pM	Au modified with thiolated PEG	González-Fernández et al. (2018a)
Protein tyrosine kinase (PTK)	Enzymatic phosphorylation	EGI**Y**DVP/probe	AuNPs	SWV/signaling-on	LOD = 10 ng mL^{-1}	Indium tin oxide modified with streptavidin	Kerman and Kraatz (2009)
Caspase 3	Enzymatic cleavage	GDGDEVDGC/probe	HRP	SWV/signaling-off	Dynamic range: 100 pM–1 nM LOD = 100 pM	Au modified with streptavidin-coated magnetic beads	Khalilzadeh et al. (2015)
Amyloid β-1–42 soluble oligomer	Affinity	RGTWEGKWK	Fc	SWV/signaling-on	Dynamic range: 480 pM–12 nM LOD = 240 pM	Au modified with thiol SAM	Li et al. (2012)
Anti-HIV antibody	Affinity	EAAEWDRVHP-K/probe	MB	ACV/signaling-off	LOD = 0.5 nM	Au modified with mixed thiol/peptide SAM	Zaitouna et al. (2015)
Human chorionic gonadotropin	Affinity	CTHSQWNKPS KPKTNMK/probe	–	EIS/signaling-on	Linear range: 0.001–0.2 IU mL^{-1} LOD = 0.6 m IU mL^{-1}	Au modified with mixed thiol/peptide SAM	Xia et al. (2017)
SW620 colorectal carcinoma cells	Affinity	DDAGNRQP/probe	–	EIS/signaling-on	Linear range: 2×10^2–2×10^8 cells/mL LOD = 79 cells/mL	Au	Han et al. (2016)
Matrix metalloproteinase-14 (PEX-14)	Affinity	VMDGYPMP/probe	–	EIS/signaling-on	Dynamic range: 0.1–7 ng mL^{-1} LOD = 0.03 ng mL^{-1}	Au modified with mixed thiol/peptide SAM	Ma et al. (2019)
Listeria monocytogenes	Affinity	NWGEAFSA/probe GVHRLANG/capture	TMB oxidation products	Potentiometry/signalling-on	Linear range: 1×10^2–1×10^6 CFU mL^{-1}	Ag ISE with PVC membrane	Lv et al. (2018)

Reproduced and adapted from Puiu, M., Bala, C., 2018a. Building switchable peptide-architectures on gold/composite surfaces: new perspectives in electrochemical bioassays. Current Opinion in Electrochemistry; with permission of Elsevier.

10.7 Conclusions and future trends

Peptide-based sensors display a net superior shelf life compared with antibodies or other proteins because of intrinsic peptide stability in harsh environments (Chen et al., 2016). Moreover, peptides are easy and costly efficient to synthesize through standard solid-phase synthesis, once the sequence information is isolated from the phage display (Liu et al., 2015; Pavan and Berti, 2012). Still, most peptide-based detection formats require the attachment of a redox/fluorescent reporter to the peptide backbone, as peptides do not directly generate a measurable signal in response to a binding event. Label-free detection involving peptide ligands/receptors is a task not yet achieved. In many cases, the binding affinity of several peptides functioning as recognition elements is not particularly selective toward structurally related targets, requiring advanced optimization steps to reduce interferences (Chen et al., 2016). The ability of peptides to decorate various nanomaterials displaying complex biological or catalytic functions seems boundless. We will probably see soon peptide sequences or peptide nanohybrids incorporated in wearable and disposable paper electrochemical devices for high-throughput analysis for point-of-care diagnosis.

Acknowledgments

This work was supported by a grant of Ministry of Research and Innovation, CNCS—UEFISCDI, project number PN-III-P4-ID-PCE-2016-0288.

References

Adler-Abramovich, L., Gazit, E., 2014. The physical properties of supramolecular peptide assemblies: from building block association to technological applications. Chemical Society Reviews 43, 6881–6893.

Arduini, F., Amine, A., Moscone, D., Palleschi, G., 2010. Biosensors based on cholinesterase inhibition for insecticides, nerve agents and aflatoxin B1 detection (review). Microchimca Acta 170, 193–214.

Bard, A.J., Faulkner, L.R., 2000. Electrochemical Methods: Fundamentals and Applications, second ed. John Wiley & Sons.

Bezer, S., Matsumoto, M., Lodewyk, M.W., Lee, S.J., Tantillo, D.J., Gagne, M.R., Waters, M.L., 2014. Identification and optimization of short helical peptides with novel reactive functionality as catalysts for acyl transfer by reactive tagging. Organic and Biomolecular Chemistry 12, 1488–1494.

Bianchi, R.C., DA Silva, E.R., Dall'Antonia, L.H., Ferreira, F.F., Alves, W.A., 2014. A nonenzymatic biosensor based on gold electrodes modified with peptide self-assemblies for detecting ammonia and urea oxidation. Langmuir 30, 11464–11473.

Chaudhry, B.R., Wilton-Ely, J.D.E.T., Tabor, A.B., Caruana, D.J., 2010. Effect of peptide orientation on electron transfer. Physical Chemistry Chemical Physics 12, 9996–9998.

Chen, Y.X.Y., Wu, J., Yin, B., Jiang, X., 2016. Click Chemistry-Mediated Nanosensors for Biochemical Assays. Theranostics 6, 969–985.

Chen, H., Huang, J., Palaniappan, A., Wang, Y., Liedberg, B., Platt, M., Tok, A.I.Y., 2016. A review on electronic bio-sensing approaches based on non-antibody recognition elements. Analyst 141, 2335–2346.

Cheng, S., Hideshima, S., Kuroiwa, S., Nakanishi, T., Osaka, T., 2015. Label-free detection of tumor markers using field effect transistor (FET)-based biosensors for lung cancer diagnosis. Sensors and Actuators B: Chemical 212, 329–334.

Ding, Y., Li, D., Li, B., Zhao, K., Du, W., Zheng, J., Yang, M., 2013. A water-dispersible, ferrocene-tagged peptide nanowire for amplified electrochemical immunosensing. Biosensors and Bioelectronics 48, 281–286.

Eckermann, A.L., Feld, D.J., Shaw, J.A., Meade, T.J., 2010. Electrochemistry of redox-active self-assembled monolayers. Coordination Chemistry Reviews 254, 1769–1802.

Ener, M.E., Gray, H.B., Winkler, J.R., 2017. Hole hopping through tryptophan in cytochrome P450. Biochemistry 56, 3531–3538.

Feldberg, S.W., 2010. Implications of Marcus–Hush theory for steady-state heterogeneous electron transfer at an inlaid disk electrode. Analytical Chemistry 82, 5176–5183.

Finklea, H.O., Hanshew, D.D., 1992. Electron-transfer kinetics in organized thiol monolayers with attached pentaammine(pyridine)ruthenium redox centers. Journal of the American Chemical Society 114, 3173–3181.

Flavel, B.S., Yu, J., Ellis, A.V., Shapter, J.G., 2009. Electroless plated gold as a support for carbon nanotube electrodes. Electrochimica Acta 54, 3191–3198.

Gatto, E., Porchetta, A., Scarselli, M., DE Crescenzi, M., Formaggio, F., Toniolo, C., Venanzi, M., 2012. Playing with peptides: how to build a supramolecular peptide nanostructure by exploiting Helix···Helix macrodipole interactions. Langmuir 28, 2817–2826.

Gatto, E., Venanzi, M., Palleschi, A., Stella, L., Pispisa, B., Lorenzelli, L., Toniolo, C., Formaggio, F., Marletta, G., 2007. Self-assembled peptide monolayers on interdigitated gold microelectrodes. Materials Science and Engineering: C 27, 1309–1312.

Gazit, E., 2007. Self-assembled peptide nanostructures: the design of molecular building blocks and their technological utilization. Chemical Society Reviews 36, 1263–1269.

Gerasimov, J.Y., Lai, R.Y., 2010. An electrochemical peptide-based biosensing platform for HIV detection. Chemical Communications 46, 395–397.

Gerasimov, J.Y., Lai, R.Y., 2011. Design and characterization of an electrochemical peptide-based sensor fabricated via "click" chemistry. Chemical Communications 47, 8688–8690.

Gobbo, P., Antonello, S., Guryanov, I., Polo, F., Soldà, A., Zen, F., Maran, F., 2016. Dipole moment effect on the electrochemical desorption of self-assembled monolayers of 310-helicogenic peptides on gold. ChemElectroChem 3, 1964.

González-Fernández, E., Staderini, M., Avlonitis, N., Murray, A.F., Mount, A.R., Bradley, M., 2018a. Effect of spacer length on the performance of peptide-based electrochemical biosensors for protease detection. Sensors and Actuators B: Chemical 255, 3040–3046.

González-Fernández, E., Staderini, M., Yussof, A., Scholefield, E., Murray, A.F., Mount, A.R., Bradley, M., 2018b. Electrochemical sensing of human neutrophil elastase and polymorphonuclear neutrophil activity. Biosensors and Bioelectronics 119, 209–214.

Gray, H.B., Winkler, J.R., 2005. Long-range electron transfer. Proceedings of the National Academy of Sciences of the United States of America 102, 3534–3539.

Groß, A., Hashimoto, C., Sticht, H., Eichler, J., 2016. Synthetic peptides as protein mimics. Frontiers in Bioengineering and Biotechnology 3, 1−16.

Guida, F., Battisti, A., Gladich, I., Buzzo, M., Marangon, E., Giodini, L., Toffoli, G., Laio, A., Berti, F., 2018. Peptide biosensors for anticancer drugs: design in silico to work in denaturizing environment. Biosensors and Bioelectronics 100, 298−303.

Guidotti, G., Brambilla, L., Rossi, D., 2017. Cell-penetrating peptides: from basic Research to clinics. Trends in Pharmacological Sciences 38, 406−424.

Han, L., Liu, P., Petrenko, V.A., Liu, A., 2016. A label-free electrochemical impedance cytosensor based on specific peptide-fused phage selected from landscape phage library. Scientific Reports 6, 22199.

Hansen, J.A., Sumbayev, V.V., Gothelf, K.V., 2007. An electrochemical sensor based on the human estrogen receptor ligand binding domain. Nano Letters 7, 2831−2834.

He, X., Zhu, Q., Liao, F., Zhu, L., Ai, Z., 2007. Differential pulse voltammetric determination and application of square-wave voltammetry of yRNA on a CPB-cellulose modified electrode. Electroanalysis 19, 1375−1381.

Heck, A., Woiczikowski, P.B., Kubař, T., Giese, B., Elstner, M., Steinbrecher, T.B., 2012. Charge transfer in model peptides: obtaining Marcus parameters from molecular simulation. The Journal of Physical Chemistry B 116, 2284−2293.

Hoyos-Nogués, M., Brosel-Oliu, S., Abramova, N., Muñoz, F.-X., Bratov, A., MAS-Moruno, C., Gil, F.-J., 2016. Impedimetric antimicrobial peptide-based sensor for the early detection of periodontopathogenic bacteria. Biosensors and Bioelectronics 86, 377−385.

Jampasa, S., Wonsawat, W., Rodthongkum, N., Siangproh, W., Yanatatsaneejit, P., Vilaivan, T., Chailapakul, O., 2014. Electrochemical detection of human papillomavirus DNA type 16 using a pyrrolidinyl peptide nucleic acid probe immobilized on screen-printed carbon electrodes. Biosensors and Bioelectronics 54, 428−434.

Juhaniewicz, J., Pawlowski, J., Sek, S., 2015. Electron transport mediated by peptides immobilized on surfaces. Israel Journal of Chemistry 55, 645−660.

Juhaniewicz, J., Sek, S., 2012. Peptide molecular junctions: distance dependent electron transmission through oligoprolines. Bioelectrochemistry 87, 21−27.

Kai, M., Takeda, K., Morita, T., Kimura, S., 2008. Distance dependence of long-range electron transfer through helical peptides. Journal of Peptide Science 14, 192−202.

Kalafatovic, D., Giralt, E., 2017. Cell-penetrating peptides: design strategies beyond primary structure and amphipathicity. Molecules 22, 1929.

Karimzadeh, A., Hasanzadeh, M., Shadjou, N., Guardia, M.D.L., 2018. Peptide based biosensors. Trends in Analytical Chemistry 107, 1−20.

Katz, E., Willner, I., 2003. Probing biomolecular interactions at conductive and semiconductive surfaces by impedance spectroscopy: routes to impedimetric immunosensors, DNA-sensors, and enzyme biosensors. Electroanalysis 15, 913−947.

Ke, X., Xenia, M., Barbara, M.N., Eugene, Z., Mitra, D., Michael, A.S., 2014. Graphene- and aptamer-based electrochemical biosensor. Nanotechnology 25, 205501.

Kerman, K., Kraatz, H.-B., 2009. Electrochemical detection of protein tyrosine kinase-catalysed phosphorylation using gold nanoparticles. Biosensors and Bioelectronics 24, 1484−1489.

Khalilzadeh, B., Shadjou, N., Eskandani, M., Charoudeh, H.N., Omidi, Y., Rashidi, M.-R., 2015. A reliable self-assembled peptide based electrochemical biosensor for detection of caspase 3 activity and apoptosis. RSC Advances 5, 58316−58326.

Kitagawa, K., Morita, T., Kimura, S., 2006. Electron transport properties of helical peptide dithiol at a molecular level: scanning tunneling microscope study. Thin Solid Films 509, 18–26.

Lakshmanan, A., Zhang, S., Hauser, C.A.E., 2012. Short self-assembling peptides as building blocks for modern nanodevices. Trends in Biotechnology 30, 155–165.

Lauz, M., Eckhardt, S., Fromm, K.M., Giese, B., 2012. The influence of dipole moments on the mechanism of electron transfer through helical peptides. Physical Chemistry Chemical Physics 14, 13785–13788.

Lee, J., Yun, J.Y., Lee, W.C., Choi, S., Lim, J., Jeong, H., Shin, D.-S., Park, Y.J., 2017. A reference electrode-free electrochemical biosensor for detecting MMP-9 using a concentric electrode device. Sensors and Actuators B: Chemical 240, 735–741.

Lequoy, P., Murschel, F., Liberelle, B., Lerouge, S., de Crescenzo, G., 2016. Controlled co-immobilization of EGF and VEGF to optimize vascular cell survival. Acta Biomaterialia 29, 239–247.

Li, H., Cao, Y., Wu, X., Ye, Z., Li, G., 2012. Peptide-based electrochemical biosensor for amyloid β 1–42 soluble oligomer assay. Talanta 93, 358–363.

Li, R., Huang, H., Huang, L., Lin, Z., Guo, L., Qiu, B., Chen, G., 2013. Electrochemical biosensor for epidermal growth factor receptor detection with peptide ligand. Electrochimica Acta 109, 233–237.

Li, Y., Afrasiabi, R., Fathi, F., Wang, N., Xiang, C., Love, R., She, Z., Kraatz, H.-B., 2014. Impedance based detection of pathogenic *E. coli* O157:H7 using a ferrocene-antimicrobial peptide modified biosensor. Biosensors and Bioelectronics 58, 193–199.

Lim, J.M., Kim, J.H., Ryu, M.Y., Cho, C.H., Park, T.J., Park, J.P., 2018. An electrochemical peptide sensor for detection of dengue fever biomarker NS1. Analytica Chimica Acta 1026, 109–116.

Liu, Q., Boyd, B.J., 2013. Liposomes in biosensors. Analyst 138, 391–409.

Liu, Q., Wang, J., Boyd, B.J., 2015. Peptide-based biosensors. Talanta 136, 114–127.

Long, Y.-T., abu-Irhayem, E., Kraatz, H.-B., 2005. Peptide electron transfer: more questions than answers. Chemistry - A European Journal 11, 5186–5194.

Loo, Y., Zhang, S., Hauser, C.A.E., 2012. From short peptides to nanofibers to macromolecular assemblies in biomedicine. Biotechnology Advances 30, 593–603.

Lowik, D.W.P.M., Leunissen, E.H.P., van den Heuvel, M., Hansen, M.B., van Hest, J.C.M., 2010. Stimulus responsive peptide based materials. Chemical Society Reviews 39, 3394–3412.

Lv, E., Ding, J., Qin, W., 2018. Potentiometric detection of Listeria monocytogenes via a short antimicrobial peptide pair-based sandwich assay. Analytical Chemistry 90, 13600–13606.

Ma, F., Zhu, Y., Chen, Y., Liu, J., Zeng, X., 2019. Labeled and non-label electrochemical peptide inhibitor-based biosensing platform for determination of hemopexin domain of matrix metalloproteinase-14. Talanta 194, 548–553.

Mandal, H.S., Kraatz, H.-B., 2006. Electron transfer across α-helical peptides: potential influence of molecular dynamics. Chemical Physics 326, 246–251.

Mandal, H.S., Kraatz, H.-B., 2012. Electron transfer mechanism in helical peptides. The Journal of Physical Chemistry Letters 3, 709–713.

Marcus, R.A., Sutin, N., 1985. Electron transfers in chemistry and biology. Biochimica et Biophysica Acta (BBA) - Reviews on Bioenergetics 811, 265–322.

Martic, S., Labib, M., Shipman, P.O., Kraatz, H.-B., 2011. Ferrocene-peptido conjugates: from synthesis to sensory applications. Dalton Transactions 40, 7264–7290.

Meng, F., Sun, H., Huang, Y., Tang, Y., Chen, Q., Miao, P., 2019. Peptide cleavage-based electrochemical biosensor coupling graphene oxide and silver nanoparticles. Analytica Chimica Acta 1047, 45–51.
Mirceski, V., Komorsky-Lovric, S., Lovric, M., 2007. Square-Wave Voltammetry: Theory and Application. Springer Science & Business Media, pp. 1–6.
Moore, K.E., Flavel, B.S., Yu, J., Abell, A.D., Shapter, J.G., 2013. Increased redox-active peptide loading on carbon nanotube electrodes. Electrochimica Acta 89, 206–211.
Nitzan, A., 2001. Electron transmission through molecules and molecular interfaces. Annual Review of Physical Chemistry 52, 681–750.
Okamoto, S., Morita, T., Kimura, S., 2009. Electron transfer through a self-assembled monolayer of a double-helix peptide with linking the terminals by ferrocene. Langmuir 25, 3297–3304.
Orlowski, G.A., Chowdhury, S., Kraatz, H.-B., 2007. Reorganization energies of ferrocene-peptide monolayers. Langmuir 23, 12765–12770.
Osteryoung, J.G., Osteryoung, R.A., 1985. Square wave voltammetry. Analytical Chemistry 57, 101A–110A.
Palaniappan, A., Goh, W.H., Fam, D.W.H., Rajaseger, G., Chan, C.E.Z., Hanson, B.J., Moochhala, S.M., Mhaisalkar, S.G., Liedberg, B., 2013. Label-free electronic detection of bio-toxins using aligned carbon nanotubes. Biosensors and Bioelectronics 43, 143–147.
Pavan, S., Berti, F., 2012. Short peptides as biosensor transducers. Analytical and Bioanalytical Chemistry 402, 3055–3070.
Pepe-Mooney, B.J., Fairman, R., 2009. Peptides as materials. Current Opinion in Structural Biology 19, 483–494.
Polo, F., Antonello, S., Formaggio, F., Toniolo, C., Maran, F., 2005. Evidence against the hopping mechanism as an important electron transfer pathway for conformationally constrained oligopeptides. Journal of the American Chemical Society 127, 492–493.
Puiu, M., Bala, C., 2018a. Building switchable peptide-architectures on gold/composite surfaces: new perspectives in electrochemical bioassays. Current Opinion in Electrochemistry.
Puiu, M., Bala, C., 2018b. Peptide-based biosensors: from self-assembled interfaces to molecular probes in electrochemical assays. Bioelectrochemistry 120, 66–75.
Puiu, M., Idili, A., Moscone, D., Ricci, F., Bala, C., 2014. A modular electrochemical peptide-based sensor for antibody detection. Chemical Communications 50, 8962–8965.
Reeves, J.H., Song, S., Bowden, E.F., 1993. Application of square wave voltammetry to strongly adsorbed quasireversible redox molecules. Analytical Chemistry 65, 683–688.
Scholz, F., 2009. Electroanalytical Methods: Guide to Experiments and Applications. Springer Berlin Heidelberg.
Shinohara, H., Kuramitz, H., Sugawara, K., 2015. Design of an electroactive peptide probe for sensing of a protein. Analytica Chimica Acta 890, 143–149.
Siddarth, P., Marcus, R.A., 1990. Comparison of experimental and theoretical electronic matrix elements for long-range electron transfer. Journal of Physical Chemistry 94, 2985–2989.
Sisido, M., Hoshino, S., Kusano, H., Kuragaki, M., Makino, M., Sasaki, H., Smith, T.A., Ghiggino, K.P., 2001. Distance dependence of photoinduced electron transfer along α-helical polypeptides. The Journal of Physical Chemistry B 105, 10407–10415.

Smith, A.M., Acquah, S.F.A., Bone, N., Kroto, H.W., Ryadnov, M.G., Stevens, M.S.P., Walton, D.R.M., Woolfson, D.N., 2005. Polar assembly in a designed protein fiber. Angewandte Chemie International Edition 44, 325–328.

Smith, K.H., Tejeda-Montes, E., Poch, M., Mata, A., 2011. Integrating top-down and self-assembly in the fabrication of peptide and protein-based biomedical materials. Chemical Society Reviews 40, 4563–4577.

Speiser, B., 2007. Linear Sweep and Cyclic Voltammetry. Encyclopedia of Electrochemistry. Wiley Online Library.

Strzemińska, I., Sainte Rose Fanchine, S., Anquetin, G., Reisberg, S., Noël, V., Pham, M.C., Piro, B., 2016. Grafting of a peptide probe for prostate-specific antigen detection using diazonium electroreduction and click chemistry. Biosensors and Bioelectronics 81, 131–137.

Ulijn, R.V., Smith, A.M., 2008. Designing peptide based nanomaterials. Chemical Society Reviews 37, 664–675.

Vieira-Pires, R.S., Morais-Cabral, J.H., 2010. 3_{10} helices in channels and other membrane proteins. The Journal of General Physiology 136, 585–592.

Wain, A.J., do, H.N.L., Mandal, H.S., Kraatz, H.-B., Zhou, F., 2008. Influence of molecular dipole moment on the redox-induced reorganization of α-helical peptide self-assembled monolayers: an electrochemical SPR investigation. Journal of Physical Chemistry C 112, 14513–14519.

Wang, L., Zhang, Y., Wu, A., Wei, G., 2017. Designed graphene-peptide nanocomposites for biosensor applications: a review. Analytica Chimica Acta 985, 24–40.

Wei, G., Su, Z., Reynolds, N.P., Arosio, P., Hamley, I.W., Gazit, E., Mezzenga, R., 2017. Self-assembling peptide and protein amyloids: from structure to tailored function in nanotechnology. Chemical Society Reviews 46, 4661–4708.

Wong, L.S., Khan, F., Micklefield, J., 2009. Selective covalent protein immobilization: strategies and applications. Chemical Reviews 109, 4025–4053.

Xia, N., Wang, X., Yu, J., Wu, Y., Cheng, S., Xing, Y., Liu, L., 2017. Design of electrochemical biosensors with peptide probes as the receptors of targets and the inducers of gold nanoparticles assembly on electrode surface. Sensors and Actuators B: Chemical 239, 834–840.

Xie, S., Zhang, J., Yuan, Y., Chai, Y., Yuan, R., 2015. An electrochemical peptide cleavage-based biosensor for prostate specific antigen detection via host-guest interaction between ferrocene and [small beta]-cyclodextrin. Chemical Communications 51, 3387–3390.

Xu, W., Jing, P., Yi, H., Xue, S., Yuan, R., 2016. Bimetallic Pt/Pd encapsulated mesoporous-hollow CeO_2 nanospheres for signal amplification toward electrochemical peptide-based biosensing for matrix metalloproteinase 2. Sensors and Actuators B: Chemical 230, 345–352.

Yin, T., Qin, W., 2013. Applications of nanomaterials in potentiometric sensors. TrAC Trends in Analytical Chemistry 51, 79–86.

Yogeswaran, U., Chen, S.-M., 2008. A review on the electrochemical sensors and biosensors composed of nanowires as sensing material. Sensors 8, 290.

Yu, H.-D., Regulacio, M.D., Ye, E., Han, M.-Y., 2013. Chemical routes to top-down nanofabrication. Chemical Society Reviews 42, 6006–6018.

Zaitouna, A.J., Lai, R.Y., 2011. Design and characterization of a metal ion-imidazole self-assembled monolayer for reversible immobilization of histidine-tagged peptides. Chemical Communications 47, 12391–12393.

Zaitouna, A.J., Lai, R.Y., 2014. An electrochemical peptide-based Ara h 2 antibody sensor fabricated on a nickel(II)-nitriloacetic acid self-assembled monolayer using a His-tagged peptide. Analytica Chimica Acta 828, 85–91.

Zaitouna, A.J., Maben, A.J., Lai, R.Y., 2015. Incorporation of extra amino acids in peptide recognition probe to improve specificity and selectivity of an electrochemical peptide-based sensor. Analytica Chimica Acta 886, 157–164.

Zhang, L., Hermans, J., 1994. 310 helix versus .alpha.-Helix: a molecular dynamics study of conformational preferences of aib and alanine. Journal of the American Chemical Society 116, 11915–11921.

Zhang, M., Zhao, J., Yang, H., Liu, P., Bu, Y., 2013. 310-Helical peptide acting as a dual relay for charge-hopping transfer in proteins. The Journal of Physical Chemistry B 117, 6385–6393.

Zhang, W., Yu, X., Li, Y., Su, Z., Jandt, K.D., Wei, G., 2017. Protein-mimetic peptide nanofibers: motif design, self-assembly synthesis, and sequence-specific biomedical applications. Progress in Polymer Science.

Zhao, J., Yang, L., Dai, Y., Tang, Y., Gong, X., Du, D., Cao, Y., 2018. Peptide-templated multifunctional nanoprobe for feasible electrochemical assay of intracellular kinase. Biosensors and Bioelectronics 119, 42–47.

Zoski, C.G. (Ed.), 2007. Handbook of Electrochemistry. Elsevier.

CHAPTER

11 Receptor-based electrochemical biosensors for the detection of contaminants in food products

Valérie Gaudin, PhD

Anses, Laboratory of Fougeres, European Union Reference Laboratory (EU-RL) for Antimicrobial and Dye Residue Control in Food-Producing Animals, Fougeres, France

11.1 Introduction

11.1.1 Context

Animal- and plant-derived food products could be contaminated by chemical (e.g., in milk (Nag, 2010a)) and bacteriological contaminants (i.e., foodborne pathogens). One definition of contaminant is "any substance not intentionally added to food which is present in such food as a result of the production (including operations carried out in crop husbandry, animal husbandry, and veterinary medicine), manufacture, processing, preparation, treatment, packing, packaging, transport or holding of such food, or as a result of environmental contamination" (1993) (Fig. 11.1). We also integrated intentionally added chemical substances such as melamine. Animals and plants could be contaminated through environmental pollution (e.g., air, water, soils), animal feed (e.g., seeds, fodder), veterinary applications (e.g., veterinary drug residues), or directly by utensils/installations (e.g., detergents and disinfectants used in milk plants for sanitation), food processing and packaging (e.g., acrylamide (Lambert et al., 2018), polycyclic aromatic hydrocarbons (PAHs) (Raza and Kim, 2018)). The contamination could occur all along the food chain, "from farm to fork."

In certain environmental conditions (i.e., moisture content, temperature, aeration), crops could be infested by molds (e.g., *Aspergillus*), which produced mycotoxins (i.e., aflatoxins, ochratoxins, citrinin, zearelanone, and trichothecenes). Feed and fodder are the main sources of contamination by mycotoxins of milk and milk products, for example (Nag, 2010b). The potential health effects of aflatoxins on human are carcinogenic, nephrotoxic, immunotoxic, teratogenic, neurotoxic, and genotoxic effects (Sakin et al., 2018). Phycotoxins (i.e., marine toxins, cyanobacterial toxins) are toxins produced by algae, especially phytoplankton,

Electrochemical Biosensors. https://doi.org/10.1016/B978-0-12-816491-4.00011-5

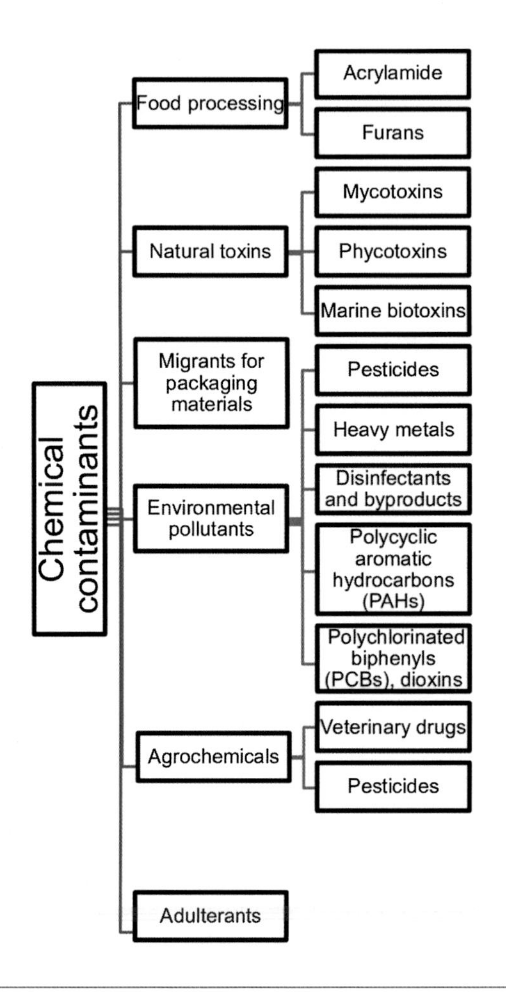

FIGURE 11.1

Sources of chemical contaminants in foodstuffs.

which could contaminate shellfish and lead to human intoxications (Alarcan et al., 2018).

Migration of elements from food packaging could also be responsible for the presence of chemical contaminants in food products (e.g., bisphenol A (BPA), phthalates, heavy metals (cobalt, lead)) (Karmaus et al., 2018; Szczepańska et al., 2018).

Human industrial activities (e.g., industrial effluents, waste chemicals, inorganic fertilizers) are also responsible for the presence of chemical contaminants (e.g., dioxins, polychlorobiphenyls (PCBs), heavy metals) in food products (Arsi and Donoghue, 2017). An Italian study (2013–2016) showed that fish, food of plant origin, and cheese are the most contributing food products to human food intake of dioxins and dioxin-like PCB (Diletti et al., 2018). Human activities could impact aquaculture production by contaminating aquaculture water (Lai et al., 2018). The opposite is also true that antibiotic use in aquaculture production could lead to contamination of environmental water. The transfer of trace elements from soils and air pollution to plants leads to the contamination of plants intended for human and animal's consumption. Chemical contaminants can accumulate in plants and in animal bodies (e.g., lipid-soluble contaminants such as pesticides accumulate in animal fat tissues). Many environmental contaminants (e.g., BPA, organochlorine pesticides, heavy metals (i.e., mercury, lead, cadmium)) could have reproductive and endocrine disruptor's effect (e.g., decrease in the fertility rate, sex inversion) (Aggerbeck and Blanc, 2018; Wielogórska et al., 2015).

Extensive agriculture (i.e., pesticides) and animal husbandry (i.e., veterinary drugs) could lead to the presence of chemical contaminants in food products from animal and plant origins. The use of veterinary drugs is necessary to treat animal diseases, especially in extensive animal husbandry. The administration of drugs to animals leads to the presence of residues in the animal body and therefore in animal by-products (e.g., meat, milk, eggs, honey) (Fischer et al., 2016). The contamination could also be due to misuse in farms or even frauds (e.g., banned substances), to contamination of soils (e.g., soils amended with composted manure (Albero et al., 2018)). The contamination of manure by antibiotic residues could contribute to the development of antimicrobial resistance of bacteria present in the manure (Zhang et al., 2018). Pesticides that are able to accumulate in environment and in living animals represent a major risk for human safety (Liu et al., 2013b). When the raw product (e.g., milk) is contaminated, the contaminants could be transferred to the transformed products (e.g., cheese) (Gajda et al., 2018). The risks for the consumers' health of veterinary drug and biocide residues are multiple: toxicological risks (e.g., carcinogenic effect), allergic risks (e.g., penicillins), antimicrobial resistance (Gaglio et al., 2016), and biocide resistance of bacteria (Fernández Márquez et al., 2017).

Some food additives are intentionally used for the food preservation (e.g., dairy products and especially cheeses). Preservatives act as antimicrobial and antifungal agents (e.g., parabens). Some are not authorized in Europe, and their misuse leads to contamination of milk and cheese (Scordino et al., 2018). Urea and melamine are potential adulterants, which are used to increase the nitrogen content (Migliorini et al., 2018). The crisis of melamine in China in 2008 is an example of adulteration of infant-formula milk, which led to the death of several children due to neurotoxic effect (Wei and Liu, 2012).

The contamination of food products by bacteriological contaminants (e.g., *Salmonella* spp., *Listeria monocytogenes, Escherichia coli*) could be due to hygiene issues at different levels of the food chain: in field cultivation (e.g., contaminated manure, water, soils) (Alegbeleye et al. 2018), in farms (Antunes et al. 2018), in slaughterhouses, in transformation plants (dairies, cheese, "charcuterie" (Lakicevic and Nastasijevic 2017), and in smoked products (e.g., salmon) (Overney et al. 2017)). Bacteriological contaminants could be responsible for foodborne diseases from low to moderate severity (e.g., gastrointestinal symptoms), and even some pathogens (e.g., *E. coli, Listeria*) could have serious consequences (e.g., neurological disorders, immunological deficiencies, abortion).

11.1.2 Regulation

There is a regulation for food (i.e., intended for humans) supplemented by regulations for animal feed. The council regulation EC/315/93 set the basic rules for the EU legislation on the contaminants in food (1993). To control human exposure and so to protect consumer health, regulatory limits (maximum levels) were set for the potential contaminants in food products (e.g., European Regulation (EC) No 1881/2006 (2006) for melamine, mycotoxins, dioxins, PCBs, heavy metals, nitrates, PAHs (2006), European Regulation (EC) No 396/2005 for pesticides in food and feed of plant and animal origin (2005), European Regulation (EC) No 37/2010 for veterinary drugs (Commission, 2009)). These levels are toxicologically acceptable. It is prohibited in Europe to use substances included in European Directive 96/22/EC (i.e., certain substances having a hormonal or thyrostatic effect and beta-agonists) (1996).

National Residue Monitoring Plans (NRMPs) are implemented in all EU member states to monitor the contamination by chemical and bacteriological contaminants of animal and plant primary productions, foodstuffs of animal origin, and animal feed. The main objective is to prevent the marketing of noncompliant food products. The concentrations should be lower than regulatory limits for authorized substances, and the residue should be absent for banned substances. A new European Union agri-food chain legislation (Commission, 2017) was published to harmonize the official controls of food products along the entire agri-food chain.

The analytical methods used for the implementation of NRMPs have to be validated according to the European regulation to ensure the quality and comparability of analytical results. Validation of analytical methods (screening and confirmation) for hormones, beta-agonists, and veterinary drugs, for example, is based on the determination of performance characteristics (European Commission. 2002). Then characteristics are compared with performance criteria, which are usually set by the regulation, but could be also set by the customer. Validation procedures for screening and confirmatory methods to control pesticides in food and feed are detailed in the guidance document SANTE 11813/2017 (2017). The other chemical and bacteriological contaminants in foodstuffs also have their own European regulation for official controls and for validation of analytical methods (e.g., comparison of an alternative method with a reference method for bacteriological contamination (ISO 16140 (2016)).

11.1.3 Requirements for screening methods

The screening of contaminants in food products is the first critical step of the control. Whatever is the type of contaminant (i.e., chemical or bacteriological), screening methods should be sensitive (detection limits lower than regulatory limits (e.g., MRL, RPA) or as low as possible if no regulatory limit), specific or selective, quick, with high sample throughput.

11.1.4 Conventional screening methods

For all chemical and bacteriological analyses, the different steps are sampling, extraction of the analyte from the matrix, cleanup of the extracts if necessary and finally detection (screening methods) or identification/quantification (confirmatory methods).

Conventional screening methods could be bioassays, immunological methods, or physicochemical methods. Biosensors are identified as innovative screening methods.

Microbiological-based methods could be used for all analytes having an inhibitory effect on bacteria (e.g., antibiotic residues) and for the screening of microbial pathogens by isolation and culture. The microbiological methods used to detect food contaminants (i.e., chemical and bacteriological) are time-consuming and laborious.

Immunological methods could be used not only for the screening of many chemical contaminants but also for the detection of bacteriological contaminants (Helali et al., 2018). Immunological methods exist for decades as in-house laboratories or commercial methods, but new developments are always in progress to improve the sensitivity and the specificity of these methods. Different kinds of immunological tests are available: ELISA (enzyme-linked immunosorbent assay) (Li et al., 2017a), radioimmunoassays (RIA) (Granja et al., 2008), receptor tests (or dipsticks) (Lei et al., 2018).

Conventional methods for the detection of pathogens are counting methods, immunological methods (i.e., ELISA), and polymerase chain reaction (PCR) (Alahi and Mukhopadhyay, 2017). The identification of foodborne pathogens is more and more performed by PCR in food products (e.g., Brucella in milk and cheese (Moslemi et al., 2018)) because this technique is more sensitive and specific (Sharma and Mutharasan, 2013). However, these methods are expensive, not portable, and use labeled reagents.

The conventional methods used to detect many chemical contaminants (e.g., pesticides, acrylamide, endocrine disrupting compounds (BPA)) in foodstuffs could be also instrumental methods (e.g., gas chromatography (GC) and liquid chromatography (LC), especially coupled to mass spectrometry (MS, MS/MS) (in tandem)) (Samsidar et al., 2018; Oracz et al., 2011; Salgueiro-González et al., 2018). During the past 5 years, multiclass methods have been developed for the simultaneous detection of several families of contaminants in food products (Rossi et al., 2018). The advantages of these instrumental methods are the wide spectrum of detection, including different classes of compounds, with detection capabilities below the regulatory limits. However, because of the high costs of instrumental methods (e.g., LC-MS/MS, GC-MS), to the need of highly skilled and trained people and to the complexity of data analyses, especially when many analytes are in the scope of the method, there is a need for easy, quick, sensitive, and cheap methods.

11.1.5 Biosensors

Biosensors are in continuous development to monitor food safety because they lead to quick, sensitive, and specific screening methods (Santoro and Ricciardi, 2016). Many improvements have been done recently in the development of biosensors, mainly because of nanotechnologies, microfluidics, and biosensing elements, especially for the screening of food contaminants. They are considered as innovative and promising tools for the detection of many contaminants in foodstuffs (e.g., food-borne pathogens (Arora et al., 2011), mycotoxins (Vidal et al., 2013), pesticides (Boulanouar et al., 2018)). Biosensors could be classified according to the biosensing element (e.g., antibody, aptamer) or according to the transducer element (e.g., optical, electrochemical, mass sensitive, thermal). Optical biosensors have been developed in recent years for the detection of food contaminants but unfortunately often too expensive for high-throughput screening. On the contrary, electrochemical biosensors make it possible to develop a promising and economically interesting approach. Examples of applications of different kinds of electrochemical biosensors for the detection of chemical and bacteriological contaminants in foodstuffs are reviewed below.

11.2 Electrochemical biosensors for the detection of contaminants in food products

Electrochemical sensing technology began in the early 1950s. Among the various transduction methodologies, electrochemical techniques are widely investigated because of their potential advantages such as low cost, miniaturization, and portable instrumentation. Because of the recent advances in the development of electrochemical biosensors, they could be considered as an alternative to conventional methodologies for the detection of many chemical and microbial contaminants in foodstuffs (Bunney et al., 2017). Technological advances (e.g., nanomaterials, microfluidics) have been made to improve the methods' sensitivity, as well as their stability and robustness. The improvement of their performance characteristics (sensitivity, specificity, quickness, low cost, ease of use) suggests that electrochemical biosensors will in the future allow controlling the food chain in real time (food quality and risk of contamination).

There are few commercial biosensors used in food and environmental analyses and especially electrochemical biosensors (Bahadır and Sezgintürk, 2015). This chapter cited commercial electrochemical biosensors for the detection of food components (e.g., glucose, lactose) but no commercial electrochemical biosensors for the screening of contaminants in food products. However, some commercial biosensors are named for the detection of chemical contaminants in environmental samples (e.g., water, soils) (e.g., NEC nitrate biosensor (Nitrate Elimination Co. Inc., USA)) for nitrates).

Even if many sensors described in the literature have not been tested with real samples, we succeeded to find some applications of electrochemical sensors to the detection of contaminants in foodstuffs. Most of the electrochemical biosensors to detect contaminants in food products are affinity biosensors. A biosensing element (e.g., antibody, molecular imprinting polymer (MIP), aptamer) is usually immobilized onto the surface of the electrode to bind selectively the analyte to be detected. Either the bioreceptor is immobilized directly onto the electrode, or the electrode is previously functionalized before immobilization. The different strategies adopted to modify electrodes with the aim of immobilizing the bioreceptor for electrochemical detection are summarized by Yang et al. (2017). Then the binding event of the bioreceptor with the target analyte should be transformed into a measurable electrochemical signal. Finally, this signal is measured before and after the binding reaction. Another strategy is to modify electrodes with nanomaterials (e.g., carbon or metal nanocomposites, quantum dots) to improve the sensitivity of the electrochemical biosensors, but also they could act as the sensitized element, with no need of using a bioreceptor.

Bioreceptors or target analyte labeled with an enzyme (e.g., alkaline phosphatase (ALP) or horseradish peroxidase (HRP)), labeled with electrochemically active species (e.g., methylene blue (MB), hydroquinone), redox species in solution (e.g., ferri/ferrocyanide ($Fe(CN)_6{}^{3-/4-}$)), metal ions, and quantum dots (QDs) could be used for signal transduction (Vasilescu and Marty, 2016). The principle of detection of the electrochemical biosensors, which use an enzyme as the biosensing element, could be based on the electrochemical detection of its catalytic activity (detection of the substrate) (e.g., HRP) or could be based on the inhibition of the enzymatic activity by the target contaminant). These later assays are so-called inhibition-based electrochemical biosensors. AchE is the most often used enzyme to develop inhibition-based biosensors for pesticide detection (Pundir and Chauhan, 2012).

Electrochemical techniques are classified according to the measured parameter (e.g., amperometric (current), potentiometric (potential)). This chapter that proposes a review of the applications of electrochemical biosensors for the detection of chemical and bacteriological contaminants in foodstuffs is organized according to this classification (Fig. 11.2).

11.2.1 Amperometry

Amperometry is based on the measurement of the current resulting from the electrochemical oxidation or reduction of an electroactive species (Thévenot et al., 2001). The working electrode of an amperometric biosensor (e.g., noble metal (e.g., gold electrode), screen-printed electrode (SPE), carbon paste electrode (CPE)) is grafted with the biosensing element (Pohanka and Skládal, 2008). When the constant potential is applied, the electroactive species are conversed at the electrode surface and the resulting current is measured. The use of a bioreceptor allows improving the

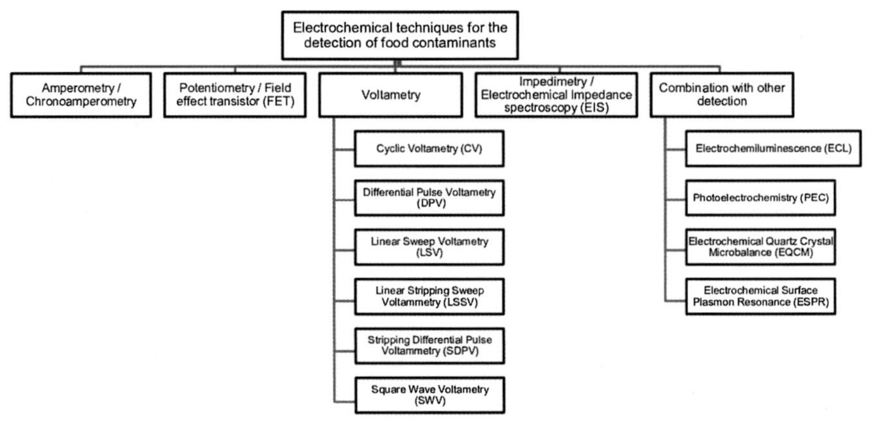

FIGURE 11.2

Different electrochemical techniques methods used for the detection of food contaminants.

selectivity of the biosensor, which is poor with classical amperometry because all the chemical species with oxidation potential below the applied voltage will be oxidized and so a current will be measured. The development of engineered enzymes is one of the explored ways to improve selectivity and sensitivity of enzyme-based biosensors, especially for pesticides (Songa and Okonkwo, 2016). Multiplex amperometric biosensors have been developed for the simultaneous high-throughput screening of multiple analytes (Wu et al., 2018; Conzuelo et al., 2014). Examples of applications of amperometric detection for the screening of contaminants in foodstuffs are presented in Table 11.1.

Different types of bioreceptors could be used to develop amperometric biosensors for the detection of contaminants in foodstuffs (i.e., enzymes, antibodies, binding proteins, MIPs) (Fig. 11.3). When no bioreceptor is needed for the detection of the target analyte, nanomaterials (e.g., carbon nanomaterials) are utilized as the sensitized element in fabricating electrochemical sensor (i.e., composite electrodes).

11.2.2 Chronoamperometry

An amperometric measurement is made while a constant potential is applied, whereas a chronoamperometric measurement is made while the applied potential is modified. The applied potential step waveform is the same (square-wave pulsed voltammetric technique). The current is monitored as a function of time. Four examples of applications of chronoamperometric measurement for the detection of contaminants in foodstuffs are presented in Table 11.1.

Magnetic nanomaterials or magnetic beads (MBs) (e.g., Fe3O4 nanoparticles (NPs)) could be used in combination with biosensing elements (e.g., antibodies, aptamer, MIPs, enzymes) to improve the detection performance of electrochemical

Table 11.1 Amperometric biosensors for the detection of various contaminants in foodstuffs classified by biosensing element.

Biosensing element	Contaminant	Matrix	Electrode/modifier of electrode surface	LOD	References
Enzymes					
AchE	4 OPPs	Milk	AuE@c-MWCNT/Fe_3O_4NP-AchE	0.1 –10 nM	Chauhan and Purdir (2011)
AchE	6 OPPs	Wheat, cabbage, apple, orange, cherry	SPCE@CoPC-AchE	10^{-5}–10^{-9} M	Crew et al. (2011)
Tyrosinase	3 CPs	Apple	GDE@tyrosinase	0.074–1.7 pM	Pita et al. (1997)
	OTA[a]	Beer	SPCE@HRP	26.77 ± 3.61 nM	Alonso-Lomillo et al. (2011)
Antibodies					
Ab + PBP	CEPH, SULFA, TTC	Milk	SPCE + MBs-Ab (or PBP) + Ag-HRP	<100 μg kg^{-1}	Conzuelo et al. (2014)
	Beta-lactams	Milk	H_2O_2E@PM-PENI + Ab-HRP	0.1 pM	Merola et al. (2015)
	Peanut allergens	Wheat flour	SPdCEs + MBs-Ab + Ab-HRP	0.07–18.0 ng mL^{-1}	Ruiz-Valdepeñas Montiel et al. (2016)
	OTA	Coffee	SPCE@ + MBs/protein G-Ab	0.32 μg L^{-1}	Jodra et al. (2015)
	AFM1[a]	Milk	SPCE@Ab + AFM1-HRP	25 pg mL^{-1}	Micheli et al. (2005)
	TTX[a]	Puffer fish	AuSPE@CDTASAMs-TTX + Ab-HRP	2.6 ng mL^{-1}	Reverté et al. (2017)
	S. pullorum	Chicken muscle	SPE@MWCNT-HRP + CNM-Ab	60–100 CFU mL^{-1}	Liu et al. (2013a)
	Histamine	Fish	SPCE@AuNP-CS/NPs-HA + Ab-HRP	1.25 pg mL^{-1}	Dong et al. (2017)

Continued

Table 11.1 Amperometric biosensors for the detection of various contaminants in foodstuffs classified by biosensing element.—*cont'd*

Biosensing element	Contaminant	Matrix	Electrode/modifier of electrode surface	LOD	References
Aptamers					
	SDMX	Milk	AuE@Apt-cDNA + C60-rGO/Tb/AuNPs + ExoNase	10 fg mL^{-1}	You et al. (2018)
	Zearalenone[a]	Beer	GCE@Au-PANI-Au-Apt + Cu@L-Glu/Pd-PtNPs-cDNA	0.45 fg mL^{-1}	Ji et al. (2019)
Hybrid Ab-Apt	VP	Fish samples	AuE@Ab/Apt + AuNPs/detection DNA + cDNA + MB	2 CFU mL^{-1}	Teng et al. (2017)
MIP					
	SDMX	Milk	AuE@PpyMIP	7.10–5 M	Turco et al. (2015)
	BPA	Honey, grape juice	AuE@MWCNTs/AuNPs	3.6 nM	Huang et al. (2011)
Other bioreceptors					
PBP	Beta-lactams	Milk	SPCE@PBP + PENI-GOase	<10 μg kg^{-1}	Setford et al. (1999)
Without bioreceptor (other electrodes materials nanocomposites)					
	Zearalenone	Malt beverage	CPE@MWCNT	0.58 ng mL^{-1}	Afzali et al. (2015)
	CAP	Milk, milk powder, honey	GCE@AuNPs/GO	0.25 μM	Karthik et al. (2016)
	Dimetridazole	Eggs, milk	SPCE@rGO/PB MCs	3.2 nM	Murugan et al. (2018)
	MP	Ugli, tomato, beetroot, broccoli	SPCE@GONRs	0.5 nM	Govindasamy et al. (2017b)
	BPA	Milk	GCE@PAMAM–Fe3O4	5 nM	Yin et al. (2011)
	H_2O_2	Milk	MnO_2/VAMWCNTs	0.8 μM	Xu et al. (2010)
	H_2O_2	Milk	GCE@GL-VS_2	26 nM	Karthik et al. (2018)

CDTA-SAMs, *carboxylate-dithiol/thioctic acid self-assembled monolayers;* c-MWCNT, *carboxylated multiwalled carbon nanotubes;* C60-rGO, *fullerene (C60)-doped graphene nanohybrid;* CoPC, *cuprous oxide and cobalt phthalocyanine;* CNM, *cellulose nitrate membrane;* Cu@L-Glu/Pd-PtNPs, *Copper modified by* L-*glutamine and palladium and platinum nanoparticles;* ExoNase, *exonuclease;* Fe_3O_4NP, *iron oxide nanoparticles;* GL-VS_2, *grass-like vanadium disulfide;* GONRs, *graphene oxide nanoribbons;* Gr, *graphene;* MnO_2/VACNTs, *MnO_2-modified vertically aligned multiwalled carbon nanotubes;* PAMAM-Fe_3O_4, *poly (amidoamine) and Fe_3O_4 magnetic nanoparticles;* PPyMIP, *molecularly imprinted overoxidized polypyrrole;* rGO/PB MCs, *reduced graphene oxide/Prussian blue microcubes;* S. pullorum, Salmonella pullorum*;* SDMX, *sulfadimethoxine;* Tb, *toluidine blue;* TTX, *tetrodotoxins;* VP, Vibrio parahaemolyticus.

[a] *Chronoamperometric measurements.*

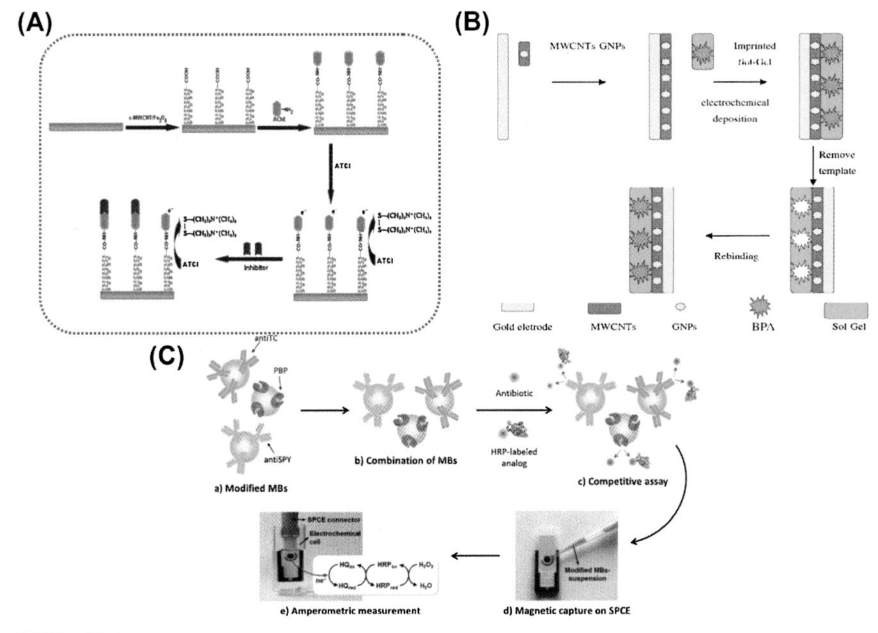

FIGURE 11.3

Three examples of amperometric biosensors developed for the detection of food contaminants, based on (A) enzyme, (B) antibodies and binding protein, and (C) MIP. (A) Amperometric biosensor based on the inhibition of acetylcholinesterase (AchE) by OPPs. (B) Amperometric biosensor based on the specific recognition of bisphenol A (BPA) by a thin film of sol-gel molecular imprinted polymer. (C) Multiplex amperometric detection of antibiotics based on magnetic beads immobilized with antibodies and binding proteins.

(A) Reprinted from Chauhan, N., Pundir, C.S., 2011. An amperometric biosensor based on acetylcholinesterase immobilized onto iron oxide nanoparticles/multi-walled carbon nanotubes modified gold electrode for measurement of organophosphorus insecticides. Analytica Chimica Acta 701, 66–74; with permission from Elsevier. (B) Reprinted from Huang, J., Zhang, X., Lin, Q., He, X., Xing, X., Huai, H., Lian, W., Zhu, H., 2011. Electrochemical sensor based on imprinted sol–gel and nanomaterials for sensitive determination of bisphenol A. Food Control 22, 786–791; with permission from Elsevier. (C). Reprinted from Conzuelo, F., Ruiz-Valdepeñas Montiel, V., Campuzano, S., Gamella, M., Torrente-Rodríguez, R.M., Reviejo, A.J., Pingarrón, J.M., 2014. Rapid screening of multiple antibiotic residues in milk using disposable amperometric magnetosensors. Analytica Chimica Acta 820, 32–38; with permission from Elsevier.

biosensors (i.e., selectivity, sensitivity) and limiting the nonspecific binding (Ríos and Zougagh, 2016).

11.2.3 Potentiometry/field-effect transistors

Potentiometric measurements consist in determining the electrical potential difference between two electrodes, the reference electrode that has a constant potential and the working electrode of which potential varies in relation to the composition of the sample. The most common potentiometric devices are pH electrodes. Potentiometric biosensors could be based on different types of working electrodes: ion-selective electrodes (ISEs) based on thin films or ion-selective membranes (ISMs) (e.g., glass membrane, enzyme) and ion-sensitive field-effect transistor (ISFET). An FET is constituted of a semiconductor channel with two electrodes. The electric field controls the current flowing in the channel. FET could detect the intrinsic charge of molecules captured by immobilized bioreceptors. An ISFET is the association of an FET with an ISM (e.g., silicon nitride (Si_3N_4), silicon dioxide (SiO_2), aluminum oxide (Al_2O_3), and tantalum pentoxide (Ta_2O_5)), which covers the transistor gate. An ISFET is used for measuring ion concentrations in solution. When an enzyme layer is immobilized on the gate, the biosensor is called enzyme FET (ENFET) (Dzyadevych et al., 2006). An enzymatic reaction produces an ion, which is monitored by the electrode (Pohanka and Skládal, 2008). When an antibody is immobilized onto the gate, it is called immunological FET (IMFET). Examples of applications of potentiometric biosensors for the screening of various contaminants are presented, respectively, in Table 11.2.

An innovative potentiometric aptasensor has been developed for the detection of bisphenol A (BPA) in grape juices (Abnous et al., 2018) (Fig. 11.4). In the absence of BPA (B)), an enzyme, the terminal deoxynucleotidyl transferase (TdT), is involved in the extension of the aptamer, which forms of a bridge between the aptamer and a complementary DNA. In this case, the redox product ($[Fe(CN)_6]^{3-/4-}$) cannot reach the surface of the electrode, and the signal is reduced. When BPA is present in a sample (A)), BPA links to its specific aptamer. So the bridge cannot be formed, the electrode is accessible to $[Fe(CN)_6]^{3-/4-}$ and so the potentiometric signal is increased.

Some enzymes (e.g., nucleases (DNase)) could be used for the amplification of the signal of electrochemical biosensors (Yao et al., 2018). In this case, the aptamer specifically binds to the target molecule. Then when the DNase is added, the aptamer is digested and so the target analyte is released. Therefore the target analyte is ready to bind again to a new aptamer on the electrode surface. For this reason, the potentiometric signal is increased by this recycling strategy.

Two different types of analyses could be performed: direct or stripping analyses. Stripping analysis is characterized by a preconcentration of a metal phase onto the electrode and the selective oxidation of each metal phase species. The analyte to be detected is electrodeposited onto the working electrode and then oxidized during the stripping step. Electrochemical stripping analysis could be based on potentiometry or voltammetry to quantify ions in solution. Potentiometric stripping analysis has been applied to the screening of heavy metals in honey (Švarc-Gajić and Stojanovic, 2014).

11.2.4 Voltammetry

The basic principles of all voltammetric techniques are the application to an electrode of a potential (E), which varies in relation to time. Because of the controlled potential, the chemical species in solution are electrolyzed (oxidized or reduced) at

Table 11.2 Potentiometric biosensors for the detection of various contaminants in foodstuffs classified by biosensing element.

Biosensing element	Contaminant	Matrix	Electrode/modifier of electrode surface	LOD	References
Enzymes					
PNCnase	Beta-lactams	Milk	PtE@PTy-PNCnase	0.3 μM; >20 μg mL^{-1} (milk)	Ismail and Adeloju (2010)
AchE	OPPs	Strawberry juice	AuFET@CNT-AchE	1.35 mg mL^{-1}	Bhatt et al. (2017)
Antibodies					
	BWAP	Cookies	FET@SAM-Ab + SDS	/	Hideshima et al. (2018)
Aptamers					
Apt + enzyme	BPA	Grape juice	AuSPE@Apt-cDNA + TdT + $[Fe(CN)6]^{3-/4-}$	15 pM	Abnous et al. (2018)
Apt + enzyme	Kanamycin	Milk	GNPE@Apt + DNase I	30.0 fg mL^{-1}	Yao et al. (2018)
	Ricin	Orange juice, milk	FG-R@Apt + CG	300 pM (10 ng mL^{-1})	White et al. (2016)
MIPs					
Five sensors	DCP	Fish	Plasticized PVC mb-MIP	0.56–1.32 μM	El-Kosasy et al. (2018)
	Dinotefuran	Cucumber	E@PVC-MIP	0.35–10.07 ng mL^{-1}	Abdel-Ghany et al. (2017)
Without bioreceptor (others electrodes materials, nanocomposites)					
	Heavy metals (Zn, Cd, Pb, Cu)	Honey	GCE@Hg	0.38–4.3 μg dm^{-3}	Švarc-Gajić and Stojanovic (2014)

AA, *acrylamide;* BWAP, *Buckwheat allergenic protein;* Cd, *cadmium;* CG, *control gate;* c-PVC, *carboxylated PVC;* Cu, *copper;* DCP, *2,4-dichlorophenol;* DVB, *divinylbenzene;* E. coli, Escherichia coli*;* EGDMA, *ethylene glycol dimethacrylate;* EGT, *electrolyte-gated transistor;* EMA, *ethyl methacrylate;* FGT, *floating gate transistor;* FGR, *right arm of the floating gate;* GCE@Hg, *mercury film electrode;* GNP, *graphene paper;* GOP, *graphene oxide paper;* MAA, *methacrylic acid;* mb, *membrane;* NAA, *nuclease-assisted amplification;* NIP, *nonimprinted polymer;* NMIP, *nanoparticulate molecularly imprinted polymers;* Pb, *lead;* PTy-PNCnase, *polytyramine-penicillinase;* QA, *quaternary ammonium;* SiO_2, *silicon dioxide;* TdT, *terminal deoxynucleotidyl transferase;* VPVC mb, *valinomycin-polyvinylchloride membrane;* Zn, *zinc.*

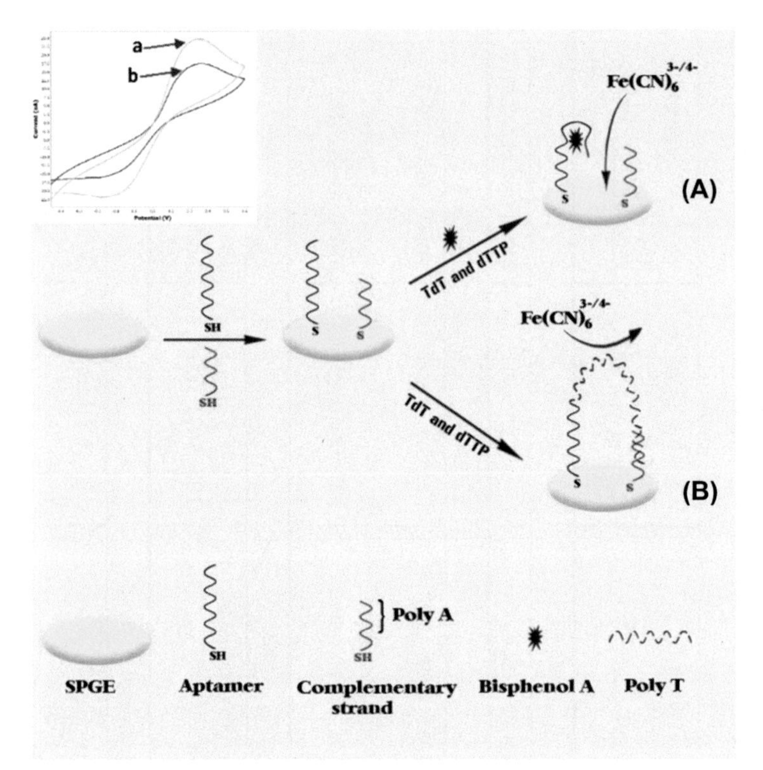

FIGURE 11.4

Principle of a potentiometric aptasensor for the detection of bisphenol A (BPA) in grape juice samples, based on the inhibition of the extension of aptamer length by BPA. The lack of a bridge in the presence of BPA allows the access of $[Fe(CN)_6]^{3-/4-}$ as redox agent to the electrode surface (A).

Reprinted from Abnous, K., Danesh, N.M., Ramezani, M., Alibolandi, M., Taghdisi, S.M., 2018. A novel electrochemical sensor for bisphenol A detection based on nontarget-induced extension of aptamer length and formation of a physical barrier. Biosensors and Bioelectronics 119, 204–208; with permission from Elsevier.

the electrode. The resulting current (1) flowing through the electrochemical cell is measured. So the basic components of a voltammetric system are a potentiostat, a computer, and the electrochemical cell, which is composed of three electrodes: one reference, one auxiliary, and one working electrode (WE) (e.g., mercury (Hg) electrodes and noble metal electrodes (e.g., gold (Au), silver (Ag))). The results are presented in the form of a voltammogram with a sigmoidal shape, which represents the variation of the potential in relation to the resulting current. The applied potential could be varied in different systematic manner (e.g., linear scan means continuous linear change in potential, pulse scan means modulation of the potential other than linear). Therefore, there are different types of voltammetric measurements: stripping voltammetry (SV) (i.e., anodic SV, cathodic SV, adsorptive SV), cyclic voltammetry (CV), pulse voltammetry (e.g., differential pulse voltammetry (DPV)), square-wave voltammetry (SWV), linear sweep voltammetry (LSV) (Farghaly et al., 2014).

Stripping voltammetry appeared to be a very sensitive technique for trace analysis because of the preconcentration step at the electrode (e.g., heavy metals). Anodic stripping voltammetry (ASV) has been applied to the detection of heavy metals (cadmium, lead), by comparison with direct potentiometric measurement. It appears that due to the drastic improvements of the sensitivity of modern ISE, a potentiometric biosensor could now compete with ASV because there is less need of preconcentration (Chumbimuni-Torres et al., 2009). The sensitivities are equivalent, whereas the selectivity of ISE is better for this specific application, even in the presence of other interfering ions in the samples.

CV is a dynamic electrochemical method for measuring redox events. In CV, the potential at WE is applied in one direction and then the direction is reversed, whereas the resulting current is measured. Pulse voltammetry, which consists of a series of potential pulses, especially DPV and SWV, is much quicker and more sensitive than the other voltammetric techniques. The voltammetric techniques DPV and SWV are extensively described for various chemical and bacteriological contaminants in the literature (Dragone et al., 2017; Chen and Shah, 2013). The advantages of voltammetric techniques are a good sensitivity, speed, and multiplexing potential (Table 11.3).

Mediator nanoparticles could be used for their excellent properties of electrocatalytic activity (i.e., enhancement of electron transfer), their low cost of production, and high stability (e.g., Prussian Blue (PB) (Li et al., 2012; Murugan et al., 2018), cobalt phthalocyanine (CoPc) (Crew et al., 2011; Prado et al., 2015).

Different types of bioreceptors could be used to develop voltammetric biosensors for the detection of contaminants in foodstuffs (i.e., Enzymes, aptamers, DNA) (Fig. 11.5).

11.2.5 Electrochemical impedance spectroscopy

Electrochemical impedance is the response of an electrochemical system to an applied potential. The principle of impedimetric sensing is based on the change in the interfacial resistance and/or capacitance (i.e., alternating voltage) at the surface of the electrode. This change could be caused by the formation of a complex between the bioreceptor (e.g., antibody, aptamer, MIP) immobilized onto the electrode surface and the target analyte in solution. Impedance technique allows developing label-free biosensors for the detection of a wide range of analytes.

No redox probe is present in capacitive biosensors because the electrode is covered by an insulating layer, whereas in faradic impedimetric biosensors a redox probe is usually present in the measuring solution (e.g., ferri/ferrocyanide system $[Fe(CN)_6]^{4-/3-}$). The electrode surface is covered by a noninsulating layer or partially covered and is able to catalyze the redox probe. Then a change in the charge transfer resistance is produced and measured.

Table 11.3 Voltammetric biosensors for the detection of various contaminants in foodstuffs classified by biosensing element.

Biosensing element	Contaminant	Matrix	Technique	Electrode/modifier of electrode surface	LOD	References
Enzymes						
PENase + L-cyst	PENI, TTC	Chicken breast, beef	CV	AgE@Au/L-cyst-Pt/AuNPs/PENase	10.5 μM PC, 15.2 μM TTC	Li et al. (2019)
TYR	BPA	Plastic feeding bottles, plastic bag for rice cooking	CV	GE@TYR/TiO_2/MWCNTs/PDDA/Nafion	0.066 μM	Kochana et al. (2015)
AchE	OPPs	Cabbage	DPV	GCE@PtPd@NCS	7.9×10^{-3} -7.1×10^{-2} pM	Ma et al. (2018)
AchE	OPPs	Cabbage	DPV	GCE@rGO/TiO_2-CS	29 nM (6.4 μg L^{-1})	Cui et al. (2018)
B-lactamase	Penicillin	Milk	DPV	CPE-B-lactamase + CoPc	0.079 μM	Prado et al. (2015)
Antibody						
	Mycotoxins	Corn samples	DPV	SPCE@PPy-ErGO/AuNPs-Ab	4.2 μg kg^{-1}	Lu et al. (2016)
	Erythromycin, tylosin	Bovine muscle	DPV	SPGrE	0.2 ng mL^{-1}	Ammida et al. (2004)
	S. ent. typhi	Milk	DPV	SPCE + MBs-Ab + AuNP-Ab	$1.5.10^3$–1.5 10^5 cfu mL^{-1}	Afonso et al. (2013)
Aptamer						
	AFB1	Peanuts	CV + EIS	PAMAM G4 on AuE + K $[Fe(CN)_6]^{3-/4-}$	0.40 ± 0.03 nM	Castillo et al. (2015)
	CAP	Milk	DPV	SPGrE@GelB-Apt	0.183 nM	Hamidi-Asl et al. (2015)
	Escherichia coli O111	Milk	DPV	AuE@cDNA/Apt + biotinylated detection DNA + ST-AP	305 CFU mL^{-1}	Luo et al. (2012)

	CAP	Honey	DPV	AuE@Apt + biotinylated detection DNA + ST-AP	0.29 ng mL^{-1}	Yan et al. (2012)
	OTA	Red wine	DPV	SPE@β-CD + cDNA-MB + RecJf	3 pg mL^{-1}	Wang et al. (2019)
	CAP	Milk	LSV	GCE@Gr/AgNPs	2 nM	Liu et al. (2017b)
	STR, CAP, TTC	Milk	SWASV	AuE@cDNA cDNA1s, QD–cDNA2 (PbS, CdS, ZnS nanoclusters)	10, 5 and 20 nM	Xue et al. (2016)
	OTA	Rice	SWV	BDDMC@PEG-Apt + K $[Fe(CN)_6]^{3-/4-}$	0.01 ng L^{-1}	Chrouda et al. (2015)
	Kanamycin, CAP	Milk	SWV	GCE@MB-Apt + cDNA-NMOF (Pb^{2+}, Cd^{2+})	0.16 pM, 0.19 pM	Chen et al., (2017)
MIP						
	CAP	Milk	CV	3D CNTs@Cu NPs-MIP	10 μM	Munawar et al. (2018)
	Olaquindox	Pork, fish	CV	GCE@AuNPs/cMWCNT	2.7 nM	Wang et al. (2017)
	Melamine	Milk, feed	CV	GCE@Au/PANI	1.39 pM	Rao et al. (2017)
	CAP	Milk, honey	DPV	GCE@GO/CKM-3/MWCNT-MIP	0.1 nM	Yang and Zhao (2015a)
	Sulfaguanidine	Honey	DPV	Au-SPE@PAM + $[Fe(CN)6]^{4-/3-}$	0.20 pg mL^{-1}	El Alami El Hassani et al. (2018b)
	N-NIP	Grilled meat	DPV	PtDE@MIP	80.9 nM	Lach et al. (2017)
	Kanamycin	Chicken/pig liver, milk	DPV	CE@CS/MMIP (MWCNTs/Fe_3O_4/PMMA)	0.023 nM	Long et al. (2015)
	TTC	Shrimp	DPV	SPCE@MIOPPy-AuNP + SDS	0.65 μM	Devkota et al. (2018)
	BPS	Mineral water	CV	GCE@GQDs/hNiNS@MIP	0.03 μM	Rao et al. (2018)

Continued

Table 11.3 Voltammetric biosensors for the detection of various contaminants in foodstuffs classified by biosensing element.—*cont'd*

Biosensing element	Contaminant	Matrix	Technique	Electrode/modifier of electrode surface	LOD	References
	OTC	Milk	DPV	PtE@PB-MIP film + GOase-OTC	300 fmol L^{-1}	Li et al. (2012)
	MP	Apple, cucumber	LSSV	GCE@MWCNT/Au-NP-MIP	0.08 μg L^{-1}	Wu et al. (2014)
	TTC	Honey	LSV	AuE@MMOF film-MIP + $[Fe(CN)6]^{-3/-4-}$	0.22 fM	Bougrini et al. (2016)
	Lindane	Orange, grape, tomato, cabbage	LSV	CuE@MWCNT-MIP	0.1 nM	Anirudhan and Alexander (2015)
	MP	Fish	SWV	m-GEC@MMIP	1.22 pg mL^{-1}	Hassan et al. (2018)
	CAP	Honey, milk	SWV	PtTFME@MIP + $K_3Fe(CN)_6$	0.39 nM	Zhao et al. (2017)
Other bioreceptors						
	Salmonella	Milk	DPV	AuE@ cDNA1 + cDNA2 + AuNP@ biotinylated detection DNA + ST-AP	6 CFU mL^{-1} (after PCR)	Zhu et al. (2014)
dsDNA	Acrylamide	Potato fries	SWV	AuSPE@hemoglobin (Hb)	0.158 μM	Asnaashari et al. (2019)
Without bioreceptor (other electrode materials, nanocomposites)						
	Acrylamide	Bake rolls, potato chips	ASV	$HgE@Ni^{2+}$	12 μg kg^{-1}	(Veselá; Šucman, 2013)
	TTC	Fish, chicken, shrimp	DPV	GCE@Fe/Zn-MMT + SDS	0.1 μM	Gan et al. (2014)

TTC	Milk	DPV	GR-SPCE	0.08 M	Hayati et al. (2016)
Melamine	Milk, tainted milk powder	DPV	SPCE@uric acid	163 ng mL^{-1}	Liao et al. (2010)
Melamine	Milk powder	DPV	AuE@CS/ZnONPs/[EMIM][Otf] Redox indicator: MB	$9.6.10^{-14}$ M	Rovina and Siddiquee (2016)
Acrylamide	Mackerel fish, cassava chips	DPV	SPE@AuNP/SiNS/ssDNA Redox indicator: MB	0.0001 ppm	Sani et al. (2018)
CAP	Milk, honey	DPV	GCE@3DRGO	0.15 μM	Zhang et al. (2016)
Sudan I	Tomato sauce	DPV	GCE@GR/βCD/PtNPs	1.6 nM	Palanisamy et al. (2017)
CAP	Milk, honey	DPV	ITOE@VSMs@CSMs	40 ng mL^{-1}	Zheng et al. (2016)
CAP	Milk, honey	SDPV	GCE@MWCNT–CTAB–PDPA	0.002 μM	Kor and Zarei (2014)
TTC, cefixime	Milk	SWV	AuSPE@AuNPs-cystein	10 μM, 1 mM	Asadollahi-Baboli and Mani-Varnosfaderani (2014)

ABPE, *acetylene black paste electrode;* BDDMC, *boron-doped diamond microcell;* β-CD-SH, *mercapto-β-cyclodextrin;* CKM-3, *mesoporous carbon;* CoPc, *cobalt phthalocyanine;* CSMs, *cylindrical surfactant micelles;* [EMIM][Otf], *ionic liquid (1-ethyl-3-methylimidazolium trifluoromethanesulfonate);* CuE, *copper electrode;* Cu-MOF-Gr, *Copper-centered metal-organic framework-graphene composites;* DON, *deoxynivalenol;* Fe/Zn-MMT, *iron/zinc cation–exchanged montmorillonite catalyst;* GEC, *graphite-epoxy composite;* GelB, *gelatine B;* GST, *glutathione-s-transferase;* m-GEC, *magnetoelectrode based on graphite-epoxy composite;* MIOPPy, *molecularly imprinted overoxidized polypyrrole;* NMOF, *nanoscale MOF;* M-NMOF, *metal ion nanoscale MOF;* CTAB-PDPA, *cetyl trimethyl ammonium bromide-poly(diphenylamine);* PMMA, *polymethyl methacrylate;* PAMAM G4, *poly (amidoamine) dendrimers of the fourth generation;* PCB72, *polychlorinated biphenyl-72;* PEG, *polyethylene glycol;* PtPd@NCS, *platinum-palladium nanoparticles encapsulated in N-doped carbon shells;* Pt TFME, *platinum thin-film microelectrode;* RecJf, *recombinant exonuclease;* S. ent. typhi, Salmonella enterica typhimurium; SiNS, *silica nanospheres;* ST-AP, *streptavidin-alkaline phosphatase;* STR, *streptomycin;* TiO_2, *titanium oxide;* TYR, *tyrosinase;* PDDA, *polycationic polymer poly(diallyl dimethyl ammonium chloride);* VSMs, *vertical silica mesochannels.*

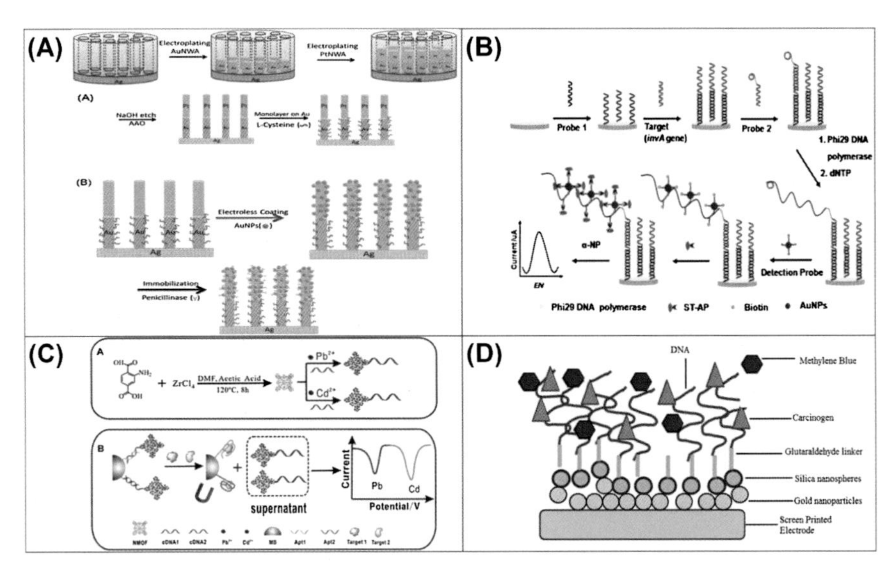

FIGURE 11.5

Different voltammetric methods implemented for the detection of contaminants in foodstuffs. (A) Fabrication process of the electrodes modified by L-cysteine (A) and penicillinase (B) on the Au-Pt multisegment nanowire array for the detection of tetracycline and penicillin in chicken breast and beef by cyclic voltammetry (CV). (B) Combination of the detection of invA target gene from *Salmonella* by capture DNA with signal amplification by RCA and detection by a DNA-AuNPs probe by DPV. (C) Simultaneous detection of kanamycin and chloramphenicol based on magnetic beads immobilized with two aptamers combined with two signal probes (cDNA-NMOF) by SWV. (D) Design of DNA toxicity biosensor for the detection of acrylamide in mackerel fish and cassava chips by differential pulse voltammetry (DPV). The DNA probe is immobilized onto silica nanospheres in the way to optimize the competitive binding with carcinogen (acrylamide) and MB.

(A) Reprinted from Li, Z., Liu, C., Sarpong, V., Gu, Z., 2019. Multisegment nanowire/nanoparticle hybrid arrays as electrochemical biosensors for simultaneous detection of antibiotics. Biosensors and Bioelectronics 126, 632–639; with permission from Elsevier. (B). Reprinted from Zhu, D., Yan, Y., Lei, P., Shen, B., Cheng, W., Ju, H., Ding, S., 2014. A novel electrochemical sensing strategy for rapid and ultrasensitive detection of Salmonella by rolling circle amplification and DNA-AuNPs probe. Anal Chim Acta 846, 44–50; with permission from Elsevier. (C) Reprinted from Chen, M., Gan, N., Zhou, Y., Li, T., Xu, Q., Cao, Y., Chen, Y., 2017. A novel aptamer-metal ions-nanoscale MOF based electrochemical biocodes for multiple antibiotics detection and signal amplification. Sensors and Actuators B: Chemical 242, 1201–1209; with permission from Elsevier. (D). Reprinted from Sani, N.D.M., Heng, L.Y., Marugan, R.S.P.M., Rajab, N.F., 2018. Electrochemical DNA biosensor for potential carcinogen detection in food sample. Food Chemistry 269, 503–510; with permission from Elsevier.

Very few publications were found using capacitive biosensors for the detection of contaminants in foodstuffs. This technique is sometimes used for the screening of foodborne pathogen bacteria. A capacitive immunosensor was developed for the detection of *E. coli* 157 (Pandey et al., 2017). This biosensor based on interdigitated microelectrodes functionalized with graphene nanocomposites and antibodies against *E. coli* was able to detect these foodborne pathogen bacteria as low as 100 cells/mL.

A variety of contaminants could be determined using impedimetric biosensors (Table 11.4).

Different types of electrodes could be used to develop impedimetric sensors: single working electrodes such as screen-printed carbon electrodes (SPCEs), metal oxide–based electrodes (e.g., indium-tin oxide (ITO)), metal electrodes (e.g., gold (Au), platinum (Pt), silver (Ag)), glassy carbon electrodes (GCEs), or a series of parallel electrodes with gaps between electrodes (i.e., interdigitated electrodes (IDEs)). Then a monolayer (self-assembled monolayer (SAM), self-assembled thiourea monolayer (SATUM)) or a multilayer (multiwalled carbon nanotubes (MWCNTs)) is deposited onto the surface to functionalize the electrode and to promote the immobilization of the biosensing element (Pänke et al., 2008). Finally a bioreceptor could be immobilized onto the functionalized electrode to selectively recognize the target analyte. Different types of bioreceptors could be used to develop impedimetric biosensors for the detection of contaminants in foodstuffs (i.e., aptamers, antibodies) (Fig. 11.6).

It is worth to highlight amplification strategies that have been developed to improve the signal and therefore to increase the sensitivity of the impedimetric biosensors (e.g., enzyme labels (e.g., HRP), redox labels (e.g., $[Fe(CN)_6]^{4-/3-}$), conductive polymer (CP) films, nanoparticles (NPs) such as carbon nanotubes (CNTs), quantum dots (QDs), gold nanoparticules (AuNPs)).

The major drawback of impedimetric techniques is that they are more sensitive to matrix effects especially when analyzing very complex matrices such as foodstuffs than classical techniques such as amperometry. However, one of the main advantages is that impedimetry is a label-free technique, particularly suited to the detection of binding events (Dragone et al., 2017).

11.2.6 Photoelectrochemical sensor

The principle of a photoelectrochemical (PEC) measurement is that the electrode is excited by a light source (input) to produce a photocurrent signal (output), which intensity is measured by the electrochemical workstation. The photoactive species onto or into the electrode are excited by the light source, so a charge transfer happens. The separation between the input and the output should increase the sensitivity because the background noise decreases.

To implement these measurements, semiconductor, especially visible light-active materials, and photoactive materials have to be synthesized to functionalize the electrodes. Different kinds of photoactive materials could be used

Table 11.4 EIS biosensors for the detection of various contaminants in foodstuffs classified by biosensing element.

Biosensing element	Contaminant	Matrix	Electrode/modifier of electrode surface ± redox probe	LOD	References
Enzyme					
AchE	Carbamates, OPP	Lettuce	AuE@AchE + $[Fe(CN)_6]^{4-/3-}$	8 ng mL^{-1}, 10 ng mL^{-1}	Malvano et al. (2017)
Urease	Urea	Milk	FTOE@PA6/PPy/ZnO/urease + $[Fe(CN)_6]^{4-/3-}$	0.11 mg L^{-1}	Migliorini et al. (2018)
Aptamer					
	AFB1	Peanuts	SPCE@PDMS film/Apt-Fe_3O_4@Au MBs + $[Fe(CN)_6]^{4-/3-}$	15 pg mL^{-1}	Wang et al. (2018a)
	Ciprofloxacin	Milk	SPCE@MWCNT-V_2O_5-CS-Apt + $[Fe(CN)_6]^{4-/3-}$	0.5 ng mL^{-1}	Hu et al. (2018)
	Kanamycin	Milk	4-CP/SPCE-Apt $[Fe(CN)_6]^{4-/3-}$	0.11 ng mL^{-1}	Sharma et al. (2017)
	Penicillin G	Milk	4-NB/SPCE-Apt + $[Fe(CN)_6]^{4-/3-}$	0.17 ng mL^{-1}	Paniel et al. (2017)
	Oxytetracycline	Milk	AuE@MCA/Ce-MOF-Apt	17.4 fg mL^{-1}	Zhou et al. (2019)
	BTX-2	Shellfish	AuE@BTX + Apt + $[Fe(CN)_6]^{4-/3-}$	106 pg mL	Eissa et al. (2015)
	E. coli O157:H7	Drinking water	3D-IDEA($TaSi_2$)@MPTES-Apt	100 cfu mL^{-1}	Brosel-Oliu et al. (2018)
	S. typhimurium	Apple juice	SPCE@Apt + $[Fe(CN)_6]^{4-/3-}$	6 cfu mL^{-1}	Bagheryan et al. (2016)
	S. typhimurium	Egg	GCE@Au/NPG-Apt + $[Fe(CN)_6]^{4-/3-}$	1 cfu mL^{-1}	Ranjbar et al. (2018)
	Salmonella	Porcine meat	GCE@GO/AuNPs + $[Fe(CN)_6]^{4-/3-}$	10–1000 cfu mL^{-1}	Ma et al. (2014)
	Salmonella	Chicken meat	GCE@GO/MWCNT-Apt + $[Fe(CN)_6]^{4-/3-}$	25 cfu mL^{-1}	Jia et al. (2016)
Antibodies					
	Sulfapyridine	Honey	AuME@MNPs/Py/Py-SA2BSA + Ab + $[Fe(CN)_6]^{4-/3-}$	0.4 ng L^{-1}	Hassani et al. (2017)
	TTC	Honey	8 AuME@TTC + MNPs/Py/Py-Ab + $[Fe(CN)_6]^{4-/3-}$	1.2 pg mL^{-1}	El Alami El Hassani et al. (2018a)
	Vomitoxin, salbutamol	Pork muscle	AuE@515-516-AlMOFs-Ab + $[Fe(CN)_6]^{4-/3-}$	0.70, 0.40 pg mL^{-1}	Liu et al. (2017a)
	AFB1	Corn	ITODE@AuNPs-PANI-Ab + $[Fe(CN)_6]^{4-/3-}$	0.05 ng mL^{-1}	Yagati et al. (2018)
	Parathion	Tomatoes, carrots	SPCE@Gr-Ab	52 pg L^{-1}	Mehta et al. (2016)

	AFB1	Olive oil	GCE@MWCNTs/IL-Ab	0.03 ng mL^{-1}	Yu et al. (2015)
	L. monocytogenes	Lettuce	IDME + AuNPs-PAb-urease + MNPs-MAb	300 cfu mL^{-1}	Chen et al. (2015)
	E. coli O157:H7	Food products	AuME@PANI/Glu-Ab	/	Chowdhury et al. (2012)
	E. coli K12	Chicken meat	IDME-Ab + $[Fe(CN)_6]^{4-/3-}$	103 cfu mL^{-1}	Helali et al. (2018)
	Fumonisins	Corn	GCE@PT/PDMA/MWCNT-Ab	0.014 FB1, 0.011 mg kg^{-1} FB2/FB3	Masikini et al. (2015)
	Okadaic acid	Mussels	SPCE@4-carboxyphenyl film	/	Hayat et al. (2012)
	SEB	Milk	AuE@PANI/IMS + $[Fe(CN)_6]^{4-/3-}$	0.017 ng mL	Wu et al. (2013)
MIP					
	Sulfaguanidine	Honey	Au-SPE@PAM + $[Fe(CN)_6]^{4-/3-}$	0.17 pg mL^{-1}	El Alami El Hassani et al. (2018b)
	N-NIP	Grilled meat	PtDE@MIP	36.9 nM	Lach et al. (2017)
Other bioreceptors					
DNA	*L. monocytogenes*	Milk	AuE@5C Pin film-DNA	1.10^{-13} M	Kashish et al. (2015)
T4 phage	*E. coli* K12	Milk	SPCMA@T4phage + MBs@T4phage	10^4 to 10^3 cfu mL^{-1}	Shabani et al. (2013)
Leucocin A	*L. monocytogenes*	/	AuIDE@AMPept (SAM)	1 cell/μL	Etayash et al. (2017)
Without bioreceptor (other electrode materials, nanocomposites)					
	Dyestuffs	/	AuE@MNCyclen-Cu^{2+}	AY25 10^{-12} M	Touzi et al. (2018)
	CAP	Milk, honey	GCE@MoS_2/f-MWCNTs + $[Fe(CN)_6]^{4-/3-}$	0.015 μM	Govindasamy et al. (2017a)

AHLs, *N-acyl homoserine lactones;* Al-MOFs, *aluminum-based MOFs;* AMPept, *antimicrobial peptide = leucocin A;* AY25, *acid yellow 25;* βCD, *β cyclodextrine;* BTX-2, *brevetoxin;* BSA, *bovine serum albumin;* 5-C, *5-carboxyindole;* 5-C Pin, *conducting polymer poly-5-carboxy indole;* Cbz-AgNPs, *carbamazepine-functionalized silver NPs;* Ce-MOF, *cerium-based metal-organic framework;* 4-CP, *4-carboxyphenyl;* E. coli, Escherichia coli; FTOE, *fluorine-doped tin oxide electrodes;* Glu, *glutaraldehyde;* IMS, *immunomagnetosome;* ITODE, *indium tin oxide disk electrodes;* L. monocytogenes, Listeria monocytogenes; Mab, *monoclonal antibody;* MCA, *melamine and cyanuric acid nanosheets;* MNCyclen, *methyl-naphthyl cyclen (1,4,7,10-tetraazacyclododecane) thin films;* Py/Py-COOH, *poly(pyrrole-co-pyrrole-2-carboxylic acid);* MoS_2/f-MWCNTs, *molybdenum disulfide nanosheets coated on functionalized MWCNTs;* MPTES, *3-mercaptopropyl-trimethoxysilane;* 4-NB, *4-nitrobenzenediazonium salt;* N-NIP, *N-nitroso-L-proline;* NPG, *nanoporous gold;* Or II, *orange II;* PA6, *polyamide 6;* PAb, *polyclonal antibody;* PAM, *molecularly imprinted electropolymer of acrylamide;* PDMS, *polydimethylsiloxane film;* PT-PDMA, *palladium telluride QDs and poly(2,5-dimethoxyaniline);* RB, *rhodamine B;* S. typhimurium, Salmonella typhimurium; SA2-BSA, *5-[4-(amino)phenylsulfonamide]-5-oxopentanoic acid-bovine serum albumin;* SH-SY5Y, *human neuroblastoma cell;* V_2O_5, *vanadium oxide;* SPCMAs, *screen-printed carbon microarrays;* SEB, *staphylococcal enterotoxin B;* $TaSi_2$, *tantalum silicide;* ZnO, *zinc oxide.*

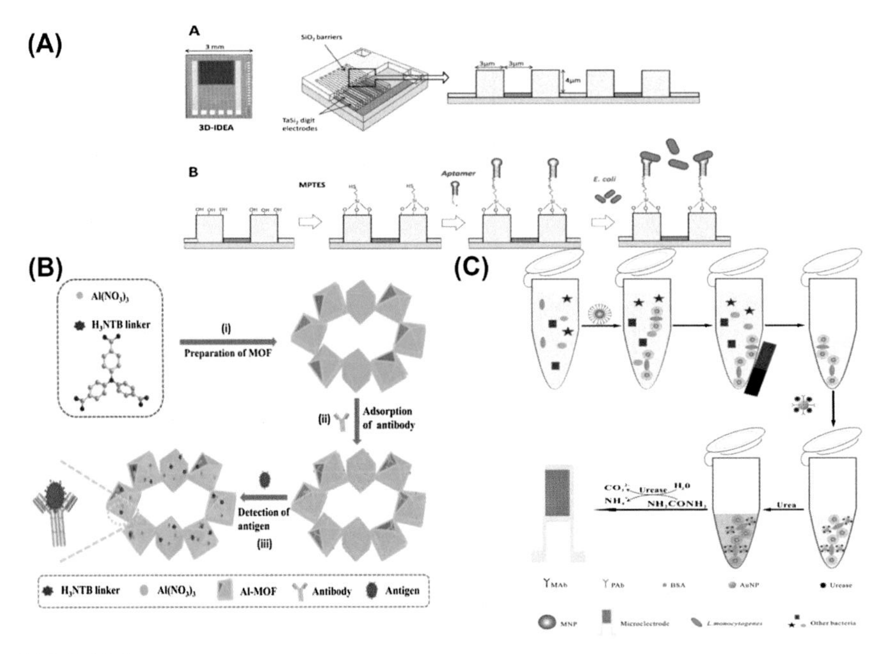

FIGURE 11.6

Different EIS methods implemented for the detection of contaminants in foodstuffs. (A) Impedimetric aptasensor for the detection of *Escherichia coli* O157 based on three-dimensional interdigitated electrode array (3D-IDEA) (A) functionalized with the aptamer (B). (B) Construction of an impedimetric immunosensor based on aluminum-metal-organic frameworks (Al-MOFs) for the detection of vomitoxin and salbutamol in wine and pork samples. (C) Development of an impedimetric immunosensor based on polyclonal and monoclonal antibodies immobilized on gold nanoparticles (AuNPs) and magnetic NPs (MNPs), respectively, for the detection of *Listeria monocytogenes*. When bacteria are present in the sample, AuNPs and MNPs form a complex from which urease can catalyze the hydrolysis of urea.

(A) Reprinted from Brosel-oliu, S., Ferreira, R., Uria, N., Abramova, N., Gargallo, R., Muñoz-Pascual, F.-X., Bratov, A., 2018. Novel impedimetric aptasensor for label-free detection of Escherichia coli O157:H7. Sensors and Actuators B: Chemical 255, 2988–2995; with permission from Elsevier. (B) Reprinted from Liu, C.-S., Sun, C.-X., Tian, J.-Y., Wang, Z.-W., Ji, H.-F., Song, Y.-P., Zhang, S., Zhang, Z.-H., He, L.-H., Du, M., 2017a. Highly stable aluminum-based metal-organic frameworks as biosensing platforms for assessment of food safety. Biosensors and Bioelectronics 91, 804–810; with permission from Elsevier. (C) Reprinted from Chen, Q., Lin, J., Gan, C., Wang, Y., Wang, D., Xiong, Y., Lai, W., Li, Y., Wang, M., 2015. A sensitive impedance biosensor based on immunomagnetic separation and urease catalysis for rapid detection of Listeria monocytogenes using an immobilization-free interdigitated array microelectrode. Biosensors and Bioelectronics 74, 504–511; with permission from Elsevier.

(e.g., plasmonic metal NPs, $BiPO_4$, multiferroic systems ($BiFeO_3$), tin-based materials Sn_3O_4, Sn_3O_4/TiO_2 composites). Tungsten disulfide (WS_2) belongs to the transition metal dichalcogenides (TMDs), which are two-dimensional semiconductor nanomaterials. WS_2 is particularly efficient for the adsorption of aptamers and single-stranded DNA (ssDNA). When the aptamer is bound to its target, the aptamer is desorbed from WS_2. The use of nanocomposites to functionalize electrodes can increase the photocurrent signal. The application of PEC sensors to the detection of contaminants in foodstuffs has emerged very recently (Table 11.5) because it is a newly emerging technique.

In some cases, a bioreceptor (eg. MIP) is immobilised onto the electrode to bind the target analyte (Fig. 11.7A) **(aptamer)**, Fig. 11.7B **(antibody)**)). In other applications, a bioreceptor is not necessary for the detection of the target analyte (Fig. 11.7C)). The photoactive material, as well as the bioreceptor, will determine the selectivity and the sensitivity of the developed PEC biosensor.

The first advantage of PEC biosensors is that it is a label-free technique. Furthermore, these sensors are characterized by simplicity, fast response, high signal-to-noise ratio, good repeatability, low background signal, high sensitivity, simple operation, and low cost.

11.2.7 Electrochemiluminescence sensor

Electrochemiluminescence (ECL) sensors are a combination of electrochemistry and measurement of visual luminescence. When a potential is applied onto an electrode, the electrode surface is excited. So an electron transfer is produced between molecules, and the resulting emitted light is measured.

ECL biosensors have been developed for the detection of bacteriological and chemical contaminants in foodstuffs (Table 11.6).

Materials widely used to build ECL sensors could be organic molecules (e.g., luminol) (Fig. 11.8A)), inorganic complexes, semiconductor nanocrystals (NCs), nanoparticles (Au NPs), and quantum dots (QDs) (e.g., CdS QDs, CdSe QDs) (Khoshbin et al., 2018). Luminol is a well-known organic ECL reagent, which is pH dependent. Ruthenium (Ru) is also currently used as an ECL reagent. Ru alone in solution has a poor ECL. However, when Ru is combined with some nanomaterials (e.g., Au NPs, QDs) or incorporated in metal-organic frameworks (MOFs) (e.g., RuMOFs: tris(2,2′-bipyridyl) dichlororuthenium(II) hexahydrate ($Ru(bpy)_3^{2+}$), tris((l,lO-phenanthroline)ru)ruthenium(II) ($Ru(phen)_3^{2+}$), Ru(bpy)32+-silica nanoparticles (RuSiNPs)) (Fig. 11.8B))) and is at a solid state, the ECL signal is increased (Hao and Wang, 2016). Indeed, some nanomaterials possess intrinsic ECL properties. A very innovative method for the detection of clenbuterol in pork liver, and kidney has been developed using upconversion nanoparticles (UCNPs) as ECL emitters (Jin et al., 2018). Reduced graphene oxide (rGO) was also used to enhance the ECL signal (Fig. 11.8C).

Table 11.5 Photoelectrochemical biosensors for the detection of various contaminants in foodstuffs classified by biosensing element.

Contaminant	Matrix	Electrode/Photoactive material	LOD	References
Enzymes				
AFB1	/	TiO_2NTs@AuNPs-AchE + AtCl	0.33 nM	Yuan et al. (2018)
Aptamers				
CAP	Milk powder	ITO@WS_2-Apt + DNase	3.6 pM	Zhou et al., (2018)
Tetracycline	Milk	ITO@$BiPO_4$/3DNGH-Apt	0.033 nM	Ge et al. (2018)
Ampicillin	Milk	ITO@$BiFeO_3$/utg-C_3N_4-Apt	0.33 pM	Ge et al. (2019)
Oxytetracycline	Chicken, raw milk	ITO@TiO_2/H-DNA@CdTe QD-Apt	0.19 nM	Li et al. (2017c)
BPA	Milk	ITO@AuNPs/ZnO-Apt	0.5 nM	Qiao et al. (2016)
SDMX	Milk	FTO@G-Bi_2S_3-Apt	0.55 nM	Okoth et al. (2016)
BPA	Milk	ITO@TiO_2/NG	0.3 fM	Zhou et al. (2016)
Kanamycin	/	FTO@GO/w-g-C_3N_4-Apt	0.2 nM	Li et al. (2014)
Acetamiprid	Apples, tomatoes	ITO@MWCNTs/rGONRs-CdTeQDs-Apt + Au NRs-DNA2	0.2 pM	Liu et al., (2016a)
Antibodies				
OTA	Milk	ITO@TiO_2/CdSe-Ab	2 pg mL^{-1}	Yang et al. (2015)
AFB1		FTO@CQDs/MnO_2 + MBs-Ab + GOase	2.1 pg mL^{-1}	Lin et al. (2017)
MIP				
2,4-D	Bean sprouts	CFP@Sn_3O_4-MIPPy	1.08×10^{-2} nM	Wang et al. (2018d)
BPA	Water (bottled)	PtE@Nb_2O_5-MIP	0.004 nM	Gao et al. (2018)
Other bioreceptors				
BoNT-LCA		ITOE@CdS QDs/SNAP25-AuNPs	1 pg mL^{-1}	Lin et al. (2018)
Without bioreceptor (other electrode materials, nanocomposites)				
BPA	Milk	FTO@TiO_2 NRAs/ZnPc	8.6 nM	Fan et al. (2018)
TBHQ	Vegetable oils	ITO@CdSe/ZnS QDs/LiTCNE	0.21 μM	Monteiro et al. (2017)

AtCl, *2-acetylsulfanylethyl (trimethyl) azanium, chloride;* AuNRs, *gold nanorods;* $BiPO_4$, *bismuth phosphate;* BoNT-LCA, *botulinum neurotoxin-light chain A;* CdSe, *cadmium selenide;* CdTe, *cadmium telluride;* CFP, *carbon fiber paper;* CN/BiOBr, *carbon nitride/ bismuth oxyhalide composites;* CQDs/MnO_2, *carbon QDs—functionalized manganese oxide nanosheets;* 2,4-D, *2,4-dichlorophenoxyacetic acid;* 3DNGH, *three-dimensional nitrogen-doped graphene hydrogel;* FTO, *fluorine-doped tin oxide glass electrode;* G-Bi_2S_3, *graphene-doped dibismuth trisulfide nanorods;* H-DNA, *hairpin DNA;* LiTCNE, *lithium tetracyanoethylenide;* MIPPy, *molecular imprinting polymer pyrrole;* rGONRs, *reduced graphene oxide nanoribbons;* N-GQDs, *nitrogen-doped graphene QDs;* Nb_2O_5, *niobium pentoxide;* NRAs, *nanorod arrays;* NTAs, *nanotube arrays;* PtE, *platinum electrode;* SDMX, *sulfadimethoxine;* Sn_3O_4, *tin oxide;* TBHQ, *tert-butyl-hydroquinone;* TiO_2NTs, *titanium dioxide nanotube array;* utg, *ultrathin graphite;* WS_2, *tungsten disulfide;* ZnPc, *zinc phthalocyanine;* ZnO, *zinc oxide nanopencils;* ZnS, *zinc sulfide.*

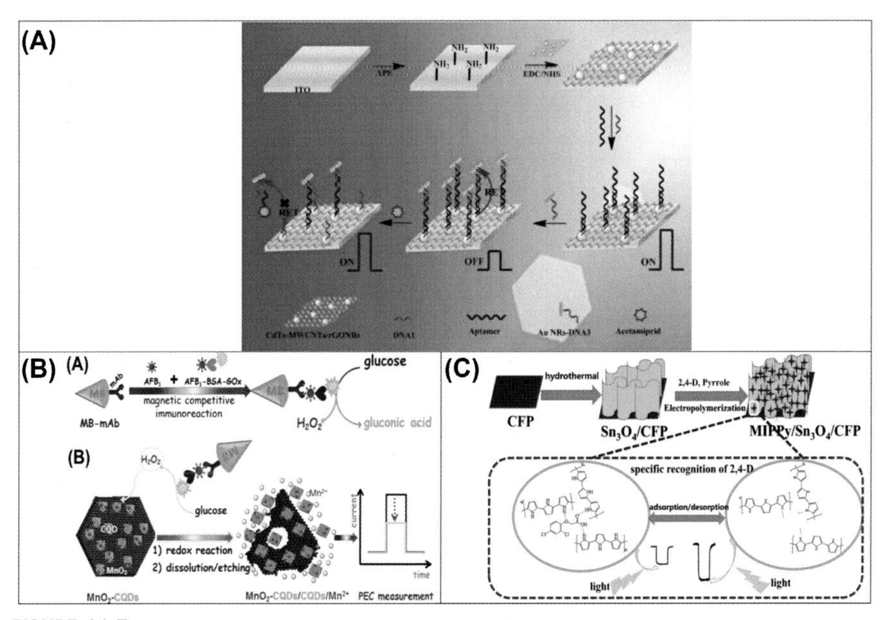

FIGURE 11.7

Different photoelectrochemical (PEC) methods implemented for the detection of contaminants in foodstuffs. (A) Fabrication of "on-off-on" PEC acetamiprid aptasensor based on the resonance energy transfer (RET). An ITO electrode is coated with a thin film of CdTe/MWCNTs/rGO nanoribbons (NRs) onto which the aptamer is immobilized. When aptamer is free for binding to cDNA3-Au nanorods (AuNRs), there is an RET from CdTe QDs to AuNRs. (B) Photoelectrochemical immunosensor for the detection of aflatoxin B1 (AFB1). A competitive immunoassay was developed using glucose oxidase (GOx) coupled to AFB1. GOx activity produces H_2O_2, which causes the dissociation of CQDs from the modified electrode (carbon QDs-coated MnO_2 nanosheets (MnO_2-CQDs)). (C) Photoelectrochemical sensor for 2,4-dichlorophenoxyacetic acid (2,4-D) detection based on carbon fiber paper (CFP) modified by Sn_3O_4 and an MIP layer.

(A) Reprinted from Liu, Q., Huan, J., Dong, X., Qian, J., Hao, N., You, T., Mao, H., Wang, K. 2016. Resonance energy transfer from CdTe quantum dots to gold nanorods using MWCNTs/rGO nanoribbons as efficient signal amplifier for fabricating visible-light-driven "on-off-on" photoelectrochemical acetamiprid aptasensor. Sensors and Actuators B: Chemical 235, 647–654; with permission from Elsevier. (B) Reprinted from Lin, Y., Zhou, Q., Tang, D., Niessner, R., Knopp, D., 2017. Signal-on photoelectrochemical immunoassay for aflatoxin B1 based on enzymatic product-etching MnO_2 nanosheets for dissociation of carbon dots. Analytical Chemistry 89, 5637–5645; with permission from ACS Publications. (C) Reprinted from Wang, J., Xu, Q., Xia, W.W., Shu, Y., Jin, D., Zang, Y., Hu, X., 2018b. High sensitive visible light photoelectrochemical sensor based on in-situ prepared flexible Sn_3O_4 nanosheets and molecularly imprinted polymers. Sensors and Actuators B: Chemical 271, 215–224; with permission from Elsevier.

Table 11.6 ECL biosensors for the detection of various contaminants in foodstuffs classified by biosensing element.

Contaminant	Matrix	Electrode/modifier of electrode surface	LOD	Reference
Enzymes				
OPP	Cabbage	GCE@MWNTs/Pt AuNPs/luminol/AchE-ChOx	From 0.08 to 29.7 nM	Miao et al. (2016a)
Aptamers				
OTA	Beer	GE@PHEMA/biotin-Apt + a-Si:H + SA-HRP + luminol	0.82 μg mL^{-1} OTA/DON, 3 ng mL^{-1} AFB1	Costantini et al. (2016)
OTA	Corn	ITO/$Ru(phen)_3^{2+}$/Apt-ssDNA + RecJf	2 pg mL^{-1}	Ni et al. (2018)
CAP	Fish	GCE@CdS NCs/Apt-ssDNA/EV (HRP)-Au-SSB + ExoI	0.034 pM (0.0079 pg mL^{-1})	Miao et al. (2016b)
E. coli	Meals	GCE@3DNGH/AgBr/luminol/CS/GA-Apt	0.17 cfu mL^{-1}	Hao et al. (2017)
VP	Seafood	GCE@GO/nanoFe$_3$O$_4$/ABEI-Ab	5 cfu mL^{-1}	Sha et al. (2016)
MG, CAP	Fish	WE1: SPCE@CdS QDs-MG Apt-Cy5 WE2: SPCE@ L-AuNPs-CAP Apt-CA	0.07 nM MG, 0.03 nM CAP	Feng et al. (2015)
Lincomycin	Meat	GCE@Au-GO + CDots-Apt + MIP	0.16 pM	Li et al. (2017b)
BPA	Milk, fruit juices	ITO@cDNA/Apt + $Ru(phen)_3^{2+}$	1.5 pM	Zhang et al. (2019)
Antibodies				
RAC	Pork meat	GCE@AuNPs/RAC + CdSe QDs/PDDA-Gr/AuNPs-Ab	2.6 pg mL^{-1}	Zhu et al. (2017)
Clenbuterol	Pork liver	GCE@AuNPs/CLB + QDs-Ab	0.0084 ng mL^{-1}	Yan et al. (2014)
AFM1	Milk	SPCE@Fe-GO/AFM1/Ab-CdTe QDs	0.3 pg mL^{-1}	Gan et al. (2013)
MIP				
MCPA	Oat, rice	AuE@AuNPs/NPs-CdSe/ZnS QDs-MIP	4.3 pM	Yang et al. (2016)
Melamine	Milk	GCE@Ru@SiNPs-MIP	0.5 pM	Lian et al. (2016)
OTA	Corn	GCE@RuSiNPs-MIP/CdTe QDs	3.0 fg mL^{-1}	Wang et al. (2016)
Clenbuterol	Pork liver and kidney	GCE@rGO/UCNPs-MIP	6.3 nM	Jin et al. (2018)
Without bioreceptor (others electrode materials, nanocomposites)				
Melamine	Milk	GCE@RuMOFs	38 pM	Feng et al. (2018)
Melamine	Milk	GCE@Nafion/CMWCNTs/RuSiNPs	0.1 pM	Chen et al. (2016)
Melamine	Milk	GCE@Nafion/RuSiNPs/P-rGO	0.1 pM	Zhou et al. (2014)
Formaldehyde	Seafood	AuE@Ru-SiNPs	6.0 nM	Chu et al. (2012)

a-Si:H, *hydrogenated amorphous silicon;* AuE, *gold electrode;* Au/g-C_3N_4, *plasmonic Au/graphitic carbon nitride composites;* AuNPs, *gold nanoparticles;* CA, *chlorogenic acid;* C-MWCNTs, *carboxylic acid–functionalized MWCNTs;* CdS, *cadmium sulfide;* CdSe, *cadmium selenide;* PDDA-Gr, *poly (diallyl dimethyl ammonium chloride)-graphene;* CdTe, *cadmium telluride;* ChOx, *choline oxidase;* CN/BiOBr, *carbon nitride/bismuth oxyhalide;* Cy5, *cyanine dye;* 3DNGH/AgBr, *AgBr nanoparticles–anchored 3D nitrogen-doped graphene hydrogel nanocomposites;* E. coli: Escherichia coli; EV, *EnVision reagent;* Fe-GO, *magnetic Fe_3O_4-graphene oxide;* GA, *glutaraldehyde;* L-AuNPs, *luminol-gold NPs;* MCPA, *2-methyl-4-chlorophenoxyacetic acid;* MG, *malachite green;* NCs, *nanocrystals;* NRAs, *nanorod arrays;* NiTAPc, *nickel tetra-amine phthalocyanine;* PHEMA, *poly(2-hydroxyethyl methacrylate;* RAC, *ractopamine;* RecJf, *RecJf exonuclease;* rGO, *reduced graphene oxide;* $Ru(bpy)_3^{2+}$, *tris(2,2′-bipyridyl) dichlororuthenium (II) hexahydrate;* RuMOFs, *Ru(bpy)32+-metal-organic frameworks;* Ru@SiNPs, *$Ru(bpy)_3^{2+}$-silica nanoparticles;* SA, *streptavidin;* SPCE, *screen-printed carbon electrode;* ssDNA, *single-stranded DNA;* SSB, *single-stranded DNA-binding protein;* UCNPs, *upconversion nanoparticles;* VP, Vibrio parahaemolyticus; WE, *working electrode;* ZnPc, *zinc phthalocyanine.*

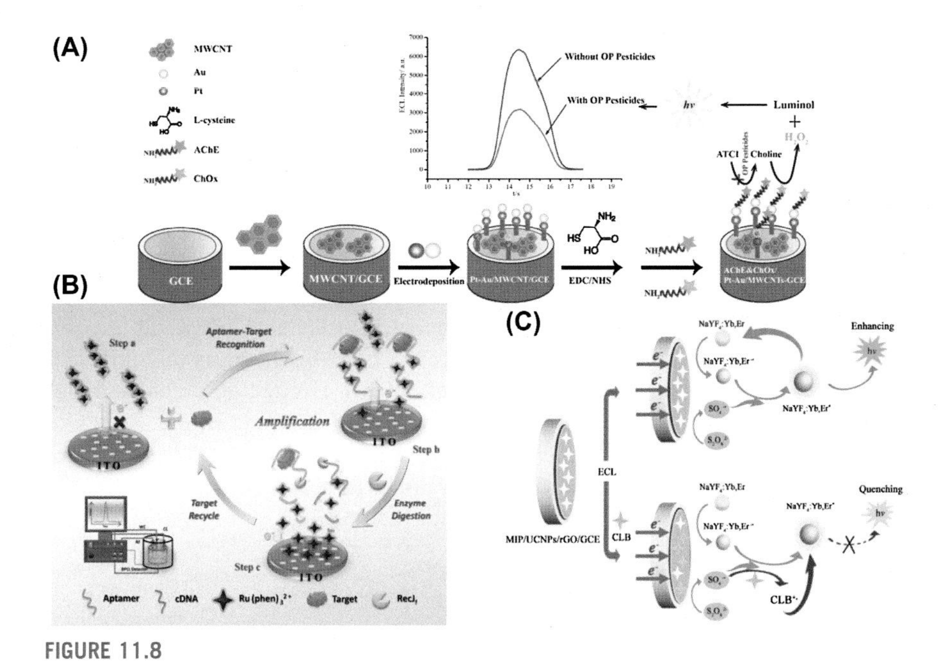

FIGURE 11.8

Different electrochemiluminescence (ECL) methods implemented for the detection of contaminants in foodstuffs. (A) Dual amplification of the ECL signal by using two enzymes (acetylcholinesterase (AChE) and choline oxidase (chOx) immobilized on a GCE functionalized with bimetallic Pt-Au/MWCNTs for organophosphate pesticides (OPs) detection. (B) ECL aptasensor for OTA detection based on ITO electrode and signal amplification by rolling-circle amplification (RCA). (C) MIP-based ECL biosensor for clenbuterol detection based on upconversion nanoparticles (UCNPs) as ECL emitters and reduced graphene oxide (rGO) to enhance the signal.

(A) Reprinted from Miao, S.S., Wu, M.S., Ma, L.Y., He, X.J., Yang, H., 2016. Electrochemiluminescence biosensor for determination of organophosphorous pesticides based on bimetallic Pt–Au/multi-walled carbon nanotubes modified electrode. Talanta, 158, 142–151; with permission from Elsevier. (B) Reprinted from Ni, J., Yang, W., Wang, Q., Luo, F., Guo, L., Qiu, B., Lin, Z., Yang, H., 2018. Homogeneous and label-free electrochemiluminescence aptasensor based on the difference of electrostatic interaction and exonuclease-assisted target recycling amplification. Biosensors and Bioelectronics 105, 182–187; with permission from Elsevier. (C) Reprinted from Jin, X., Fang, G., Pan, M., Yang, Y., Bai, X., Wang, S., 2018. A molecularly imprinted electrochemiluminescence sensor based on upconversion nanoparticles enhanced by electrodeposited rGO for selective and ultrasensitive detection of clenbuterol. Biosensors and Bioelectronics 102, 357–364; with permission from Elsevier.

11.2.8 Electrochemical surface plasmon resonance

The combination of electrochemistry with localized surface plasmon resonance (LSPR) produced by the excitation of noble metallic nanoparticles (e.g., gold, silver, copper) or metal/semiconductor composites (e.g., chitosan-AuNPs, graphene-AuNPs) could enhance the performance (sensitivity) of electrochemical sensors

(Wang et al., 2018b). Very few examples of applications to the detection of contaminants in foodstuffs have been found. One is the detection of chloramphenicol in fish samples by an SPR electrochemiluminescence assay based on aptamer recognition (Miao et al., 2016b). Gold nanoparticles labeled with a single-stranded DNA-binding protein (SSB), the enzyme HRP, and an ECL reagent (EV−Au−SSB) are responsible for the enhancement of ECL intensity of CdSNCs (cadmium sulfide nanocrystals) by SPR. A second example is an ECL biosensor for the detection of fumonisin B1 in milk band corn samples (Zhang et al., 2017). Gold nanoparticles (AuNPs) labeled with the MIP were used to enhance the ECL signal. A detection limit of 0.35 pg mL^{-1} was reported. A photoelectrochemical (PEC) aptasensor was developed for the detection of bisphenol A in milk, on the basis of gold nanoparticles labeled with an aptamer (Qiao et al., 2016). Excited gold nanoparticles enhance the photoelectrochemical signal by means of LSPR and so the sensitivity of the assay. The detection limit was determined at 0.5 nM. Another PEC aptasensor using Au NPs as PEC signal enhancer was developed for the screening of kanamycin in milk (Liu et al., 2018a) with a detection limit of 0.05 nM.

11.2.9 Electrochemical quartz crystal microbalance

QCM alone is not enough sensitive to the detection of small molecules. The combination of an electrochemical sensor to quartz crystal microbalance (QCM) could enhance the sensitivity. Electrochemical QCM (EQCM) produces real-time kinetic curves of mass change versus time. EQCM is very underdeveloped in the field of food safety. An Indian team has worked on the detection of sulfadiazine in milk based an antibody immobilized onto a gold quartz crystal. The limit of detection was 10 μg kg^{-1} in milk (Mishra and Bhand, 2012). A similar method was developed for streptomycin in milk (Mishra et al., 2015).

11.3 Perspectives/future developments

As it was observed by Dragone et al. (2017) for the detection of chemical contaminants in environmental samples, we can conclude that pulse voltammetric techniques (more sensitive, quicker) (DPV, SWV, and CV), as well as amperometry and electrochemical impedance spectroscopy (EIS), are very frequently used for the detection of chemical and microbiological contaminants in foodstuffs.

When looking at the recent publications on electrochemical biosensors from the 5 to 10 past years, most of the improvements of biosensor performance are due to the development of novel biosensing elements and the miniaturization of the biosensors (e.g., microfluidics, nanofluidics, lab on a chip (LOC)), as well as the advances in nanotechnologies (i.e., improvement of sensitivity) and the development of smartphones. Electrochemical biosensors could provide continuous monitoring systems (point-of-care (POC) testing) that are easy to use, sensitive, and quick.

Most often the technological advances in biosensors are first developed for the analysis of molecules in the human clinical field. Indeed, the needs are very similar to develop rapid, specific, inexpensive, and automated methods. So to see the trends of possible developments in the field of contaminant detection in foodstuffs, we can go and see what is happening recently in the clinical field (Campuzano et al., 2018). The perspectives for the future development of electrochemical biosensors for the detection of chemical and bacteriological contaminants in foodstuffs are highlighted in Fig. 11.10 and developed below.

11.3.1 Developing new biosensing elements

In the past, antibodies and enzymes were widely used to develop biosensors for the detection of contaminants. There is a need to develop synthetic materials with a high stability, low cost of production, and high selectivity and sensitivity. Nowadays, MIPs and aptamers are more and more used. The advantages of electrochemical sensors based on MIPs as the biosensing element are their high selectivity and sensitivity, better stability than antibodies, more resistant to drastic conditions (e.g., solvents), facile preparation, and low cost, reusability, versatility of the target analytes (antibiotics, pesticides) (Gui et al., 2018). Aptamers are also promising biosensing elements, which have to be more developed for biological and chemical contaminants because for the moment they are more developed in the clinical field than in the field of detection of food contaminants (Rhouati et al., 2013).

Enzymatic amperometric biosensors could be based on different enzymes: sugar oxidases (e.g., glucose oxidase), peroxidases (HRP), nitrate reductases, and blue multicopper oxidases (Bollella and Gorton, 2018). To improve sensitivity and selectivity of the enzymatic electrochemical biosensors, engineered enzymes (recombinant proteins) are developed to improve the recognition analyte/enzyme and to increase the electron transfer between the enzyme and the electrode. Many works are also performed on immobilization strategies of enzymes to control the enzyme orientation and to increase enzyme stability, by engineering enzymes and using nanomaterials.

11.3.2 Miniaturization: micro/nanofluidic platforms and micro/nanoelectrodes

To enhance the analytical performances of electrochemical biosensors, a lot of work has already been done to miniaturize and improve the fluidic systems and the electrodes.

A review presented different kinds of microfluidic devices (i.e., micro total Analysis systems (μTAS)), their components, and their fabrication, especially for their application in food safety (Atalay et al., 2011). The use of microfluidic devices allows to miniaturize the biosensor systems, to reduce analysis time, and so to enhance the portability and the POC testing for field use. Moreover, the electrochemical biosensors developed in combination with microfluidics are cost-effective, disposable, and with a multiplexing potential. Microfluidic analytical devices are already implemented in the field of food safety monitoring but not so much compared with clinical diagnostics (Weng and Neethirajan, 2017). The advantages of flow injection immunoanalysis are rapidity, sensitivity, automation, real-time monitoring, and high throughput. One drawback with the classical microfluidics is linked to complex matrices such as foodstuffs, which could block the microfluidic systems. A previous sample preparation should be needed before applying samples into the microfluidic system.

The main advances occurred during the past two years for the development of electrochemical biosensors have been reviewed (Pereira da Silva Neves et al., 2018). The technology progresses in microfabrication and electronics led to produce cost-effective, miniaturized, disposable, and portable sensors.

Apart from traditional physical materials (i.e., glass, silicon), one of the promising alternatives is printed papers (e.g., cellulosic paper) to develop low-cost POC tests because they are cost-effective and disposable. These sensors are named microfluidic paper—based sensors (μPADs). The passive liquid transport by capillarity is an advantage when compared with traditional microfluidics, especially when analyzing complex matrices such as foodstuffs because there is no risk of fluidics clogging (Chinnadayyala et al., 2018). These sensors, named μPADs, are also very interesting because they have a potential for multiplexing detection of analytes. The advantages of two-dimensional (2D) or three-dimensional (3D) (i.e., origami papers) paper-based biosensors are cost-effective fluidics, small sample volume, quick, sustainable material, capillary action of the paper, and relevant for POC testing development. A review of applications of paper-based electrochemical biosensors in different fields (i.e., clinical, environmental, food safety) during the five past years has been published (Bee Chin Lee et al., 2018). The first multiplex 3D origami paper—based device for the simultaneous detection of several classes of pesticides based on three different enzymes was developed with satisfactory performance characteristics (e.g., quick, cost-effective, sensitive (ppb level), accurate) (Arduini et al., 2018). Compared with traditional PADs, an innovative and promising hollow-channel paper device (HC-PAD) allows to control the flow rate paper analytical devices and to develop multiplex assays (Wang et al., 2018e).

A comprehensive review presents the advances in microfluidics, which allow to develop quick, sensitive, cost-effective, portable and high-throughput biosensors, and especially multiplexed biosensors (Liao et al., 2018). Independent units in the microfluidic systems could be a way to completely avoid the cross-talks between the different electrodes (e.g., parallel assay channels, independent cells). The fabrication and the advantages of arrays consisting of several SPEs in parallel (e.g., by inkjet printing) and paper-based devices (e.g., paper-based digital microfluidic chip) are also discussed. For a multiplexing detection of analytes, which is most commonly needed for contaminants in foodstuffs, it is worth to highlight that multielectrodes are of great interest. Arrays in which each electrode has a different antibody attached or label reporting antibodies with distinct labels are used to develop multiplex assays.

Furthermore, the integration of polymer-based substrates (e.g., polydimethylsiloxane (PDMS)) to develop electrochemical biosensors has to be highlighted. The drawbacks of traditional potentiometric ISEs could be solved by the fabrication of polymeric ISMs (e.g., poly(vinyl chloride) (PVC)) (Cuartero and Crespo, 2018) and CP film ((e.g., polypyrrole (PPy), poly(3-octylthiophene) (POT), polyaniline (PANI), and poly(3,4- ethylenedioxythiophene (PEDOT)). The ISMs are deposited onto the CP film.

Among the different electrode patterning techniques, screen-printing and ink jet printing belong to low-cost manufacturing techniques. SPEs appeared in the 1990s. Their advantages have to be highlighted: low cost, versatility of applications, miniaturization, portability, POC testing, multiplexing, disposable, and possible coupling with magnetic beads (Campuzano et al., 2018). For a multiplex detection, several biosensing elements (e.g., enzymes) could be deposited one by one onto the surface of the electrode arrays by using a multinozzle inkjet system (i.e., "drop-on-demand" inkjet printing) (Li et al., 2018).

Concerning the electrodes, some developments have been done in the past years by developing new materials to produce electrodes (Dragone et al., 2017). One of them is the LOC technology for which a comprehensive review of the applications especially to the development of electrochemical biosensors has been published (Lafleur et al., 2016). Even if LOC technology related to the development of electrochemical biosensors for the detection of contaminants in foodstuffs is still in its infancy, this combination presents an indisputable potential (e.g., mycotoxins (Guo et al., 2015)).

Nanoelectrodes (i.e., below 100 nm) owing to the miniaturization can be integrated into LOC devices and allow developing rapid and sensitive methods. They could be used as an individual electrode or in an array of electrode nanobiosensors (Abi et al., 2018). The proof-of-concept of an array of IDEs within a microfluidic cartridge was done for the impedimetric detection of blood biomarkers (Lakey et al., 2019). This system could be used to develop low-cost and POC testing for all target analytes for which a biosensing element is available.

Biomedical microelectromechanical system (Bio-MEMS), which belongs to the LOC technology, was first applied to clinical diagnostics (Jivani et al., 2016). There are different modes of fabrication. Thin films could be deposited onto substrate materials (e.g., silicon, glass, or plastics). A pattern could be transferred onto a material (i.e., photolithography), or defined patterns could be etched onto the surface of materials or thin films. This technology could also profit to the detection of contaminants in foodstuffs (El Alami El Hassani et al., 2018a; Hassani et al., 2017).

11.3.3 Amplification strategies

Some recent trends in amplification strategies applicable to electrochemical biosensors have been reviewed (Liu et al., 2018b).

11.3.3.1 Nanomaterials

It is worth to highlight that to improve the analytical performances of electrochemical biosensors (i.e., sensitivity), nanomaterials are used to modify surface architectures and functions of electrodes. Atomic layer deposition (ALD) is one of the recently developed techniques to precisely deposit thin film of nanomaterials on the electrode surfaces with good conformality and uniformity for the immobilization for a biosensing element (Graniel et al., 2018). Nanomaterials could be hybridized with other materials to obtain synergistic effects or could be associated to biosensing elements to improve the selectivity.

In combination or not with biosensing elements, nanomaterials are used to develop more selective and sensitive biosensors for the detection of biological and chemical contaminants in foodstuffs (Lim and Kim, 2016; Socas-Rodríguez et al., 2017; Liu et al., 2014).

There are different kinds of nanomaterials: carbon-based nanomaterials (e.g., CNTs, graphene (GR), or graphene oxide (GO)-based nanoparticles (NPs)) (Lawal, 2018), metallic nanoparticles (e.g., noble-metal NPs (Au, Ag, Pt), metal oxides (TiO_2, MnO_2, CuO, ZnO), quantum dots (QDs) (i.e., semiconductor NPs (e.g., CdSe/ZnS core shell)), uncommon nanomaterials (e.g., nanohorns, fullerenes (the most known C60 made up of 60 closely packed carbon atoms), membranes, and CPs. The CNTs are constituted of one, two, or several layers of atoms of carbon, which form a tube (i.e., single-walled carbon nanotubes (SWCNTs), double-walled carbon nanotubes, and multiwalled carbon nanotubes (MWCNTs)). Fullerenes are stable single-walled cage molecules (Kurbanoglu et al., 2017) that work as acceptors. The properties and functionalities of the different classes of nanomaterials are summarized in a review, which especially deals with biosensors developed to assess food safety and quality (Warriner et al., 2014). Some advantages of each type of nanomaterials are presented in a comprehensive review for biological and biomedical applications (Maduraiveeran et al., 2018). The main advantages of nanomaterials are that they have high conductivity, and they are able to selectively adsorb high quantity of biosensing elements, because of a huge surface of binding (Cho et al., 2018).

Nanomaterials could be directly used to modify the surface chemistry of the electrodes and/or biomolecules (e.g., antibodies, aptamers). They can act as an immobilization support (or carriers) for biosensing elements, as a signal amplifier (i.e., by enhancement of the electron transfer), as artificial biosensing elements or signal tracer (i.e., artificial enzymes or nanozymes (e.g., catalase, oxidase, peroxidase like activities of NPs) (Jv et al., 2010; Jiang et al., 2017)), as redox tracers, or as magnetic separators (e.g., magnetic beads) to develop electrochemical biosensors (Rhouati et al., 2017; Abi et al., 2018).

Chitosan (CS) could be codoped with electroactive materials to improve the dispersibility of nanomaterials and the stability of the modified surface (Gui et al., 2018). Other kind of materials could play the same rule (e.g., ionic liquids (ILs), some metal NPs).

11.3.3.1.1 The use of nanomaterials as immobilization support

Nanomaterials are innovative materials that could be used to modify the surface architecture of electrodes. A good immobilization of the biosensing element onto the surface of the electrode is very crucial for the development of a biosensor. Indeed, the biosensing element should be firmly attached to avoid leakage (i.e., stability), properly oriented to recognize its target, and should not be denaturized. Nanomaterials have a strong adsorption capability, reduce nonspecific adsorption, and improve the stability of the electrode surface. Nanomaterials (e.g., magnetic NPs, magnetic beads) could be used for purification and preconcentration of target analytes (Conzuelo et al., 2014). Their use provides a huge surface area, which allows binding more biosensing elements than a classical electrode alone and so more surface for bioreactions (Campuzano et al., 2018). The combination of magnetic separation with the electrochemical detection allows improving the biosensors sensitivity.

Ordered mesoporous carbon (OMC) materials are three-dimensional (3D) carbon nanostructures with good conductivity and intrinsic electrocatalytic capacities. The surface of electrodes could be modified by OMC materials to allow the immobilization of large quantities of electrochemical mediators or of biosensing elements (e.g., MIP, aptamers, enzymes) (Walcarius, 2017). OMC materials have been applied to the development of electrochemical biosensors for the detection of contaminants in foodstuffs (e.g., dimetridazole (Yang and Zhao, 2015b), chloramphenicol (Zhu et al., 2015)).

11.3.3.1.2 The use of nanomaterials as signal amplification strategies

The electrochemical signals could be enhanced by nanomaterials because the surface of interaction is increased, the electron transfer (i.e., conductivity) is improved, and so conjugation and redox reactions are facilitated. Furthermore the spacing of bioreceptors could be increased, and the catalytic activity of enzyme amplified (Campuzano et al., 2018; Kurbanoglu et al., 2017). The consequence is a synergistic effect, which improves sensitivity and potentially increases selectivity and stability of electrochemical biosensors. The effects of the modification of paper-based electrodes by different nanomaterials were studied. The sensitivity of a developed electrochemical biosensor is improved by the use of nanomaterial (i.e., AuNPs, GO, CNFs), depending on the type of nanomaterial, and by the use of hybrid nanocomposites (Sanchez et al., 2018).

Several electroactive materials could be used at the same time in parallel to develop a much more sensitive electrochemical biosensor because they can have synergistic effect. These biosensors are using dual or even multielectroactive nanomaterials (e.g., GR-AuNPs, MnO_2-MWCNTs, GO- MWCNTs, AuNPs-GO) to modify the electrodes that could significantly improve the sensitivity of electrochemical signal. Carbon and metallic nanomaterials (metal oxide NPs) could be used together to immobilize biosensing elements (e.g., enzymes) to produce a synergistic effect toward enzymatic catalysis. For example, a ternary composite has been prepared with titanate nanotubes (TiO_2NTs), polyaniline (PANI), and gold nanoparticles (AuNPs) applied onto a GCE to enhance the catalytic current of hydrogen peroxide produced by the enzyme HRP (Liu et al., 2016b)).

CPs or nanocomposites of CPs could be deposited onto the electrode by inkjet printing to form CP-based electrodes (Põldsalu et al., 2018). A synergistic effect of nanomaterials and CPs is observed (Wang et al., 2018c).

MOFs are constituted of electroactive metal ions (e.g., cadmium, copper, silver, mercury) connected by conductive organic ligands (Elsaidi et al., 2018). MOFs present outstanding optical and electromagnetic properties. They are flexible and porous nanostructures with large surface area and active sites, which can be used for target analyte separation and electrocatalysis. Some of the MOFs, the redox-active MOFs (ra-MOFs), are applicable to the development of more sensitive electrochemical biosensors (Liu and Yin, 2016). MOF composites are a combination of MOFs with guest molecules integrated into the nanopores (e.g., enzymes, conductive substrates, NPs) (Kempahanumakkagari et al., 2018). The increase of the electrochemical signal is higher when highly conductive materials (e.g., electrochemically reduced graphene oxide (ERGO), CNTs, noble NPs) are used in combination with MOFs. An MOF composite film is deposited onto the electrode surface to promote electrochemical reactions.

Nanowire sensors are constituted of an assembly of an active material (e.g., carbon nanotubes, silicon, noble metals (gold, silver), conducting polymers), which forms a nanostructure (Hernandez-Vargas et al., 2018). Nanowire sensors are used to modify the surface of the electrodes to enhance the electron transfer and so to improve the sensitivity of the developed sensor. They are easy to prepare and present a large surface for immobilization. Organic-inorganic hybrid nanoflowers (HNFs), which are an assembly of metal ions with organic substances (e.g., enzymes, antibodies), offer a better catalytic activity, stability, and biocompatibility than enzymes or antibodies alone (Zhu et al., 2018) (e.g., ractopamine (Zhang et al., 2015)). The recent advances of two-dimensional nanomaterials (2DNMs) are presented in a comprehensive review regarding the development of electrochemical biosensors based on 2DNMs (i.e., graphene- and molybdenum disulfide (MoS_2)-based nanocomposites) for the detection of environmental pollutants (Su et al., 2018).

In the few past years, many advances have been performed in the development of silica-based nanomaterials (i.e., silica-based NPs and ordered mesoporous silica thin films), which are very interesting tools as electrode modifiers (Walcarius, 2018). They are particularly interesting when combined with MIPs integrated in silica thin films because the recognition of target analytes by MIP is enhanced.

Multiplex methods for the simultaneous detection of several contaminants could be developed if nanomaterials (e.g., AuNPs) with different sizes are produced (Alim et al., 2018).

The use of nanomaterials to develop electrochemical biosensors to detect contaminants in foodstuffs is, however, still in its infancy. Attention has to be paid to nanomaterials because they could become environmental pollutants (i.e., toxic substances). A lot of work has to be done on the interferences of nanomaterials with complex matrices such as food (Liu et al., 2014).

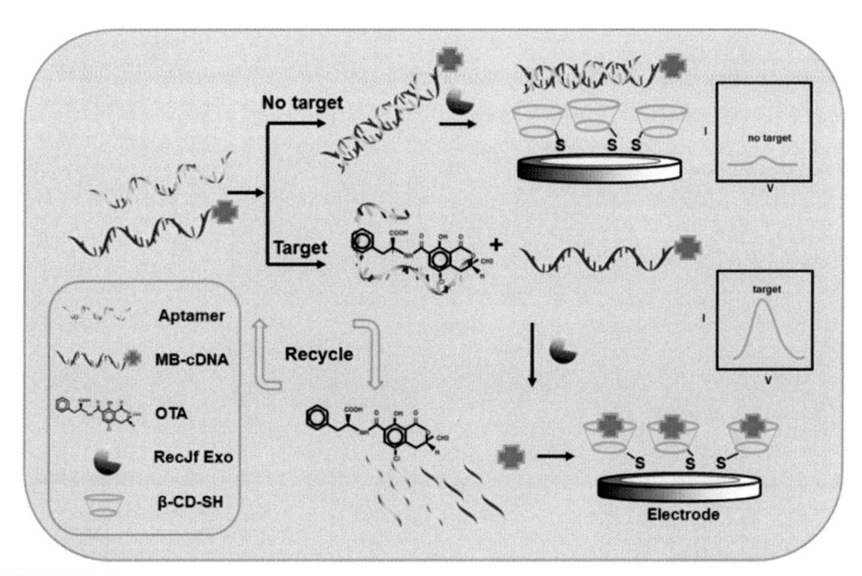

FIGURE 11.9

Basic principle of the rolling-circle amplification (RCA). One of the techniques of amplification of nucleic acids to enhance the electrochemical signal and so to improve the sensitivity.

Reprinted from Wang, Y., Ning, G., Wu, Y., Wu, S., Zeng, B., Liu, G., He, X., Wang, K., 2019. Facile combination of beta-cyclodextrin host-guest recognition with exonuclease-assistant signal amplification for sensitive electrochemical assay of ochratoxin A. Biosensors and Bioelectronics 124–125, 82–88; with permission from Elsevier.

11.3.3.2 Nucleic acid amplification strategy (rolling-circle amplification)

Among the isothermal exponential amplification techniques, loop-mediated amplification (LAMP), recombinase polymerase amplification (RPA), and rolling-circle amplification (RCA) could be used in combination with electrochemical biosensors (Fig. 11.9). These amplification strategies are based on the combination of strand displacement polymerases (e.g., nucleases) with designed primers or probes (Yan et al., 2016). The principle of RCA is that the aptamer forms a stem-loop structure with a DNA probe bound onto the electrode. When the aptamer binds to its target analyte, the stem-loop structure is denaturized and the DNA strand is released. Then the nucleases (e.g., DNase, RecJf exonuclease) specifically digest the DNA strand. The aptamer is released and therefore free to bind again to the DNA probe. A new cycle starts when the aptamer binds again with the target analyte, which leads to the amplification of the electrochemical signal.

These nucleic acid amplification strategies have been applied to the detection of contaminants in foodstuffs (e.g., ochratoxin A (Yang et al., 2014; Wang et al., 2019), *E. coli* O157:H7 and *Salmonella enterica* (Huang et al., 2018), *Vibrio parahaemolyticus* (Teng et al., 2017), *Salmonella* (Zhu et al., 2014).

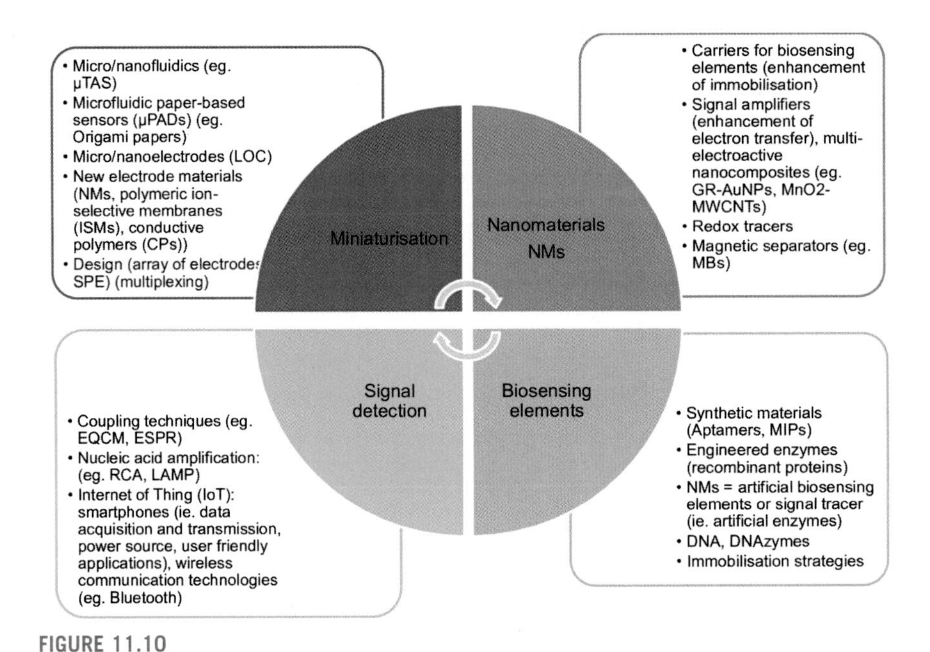

FIGURE 11.10

Perspectives in the development of electrochemical biosensors for the detection of chemical and bacteriological contaminants in foodstuffs.

11.3.4 Smartphone-based biosensors for on field use ("point-of-care testing")

Because of the progresses in Internet-of-Things (IoT) technology and the phenomenous development of smartphones (i.e., ubiquitous), there is a growing interest to develop electrochemical biosensors in combination with smartphones by the way of wireless communication technologies (e.g., Bluetooth) (Pereira da Silva Neves et al., 2018). Wireless chemical sensors (i.e., potentiometric, amperometric, and voltammetric) for the detection of chemical contaminants in environmental samples have been reviewed (Kassal et al., 2018).

Even if the use of smartphones is for the moment most developed for clinical assays for POC testing, their use is also promising to develop electrochemical biosensors for the detection of contaminants in foodstuffs. Smartphones could be divided into several categories depending on how the smartphone is interfaced with the biosensor for diagnostics and what is its function (Russell and de la Rica, 2018; Kanchi et al., 2018): real-time acquisition of data coming from the biosensor (Kassal et al., 2018), transmission of data to a user or a client, control of the biosensor, power source for the biosensor, readout device (i.e., camera as an electrochemiluminescence reader (Roda et al., 2016)). Specific lab-on smartphone and user-friendly applications are developed to interpret the signal reported by an electrochemical reader.

The combination of POC devices with smartphones for food safety monitoring (i.e., foodborne bacteria, chemical contaminants (pesticides, antibiotics)) was reviewed (Wu et al., 2017). This chapter demonstrates the potential of POC devices for the surveillance of contaminants in foodstuffs, the advances already performed in this field, and finally the perspectives to improve these devices and their performances (i.e., rapid and multiplexed methods, commercial systems, robustness for field applications, user-friendly applications for smartphones).

Indeed at that time, there are no commercial biosensors interfaced with smartphones in the field of the detection of contaminants in foodstuffs. Even though it is a very promising trend, we are only at the early stage of research and development, at best applied research.

The development of electrochemical biosensors to detect contaminants in foodstuffs and in environment has been continuously increasing since 10 years and will continue to grow. Indeed the demand from the consumers regarding the guarantee of food safety and even the possibility to checking themselves the absence of contaminants (e.g., allergens, antibiotics) in their food is growing. People are more and more aware about the potential presence of these contaminants and about the risks. The development of nanotechnologies will help producing electrochemical biosensors, which could answer to this demand as sensitive and cost-effective POC devices. The development of multiplex biosensors is necessary especially when different families of contaminants have to be detected in all-in-one test, to be able to compete with multiresidue physicochemical methods, which have progressed a lot also during the past 10 years.

Even if the future prospects of electrochemical biosensors for the detection of contaminants in foodstuffs is very promising, there are some remaining challenges. In the field of microbiological and chemical contaminants, the new biosensing elements (i.e., aptamers, MIPs) are not enough developed and commercialized to be able to develop relevant electrochemical biosensors. Furthermore to control contaminants in foodstuffs, there is a need of multiplex methods. Therefore the future developed biosensing elements should be selective for at least one family of compounds and not only for one substance in most of the cases (e.g., penicillin family instead of penicillin G alone). The biosensing elements should be more stable to be able to produce commercial biosensors with a long-term storage (Rotariu et al., 2016). Another solution to remove the issue of stability is to produce cost-effective and disposable biosensors.

The application of portable POC devices in food safety on the field or at the farm is also linked to the development of self-powered devices. It is already a search path for clinical diagnostics (Montes-Cebrián et al., 2018). A self-powered system named "plug-and-power' system was developed for glucose detection in human blood. According to the authors, this system should be applicable to electrochemical sensors for the detection of other analytes than glucose and in other matrices than blood. A lot of work has to be done to transform some of these prototypes biosensors into commercial products able to detect contaminants in foodstuffs in real conditions on field or in the laboratories. The transfer of electrochemical biosensors from research to the market needs the collaboration of different disciplines (e.g., biology, organic chemistry, electrochemistry, physics, informatics).

None of the reported biosensors has been validated as it is required by the EU regulation. About 99% of the electrochemical biosensors presented here as examples have been developed in buffer and generally tested with few real samples. Therefore the step of a fully validation with real samples from different origins should be performed. The robustness of the methods has to be proved.

List of abbreviations

Ab	Antibody
AchE	Acetylcholinesterase
AFM1/AFB1	Aflatoxin M1/B1
Apt	Aptamer
ASV	Adsorptive stripping voltammetry
AuE	Gold electrode
Au IDE	Interdigitated gold electrode
Au ME	Gold microelectrode
AuNPs	Gold nanoparticles
AuSPE	Screen-printed gold electrode
BPA/BPS	Bisphenol A/S
cDNA	Capture/complementary DNA
CA	Chronoamperometric measurements
CAP	Chloramphenicol
CdTe QDs	Cadmium telluride quantum dots
CE	Carbon electrode
CEPH	Cephalosporins
CMOS	Complementary metal oxide semiconductor
CNM	Cellulose nitrate membrane
CNT/F	Carbon nanotubes/fibers
CPE	Carbon paste electrode
CPs	Carbamate pesticides
CS	Chitosan
3D-Cu NPs	Three-dimensional copper nanoparticles
CV	Cyclic voltammetry
DPV	Differential pulse voltammetry
dsDNA	double-stranded DNA
ErGO	Electrochemically reduced GO
ExoNase	Exonuclease (Exo I)
FB1	Fumonisin B1
$[Fe(CN)6]^{-3/-4}$	Hexacyanoferrate/hexacyanoferrite
G©E	Glassy (carbon) electrode
GDE	Graphite-disk electrode
GO	Graphene oxide
GOase	Glucose oxidase
GONRs	Graphene oxide nanoribbons
Gr	Graphene

H_2O_2E	Electrode for hydrogen peroxide
HgE	Mercury drop electrode
hNiNS	N\Hollow nickel nanospheres
HRP	Horseradish peroxidase
PCR	Polymerase chain reaction
IDE(A)	Interdigitated electrode (array)
IDME	Interdigitated microelectrodes
ISE	Ion-selective electrode
ITOE	Indium tin oxide electrode
LSV	Linear sweep voltammetry
LSSV	Linear stripping sweep voltammetry
MB	Methylene blue
MBs	Magnetic beads
MIP	Molecular imprinting polymer
MMIP	Magnetic MIP
MNPs	Magnetic NPs
MOF	Metal-organic framework
MP	Methyl parathion
MWCNTs	Multiwalled carbon nanotubes
OPP	Organophosphorous pesticides
OTA	Ochratoxin A
PAMAM	Poly(amidoamine)
PANI	Polyaniline nanofibers
PAP	Peanut allergenic proteins
PB	Prussian blue
PBP	Penicillin-binding protein
PENI	Penicillin G
PM	Polymeric membrane
PPyMIP	Molecularly imprinted overoxidized polypyrrole
Pt(D)E	Platinum (disk) electrode
PVC	Polyvinyl chloride
QD	Quantum dot
RCA	Rolling-chain amplification
rGO/PB MCs	Reduced graphene oxide/Prussian blue microcubes
SAMs	Self-assembled (SA) monolayer
SATUM	SA thiourea monolayer
SDPV	Stripping differential pulse voltammetry
SDS	Sodium dodecyl sulfate
SPCE	Screen-printed carbon electrode
SPdCEs	Dual SPCE
SPGrE	Screen-printed graphite electrode
SP IDME	Screen-printed interdigitated microelectrode
SULFA	sulfonamides
SWCNTs	Single-walled carbon nanotubes
SWV	Square-wave voltammetry
SWASV	Square-wave anodic stripping voltammetry
TTC	Tetracyclines
VAMWCNTs	Vertically aligned multiwalled carbon nanotubes

References

Abdel-Ghany, M.F., Hussein, L.A., EL Azab, N.F., 2017. Novel potentiometric sensors for the determination of the dinotefuran insecticide residue levels in cucumber and soil samples. Talanta 164, 518–528.

Abi, A., Mohammadpour, Z., Zuo, X., Safavi, A., 2018. Nucleic acid-based electrochemical nanobiosensors. Biosensors and Bioelectronics 102, 479–489.

Abnous, K., Danesh, N.M., Ramezani, M., Alibolandi, M., Taghdisi, S.M., 2018. A novel electrochemical sensor for bisphenol A detection based on nontarget-induced extension of aptamer length and formation of a physical barrier. Biosensors and Bioelectronics 119, 204–208.

Afonso, A.S., Perez-Lopez, B., Faria, R.C., Mattoso, L.H., Hernandez-Herrero, M., Roig-Sagues, A.X., Maltez-da Costa, M., Merkoci, A., 2013. Electrochemical detection of Salmonella using gold nanoparticles. Biosensors and Bioelectronics 40, 121–126.

Afzali, D., Padash, M., Mostafavi, A., 2015. Determination of trace amounts of zearalenone in beverage samples with an electrochemical sensor. Mycotoxin Research 31, 203–208.

Aggerbeck, M., Blanc, E.B., 2018. Role of mixtures of organic pollutants in the development of metabolic disorders via the activation of xenosensors. Current Opinion in Toxicology 8, 57–65.

Alahi, M., Mukhopadhyay, S., 2017. Detection methodologies for pathogen and toxins: a review. Sensors 17, 1885.

Alarcan, J., Biré, R., Le Hégarat, L., Fessard, V., 2018. Mixtures of lipophilic phycotoxins: exposure data and toxicological assessment. Marine Drugs 16, 46.

Albero, B., Tadeo, J.L., Escario, M., Miguel, E., Pérez, R.A., 2018. Persistence and availability of veterinary antibiotics in soil and soil-manure systems. The Science of the Total Environment 643, 1562–1570.

Alegbeleye, O.O., Singleton, I., Sant'Ana, A.S., 2018. Sources and contamination routes of microbial pathogens to fresh produce during field cultivation: a review. Food Microbiology 73, 177–208.

Alim, S., Vejayan, J., Yusoff, M.M., Kafi, A.K.M., 2018. Recent uses of carbon nanotubes & gold nanoparticles in electrochemistry with application in biosensing: a review. Biosensors and Bioelectronics.

Alonso-Lomillo, M.A., Domínguez-Renedo, O., Román, L.D.T.-D., Arcos-Martínez, M.J., 2011. Horseradish peroxidase-screen printed biosensors for determination of Ochratoxin A. Analytica Chimica Acta 688, 49–53.

Ammida, N.H.S., Volpe, G., Draisci, R., Delliquadri, F., Palleschi, L., Palleschi, G., 2004. Analysis of erythromycin and tylosin in bovine muscle using disposable screen printed electrodes. Analyst 129, 15–19.

Anirudhan, T.S., Alexander, S., 2015. Design and fabrication of molecularly imprinted polymer-based potentiometric sensor from the surface modified multiwalled carbon nanotube for the determination of lindane (γ-hexachlorocyclohexane), an organochlorine pesticide. Biosensors and Bioelectronics 64, 586–593.

Antunes, P., Campos, J., Mourão, J., Pereira, J., Novais, C., Peixe, L., 2018. Inflow water is a major source of trout farming contamination with Salmonella and multidrug resistant bacteria. Science of The Total Environment 642, 1163–1171.

Arduini, F., Cinti, S., Caratelli, V., Amendola, L., Palleschi, G., Moscone, D., 2018. Origami multiple paper-based electrochemical biosensors for pesticide detection. Biosensors and Bioelectronics.

Arora, P., Sindhu, A., Dilbaghi, N., Chaudhury, A., 2011. Biosensors as innovative tools for the detection of food borne pathogens. Biosensors and Bioelectronics 28, 1–12.

Arsi, K., Donoghue, D.J., 2017. Chapter 19 - chemical contamination of poultry meat and eggs. In: Schrenk, D., Cartus, A. (Eds.), Chemical Contaminants and Residues in Food, second ed. Woodhead Publishing.

Asadollahi-Baboli, M., MANI-Varnosfaderani, A., 2014. Rapid and simultaneous determination of tetracycline and cefixime antibiotics by mean of gold nanoparticles-screen printed gold electrode and chemometrics tools. Measurement 47, 145–149.

Asnaashari, M., Kenari, R.E., Farahmandfar, R., Abnous, K., Taghdisi, S.M., 2019. An electrochemical biosensor based on hemoglobin-oligonucleotides-modified electrode for detection of acrylamide in potato fries. Food Chemistry 271, 54–61.

Atalay, Y.T., Vermeir, S., Witters, D., Vergauwe, N., Verbruggen, B., Verboven, P., Nicolaï, B.M., Lammertyn, J., 2011. Microfluidic analytical systems for food analysis. Trends in Food Science & Technology 22, 386–404.

Bagheryan, Z., Raoof, J.-B., Golabi, M., Turner, A.P.F., Beni, V., 2016. Diazonium-based impedimetric aptasensor for the rapid label-free detection of *Salmonella typhimurium* in food sample. Biosensors and Bioelectronics 80, 566–573.

Bahadir, E.B., Sezgintürk, M.K., 2015. Applications of commercial biosensors in clinical, food, environmental, and biothreat/biowarfare analyses. Analytical Biochemistry 478, 107–120.

Bee Chin Lee, V., Faizah Mohd-Naim, N., Tamiya, E., Ahmed, M.U., 2018. Trends in Paper-Based Electrochemical Biosensors: From Design to Application.

Bhatt, V.D., Joshi, S., Becherer, M., Lugli, P., 2017. Flexible, low-cost sensor based on electrolyte gated carbon nanotube field effect transistor for organo-phosphate detection. Sensors 17.

Bollella, P., Gorton, L., 2018. Enzyme based amperometric biosensors. Current Opinion in Electrochemistry 10, 157–173.

Bougrini, M., Florea, A., Cristea, C., Sandulescu, R., Vocanson, F., Errachid, A., Bouchikhi, B., El Bari, N., Jaffrezic-Renault, N., 2016. Development of a novel sensitive molecularly imprinted polymer sensor based on electropolymerization of a microporous-metal-organic framework for tetracycline detection in honey. Food Control 59, 424–429.

Boulanouar, S., Mezzache, S., Combès, A., Pichon, V., 2018. Molecularly imprinted polymers for the determination of organophosphorus pesticides in complex samples. Talanta 176, 465–478.

Brosel-Oliu, S., Ferreira, R., Uria, N., Abramova, N., Gargallo, R., Muñoz-Pascual, F.-X., Bratov, A., 2018. Novel impedimetric aptasensor for label-free detection of *Escherichia coli* O157:H7. Sensors and Actuators B: Chemical 255, 2988–2995.

Bunney, J., Williamson, S., Atkin, D., Jeanneret, M., Cozzolino, D., Chapman, J., Power, A., Chandra, S., 2017. The use of electrochemical biosensors in food analysis. Current Research in Nutrition and Food Science 5, 183–195.

Campuzano, S., Yáñez-Sedeño, P., Pingarrón, J.M., 2018. Current trends and challenges in bioelectrochemistry for non-invasive and early diagnosis. Current Opinion in Electrochemistry.

Castillo, G., Spinella, K., Poturnayová, A., Šnejdárková, M., Mosiello, L., Hianik, T., 2015. Detection of aflatoxin B1 by aptamer-based biosensor using PAMAM dendrimers as immobilization platform. Food Control 52, 9–18.

Chauhan, N., Pundir, C.S., 2011. An amperometric biosensor based on acetylcholinesterase immobilized onto iron oxide nanoparticles/multi-walled carbon nanotubes modified gold electrode for measurement of organophosphorus insecticides. Analytica Chimica Acta 701, 66–74.

Chen, A., Shah, B., 2013. Electrochemical sensing and biosensing based on square wave voltammetry. Analytical Methods 5, 2158–2173.

Chen, Q., Lin, J., Gan, C., Wang, Y., Wang, D., Xiong, Y., Lai, W., Li, Y., Wang, M., 2015. A sensitive impedance biosensor based on immunomagnetic separation and urease catalysis for rapid detection of Listeria monocytogenes using an immobilization-free interdigitated array microelectrode. Biosensors and Bioelectronics 74, 504–511.

Chen, X., Lian, S., Ma, Y., Peng, A., Tian, X., Huang, Z., Chen, X., 2016. Electrochemiluminescence sensor for melamine based on a Ru(bpy)32+-doped silica nanoparticles/carboxylic acid functionalized multi-walled carbon nanotubes/Nafion composite film modified electrode. Talanta 146, 844–850.

Chen, M., Gan, N., Zhou, Y., Li, T., Xu, Q., Cao, Y., Chen, Y., 2017. A novel aptamer- metal ions- nanoscale MOF based electrochemical biocodes for multiple antibiotics detection and signal amplification. Sensors and Actuators B: Chemical 242, 1201–1209.

Chinnadayyala, S.R., Park, J., Le, H.T.N., Santhosh, M., Kadam, A.N., Cho, S., 2018. Recent advances in microfluidic paper-based electrochemiluminescence analytical devices for point-of-care testing applications. Biosensors and Bioelectronics.

Cho, I.H., Lee, J., Kim, J., Kang, M.S., Paik, J.K., Ku, S., Cho, H.M., Irudayaraj, J., Kim, D.H., 2018. Current technologies of electrochemical immunosensors: perspective on signal amplification. Sensors 18.

Chowdhury, A.D., De, A., Chaudhuri, C.R., Bandyopadhyay, K., Sen, P., 2012. Label free polyaniline based impedimetric biosensor for detection of *E. coli* O157:H7 Bacteria. Sensors and Actuators B: Chemical 171–172, 916–923.

Chrouda, A., Sbartai, A., Baraket, A., Renaud, L., Maaref, A., Jaffrezic-Renault, N., 2015. An aptasensor for ochratoxin A based on grafting of polyethylene glycol on a boron-doped diamond microcell. Analytical Biochemistry 488, 36–44.

Chu, L., Zou, G., Zhang, X., 2012. Electrogenerated chemiluminescence sensor for formaldehyde based on Ru(bpy)32+-doped silica nanoparticles modified Au electrode. Materials Science and Engineering: C 32, 2169–2174.

Chumbimuni-Torres, K.Y., Calvo-Marzal, P., Wang, J., 2009. Comparison between potentiometric and stripping voltammetric detection of trace metals: measurements of cadmium and lead in the presence of thalium, indium, and tin. Electroanalysis 21, 1939–1943.

Commission, E., 2009. Council regulation (EU) (2010) N 37/2010 of 22 December 2009 on pharmacologically active substances and their classification regarding maximum residue limits in foodstuffs of animal origin. Official Journal of European Union.

Commission, E., 2017. Regulation (EU) 2017/625 of the European Parliament and of the Council of 15 March 2017 on Official Controls and Other Official Activities Performed to Ensure the Application of Food and Feed Law, Rules on Animal Health and Welfare, Plant Health and Plant Protection Products.

2006. Commission Regulation (EC) No 1881/2006 of 19 December 2006 Setting Maximum Levels for Certain Contaminants in Foodstuffs.

Conzuelo, F., Ruiz-Valdepeñas Montiel, V., Campuzano, S., Gamella, M., Torrente-Rodríguez, R.M., Reviejo, A.J., Pingarrón, J.M., 2014. Rapid screening of multiple antibiotic residues in milk using disposable amperometric magnetosensors. Analytica Chimica Acta 820, 32–38.

Costantini, F., Sberna, C., Petrucci, G., Reverberi, M., Domenici, F., Fanelli, C., Manetti, C., De Cesare, G., Derosa, M., Nascetti, A., Caputo, D., 2016. Aptamer-based sandwich assay for on chip detection of Ochratoxin A by an array of amorphous silicon photosensors. Sensors and Actuators B: Chemical 230, 31–39.

1996. Council Directive 96/22/EC of 29 April 1996 Concerning the Prohibition on the Use in Stockfarming of Certain Substances Having a Hormonal or Thyrostatic Action and of Beta-Agonists, and Repealing Directives 81/602/EEC, 88/146/EEC and 88/299/EEC. Communities, O. J. O. T. E..

1993. Council Regulation (EEC) No 315/93 of 8 February 1993 Laying Down Community Procedures for Contaminants in Food.

Crew, A., Lonsdale, D., Byrd, N., Pittson, R., Hart, J.P., 2011. A screen-printed, amperometric biosensor array incorporated into a novel automated system for the simultaneous determination of organophosphate pesticides. Biosensors and Bioelectronics 26, 2847–2851.

Cuartero, M., Crespo, G.A., 2018. All-solid-state potentiometric sensors: a new wave for in situ aquatic research. Current Opinion in Electrochemistry 10, 98–106.

Cui, H.-F., Wu, W.-W., Li, M.-M., Song, X., Lv, Y., Zhang, T.-T., 2018. A highly stable acetylcholinesterase biosensor based on chitosan-TiO_2-graphene nanocomposites for detection of organophosphate pesticides. Biosensors and Bioelectronics 99, 223–229.

Devkota, L., Nguyen, L.T., Vu, T.T., Piro, B., 2018. Electrochemical determination of tetracycline using AuNP-coated molecularly imprinted overoxidized polypyrrole sensing interface. Electrochimica Acta 270, 535–542.

Diletti, G., Scortichini, G., Abete, M.C., Binato, G., Candeloro, L., Ceci, R., Chessa, G., Conte, A., DI Sandro, A., Esposito, M., Fedrizzi, G., Ferrantelli, V., Ferretti, E., Menotta, S., Nardelli, V., Neri, B., Piersanti, A., Roberti, F., Ubaldi, A., Brambilla, G., 2018. Intake estimates of dioxins and dioxin-like polychlorobiphenyls in the Italian general population from the 2013-2016 results of official monitoring plans in food. The Science of the Total Environment 627, 11–19.

Dong, X.-X., Yang, J.-Y., Luo, L., Zhang, Y.-F., Mao, C., Sun, Y.-M., Lei, H.-T., Shen, Y.-D., Beier, R.C., Xu, Z.-L., 2017. Portable amperometric immunosensor for histamine detection using Prussian blue-chitosan-gold nanoparticle nanocomposite films. Biosensors and Bioelectronics 98, 305–309.

Dragone, R., Grasso, G., Muccini, M., Toffanin, S., 2017. Portable bio/chemosensoristic devices: innovative systems for environmental health and food safety diagnostics. Frontiers in Public Health 5.

Dzyadevych, S., Soldatkin, A., EL'skaya, A., Martelet, C., Jaffrezic-Renault, N., 2006. Enzyme biosensors based on ion- selective field effect transistors. Analytica Chimica Acta 568, 248–258.

Eissa, S., Siaj, M., Zourob, M., 2015. Aptamer-based competitive electrochemical biosensor for brevetoxin-2. Biosensors and Bioelectronics 69, 148–154.

El Alami El Hassani, N., Baraket, A., Boudjaoui, S., Neto, E.T.T., Bausells, J., EL Bari, N., Bouchikhi, B., Elaissari, A., Errachid, A., Zine, N., 2018a. Development and application of a novel electrochemical immunosensor for tetracycline screening in honey using a fully integrated electrochemical Bio-MEMS. Biosensors and Bioelectronics.

El Alami El Hassani, N., Llobet, E., Popescu, L.-M., Ghita, M., Bouchikhi, B., El Bari, N., 2018b. Development of a highly sensitive and selective molecularly imprinted electrochemical sensor for sulfaguanidine detection in honey samples. Journal of Electroanalytical Chemistry 823, 647–655.

EL-Kosasy, A.M., Kamel, A.H., Hussin, L.A., Ayad, M.F., Fares, N.V., 2018. Mimicking new receptors based on molecular imprinting and their application to potentiometric assessment of 2,4-dichlorophenol as a food taint. Food Chemistry 250, 188–196.

Elsaidi, S.K., Mohamed, M.H., Banerjee, D., Thallapally, P.K., 2018. Flexibility in metal–organic frameworks: a fundamental understanding. Coordination Chemistry Reviews 358, 125–152.

Etayash, H., Thundat, T., Kaur, K., 2017. Bacterial detection using peptide-based platform and impedance spectroscopy. In: Prickril, B., Rasooly, A. (Eds.), Biosensors and Biodetection: Methods and Protocols, Electrochemical, Bioelectronic, Piezoelectric, Cellular and Molecular Biosensors, vol. 2. Springer, New York, NY.

European Commission, 2002. Commission Decision (EC) N° 2002/657 of 12 August 2002 implementing Council Directive 96/23/EC concerning the performance of analytical methods and interpretation of results. Official Journal of European Communities.

Fan, Z., Fan, L., Shuang, S., Dong, C., 2018. Highly sensitive photoelectrochemical sensing of bisphenol A based on zinc phthalocyanine/TiO_2 nanorod arrays. Talanta 189, 16–23.

Farghaly, O., ABDEL Hameed, R.S., ABU-Nawwas, A.A.H., 2014. Analytical Application Using Modern Electrochemical Techniques.

Feng, X., Gan, N., Zhang, H., Yan, Q., Li, T., Cao, Y., Hu, F., Yu, H., Jiang, Q., 2015. A novel "dual-potential" electrochemiluminescence aptasensor array using CdS quantum dots and luminol-gold nanoparticles as labels for simultaneous detection of malachite green and chloramphenicol. Biosensors and Bioelectronics 74, 587–593.

Feng, D., Wu, Y., Tan, X., Chen, Q., Yan, J., Liu, M., Ai, C., Luo, Y., Du, F., Liu, S., Han, H., 2018. Sensitive detection of melamine by an electrochemiluminescence sensor based on tris(bipyridine)ruthenium(II)-functionalized metal-organic frameworks. Sensors and Actuators B: Chemical 265, 378–386.

Fernández Márquez, M.L., Grande Burgos, M.J., López Aguayo, M.C., Pérez Pulido, R., Gálvez, A., Lucas, R., 2017. Characterization of biocide-tolerant bacteria isolated from cheese and dairy small-medium enterprises. Food Microbiology 62, 77–81.

Fischer, W.J., Schilter, B., Tritscher, A.M., Stadler, R.H., 2016. Contaminants of Milk and Dairy Products: Contamination Resulting from Farm and Dairy Practices. Reference Module in Food Science. Elsevier.

Gaglio, R., Couto, N., Marques, C., De Fatima Silva Lopes, M., Moschetti, G., Pomba, C., Settanni, L., 2016. Evaluation of antimicrobial resistance and virulence of enterococci from equipment surfaces, raw materials, and traditional cheeses. International Journal of Food Microbiology 236, 107–114.

Gajda, A., Nowacka-Kozak, E., Gbylik-Sikorska, M., Posyniak, A., 2018. Tetracycline antibiotics transfer from contaminated milk to dairy products and the effect of the skimming step and pasteurisation process on residue concentrations. Food Additives & Contaminants: Part A 35, 66–76.

Gan, N., Zhou, J., Xiong, P., Hu, F., Cao, Y., Li, T., Jiang, Q., 2013. An ultrasensitive electrochemiluminescent immunoassay for aflatoxin M1 in milk, based on extraction by magnetic graphene and detection by antibody-labeled CdTe quantumn dots-carbon nanotubes nanocomposite. Toxins 5, 865.

Gan, T., Shi, Z., Sun, J., Liu, Y., 2014. Simple and novel electrochemical sensor for the determination of tetracycline based on iron/zinc cations—exchanged montmorillonite catalyst. Talanta 121, 187—193.

Gao, P., Wang, H., Li, P., Gao, W., Zhang, Y., Chen, J., Jia, N., 2018. In-site synthesis molecular imprinting Nb_2O_5-based photoelectrochemical sensor for bisphenol A detection. Biosensors and Bioelectronics.

Ge, L., Li, H., Du, X., Zhu, M., Chen, W., Shi, T., Hao, N., Liu, Q., Wang, K., 2018. Facile one-pot synthesis of visible light-responsive $BiPO_4$/nitrogen doped graphene hydrogel for fabricating label-free photoelectrochemical tetracycline aptasensor. Biosensors and Bioelectronics 111, 131—137.

Ge, L., Xu, Y., Ding, L., You, F., Liu, Q., Wang, K., 2019. Perovskite-type $BiFeO_3$/ultrathin graphite-like carbon nitride nanosheets p-n heterojunction: boosted visible-light-driven photoelectrochemical activity for fabricating ampicillin aptasensor. Biosensors and Bioelectronics 124—125, 33—39.

Govindasamy, M., Chen, S.-M., Mani, V., Devasenathipathy, R., Umamaheswari, R., Joseph Santhanaraj, K., Sathiyan, A., 2017a. Molybdenum disulfide nanosheets coated multiwalled carbon nanotubes composite for highly sensitive determination of chloramphenicol in food samples milk, honey and powdered milk. Journal of Colloid and Interface Science 485, 129—136.

Govindasamy, M., Umamaheswari, R., Chen, S.-M., Mani, V., Su, C., 2017b. Graphene oxide nanoribbons film modified screen-printed carbon electrode for real-time detection of methyl parathion in food samples. Journal of the Electrochemical Society 164, B403—B408.

Graniel, O., Weber, M., Balme, S., Miele, P., Bechelany, M., 2018. Atomic layer deposition for biosensing applications. Biosensors and Bioelectronics.

Granja, R.H.M.M., Montes Nino, A.M., Rabone, F., Montes Nino, R.E., Cannavan, A., Gonzalez Salerno, A., 2008. Validation of radioimmunoassay screening methods for Î²-agonists in bovine liver according to Commission Decision 2002/657/EC. Food Additives & Contaminants: Part A 25, 1475—1481.

Gui, R., Jin, H., Guo, H., Wang, Z., 2018. Recent advances and future prospects in molecularly imprinted polymers-based electrochemical biosensors. Biosensors and Bioelectronics 100, 56—70.

Guo, L., Feng, J., Fang, Z., Xu, J., Lu, X., 2015. Application of microfluidic "lab-on-a-chip" for the detection of mycotoxins in foods. Trends in Food Science & Technology 46, 252—263.

Hamidi-Asl, E., Dardenne, F., Blust, R., DE Wael, K., 2015. An improved electrochemical aptasensor for chloramphenicol detection based on aptamer incorporated gelatine. Sensors 15, 7605—7618.

Hao, N., Wang, K., 2016. Recent development of electrochemiluminescence sensors for food analysis. Analytical and Bioanalytical Chemistry 408, 7035—7048.

Hao, N., Zhang, X., Zhou, Z., Hua, R., Zhang, Y., Liu, Q., Qian, J., Li, H., Wang, K., 2017. AgBr nanoparticles/3D nitrogen-doped graphene hydrogel for fabricating all-solid-state luminol-electrochemiluminescence *Escherichia coli* aptasensors. Biosensors and Bioelectronics 97, 377—383.

Hassan, A.H.A., Moura, S.L., Ali, F.H.M., Moselhy, W.A., Taboada Sotomayor, M.D.P., Pividori, M.I., 2018. Electrochemical sensing of methyl parathion on magnetic molecularly imprinted polymer. Biosensors and Bioelectronics 118, 181—187.

Hassani, N.E.A.E., Baraket, A., Neto, E.T.T., Lee, M., Salvador, J.P., Marco, M.P., Bausells, J., Bari, N.E., Bouchikhi, B., Elaissari, A., Errachid, A., Zine, N., 2017. Novel strategy for sulfapyridine detection using a fully integrated electrochemical Bio-MEMS: application to honey analysis. Biosensors and Bioelectronics 93, 282–288.

Hayat, A., Barthelmebs, L., Marty, J.-L., 2012. Electrochemical impedimetric immunosensor for the detection of okadaic acid in mussel sample. Sensors and Actuators B: Chemical 171–172, 810–815.

Hayati, F., Asiye, A.A., Sevda, A., Dilek, O., Birsen, D., 2016. Determination of tetracycline on the surface of a high- performance graphene modified screen-printed carbon electrode in milk and honey samples. Current Nanoscience 12, 527–533.

Helali, S., Sawelem Eid Alatawi, A., Abdelghani, A., 2018. Pathogenic *Escherichia coli* biosensor detection on chicken food samples. Journal of Food Safety 38, e12510.

Hernandez-Vargas, G., Sosa-Hernández, J., Saldarriaga-Hernandez, S., Villalba-Rodríguez, A., Parra-Saldivar, R., Iqbal, H., 2018. Electrochemical biosensors: a solution to pollution detection with reference to environmental contaminants. Biosensors 8, 29.

Hideshima, S., Saito, M., Fujita, K., Harada, Y., Tsuna, M., Sekiguchi, S., Kuroiwa, S., Nakanishi, T., Osaka, T., 2018. Label-free detection of allergens in food via surfactant-induced signal amplification using a field effect transistor-based biosensor. Sensors and Actuators B: Chemical 254, 1011–1016.

Hu, X., Goud, K.Y., Kumar, V.S., Catanante, G., Li, Z., Zhu, Z., Marty, J.L., 2018. Disposable electrochemical aptasensor based on carbon nanotubes-V_2O_5-chitosan nanocomposite for detection of ciprofloxacin. Sensors and Actuators B: Chemical 268, 278–286.

Huang, J., Zhang, X., Lin, Q., He, X., Xing, X., Huai, H., Lian, W., Zhu, H., 2011. Electrochemical sensor based on imprinted sol–gel and nanomaterials for sensitive determination of bisphenol A. Food Control 22, 786–791.

Huang, T.-T., Liu, S.-C., Huang, C.-H., Lin, C.-J., Huang, S.-T., 2018. An integrated real-time electrochemical LAMP device for pathogenic bacteria detection in food. Electroanalysis 30, 2397–2404.

Ismail, F., Adeloju, S.B., 2010. Galvanostatic entrapment of penicillinase into polytyramine films and its utilization for the potentiometric determination of penicillin. Sensors 10, 2851.

Ji, X., Yu, C., Wen, Y., Chen, J., Yu, Y., Zhang, C., Gao, R., Mu, X., He, J., 2019. Fabrication of pioneering 3D sakura-shaped metal-organic coordination polymers Cu@L-Glu phenomenal for signal amplification in highly sensitive detection of zearalenone. Biosensors and Bioelectronics.

Jia, F., Duan, N., Wu, S., Dai, R., Wang, Z., Li, X., 2016. Impedimetric Salmonella aptasensor using a glassy carbon electrode modified with an electrodeposited composite consisting of reduced graphene oxide and carbon nanotubes. Microchimica Acta 183, 337–344.

Jiang, C., Zhu, J., Li, Z., Luo, J., Wang, J., Sun, Y., 2017. Chitosan–gold nanoparticles as peroxidase mimic and their application in glucose detection in serum. RSC Advances 7, 44463–44469.

Jin, X., Fang, G., Pan, M., Yang, Y., Bai, X., Wang, S., 2018. A molecularly imprinted electrochemiluminescence sensor based on upconversion nanoparticles enhanced by electrodeposited rGO for selective and ultrasensitive detection of clenbuterol. Biosensors and Bioelectronics 102, 357–364.

Jivani, R.R., Lakhtaria, G.J., Patadiya, D.D., Patel, L.D., Jivani, N.P., Jhala, B.P., 2016. Biomedical microelectromechanical systems (BioMEMS): revolution in drug delivery and analytical techniques. Saudi Pharmaceutical Journal 24, 1–20.

Jodra, A., Hervás, M., López, M.Á., Escarpa, A., 2015. Disposable electrochemical magneto immunosensor for simultaneous simplified calibration and determination of Ochratoxin A in coffee samples. Sensors and Actuators B: Chemical 221, 777–783.

Jv, Y., Li, B., Cao, R., 2010. Positively-charged gold nanoparticles as peroxidase mimic and their application in hydrogen peroxide and glucose detection. Chemical Communications 46.

Kanchi, S., Sabela, M.I., Mdluli, P.S., Inamuddin, Bisetty, K., 2018. Smartphone based bioanalytical and diagnosis applications: a review. Biosensors and Bioelectronics 102, 136–149.

Karmaus, A.L., Osborn, R., Krishan, M., 2018. Scientific advances and challenges in safety evaluation of food packaging materials: workshop proceedings. Regulatory Toxicology and Pharmacology 98, 80–87.

Karthik, R., Govindasamy, M., Chen, S.-M., Mani, V., Lou, B.-S., Devasenathipathy, R., Hou, Y.-S., Elangovan, A., 2016. Green synthesized gold nanoparticles decorated graphene oxide for sensitive determination of chloramphenicol in milk, powdered milk, honey and eye drops. Journal of Colloid and Interface Science 475, 46–56.

Karthik, R., Vinoth Kumar, J., Chen, S.-M., Sundaresan, P., Mutharani, B., Chi Chen, Y., Muthuraj, V., 2018. Simple sonochemical synthesis of novel grass-like vanadium disulfide: a viable non-enzymatic electrochemical sensor for the detection of hydrogen peroxide. Ultrasonics Sonochemistry 48, 473–481.

Kashish, Soni, D.K., Mishra, S.K., Prakash, R., Dubey, S.K., 2015. Label-free impedimetric detection of Listeria monocytogenes based on poly-5-carboxy indole modified ssDNA probe. Journal of Biotechnology 200, 70–76.

Kassal, P., Steinberg, M.D., Steinberg, I.M., 2018. Wireless chemical sensors and biosensors: a review. Sensors and Actuators B: Chemical 266, 228–245.

Kempahanumakkagari, S., Vellingiri, K., Deep, A., Kwon, E.E., Bolan, N., Kim, K.-H., 2018. Metal–organic framework composites as electrocatalysts for electrochemical sensing applications. Coordination Chemistry Reviews 357, 105–129.

Khoshbin, Z., Verdian, A., Housaindokht, M.R., Izadyar, M., Rouhbakhsh, Z., 2018. Aptasensors as the future of antibiotics test kits-A case study of the aptamer application in the chloramphenicol detection. Biosensors and Bioelectronics.

Kochana, J., Wapiennik, K., Kozak, J., Knihnicki, P., Pollap, A., Woźniakiewicz, M., Nowak, J., Kościelniak, P., 2015. Tyrosinase-based biosensor for determination of bisphenol A in a flow-batch system. Talanta 144, 163–170.

Kor, K., Zarei, K., 2014. Electrochemical determination of chloramphenicol on glassy carbon electrode modified with multi-walled carbon nanotube–cetyltrimethylammonium bromide–poly(diphenylamine). Journal of Electroanalytical Chemistry 733, 39–46.

Kurbanoglu, S., Ozkan, S.A., Merkoçi, A., 2017. Nanomaterials-based enzyme electrochemical biosensors operating through inhibition for biosensing applications. Biosensors and Bioelectronics 89, 886–898.

Lach, P., Sharma, P.S., Golebiewska, K., Cieplak, M., D'souza, F., Kutner, W., 2017. Molecularly imprinted polymer chemosensor for selective determination of an N-Nitroso-l-proline food toxin. Chemistry - A European Journal 23, 1942–1949.

Lafleur, J.P., Jönsson, A., Senkbeil, S., Kutter, J.P., 2016. Recent advances in lab-on-a-chip for biosensing applications. Biosensors and Bioelectronics 76, 213–233.

Lai, W.W.-P., Lin, Y.-C., Wang, Y.-H., Guo, Y.L., Lin, A.Y.-C., 2018. Occurrence of emerging contaminants in aquaculture waters: cross-contamination between aquaculture systems and surrounding waters. Water, Air, & Soil Pollution 229, 249.

Lakey, A., Ali, Z., Scott, S.M., Chebil, S., Youssoufi, H.K., Hunor, S., Ohlander, A., Kuphal, M., Marti, J.S., 2019. Impedimetric array in polymer microfluidic cartridge for low cost point-of-care diagnostics. Biosensors and Bioelectronics.

Lakicevic, B., Nastasijevic, I., 2017. Listeria monocytogenes in retail establishments: contamination routes and control strategies. Food Reviews International 33, 247–269.

Lambert, M., Inthavong, C., Hommet, F., Leblanc, J.-C., Hulin, M., Guérin, T., 2018. Levels of acrylamide in foods included in 'the first French total diet study on infants and toddlers'. Food Chemistry 240, 997–1004.

Lawal, A.T., 2018. Progress in utilisation of graphene for electrochemical biosensors. Biosensors and Bioelectronics 106, 149–178.

Lei, X., Xu, L., Song, S., Liu, L., Kuang, H., 2018. Development of an ultrasensitive ic-ELISA and immunochromatographic strip assay for the simultaneous detection of florfenicol and thiamphenicol in eggs. Food and Agricultural Immunology 29, 254–266.

Li, J., Li, Y., Zhang, Y., Wei, G., 2012. Highly sensitive molecularly imprinted electrochemical sensor based on the double amplification by an inorganic prussian blue catalytic polymer and the enzymatic effect of glucose oxidase. Analytical Chemistry 84, 1888–1893.

Li, R., Liu, Y., Cheng, L., Yang, C., Zhang, J., 2014. Photoelectrochemical aptasensing of kanamycin using visible light-activated carbon nitride and graphene oxide nanocomposites. Analytical Chemistry 86, 9372–9375.

Li, L., Peng, A.-H., Lin, Z.-Z., Zhong, H.-P., Chen, X.-M., Huang, Z.-Y., 2017a. Biomimetic ELISA detection of malachite green based on molecularly imprinted polymer film. Food Chemistry 229, 403–408.

Li, S., Liu, C., Yin, G., Zhang, Q., Luo, J., Wu, N., 2017b. Aptamer-molecularly imprinted sensor base on electrogenerated chemiluminescence energy transfer for detection of lincomycin. Biosensors and Bioelectronics 91, 687–691.

Li, Y., Tian, J., Yuan, T., Wang, P., Lu, J., 2017c. A sensitive photoelectrochemical aptasensor for oxytetracycline based on a signal "switch off-on" strategy. Sensors and Actuators B: Chemical 240, 785–792.

Li, L., Pan, L., Ma, Z., Yan, K., Cheng, W., Shi, Y., Yu, G., 2018. All inkjet-printed amperometric multiplexed biosensors based on nanostructured conductive hydrogel electrodes. Nano Letters 18, 3322–3327.

Li, Z., Liu, C., Sarpong, V., Gu, Z., 2019. Multisegment nanowire/nanoparticle hybrid arrays as electrochemical biosensors for simultaneous detection of antibiotics. Biosensors and Bioelectronics 126, 632–639.

Lian, S., Huang, Z., Lin, Z., Chen, X., Oyama, M., Chen, X., 2016. A highly selective melamine sensor relying on intensified electrochemiluminescence of the silica nanoparticles doped with [Ru(bpy)3]2+/molecularly imprinted polymer modified electrode. Sensors and Actuators B: Chemical 236, 614–620.

Liao, C.W., Chen, Y.R., Chang, J.L., Zen, J.-M., 2010. A sensitive electrochemical approach for melamine detection using a disposable screen printed carbon electrode. Electroanalysis.

Liao, Z., Wang, J., Zhang, P., Zhang, Y., Miao, Y., Gao, S., Deng, Y., Geng, L., 2018. Recent advances in microfluidic chip integrated electronic biosensors for multiplexed detection. Biosensors and Bioelectronics.

Lim, M.C., Kim, Y.R., 2016. Analytical applications of nanomaterials in monitoring biological and chemical contaminants in food. Journal of Microbiology and Biotechnology 26, 1505–1516.

Lin, Y., Zhou, Q., Tang, D., Niessner, R., Knopp, D., 2017. Signal-on photoelectrochemical immunoassay for aflatoxin B1 based on enzymatic product-etching MnO2 nanosheets for dissociation of carbon dots. Analytical Chemistry 89, 5637–5645.

Lin, P., Liu, D., Wei, W., Guo, J., Ke, S., Zeng, X., Chen, S., 2018. A novel protein binding strategy for energy-transfer-based photoelectrochemical detection of enzymatic activity of botulinum neurotoxin A. Electrochemistry Communications 97, 114–118.

Liu, W., Yin, X.-B., 2016. Metal–organic frameworks for electrochemical applications. TRAC Trends in Analytical Chemistry 75, 86–96.

Liu, G., Chai, C., Yao, B., 2013a. Rapid evaluation of Salmonella pullorum contamination in chicken based on a portable Amperometric sensor. Journal of Biosensors & Bioelectronics 4, 137.

Liu, S., Zheng, Z., Li, X., 2013b. Advances in pesticide biosensors: current status, challenges, and future perspectives. Analytical and Bioanalytical Chemistry 405, 63–90.

Liu, Q., Zhou, Q., Jiang, G., 2014. Nanomaterials for analysis and monitoring of emerging chemical pollutants. Trends in Analytical Chemistry 58, 10–22.

Liu, Q., Huan, J., Dong, X., Qian, J., Hao, N., You, T., Mao, H., Wang, K., 2016a. Resonance energy transfer from CdTe quantum dots to gold nanorods using MWCNTs/rGO nanoribbons as efficient signal amplifier for fabricating visible-light-driven "on-off-on" photoelectrochemical acetamiprid aptasensor. Sensors and Actuators B: Chemical 235, 647–654.

Liu, X., Zhu, J., Huo, X., Yan, R., Wong, D.K.Y., 2016b. An intimately bonded titanate nanotube-polyaniline-gold nanoparticle ternary composite as a scaffold for electrochemical enzyme biosensors. Analytica Chimica Acta 911, 59–68.

Liu, C.-S., Sun, C.-X., Tian, J.-Y., Wang, Z.-W., Ji, H.-F., Song, Y.-P., Zhang, S., Zhang, Z.-H., He, L.-H., Du, M., 2017a. Highly stable aluminum-based metal-organic frameworks as biosensing platforms for assessment of food safety. Biosensors and Bioelectronics 91, 804–810.

Liu, S., Lai, G., Zhang, H., Yu, A., 2017b. Amperometric aptasensing of chloramphenicol at a glassy carbon electrode modified with a nanocomposite consisting of graphene and silver nanoparticles. Microchimica Acta 184, 1445–1451.

Liu, X., Liu, P., Tang, Y., Yang, L., Li, L., Qi, Z., Li, D., Wong, D.K.Y., 2018a. A photoelectrochemical aptasensor based on a 3D flower-like TiO2-MoS2-gold nanoparticle heterostructure for detection of kanamycin. Biosensors and Bioelectronics 112, 193–201.

Liu, Y., Liu, Y., Qiao, L., Liu, Y., Liu, B., 2018b. Advances in signal amplification strategies for electrochemical biosensing. Current Opinion in Electrochemistry.

Long, F., Zhang, Z., Yang, Z., Zeng, J., Jiang, Y., 2015. Imprinted electrochemical sensor based on magnetic multi-walled carbon nanotube for sensitive determination of kanamycin. Journal of Electroanalytical Chemistry 755, 7–14.

Lu, L., Seenivasan, R., Wang, Y.-C., Yu, J.-H., Gunasekaran, S., 2016. An electrochemical immunosensor for rapid and sensitive detection of mycotoxins fumonisin B1 and Deoxynivalenol. Electrochimica Acta 213, 89–97.

Luo, C., Lei, Y., Yan, L., Yu, T., Li, Q., Zhang, D., Ding, S., Ju, H., 2012. A rapid and sensitive aptamer-based electrochemical biosensor for direct detection of Escherichia coli O111. Electroanalysis 24, 1186–1191.

Ma, X., Jiang, Y., Jia, F., Yu, Y., Chen, J., Wang, Z., 2014. An aptamer-based electrochemical biosensor for the detection of Salmonella. Journal of Microbiological Methods 98, 94–98.

Ma, L., Zhou, L., He, Y., Wang, L., Huang, Z., Jiang, Y., Gao, J., 2018. Hierarchical nanocomposites with an N-doped carbon shell and bimetal core: novel enzyme nanocarriers for electrochemical pesticide detection. Biosensors and Bioelectronics.

Maduraiveeran, G., Sasidharan, M., Ganesan, V., 2018. Electrochemical sensor and biosensor platforms based on advanced nanomaterials for biological and biomedical applications. Biosensors and Bioelectronics 103, 113–129.

Malvano, F., Albanese, D., Pilloton, R., Di Matteo, M., Crescitelli, A., 2017. A new label-free impedimetric affinity sensor based on cholinesterases for detection of organophosphorous and carbamic pesticides in food samples: impedimetric versus amperometric detection. Food and Bioprocess Technology 10, 1834–1843.

Masikini, M., Mailu, S., Tsegaye, A., Njomo, N., Molapo, K., Ikpo, C., Sunday, C., Rassie, C., Wilson, L., Baker, P., Iwuoha, E., 2015. A fumonisins immunosensor based on polyanilino-carbon nanotubes doped with palladium telluride quantum dots. Sensors 15, 529.

Mehta, J., Vinayak, P., Tuteja, S.K., Chhabra, V.A., Bhardwaj, N., Paul, A.K., Kim, K.-H., Deep, A., 2016. Graphene modified screen printed immunosensor for highly sensitive detection of parathion. Biosensors and Bioelectronics 83, 339–346.

Merola, G., Martini, E., Tomassetti, M., Campanella, L., 2015. Simple and suitable immunosensor for β-lactam antibiotics analysis in real matrixes: milk, serum, urine. Journal of Pharmaceutical and Biomedical Analysis 106, 186–196.

Miao, S.S., Wu, M.S., Ma, L.Y., He, X.J., Yang, H., 2016a. Electrochemiluminescence biosensor for determination of organophosphorous pesticides based on bimetallic Pt-Au/multi-walled carbon nanotubes modified electrode. Talanta 158, 142–151.

Miao, Y.-B., Ren, H.-X., Gan, N., Zhou, Y., Cao, Y., Li, T., Chen, Y., 2016b. A triple-amplification SPR electrochemiluminescence assay for chloramphenicol based on polymer enzyme-linked nanotracers and exonuclease-assisted target recycling. Biosensors and Bioelectronics 86, 477–483.

Micheli, L., Grecco, R., Badea, M., Moscone, D., Palleschi, G., 2005. An electrochemical immunosensor for aflatoxin M1 determination in milk using screen-printed electrodes. Biosensors and Bioelectronics 21, 588–596.

Migliorini, F.L., Sanfelice, R.C., Mercante, L.A., Andre, R.S., Mattoso, L.H.C., Correa, D.S., 2018. Urea impedimetric biosensing using electrospun nanofibers modified with zinc oxide nanoparticles. Applied Surface Science 443, 18–23.

Mishra, G.K., Bhand, S., 2012. FIA-EQCN biosensor for analysis of sulphadiazine residues in milk. In: 2012 Sixth International Conference on Sensing Technology (ICST), 18-21 Dec. 2012, pp. 672–676.

Mishra, G.K., Sharma, A., Bhand, S., 2015. Ultrasensitive detection of streptomycin using flow injection analysis-electrochemical quartz crystal nanobalance (FIA-EQCN) biosensor. Biosensors and Bioelectronics 67, 532–539.

Monteiro, T.O., Tanaka, A.A., Damos, F.S., Luz, R.D.C.S., 2017. Photoelectrochemical determination of tert-butylhydroquinone in edible oil samples employing CdSe/ZnS quantum dots and LiTCNE. Food Chemistry 227, 16–21.

Montes-Cebrián, Y., Del torno-de Román, L., Álvarez-Carulla, A., Colomer-Farrarons, J., Minteer, S.D., Sabaté, N., Miribel-Català, P.L., Esquivel, J.P., 2018. 'Plug-and-Power' Point-of-Care diagnostics: a novel approach for self-powered electronic reader-based portable analytical devices. Biosensors and Bioelectronics 118, 88–96.

Moslemi, E., Soltandalal, M.M., Beheshtizadeh, M.R., Taghavi, A., Kheiri Manjili, H., Mahmoudi Lamouki, R., Izadi, A., 2018. Detection of Brucella spp. in dairy products by real-time PCR. Archives of Clinical Infectious Diseases 13, e12673.

Munawar, A., Tahir, M.A., Shaheen, A., Lieberzeit, P.A., Khan, W.S., Bajwa, S.Z., 2018. Investigating nanohybrid material based on 3D CNTs@Cu nanoparticle composite and imprinted polymer for highly selective detection of chloramphenicol. Journal of Hazardous Materials 342, 96−106.

Murugan, K., Muthumariappan, A., Chen, S.-M., Sakthivel, K., Govindasamy, M., Mani, V., Ajmal Ali, M., Al-Hemaid, M.A.,F., Elshikh, M.S., 2018. One-Pot Biosynthesis of Reduced Graphene Oxide/Prussian Blue Microcubes Composite and its Sensitive Detection of Prophylactic Drug Dimetridazole.

Nag, S.K., 2010a. 5 - pesticides, veterinary residues and other contaminants in milk. In: Griffiths, M.W. (Ed.), Improving the Safety and Quality of Milk. Woodhead Publishing.

Nag, S.K., 2010b. 6 - contaminants in milk: routes of contamination, analytical techniques and methods of control. In: Griffiths, M.W. (Ed.), Improving the Safety and Quality of Milk. Woodhead Publishing.

2016. NF EN FDIS 16140-2 Microbiology of the Food Chain — Method Validation — Part 2: Protocol for the Validation of Alternative (Proprietary) Methods against a Reference Method.

Ni, J., Yang, W., Wang, Q., Luo, F., Guo, L., Qiu, B., Lin, Z., Yang, H., 2018. Homogeneous and label-free electrochemiluminescence aptasensor based on the difference of electrostatic interaction and exonuclease-assisted target recycling amplification. Biosensors and Bioelectronics 105, 182−187.

Okoth, O.K., Yan, K., Liu, Y., Zhang, J., 2016. Graphene-doped Bi2S3 nanorods as visible-light photoelectrochemical aptasensing platform for sulfadimethoxine detection. Biosensors and Bioelectronics 86, 636−642.

Oracz, J., Nebesny, E., Żyżelewicz, D., 2011. New trends in quantification of acrylamide in food products. Talanta 86, 23−34.

Overney, A., Jacques-André-Coquin, J., Ng, P., Carpentier, B., Guillier, L., Firmesse, O., 2017. Impact of environmental factors on the culturability and viability of Listeria monocytogenes under conditions encountered in food processing plants. International Journal of Food Microbiology 244, 74−81.

Palanisamy, S., Kokulnathan, T., Chen, S.-M., Velusamy, V., Ramaraj, S.K., 2017. Voltammetric determination of Sudan I in food samples based on platinum nanoparticles decorated on graphene-β-cyclodextrin modified electrode. Journal of Electroanalytical Chemistry 794, 64−70.

Pandey, A., Gurbuz, Y., Ozguz, V., Niazi, J.H., Qureshi, A., 2017. Graphene-interfaced electrical biosensor for label-free and sensitive detection of foodborne pathogenic *E. coli* O157:H7. Biosensors and Bioelectronics 91, 225−231.

Paniel, N., Istamboulié, G., Triki, A., Lozano, C., Barthelmebs, L., Noguer, T., 2017. Selection of DNA aptamers against penicillin G using Capture-SELEX for the development of an impedimetric sensor. Talanta 162, 232−240.

Pänke, O., Balkenhohl, T., Kafka, J., Schäfer, D., Lisdat, F., 2008. Impedance spectroscopy and biosensing. In: Renneberg, R., Lisdat, F. (Eds.), Biosensing for the 21st Century. Springer Berlin Heidelberg, Berlin, Heidelberg.

Pereira Da Silva Neves, M.M., González-García, M.B., Hernández-Santos, D., Fanjul-Bolado, P., 2018. Future trends in the market for electrochemical biosensing. Current Opinion in Electrochemistry 10, 107−111.

Pita, M.T.P., Reviejo, A.J., De Villena, F.J.M., Pingarrón, J.M., 1997. Amperometric selective biosensing of dimethyl- and diethyldithiocarbamates based on inhibition processes in a medium of reversed micelles. Analytica Chimica Acta 340, 89–97.

Pohanka, M., Skládal, P., 2008. Electrochemical biosensors - principles and applications. Journal of Applied Biomedicine 6, 57–64.

Põldsalu, I., Johanson, U., Tamm, T., Punning, A., Greco, F., Peikolainen, A.-L., Kiefer, R., Aabloo, A., 2018. Mechanical and electro-mechanical properties of EAP actuators with inkjet printed electrodes. Synthetic Metals 246, 122–127.

Prado, T.M.D., Foguel, M.V., Gonçalves, L.M., Sotomayor, M.D.P.T., 2015. β-Lactamase-based biosensor for the electrochemical determination of benzylpenicillin in milk. Sensors and Actuators B: Chemical 210, 254–258.

Pundir, C.S., Chauhan, N., 2012. Acetylcholinesterase inhibition-based biosensors for pesticide determination: a review. Analytical Biochemistry 429, 19–31.

Qiao, Y., Li, J., Li, H., Fang, H., Fan, D., Wang, W., 2016. A label-free photoelectrochemical aptasensor for bisphenol A based on surface plasmon resonance of gold nanoparticle-sensitized ZnO nanopencils. Biosensors and Bioelectronics 86, 315–320.

Ranjbar, S., Shahrokhian, S., Nurmohammadi, F., 2018. Nanoporous gold as a suitable substrate for preparation of a new sensitive electrochemical aptasensor for detection of *Salmonella typhimurium*. Sensors and Actuators B: Chemical 255, 1536–1544.

Rao, H., Chen, M., Ge, H., Lu, Z., Liu, X., Zou, P., Wang, X., He, H., Zeng, X., Wang, Y., 2017. A novel electrochemical sensor based on Au@PANI composites film modified glassy carbon electrode binding molecular imprinting technique for the determination of melamine. Biosensors and Bioelectronics 87, 1029–1035.

Rao, H., Zhao, X., Liu, X., Zhong, J., Zhang, Z., Zou, P., Jiang, Y., Wang, X., Wang, Y., 2018. A novel molecularly imprinted electrochemical sensor based on graphene quantum dots coated on hollow nickel nanospheres with high sensitivity and selectivity for the rapid determination of bisphenol S. Biosensors and Bioelectronics 100, 341–347.

Raza, N., Kim, K.-H., 2018. Quantification techniques for important environmental contaminants in milk and dairy products. TRAC Trends in Analytical Chemistry 98, 79–94.

Reverté, L., Campbell, K., Rambla-Alegre, M., Elliott, C.T., Diogène, J., Campàs, M., 2017. Immunosensor array platforms based on self-assembled dithiols for the electrochemical detection of tetrodotoxins in puffer fish. Analytica Chimica Acta 989, 95–103.

Rhouati, A., Yang, C., Hayat, A., Marty, J.-L., 2013. Aptamers: a promising tool for ochratoxin a detection in food analysis. Toxins 5, 1988.

Rhouati, A., Bulbul, G., Latif, U., Hayat, A., Li, Z.-H., Marty, J., 2017. Nano-Aptasensing in mycotoxin analysis: recent updates and progress. Toxins 9, 349.

Ríos, Á., Zougagh, M., 2016. Recent advances in magnetic nanomaterials for improving analytical processes. TRAC Trends in Analytical Chemistry 84, 72–83.

Roda, A., Michelini, E., Zangheri, M., DI Fusco, M., Calabria, D., Simoni, P., 2016. Smartphone-based biosensors: a critical review and perspectives. TRAC Trends in Analytical Chemistry 79, 317–325.

Rossi, R., Saluti, G., Moretti, S., Diamanti, I., Giusepponi, D., Galarini, R., 2018. Multiclass methods for the analysis of antibiotic residues in milk by liquid chromatography coupled to mass spectrometry: a review. Food Additives & Contaminants: Part A 35, 241–257.

Rotariu, L., Lagarde, F., Jaffrezic-Renault, N., Bala, C., 2016. Electrochemical biosensors for fast detection of food contaminants – trends and perspective. TRAC Trends in Analytical Chemistry 79, 80–87.

Rovina, K., Siddiquee, S., 2016. Electrochemical sensor based rapid determination of melamine using ionic liquid/zinc oxide nanoparticles/chitosan/gold electrode. Food Control 59, 801–808.

Ruiz-Valdepeñas Montiel, V., Torrente-Rodríguez, R., Campuzano, S., Pellicanò, A., Reviejo, Á., Cosio, M., Pingarrón, J., 2016. Simultaneous determination of the main peanut allergens in foods using disposable amperometric magnetic beads-based immunosensing platforms. Chemosensors 4, 11.

Russell, S.M., De La Rica, R., 2018. Policy considerations for mobile biosensors. ACS Sensors 3, 1059–1068.

Sakin, F., Tekeli, I.O., Yipel, M., Kürekci, C., 2018. Occurrence and health risk assessment of aflatoxins and ochratoxin a in Sürk, a Turkish dairy food, as studied by HPLC. Food Control 90, 317–323.

Salgueiro-González, N., Castiglioni, S., Zuccato, E., Turnes-Carou, I., López-Mahía, P., Muniategui-Lorenzo, S., 2018. Recent advances in analytical methods for the determination of 4-alkylphenols and bisphenol A in solid environmental matrices: a critical review. Analytica Chimica Acta 1024, 39–51.

Samsidar, A., Siddiquee, S., Shaarani, S.M., 2018. A review of extraction, analytical and advanced methods for determination of pesticides in environment and foodstuffs. Trends in Food Science & Technology 71, 188–201.

Sanchez, A., Núnez-Bajo, E., Fernández-Abedul, M.T., Blanco-López, M.C., Costa-García, A., 2018. Optimization and characterization of nanostructured paper-based electrodes. Electrochimica Acta.

Sani, N.D.M., Heng, L.Y., Marugan, R.S.P.M., Rajab, N.F., 2018. Electrochemical DNA biosensor for potential carcinogen detection in food sample. Food Chemistry 269, 503–510.

SANTE/11813/2017 Guidance document on analytical quality control and method validation procedures for pesticide residues and analysis in food and feed. European Commission. Directorate general for health and food safety. Safety of the Food Chain Pesticides and Biocides.

Santoro, K., Ricciardi, C., 2016. Biosensors. In: Caballero, B., Finglas, P.M., Toldrá, F. (Eds.), Encyclopedia of Food and Health. Academic Press, Oxford.

Scordino, M., Lazzaro, F., Borzì, M.A., Sabatino, L., Traulo, P., Gagliano, G., 2018. Dehydroacetic acid in cheese and cheese coating, results of official control in Italy. Food Additives and Contaminants: Part B 11, 75–81.

Setford, S.J., Van Es, R.M., Blankwater, Y.J., Kröger, S., 1999. Receptor binding protein amperometric affinity sensor for rapid β-lactam quantification in milk. Analytica Chimica Acta 398, 13–22.

Sha, Y., Zhang, X., Li, W., Wu, W., Wang, S., Guo, Z., Zhou, J., Su, X., 2016. A label-free multi-functionalized graphene oxide based electrochemiluminscence immunosensor for ultrasensitive and rapid detection of Vibrio parahaemolyticus in seawater and seafood. Talanta 147, 220–225.

Shabani, A., Marquette, C.A., Mandeville, R., Lawrence, M.F., 2013. Magnetically-assisted impedimetric detection of bacteria using phage-modified carbon microarrays. Talanta 116, 1047–1053.

Sharma, H., Mutharasan, R., 2013. Review of biosensors for foodborne pathogens and toxins. Sensors and Actuators B: Chemical 183, 535–549.

Sharma, A., Istamboulie, G., Hayat, A., Catanante, G., Bhand, S., Marty, J.L., 2017. Disposable and portable aptamer functionalized impedimetric sensor for detection of kanamycin residue in milk sample. Sensors and Actuators B: Chemical 245, 507–515.

Socas-Rodríguez, B., González-Sálamo, J., Hernández-Borges, J., Rodríguez-Delgado, M.Á., 2017. Recent applications of nanomaterials in food safety. TRAC Trends in Analytical Chemistry 96, 172–200.

Songa, E.A., Okonkwo, J.O., 2016. Recent approaches to improving selectivity and sensitivity of enzyme-based biosensors for organophosphorus pesticides: a review. Talanta 155, 289–304.

Su, S., Chen, S., Fan, C., 2018. Recent advances in two-dimensional nanomaterials-based electrochemical sensors for environmental analysis. Green Energy & Environment 3, 97–106.

Švarc-Gajić, J., Stojanovic, Z., 2014. Direct determination of heavy metals in honey by potentiometric stripping analysis. International Journal of Food Processing Technology 1, 1–6.

Szczepańska, N., Kudłak, B., Namieśnik, J., 2018. Recent advances in assessing xenobiotics migrating from packaging material – a review. Analytica Chimica Acta 1023, 1–21.

Teng, J., Ye, Y., Yao, L., Yan, C., Cheng, K., Xue, F., Pan, D., Li, B., Chen, W., 2017. Rolling circle amplification based amperometric aptamer/immuno hybrid biosensor for ultrasensitive detection of Vibrio parahaemolyticus. Microchimica Acta 184, 3477–3485.

Thévenot, D.R., Toth, K., Durst, R.A., Wilson, G.S., 2001. Electrochemical biosensors: recommended definitions and classification. Biosensors and Bioelectronics 16, 121–131.

Touzi, H., Chevalier, Y., Bessueille, F., Ben Ouada, H., Jaffrezic-Renault, N., 2018. Detection of dyestuffs with an impedimetric sensor based on Cu2+-methyl-naphthyl cyclen complex functionalized gold electrodes. Sensors and Actuators B: Chemical 273, 1211–1221.

Turco, A., Corvaglia, S., Mazzotta, E., 2015. Electrochemical sensor for sulfadimethoxine based on molecularly imprinted polypyrrole: study of imprinting parameters. Biosensors and Bioelectronics 63, 240–247.

Vasilescu, A., Marty, J.-L., 2016. Electrochemical aptasensors for the assessment of food quality and safety. TRAC Trends in Analytical Chemistry 79, 60–70.

Veselá, H., Šucman, E., 2013. Determination of acrylamide in food using adsorption stripping voltammetry. Czech Journal of Food Sciences 31, 401–406.

Vidal, J.C., Bonel, L., Ezquerra, A., Hernández, S., Bertolín, J.R., Cubel, C., Castillo, J.R., 2013. Electrochemical affinity biosensors for detection of mycotoxins: a review. Biosensors and Bioelectronics 49, 146–158.

Walcarius, A., 2017. Recent trends on electrochemical sensors based on ordered mesoporous carbon. Sensors 17, 1863.

Walcarius, A., 2018. Silica-based electrochemical sensors and biosensors: recent trends. Current Opinion in Electrochemistry 10, 88–97.

Wang, Q., Chen, M., Zhang, H., Wen, W., Zhang, X., Wang, S., 2016. Enhanced electrochemiluminescence of RuSi nanoparticles for ultrasensitive detection of ochratoxin A by energy transfer with CdTe quantum dots. Biosensors and Bioelectronics 79, 561–567.

Wang, H., Yao, S., Liu, Y., Wei, S., Su, J., Hu, G., 2017. Molecularly imprinted electrochemical sensor based on Au nanoparticles in carboxylated multi-walled carbon nanotubes for sensitive determination of olaquindox in food and feedstuffs. Biosensors and Bioelectronics 87, 417–421.

Wang, C., Qian, J., An, K., Ren, C., Lu, X., Hao, N., Liu, Q., Li, H., Huang, X., Wang, K., 2018a. Fabrication of magnetically assembled aptasensing device for label-free determination of aflatoxin B1 based on EIS. Biosensors and Bioelectronics 108, 69–75.

Wang, C., Shi, Y., Yang, D.-R., Xia, X.-H., 2018b. Combining plasmonics and electrochemistry at the nanoscale. Current Opinion in Electrochemistry 7, 95–102.

Wang, G., Morrin, A., Li, M., Liu, N., Luo, X., 2018c. Nanomaterial-doped conducting polymers for electrochemical sensors and biosensors. Journal of Materials Chemistry B 6, 4173–4190.

Wang, J., Xu, Q., Xia, W.W., Shu, Y., Jin, D., Zang, Y., Hu, X., 2018d. High sensitive visible light photoelectrochemical sensor based on in-situ prepared flexible Sn3O4 nanosheets and molecularly imprinted polymers. Sensors and Actuators B: Chemical 271, 215–224.

Wang, Y., Zhang, L., Kong, Q., Ge, S., Yu, J., 2018e. Time-resolution addressable photoelectrochemical strategy based on hollow-channel paper analytical devices. Biosensors and Bioelectronics 120, 64–70.

Wang, Y., Ning, G., Wu, Y., Wu, S., Zeng, B., Liu, G., He, X., Wang, K., 2019. Facile combination of beta-cyclodextrin host-guest recognition with exonuclease-assistant signal amplification for sensitive electrochemical assay of ochratoxin A. Biosensors and Bioelectronics 124–125, 82–88.

Warriner, K., Reddy, S.M., Namvar, A., Neethirajan, S., 2014. Developments in nanoparticles for use in biosensors to assess food safety and quality. Trends in Food Science & Technology 40, 183–199.

Wei, Y., Liu, D., 2012. Review of melamine scandal:still a long way ahead. Toxicology and Industrial Health 28, 579–582.

Weng, X., Neethirajan, S., 2017. Ensuring food safety: quality monitoring using microfluidics. Trends in Food Science & Technology 65, 10–22.

White, S.P., Sreevatsan, S., Frisbie, C.D., Dorfman, K.D., 2016. Rapid, selective, label-free aptameric capture and detection of ricin in potable liquids using a printed floating gate transistor. ACS Sensors 1, 1213–1216.

Wielogórska, E., Elliott, C.T., Danaher, M., Connolly, L., 2015. Endocrine disruptor activity of multiple environmental food chain contaminants. Toxicology in Vitro 29, 211–220.

Wu, L., Gao, B., Zhang, F., Sun, X., Zhang, Y., Li, Z., 2013. A novel electrochemical immunosensor based on magnetosomes for detection of staphylococcal enterotoxin B in milk. Talanta 106, 360–366.

Wu, B., Hou, L., Du, M., Zhang, T., Wang, Z., Xue, Z., Lu, X., 2014. A molecularly imprinted electrochemical enzymeless sensor based on functionalized gold nanoparticle decorated carbon nanotubes for methyl-parathion detection. RSC Advances 4, 53701–53710.

Wu, M.Y.-C., Hsu, M.-Y., Chen, S.-J., Hwang, D.-K., Yen, T.-H., Cheng, C.-M., 2017. Point-of-Care detection devices for food safety monitoring: proactive disease prevention. Trends in Biotechnology 35, 288–300.

Wu, D., Rios-Aguirre, D., Chounlakone, M., Camacho-Leon, S., Voldman, J., 2018. Sequentially multiplexed amperometry for electrochemical biosensors. Biosensors and Bioelectronics 117, 522–529.

Xu, B., Ye, M.-L., Yu, Y.-X., Zhang, W.-D., 2010. A highly sensitive hydrogen peroxide amperometric sensor based on MnO2-modified vertically aligned multiwalled carbon nanotubes. Analytica Chimica Acta 674, 20–26.

Xue, J., Liu, J., Wang, C., Tian, Y., Zhou, N., 2016. Simultaneous electrochemical detection of multiple antibiotic residues in milk based on aptamers and quantum dots. Analytical Methods 8, 1981–1988.

Yagati, A., Chavan, S., Baek, C., Lee, M.-H., Min, J., 2018. Label-free impedance sensing of aflatoxin B1 with polyaniline nanofibers/Au nanoparticle electrode array. Sensors 18, 1320.

Yan, L., Luo, C., Cheng, W., Mao, W., Zhang, D., Ding, S., 2012. A simple and sensitive electrochemical aptasensor for determination of Chloramphenicol in honey based on target-induced strand release. Journal of Electroanalytical Chemistry 687, 89–94.

Yan, P., Tang, Q., Deng, A., Li, J., 2014. Ultrasensitive detection of clenbuterol by quantum dots based electrochemiluminescent immunosensor using gold nanoparticles as substrate and electron transport accelerator. Sensors and Actuators B: Chemical 191, 508–515.

Yan, M., Bai, W., Zhu, C., Huang, Y., Yan, J., Chen, A., 2016. Design of nuclease-based target recycling signal amplification in aptasensors. Biosensors and Bioelectronics 77, 613–623.

Yang, G., Zhao, F., 2015a. Electrochemical sensor for chloramphenicol based on novel multi-walled carbon nanotubes@molecularly imprinted polymer. Biosensors and Bioelectronics 64, 416–422.

Yang, G., Zhao, F., 2015b. Electrochemical sensor for dimetridazole based on novel gold nanoparticles@molecularly imprinted polymer. Sensors and Actuators B: Chemical 220, 1017–1022.

Yang, M., Jiang, B., Xie, J., Xiang, Y., Yuan, R., Chai, Y., 2014. Electrochemiluminescence recovery-based aptasensor for sensitive Ochratoxin A detection via exonuclease-catalyzed target recycling amplification. Talanta 125, 45–50.

Yang, J., Gao, P., Liu, Y., Li, R., Ma, H., Du, B., Wei, Q., 2015. Label-free photoelectrochemical immunosensor for sensitive detection of Ochratoxin A. Biosensors and Bioelectronics 64, 13–18.

Yang, Y., Fang, G., Wang, X., Liu, G., Wang, S., 2016. Imprinting of molecular recognition sites combined with π-donor–acceptor interactions using bis-aniline-crosslinked Au–CdSe/ZnS nanoparticles array on electrodes: development of electrochemiluminescence sensor for the ultrasensitive and selective detection of 2-methyl-4-chlorophenoxyacetic acid. Biosensors and Bioelectronics 77, 1134–1143.

Yang, Y., Yin, S., Li, Y., Lu, D., Zhang, J., Sun, C., 2017. Application of aptamers in detection and chromatographic purification of antibiotics in different matrices. TRAC Trends in Analytical Chemistry 95, 1–22.

Yao, Y., Jiang, C., Ping, J., 2018. Flexible freestanding graphene paper-based potentiometric enzymatic aptasensor for ultrasensitive wireless detection of kanamycin. Biosensors and Bioelectronics.

Yin, H., Cui, L., Chen, Q., Shi, W., Ai, S., Zhu, L., Lu, L., 2011. Amperometric determination of bisphenol A in milk using PAMAM–Fe3O4 modified glassy carbon electrode. Food Chemistry 125, 1097–1103.

You, H., Bai, L., Yuan, Y., Zhou, J., Bai, Y., Mu, Z., 2018. An amperometric aptasensor for ultrasensitive detection of sulfadimethoxine based on exonuclease-assisted target recycling and new signal tracer for amplification. Biosensors and Bioelectronics 117, 706–712.

Yu, L., Zhang, Y., Hu, C., Wu, H., Yang, Y., Huang, C., Jia, N., 2015. Highly sensitive electrochemical impedance spectroscopy immunosensor for the detection of AFB1 in olive oil. Food Chemistry 176, 22–26.

Yuan, Q., He, C., Mo, R., He, L., Zhou, C., Hong, P., Sun, S., Li, C., 2018. Detection of AFB1 via TiO_2 nanotubes/Au nanoparticles/enzyme photoelectrochemical biosensor. Coatings 8, 90.

Zhang, Z., Zhang, Y., Song, R., Wang, M., Yan, F., He, L., Feng, X., Fang, S., Zhao, J., Zhang, H., 2015. Manganese(II) phosphate nanoflowers as electrochemical biosensors for the high-sensitivity detection of ractopamine. Sensors and Actuators B: Chemical 211, 310–317.

Zhang, X., Zhang, Y.-C., Zhang, J.-W., 2016. A highly selective electrochemical sensor for chloramphenicol based on three-dimensional reduced graphene oxide architectures. Talanta 161, 567–573.

Zhang, W., Xiong, H., Chen, M., Zhang, X., Wang, S., 2017. Surface-enhanced molecularly imprinted electrochemiluminescence sensor based on Ru@SiO2 for ultrasensitive detection of fumonisin B1. Biosensors and Bioelectronics 96, 55–61.

Zhang, R., Gu, J., Wang, X., Li, Y., Zhang, K., Yin, Y., Zhang, X., 2018. Contributions of the microbial community and environmental variables to antibiotic resistance genes during co-composting with swine manure and cotton stalks. Journal of Hazardous Materials 358, 82–91.

Zhang, H., Luo, F., Wang, P., Guo, L., Qiu, B., Lin, Z., 2019. Signal-on electrochemiluminescence aptasensor for bisphenol a based on hybridization chain reaction and electrically heated electrode. Biosensors and Bioelectronics.

Zhao, X., Zhang, Q., Chen, H., Liu, G., Bai, W., 2017. Highly sensitive molecularly imprinted sensor based on platinum thin-film microelectrode for detection of chloramphenicol in food samples. Electroanalysis 29, 1918–1924.

Zheng, W., Yan, F., Su, B., 2016. Electrochemical determination of chloramphenicol in milk and honey using vertically ordered silica mesochannels and surfactant micelles as the extraction and anti-fouling element. Journal of Electroanalytical Chemistry 781, 383–388.

Zhou, L., Huang, J., Yang, L., Li, L., You, T., 2014. Enhanced electrochemiluminescence based on Ru(bpy)32+-doped silica nanoparticles and graphene composite for analysis of melamine in milk. Analytica Chimica Acta 824, 57–63.

Zhou, L., Jiang, D., Du, X., Chen, D., Qian, J., Liu, Q., Hao, N., Wang, K., 2016. Femtomolar sensitivity of bisphenol A photoelectrochemical aptasensor induced by visible light-driven TiO2 nanoparticle-decorated nitrogen-doped graphene. Journal of Materials Chemistry B 4, 6249–6257.

Zhou, Y., Sui, C., Yin, H., Wang, Y., Wang, M., Ai, S., 2018. Tungsten disulfide (WS2) nanosheet-based photoelectrochemical aptasensing of chloramphenicol. Microchimica Acta 185, 453.

Zhou, N., Ma, Y., Hu, B., He, L., Wang, S., Zhang, Z., Lu, S., 2019. Construction of Ce-MOF@COF hybrid nanostructure: label-free aptasensor for the ultrasensitive detection of oxytetracycline residues in aqueous solution environments. Biosensors and Bioelectronics 127, 92–100.

Zhu, D., Yan, Y., Lei, P., Shen, B., Cheng, W., Ju, H., Ding, S., 2014. A novel electrochemical sensing strategy for rapid and ultrasensitive detection of Salmonella by rolling circle amplification and DNA-AuNPs probe. Analytica Chimica Acta 846, 44–50.

Zhu, M., Zhang, Y., Ye, J., Haijun, D., 2015. Sensitive and Selective Determination of Chloramphenicol on Ordered Mesoporous Carbon/Nafion Composite Film.

Zhu, Q., Liu, H., Zhang, J., Wu, K., Deng, A., Li, J., 2017. Ultrasensitive QDs based electrochemiluminescent immunosensor for detecting ractopamine using AuNPs and Au nanoparticles@PDDA-graphene as amplifier. Sensors and Actuators B: Chemical 243, 121–129.

Zhu, J., Wen, M., Wen, W., Du, D., Zhang, X., Wang, S., Lin, Y., 2018. Recent progress in biosensors based on organic-inorganic hybrid nanoflowers. Biosensors and Bioelectronics.

Index

'*Note:* Page numbers followed by "f" indicate figures and "t" indicates tables.'

A

Acetylthiocholine, 93–94
Active ester chemistry, 56–58
Adenosine-5'-triphosphate (ATP), 219–220, 220f
Adsorption, 193
Affinity-based biosensors, 62–69
  antibodies-based biosensors, 63–65
    immunoglobulin A (IgA), 63–64
    immunoglobulin D (IgD), 63–64
    immunoglobulin E (IgE), 63–64
    immunoglobulin G (IgG), 63–64
    immunoglobulin M (IgM), 63–64
  nucleic acid biosensors, 65–69
    DNA sensors, 65–66
    electrochemical DNA sensor, 66
    electrostatic adsorption, 66
Affinity immobilization, 58
Ag nanoparticles (AgNPs), 14
Alcohol oxidases (AOx), 180
Amine removal, 259
Amperometric method, 13–14, 14f
Amperometry, 313–314, 315t–316t
Antibiotics, 214–218, 219f
Aptamers, 85–90
  cancer cells, 232–233, 233f, 234t–236t
  microorganisms, 232–233, 234t–236t
  proteins, 224–226, 226f, 227t–231t
    differential pulse stripping voltammetry (DPSV), 225–226
    Rhodamin B (RhB), 225
    rolling-circle amplification (RCA), 224
    thrombin-binding aptamer (TBA), 224–225
    thrombin (TB), 224
  small molecules, 214–223, 215t–217t
    adenosine-5'-triphosphate (ATP), 219–220, 220f
    antibiotics, 214–218, 219f
    dopamine, 218–219
    drugs, 222–223, 224f
    heavy metals ion, 214
    pesticides, 221
    toxins, 221, 222f
Atomic force microscope (AFM), 32
Au@Pd core-shell nanostructure, 112
Avidin-biotin complex, 257

B

Biomolecule-contained nanocomposite, 123–124, 124f
Biomolecules, 84–85
Biomolecules immobilization techniques, 58–62
  active ester chemistry, 56–58
  affinity-based biosensors, 62–69
  affinity immobilization, 58
  chemisorption, 55–56
  entrapment, 54–55
  enzymatic sensors, 58–62
  physisorption, 53–54
  whole-cell biosensors, 69–70, 70f
Biorecognition components, 11
Biosensing surface, 77
Biosensor, 168–169
Biosensors
  applications, 7, 7f
  electrochemical biosensors. *See* Electrochemical biosensors
  electrochemical techniques, 7–9
  glucose oxidase (GOx), 3
  research challenges, 8–9
  signalization, 2–3
  transducer-based output signals, 3
  transducers, 2–3
Bisphenol A (BPA), 318

C

Carbon nanomaterials (CNMs)
  application, 84–85
  aptamer, 85–90
  biomolecules, 84–85
  biosensing surface, 77
  biosensor, 85
  carbon nanotubes, 79–80
  DNA biosensors, 85–90, 90f–91f
  fabrication of biosensors, 77–78
  fullerene, 81
  graphene, 80–81
  graphene oxide (GO), 80–81
  hexagonal configuration, 78–79
  higher currents, 84
  higher stability and resistance to passivation, 84
  immunosensors, 90–94, 94f

Carbon nanomaterials (CNMs) (*Continued*)
lower detection potentials, 84
modification, 83–84
onedimensional (1D) objects, 79
properties, 79–81
synthesis methods, 81–83
two-dimensional (2D) sheet, 78–79
Carbon nanotubes, 79–80
Catalysis, 103
Chemisorption, 55–56
Chitosan, 120
Choline oxidase, 180
Chronoamperometry, 17–18, 17f, 314–317
Chronocoulometry, 18–19, 19f
Concanavalin A (ConA), 88
Conductive polymer nanowire, 126
Conjugate nanomaterials, 116–118
Controlled potential adsorption, 256
Core-shell nanomaterials, 110–112, 111f, 113f
Cottrell's equation, 17–18
Covalent attachment, 256
Covalent bonding, 60, 194–195
C-reactive protein (CRP), 35
Cross-linking, 196–198
Cyclic voltammetry (CV), 14–17, 16f, 110
Cystine-based cyclic bisureas (CBU)-gold nanoparticle (AuNP), 62
Cytochrome c oxidase, 181–183
Cytochrome P-450, 188–189

D

Daunorubicin (DNR), 30
Dehydrogenases, 183
Detect double-stranded DNA (dsDNA), 51
Detection limit, 5
Dichalcogenides, 127
Differential pulse stripping voltammetry (DPSV), 225–226
Differential pulse voltammetry (DPV), 14
Dioxygenases, 188
Direct detection biosensors, 136–140
direct electrochemical biosensors, 138–139
direct optical biosensors, 137
direct piezoelectric (mechanical) biosensors, 139–140
Direct detection methods, 144–148
Direct electrochemical biosensors, 138–139
Direct electron transfer (DET), 175–176
Direct optical biosensors, 137
Direct piezoelectric (mechanical) biosensors, 139–140
Dissipation monitoring, 34–35, 35f
DNA alkylation, 259
DNA biosensors, 85–90, 90f–91f
DNA-mediated charge transport platforms, 149–153, 150f
Dopamine, 218–219
Double-analyte electrochemical aptasensors, 114
Dynamic chronoamperometry method, 14

E

Efficient electrochemical-SPR (EC-SPR), 29, 29f
Electrochemical biosensors
accuracy, 6
characteristics, 4–7
detection limit, 5
linear dynamic range, 5
linearity, 4
operational stability, 6–7
recovery time, 5–6
repeatability, 6
reproducibility, 6
response time, 5–6
ruggedness, 6
selectivity, 5
sensitivity, 5
storage stability, 6–7
Electrochemical detection techniques
amperometric method, 13–14, 14f
biorecognition components, 11
chronoamperometry, 17–18, 17f
chronocoulometry, 18–19, 19f
cyclic voltammetry, 15–17, 16f
daunorubicin (DNR), 30
dissipation monitoring, 34–35, 35f
electrochemical impedance spectroscopy, 25–26, 25f, 26t, 27f–28f
ellipsometry, 30–32, 31f
field-effect transistor (FET), 19–23, 21f–24f
impedimetric methods, 11
nanotechnology and bioelectronics, 12
potentiometric devices, 14–15
reference electrode, 11–12
scanning probe microscopy (SPM), 32–34, 32f–34f
screenprinted electrodes (SPEs), 13, 13f
surface plasmon resonance (SPR) optical sensor, 27–30
three-electrode electrochemical cell, 11–12, 12f
waveguide-based techniques, 26–27
Electrochemical DNA biosensors, 66, 144–146, 259–271
direct detection methods, 144–148

DNA-mediated charge transport platforms, 149–153, 150f
electrochemical impedance spectroscopy, 146–148
electrostatic mode, 150–151
enzyme label–based electrochemical detection, 153–154
field-effect transistors, 148
groove mode, 153
indirect detection method, 148–156
intercalation mode, 151–153
nanoparticles label–based electrochemical detection, 155–156
polymerase chain reaction (PCR), 145
redox mediator–based electrochemical detection, 148–149, 149f
Electrochemical impedance spectroscopy, 25–26, 25f, 26t, 27f–28f, 146–148, 321–327, 328t–329t
Electrochemical quartz crystal microbalance (EQCM), 34–35, 336
Electrochemical reduction, 82–83
Electrochemical signal amplification, 100–101
catalytic reaction-based signal amplification, 100–101
mediators, 101
Electrochemical surface plasmon resonance, 335–336
Electrochemiluminescence (ECL) sensors, 331, 334t, 335f
Electrodeposition, 49–50
Electrode surface immobilization, 255–257
Electrostatic adsorption, 66
Electrostatic bond external junction, 258
Electrostatic mode, 150–151
Electrotransducer surface, 100
Ellipsometry, 30–32, 31f
Encapsulation, 195
Entrapment, 54–55, 195
Enzymatic behavior, 102
Enzymatic sensors
covalent bonding, 60
cystine-based cyclic bisureas (CBU)-gold nanoparticle (AuNP), 62
glucose oxidase (GOx), 59, 62
horseradish peroxidase (HRP), 60
hydrophilicity, 61–62
hydrophobic core, 59
nanotechnology, 60–61
porous entrapment method, 59
Enzyme-based electrochemical biosensors
alcohol oxidases (AOx), 180
biosensor, 168–169
choline oxidase, 180
classification, 169–176
cytochrome c oxidase, 181–183
cytochrome P-450, 188–189
dehydrogenases, 183
dioxygenases, 188
direct electron transfer (DET), 175–176
glucose oxidase (GOx), 177
horseradish peroxidase (HRP), 186
hydrolases, 171
hydroxylases, 188
immobilization of enzymes, 192–198
lactate oxidases (Lox), 180
molybdopterin, 172
monooxygenases, 188
nicotine amide adenine dinucleotide, 171–172
oxidases, 177
oxidoreductases, 170–171
oxidoreductase subclasses, 176–179
oxygenases, 188
peroxidases, 185
reductases, 189–190
sequence enzyme, 187
steroid oxidases, 180
Enzyme label–based electrochemical detection, 153–154
Epidermal growth factor receptor (EGFR), 87
Epithelial cell adhesion molecule (EpCAM), 87
Equivalent circuits, 25
Esterification, 83

F

Fabrication of biosensors, 77–78
Fc-tagged peptides (Fc-PNW), 289
Field-effect transistor (FET), 19–23, 21f–24f, 148
Flavin adenine dinucleotide (FAD), 104
Folate receptor (FR), 109–110
Fullerene, 81

G

Glucose oxidase (GOx), 3, 59, 62, 177
G-quadruplex structure, 265–268
Graphene, 80–81
Graphene oxide (GO), 51, 80–81, 288
Groove mode, 153

H

Heavy metal ions, 214, 268–271
Hexagonal boron nitride (h-BN) nanosheets, 127
Hexagonal configuration, 78–79
Hollow mesoporous silica microspheres (HSMs), 125

Horseradish peroxidase (HRP), 60, 186
Human chorionic gonadotropin antibody (HGC), 123
Human epidermal growth factor receptor 2 (HER2), 87
Hybridization, 255
Hybridization chain reactions (HCRs), 109–110
Hydrolases, 171
Hydrophilicity, 61–62
Hydrophobic core, 59
Hydroxylases, 188

I

Immobilization of biomolecules, 100
Immobilization of enzymes, 192–198
- adsorption, 193
- covalent bonding, 194–195
- cross-linking, 196–198
- encapsulation, 195
- entrapment, 195
- kinetics of, 198–200, 199f

Immunoglobulin A (IgA), 63–64
Immunoglobulin D (IgD), 63–64
Immunoglobulin E (IgE), 63–64
Immunoglobulin G (IgG), 63–64
Immunoglobulin M (IgM), 63–64
Immunosensors, 90–94, 94f
Impedance, 25
Impedimetric methods, 11
Indirect detection biosensors
- indirect electrochemical biosensors, 142–144, 143f
- indirect optical biosensors, 140–142, 141f–142f

Indirect detection method, 148–156
Inert electrode surface, 46–47
Intercalation mode, 151–153
Intercalation of compounds, 257–258, 257f
Intrinsic and distinctive properties, 99
Ion-selective electrodes (ISEs), 14–15, 295
Ion-selective field-effect transistor (ISFET), 21–22
Isoelectric point (IEP), 23, 120–121

J

Junction gate field-effect transistor (JFET), 21

L

Lactate oxidases (Lox), 180
Layered double hydroxides (LDHs), 111
Limit of detection (LOD), 87
Linear dynamic range, 5
Linearity, 4
Linear sweep voltammetry (LSV), 14
Listeria monocytogenes (LM), 295
Lower detection potentials, 84

M

Maltose-binding protein (MBP), 22
Mesoporous silica (MPS), 93–94
Mesoporous silica nanoparticles (MSNs), 124
Metal and metal oxide, 103–110, 105f–108f
Metallic nanoparticles, 103
Metal-organic frameworks (MOFs), 51–53
Metal-oxide-semiconductor field-effect transistor (MOSFET), 21
Methylene blue (MB), 89
MOFs. *See* Metal-organic frameworks (MOFs)
Molecule grooves, 258
Molybdopterin, 172
Monitoring biosensor signals
- bioreceptor, 135
- direct detection biosensors, 136–140
- electrochemical DNA biosensors. *See* Electrochemical DNA biosensors
- indirect detection biosensors, 140–144
- signal monitoring, 135–136, 136f

Monooxygenases, 188
Multimetallic nanocomposites, 120–123, 122f
Multi-walled carbon nanotubes (MWCNT), 17, 79
Mycobacterium tuberculosis (Mtb), 114–115

N

Nanofibers, 125–126, 127f
Nanoparticles label–based electrochemical detection, 155–156
Nanosheets, 127–130, 128f
Nanowires, 125–126, 127f
Naphthalene acetic acid, 92
Nicotine amide adenine dinucleotide, 171–172
Noble metal nanoparticles, 103
Noncarbon nanomaterials
- classification, 102–130
- composites, 116–125
  - biomolecule-contained nanocomposite, 123–124, 124f
  - multimetallic nanocomposites, 120–123, 122f
  - polymer-based nanocomposites, 118–120, 119f
  - semiconductor material–based nanocomposites, 124–125
- core-shell, 110–112, 111f, 113f
- electrochemical signal amplification, 100–101
- electrotransducer surface, 100
- enzymatic behavior, 102
- immobilization of biomolecules, 100

intrinsic and distinctive properties, 99
metal and metal oxide, 103–110, 105f–108f
nanofibers, 125–126, 127f
nanosheets, 127–130, 128f
nanowires, 125–126, 127f
quantum dots, 113–116, 115f
signalproducing probes, 101–102
Nonspecific adsorption (NSA), 46–47
Normal pulse voltammetry (NPV), 14
Nucleic acid biosensors, 65–69
chemical sensor, 253
deoxyribonucleic acid's (DNA), 253
DNA hybridization, 254
DNA immobilization, 255–257
avidin-biotin complex, 257
controlled potential adsorption, 256
covalent attachment, 256
physical adsorption, 255–256
DNA sensors, 65–66
electrochemical DNA biosensors, 259–271
electrode surface immobilization, 255–257
genetic defects and clinical applications, 262–265
G-quadruplex structure, 265–268
heavy metal ions, 268–271
hybridization, 255
molecules and ions, 257–258
electrostatic bond external junction, 258
intercalation of compounds, 257–258, 257f
molecule grooves, 258
mutations, 258–259
amine removal from, 259
DNA alkylation, 259
DNA oxidation, 259
types, 258–259
ultraviolet radiation, 259
ribonucleic acid's (RNA), 253
single-stranded oligonucleotide (ssDNA), 253–254
stabilizing ligands, 265–268

O

Onedimensional (1D) objects, 79
Operational stability, 6–7
Oxidases, 177
Oxidoreductases, 170–171
Oxidoreductase subclasses, 176–179
Oxygenases, 188

P

Passive probe configuration, 32–33
Peptide-based electrochemical biosensors
amino acids (AAs), 277–278
antimicrobial peptides (AMPs), 278
biomolecules, 298t–299t
creating interfaces, 278–279
drawbacks, 296–297
earnings, 296–297
electrochemical techniques, 283–296
alternating current voltammetry, 292–294, 294f
cyclic voltammetry, 284–288, 286f
differential pulse voltammetry, 289–291, 290f
electrochemical impedance spectroscopy, 292–294, 294f
linear sweep voltammetry, 284–288, 286f
potentiometry, 294–296, 297f
square wave voltammetry, 289–291, 290f
electrodes coating strategies, 278–279
electron transfer, 280–283, 281f
electron transfer (ET), 277–278
interrogation modes, 279–280
peptide-modified surfaces, 279–280
Peroxidases, 185
Pesticides, 221
Photoelectrochemical (PEC) sensor, 327–331, 332t
Physical adsorption, 255–256
Physisorption, 53–54
Polymerase chain reaction (PCR), 145
Polymer-based nanocomposites, 118–120, 119f
Poly(vinyl chloride) (PVC), 295
Porous entrapment method, 59
Potentiometric devices, 14–15
Potentiometry/field-effect transistors, 317–318, 319t
Proteins, aptamer, 224–226, 226f, 227t–231t
differential pulse stripping voltammetry (DPSV), 225–226
Rhodamin B (RhB), 225
rolling-circle amplification (RCA), 224
thrombin-binding aptamer (TBA), 224–225
thrombin (TB), 224

Q

Quantum dots, 113–116, 115f
Quartz crystal microbalance (QCM), 34–35

R

Receptor biosensors, food products contaminants
biosensors, 312
chemical contaminants, 308–309, 308f
conventional screening methods, 311–312
electrochemical biosensors, 312–336

Receptor biosensors, food products contaminants (*Continued*)
amperometry, 313–314, 315t–316t
bisphenol A (BPA), 318
chronoamperometry, 314–317
electrochemical impedance spectroscopy, 321–327, 328t–329t
electrochemical quartz crystal microbalance, 336
electrochemical surface plasmon resonance, 335–336
electrochemiluminescence (ECL) sensors, 331, 334t, 335f
photoelectrochemical (PEC) sensor, 327–331, 332t
potentiometry/field-effect transistors, 317–318, 319t
voltammetry, 318–321, 322t–325t, 326f
future perspectives, 336–346
amplification strategies, 339–343
immobilization support, 341
micro/nanoelectrodes, 337–339
micro/nanofluidic platforms, 337–339
miniaturization, 337–339
nanomaterials, 340–342
new biosensing elements, 337
nucleic acid amplification strategy, 343
signal amplification strategies, 341–342
human industrial activities, 308–309
migration of elements, 308
regulation, 310
screening methods, 310–311
smartphone-based biosensors, 344–346
Recovery time, 5–6
Redox mediator–based electrochemical detection, 148–149, 149f
Reduced graphene oxide (rGO), 51
Reductases, 189–190
Reference electrode, 11–12
Repeatability, 6
Reproducibility, 6
Response time, 5–6
Reversible redox reaction, 17–18
Rhodamin B (RhB), 225
Ribonucleic acid's (RNA), 253
Rolling-circle amplification (RCA), 224

S

Sambucus nigra agglutinin (SNA), 123
SAMs. *See* Self-assembled monolayers (SAMs)
Scanning probe microscopy (SPM), 32–34, 32f–34f
Screen-printed electrode (SPE), 13, 13f, 87
Self-assembled monolayers (SAMs), 47–53, 48f
Semiconductor material–based nanocomposites, 124–125
Sensors
biosensors, 2–3, 2f
characterization, 1
classification, 1–2
definition, 1
electrochemical biosensors, 3–4, 4f
Sequence enzyme, 187
Signaling-on approach, 289
Signalization, 2–3
Signal producing probes, 101–102
Silicon carbide (SiC), 82–83
Silver nanoparticles (AgNPs), 288
Single-nucleotide polymorphisms (SNPs), 115–116
Single-stranded oligonucleotide (ssDNA), 253–254
Single-walled carbon nanotubes(SWCNT), 79
Small molecules, 214–223, 215t–217t
adenosine-5'-triphosphate (ATP), 219–220, 220f
antibiotics, 214–218, 219f
dopamine, 218–219
drugs, 222–223, 224f
heavy metals ion, 214
pesticides, 221
toxins, 221, 222f
Smartphone-based biosensors, 344–346
Square-wave voltammetry (SWV), 14
Steroid oxidases, 180
Storage stability, 6–7
Surface modification methods
biomolecules immobilization techniques, 53–58, 54f
conducting polymers, 50
detect double-stranded DNA (dsDNA), 51
electrodeposition, 49–50
graphene, 51
graphene oxide (GO), 51
inert electrode surface, 46–47
metal-organic frameworks (MOFs), 51–53
nanomaterials, 50–51
needs, 46
nonspecific adsorption (NSA), 46–47
reduced graphene oxide (rGO), 51
self-assembled monolayers (SAMs), 47–53, 48f
specific *vs.* nonspecific binding, 46–47
strategies, 47–53
thrombin (TB), 53

Surface plasmon resonance (SPR) optical sensor, 27–30
SWV method, 116
Synthesis methods, 81–83

T

4-Tertbutyl catechol, 93
Three-electrode electrochemical cell, 11–12, 12f
Thrombin-binding aptamer (TBA), 224–225
Thrombin (TB), 53, 224
Titration devices, 15
Toxins, 221, 222f
Transducer-based output signals, 3
Transducers, 2–3
Two-dimensional (2D) sheet, 78–79

U

Ultraviolet radiation, 259

V

Voltammetry, 318–321, 322t–325t, 326f
Voltammograms, 90

W

Warburg impedance, 25
Waveguide-based techniques, 26–27
Whole-cell biosensors, 69–70, 70f

Printed in the United States
By Bookmasters